BIOCHEMISTRY OF COLLAGEN

BIOCHEMISTRY OF COLLAGEN

Edited by

G. N. Ramachandran

Indian Institute of Science
Bangalore, India

and

A. H. Reddi

The Ben May Laboratory for Cancer Research
University of Chicago, Illinois

SPRINGER SCIENCE+BUSINESS MEDIA, LLC

Library of Congress Cataloging in Publication Data

Main entry under title:

Biochemistry of collagen.

Includes bibliographies and index.
1. Collagen. 2. Biological chemistry. I. Ramachandran, G. N., 1922- II. Reddi, A. H., 1942-
QP552.C6B56 591.1'9245 76-7075
ISBN 978-1-4757-4604-4 ISBN 978-1-4757-4602-0 (eBook)
DOI 10.1007/978-1-4757-4602-0

Originally published by Plenum Press, New York in 1976
Softcover reprint of the hardcover 1st edition 1976

Preface

Collagen is a fascinating protein not only because of its ubiquitous occurrence in multicellular animals, but also because of its unique chemical structure. As the predominant constituent in bone, cartilage, skin, tendon, and tooth, it is not surprising that collagen is of interest to anatomists, biochemists, biomedical engineers, cell biologists, dermatologists, dental surgeons, leather chemists, orthopedic surgeons, physiologists, physicians, zoologists, and a host of others.

This book was planned to provide an up-to-date comprehensive survey of all aspects of biochemistry of collagen. The recent discovery of genetically distinct collagens with tissue specificity has opened a new era in collagen biochemistry, and Karl Piez discusses this in the opening chapter on primary structure. In the next chapter, Ramachandran and Ramakrishnan deal with the molecular structure of collagen, placing special emphasis on the conformational aspects of its polypeptide chains. Following the consideration of primary and secondary structure of collagen, the three-dimensional arrangement of collagen molecules in the fibrils is covered by Miller in Chapter 3. Collagen is generally in the insoluble state in the living organism due to the cross-linking of individual molecules, and Tanzer describes the various aspects of this cross-linkage in Chapter 4.

The biosynthesis of collagen is discussed in depth by Prockop and his colleagues. Chapter 5 investigates the numerous posttranslational modifications of collagen, including the hydroxylations of lysine and proline, and the glycosylation and deamination of lysine, which give rise to cross-link precursors. It is axiomatic that rates of both synthesis and degradation determine the steady-state levels of most proteins. In addition, mechanisms must exist for the remodeling of collagen, an important component of connective and other tissues both in normal and diseased states and during development. Enzymatic degradation by collagenases is an integral part of the biochemistry of collagen and is reviewed by J. Gross.

Collagenous proteins are antigenic, and Timpl discusses this aspect in his chapter on the immunology of collagens. The myriad of enzymatic steps involved in the intracellular biosynthesis of procollagen and its conversion to collagen fibrils extracellularly points to the possibility that biochemical lesions in this step-wise process may lead to pathological states. In fact, this is the topic for discussion in Chapter 8 by Lapiere and Nusgens. Cell differentiation is one of the central problems in biology and medicine, and Reddi's chapter deals with the role of collagen in differentiation of cells. Finally, Bhatnagar and Rapaka survey the synthetic polypeptide models of collagen and their applications in biochemistry and medicine.

We would like to thank our contributors for their enthusiastic cooperation in this collaborative venture. Our thanks are also due to Mr. Robert N. Ubell and his colleagues at Plenum Press for their dedicated effort in the publication of this book.

Bangalore, India G. N. RAMACHANDRAN
Chicago, Illinois A. H. REDDI

Contributors

RICHARD A. BERG, Department of Biochemistry, College of Medicine and Dentistry of New Jersey, Rutgers Medical School, Piscataway, New Jersey

RAJENDRA S. BHATNAGAR, Laboratory of Connective Tissue Biochemistry, School of Dentistry, University of California, San Francisco, California

JEROME GROSS, The Developmental Biology Laboratory, Department of Medicine, Massachusetts General Hospital and the Harvard Medical School, Boston, Massachusetts

KARI I. KIVIRIKKO, Department of Biochemistry, College of Medicine and Dentistry of New Jersey, Rutgers Medical School, Piscataway, New Jersey. Present address: Department of Medical Chemistry, University of Oulu, Oulu, Finland

CHARLES M. LAPIÈRE, Service de Dermatologie, Hôpital de Bavière, Université de Liège, Liège, Belgium

A. MILLER, Laboratory of Molecular Biophysics, Zoology Department, Oxford University, Oxford, England

BETTY NUSGENS, Service de Dermatologie, Hôpital de Bavière, Université de Liège, Liège, Belgium

KARL A. PIEZ, Laboratory of Biochemistry, National Institute of Dental Research, National Institutes of Health, Bethesda, Maryland

DARWIN J. PROCKOP, Department of Biochemistry, College of Medicine and Dentistry of New Jersey, Rutgers Medical School, Piscataway, New Jersey

G. N. RAMACHANDRAN, Molecular Biophysics Unit, Indian Institute of Science, Bangalore, India

C. RAMAKRISHNAN, Molecular Biophysics Unit, Indian Institute of Science, Bangalore, India

RAO S. RAPAKA, Laboratory of Connective Tissue Biochemistry, School of Dentistry, University of California, San Francisco, California

A. H. REDDI, The Ben May Laboratory for Cancer Research, University of Chicago, Chicago, Illinois

MARVIN LAWRENCE TANZER, Department of Biochemistry, University of Connecticut Health Center, Farmington, Connecticut

RUPERT TIMPL, Max-Planck-Institut für Biochemie, Martinsried b. München, Germany

JOUNI UITTO, Department of Biochemistry, College of Medicine and Dentistry of New Jersey, Rutgers Medical School, Piscataway, New Jersey. Present address: Division of Dermatology, Department of Medicine, Washington University School of Medicine, St. Louis, Missouri

Contents

1. Primary Structure

KARL A. PIEZ

2. Molecular Structure

G. N. RAMACHANDRAN AND C. RAMAKRISHNAN

3. Molecular Packing in Collagen Fibrils

A. MILLER

4. Cross-Linking

MARVIN LAWRENCE TANZER

5. Intracellular Steps in the Biosynthesis of Collagen

DARWIN J. PROCKOP, RICHARD A. BERG, KARI I. KIVIRIKKO, AND JOUNI UITTO

6. Aspects of the Animal Collagenases

JEROME GROSS

7. Immunological Studies on Collagen

RUPERT TIMPL

8. Collagen Pathology at the Molecular Level

CHARLES M. LAPIÈRE AND BETTY NUSGENS

9. Collagen and Cell Differentiation

A. H. REDDI

10. Synthetic Polypeptide Models of Collagen: Synthesis and Applications

RAJENDRA S. BHATNAGAR AND RAO S. RAPAKA

1
Primary Structure

KARL A. PIEZ

I. Introduction

The primary structure of a protein is defined as the sequence of the amino acid residues in its polypeptide chain(s). This definition will be expanded here to include all covalent structures such as cross-links, attached carbohydrates, and modifications to amino acid side chains. However, it should be remembered that the sequence of amino acids in α-amino peptide linkage is the product of a structural gene, while all other aspects of covalent structure result from posttranslational events. Thse two aspects of covalent structure require that two quite different concepts be invoked in understanding the biological regulation of structure–function relationships.

This is particularly true in the case of collagen where posttranslational modifications are varied and extensive. They include conversion of proline to 4-hydroxyproline and to 3-hydroxyproline, conversion of lysine to hydroxylysine, glycosylation of hydroxylysine, oxidative deamination of lysine and hydroxylysine to yield the cross-link-precursor aldehydes allysine and hydroxyallysine, and proteolytic cleavage of procollagen to collagen. At least seven different enzymes are involved. Biosynthetic and functional aspects of these alterations are considered elsewhere in this book. The chemical aspects will be covered in this chapter except for cross-link chemistry which is more conveniently discussed separately (Chapter 4).

The most recent comprehensive summaries of collagen amino acid sequences are in the reviews by Traub and Piez (1971) and Gallop *et al.*,

KARL A. PIEZ · Laboratory of Biochemistry, National Institute of Dental Research, National Institutes of Health, Bethesda, Maryland 20014.

(1972), the *Atlas of Protein Sequence and Structure* (Dayhoff, 1972), and a paper by Hulmes *et al.* (1973). All data available through March, 1975, as well as some data that are in press or unpublished will be summarized here. Where sequence data are not available and amino acid composition or other chemical data provides some insight into primary structure, that information will be discussed. Because of the regular triple-chain collagen helix and the resulting constant residue spacing (Chapter 2), primary structure information can be obtained by electron microscopy. These results will also be reviewed here.

As will be seen, a large amount of new and important information has become available in the last few years. The complete sequence of 1052 residues in an $\alpha 1$ chain, which is presently the largest polypeptide chain sequenced, can be assembled. Another important body of information relates to the demonstration that higher animals have at least several genetically distinct collagens that have some degree of tissue specificity. Thus, the collagen that has received the most study (type I) is the major collagen of skin and the only collagen in bone and tendon. It has the chain composition $[\alpha 1(\mathrm{I})]_2\alpha 2$ where $\alpha 1(\mathrm{I})$ and $\alpha 2$ are homologous. The collagen specific to hyaline cartilage (type II) has three identical chains and is designated $[\alpha 1(\mathrm{II})]_3$ (Miller and Matukas, 1969; Miller, 1973). Similarly, the collagen that is a minor but important constituent in skin and perhaps a major constituent in large blood vessels (type III) is designated $[\alpha 1(\mathrm{III})]_3$ (Miller *et al.,* 1971). The collagen of basement membrane is referred to as type IV. As reviewed by Kefalides (1973), the amino acid composition and limited studies on cyanogen bromide peptides strongly suggest that basement membranes such as lens capsule, Descemet's membrane, and glomerulus contain a collagen that is genetically distinct from types I, II, and III. The polypeptide chains apparently have a collagenous portion similar in size to other collagens but may also have a large noncollagenous portion that contains a heteropolysaccharide and disulfide links (Olson *et al.,* 1973). This structure may be similar to and possibly derived from procollagen (Section VII). However, there is not complete agreement about the relationship of basement membrane collagen to other collagens (Hudson and Spiro, 1972*a*).

II. Methods

A. Separation of Collagen Types

Type I collagen is readily obtained in soluble form from many vertebrate species by extraction of skin or other tissues (see Piez, 1967).

Type II collagen is apparently always insoluble, but good yields can be obtained from several cartilages if a lathyrogen is used to inhibit cross-linking (Miller and Matukas, 1969; Trelstad *et al.*, 1970; Miller, 1971*a*). Limited digestion with a protease also solubilizes type II collagen effectively by cleaving off the nonhelical ends where cross-links originate (Strawich and Nimni, 1971; Miller, 1972). Although small amounts of type III collagen can be obtained by extraction (Byers *et al.*, 1974), it has been obtained in preparative amounts only after partial pepsin digestion (Epstein, 1974; Chung and Miller, 1974; Chung *et al.*, 1974; Trelstad, 1974). This type of preparation is suitable for many studies but of course not for the characterization of the cross-link regions.

Chromatographic procedures using CM-cellulose and molecular-sieve materials for the separation and purification of denatured α chains have been reviewed (Piez, 1967). The three α1 chain types are not readily resolved from each other by these methods. α1(III), however, is present normally in extracts as the disulfide cross-linked trimer, unlike α1(I) and α1(II) which are devoid of cysteine and cystine, and can be chromatographed both as the trimer and monomer (Epstein, 1974; Chung *et al.*, 1974; Byers *et al.*, 1974). Also, the three collagen types can be separated as native proteins by fractional salt precipitation. In cold neutral solution, type III precipitates at about 1.5 M NaCl (Epstein, 1974; Chung *et al.*, 1974), type I at about 2.2 M NaCl, and type II at about 4.4 M NaCl (Trelstad *et al.*, 1972). Separation in the native form is preferable to separation of denatured components since native collagen is much more stable than denatured collagen.

A procedure for the separation of denatured α chains on DEAE-cellulose has been reported (Trelstad *et al.*, 1972). It seems to have high resolution, but its utility has not yet been demonstrated.

Recent studies on collagen from basement membranes (type IV collagen) suggest that it is closely related to other collagens but has distinctive features. It has not been obtained in preparative amounts except by degradative procedures. Since the starting material is a tissue such as lens capsule or glomerulus, large amounts are difficult to make. The procedures used and the nature of the preparations obtained have been reviewed by Kefalides (1973).

B. *Separation of Cyanogen Bromide Peptides*

Cyanogen bromide cleavage of collagen in solution or in the insoluble state and the separation by ion-exchange and molecular-sieve chromatography of the peptides derived from type I collagen have been reviewed

(Traub and Piez, 1971). The same general procedures have been successfully used to characterize all the cyanogen bromide peptides from type II collagen and many from type III collagen (see Section III). SDS-gel electrophoresis is particularly effective for analytical determination of collagen type (Byers *et al.*, 1974).

C. *Amino Acid Sequencing*

The standard approach to the sequencing of a polypeptide chain is to prepare small peptides by a specific cleavage method, sequence the peptides, and then order them by using another cleavage procedure to prepare overlapping peptides. In the case of collagen, cyanogen bromide cleavage has proven to be extremely useful for the first step. Although the peptides are often of protein size, other specific cleavage procedures can be more effectively used than is possible with whole α chains. Ordering the cyanogen bromide peptides has in part been possible by the isolation and characterization of peptides arising from incomplete cyanogen bromide cleavage and thereby containing the sequences of two peptides. Unique to collagen is ordering by electron microscopy of renatured peptides, which in the SLS form (Section VIA) have a characteristic band pattern that can be matched to whole collagen SLS, and by taking advantage of the specific activity gradient along the chains after pulse labeling. These procedures have been reviewed (Traub and Piez, 1971), and recent examples can be found in the references quoted below (Sections III and IV).

The major procedures for sequencing peptides from collagen, as from other proteins, has been Edman degradation. The automatic sequencer has been particularly valuable. In some cases, polypeptides of more than 200 residues have been sequenced completely on the automatic instrument by utilizing appropriate overlapping fragments. Single runs of 40–60 residues are often possible. Details may be found in the references to the sequences (Section IV).

III. *Cyanogen Bromide Peptide Patterns*

The cyanogen bromide peptides of the α chains of collagen are not only useful as the first step in amino acid sequencing, but the peptide patterns serve as a map or "fingerprint" to characterize and compare α

chains. The distribution of methionine residues, as deduced from the size and order of the cyanogen bromide peptides, is therefore important information.

An earlier review summarized the data for the α1(I) and α2 chains of rat, calf, baboon, human, and chick (Traub and Piez, 1971). Similar data for guinea pig (Clark and Bornstein, 1972), rabbit (Becker and Timpl, 1972), and additional data on bovine (steer) type I collagen (Volpin and Veis, 1971, 1973) are now available. The size and order of all but two small cyanogen bromide peptides from α1(II) of chick cartilage (Miller, 1971*b*, 1972; Miller *et al.*, 1973) and human and bovine cartilage (Miller and Lunde, 1973) have also appeared. The order of the peptides in the

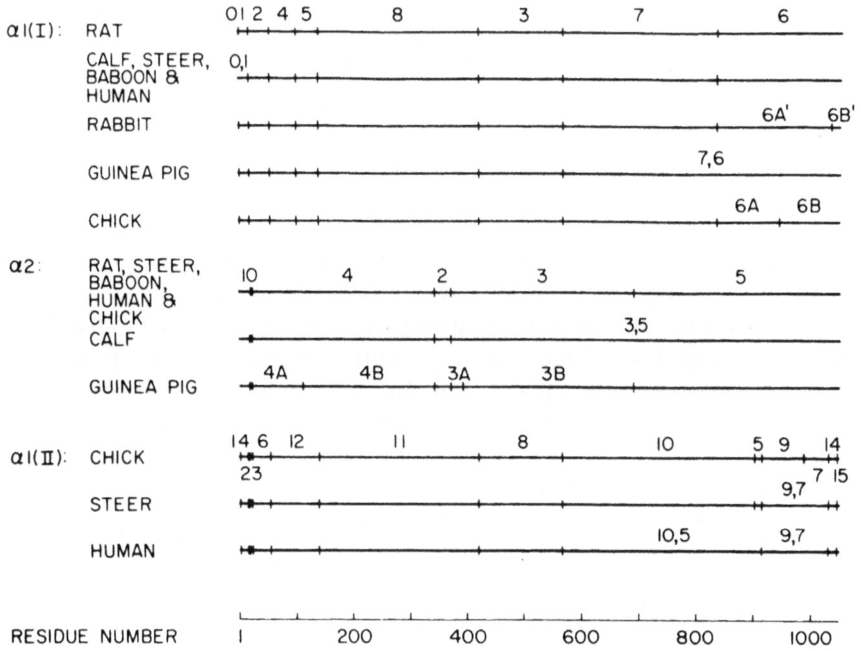

FIGURE 1. Distribution of methionine residues in several α chains of collagen. The short vertical lines show the positions of methionine in the amino acid sequence. The peptides arising from cleavage of these residues by CNBr are numbered according to their positions in chromatographic effluents using as standards rat type I collagen for the α1 (I) and α2 chains and chick type II collagen for the α1(II) chains. Where a methionine is missing in another species, the numbers designating the two peptides in the standard are both used, separated by a comma. When there is an extra methionine, the two peptides are designated A and B. The three α chain types are sufficiently different to require different numbering systems.

α2 chain of guinea pig (Clark *et al.*, 1975) and the position of the final two peptides (2 and 3) in α1(II) (Butler *et al.*, 1976) have been determined and will be reported shortly.

The available data are summarized in Figure 1. Although calf and steer represent the same species, they are listed separately in the figure since the α2 chain of calf is reported (Fietzek *et al.*, 1970) to have one less methionine than steer (Volpin and Veis, 1971, 1973). This is presumably a strain rather than an age difference. The large CNBr peptides from pig skin α1(I) and α2 (not shown) are reported to be the same as from calf (Heinrich *et al.*, 1971).

The close similarity of the methionine distribution among the α1(I) chains and among the α2 chains is readily apparent, although the two groups differ markedly. The α1(II) chains also form a distinct group, although the methionine distribution is closer to that found for the α1(I) chain than to the α2 chain. The α1(I) and α1(II) chains both have methionine residues at positions 19, 55, 139, 418, and 567. This justifies the designation of the α chain in cartilage collagen as α1(II).

The cyanogen bromide peptides from the α1(III) chain of human skin collagen have been partially characterized (Chung *et al.*, 1974) but not ordered. The data suggest that this chain also is closely related to α1(I). Some of the peptides appear to be homologous to α1(I) peptides, but others are sufficiently different to suggest a quite different distribution of methionine. The presence of a cyanogen bromide peptide without homoserine and with two sulfhydryl groups places the interchain disulfide bonds in type III collagen in the C-terminal region. This region is probably helical since it is pepsin resistant, but it is also possible that it has some other stable configuration or simply lacks a pepsin-susceptible peptide bond.

Cyanogen bromide peptides from type IV collagen have been isolated and compositions reported (Kefalides, 1973; Kefalides *et al.*, 1974). Although some are similar to peptides from α1(I) and α1(II), there are not sufficient data to determine the degree of homology with any certainty. The designation of the α chain in type IV collagen as α1(IV) is therefore tentative.

IV. Amino Acid Sequences

A. Terminal Regions

The N- and C-terminal ends of the α chains of type I collagen consist of sequences of 10–25 residues that do not contain glycine in every third position and therefore cannot be triple helical like the body of the molecule. These regions are the site of lysine residues from which cross-links originate (Chapter 4); they contain major antigenic determinants (Chapter 7); and they may play other important roles such as a directing function in fibril formation.

Partial or complete sequences for the N-terminal nonhelical regions of the α1(I) chain of six species (chick, rat, rabbit, calf, baboon, and human) have been previously summarized (Traub and Piez, 1971; Gallop *et al.*, 1972). Additional data have appeared which show that the four missing residues in rat skin α1(I), as usually obtained, are lost as a result of proteolysis during isolation (Stoltz *et al.*, 1972). The complete sequence of the N-terminal region in calf α1(I) (Rauterberg *et al.*, 1972*a*) and a partial sequence for the same region in rabbit α1(I) (Timpl *et al.*, 1973; Fietzek, personal communication) have also been reported. The amino acid composition of pig skin α1(I) shows it to have a similar N-terminal region except that the tyrosine residue is missing (Allam and Heidemann, 1974). These data, together with the earlier data, are shown in Figure 2. The cyanogen bromide peptides from which these sequences or compositions were derived are α1(I)CB-0 and α1(I)CB-1 (or α1(I)CB-0,1). They include the first three residues (Gly-Pro-Met) of the helical region.

Sequences from the N-terminal end of the α2 chain of the same six species have also been summarized (Traub and Piez, 1971; Gallop *et al.*, 1972). Additional data that have appeared recently are a partial sequence for the rabbit α2 chain (Timpl *et al.*, 1973; Fietzek, personal communication), the complete sequence for the calf α2 chain (Fietzek *et al.*, 1974*a*), and amino acid composition data for the pig α2 chain (Allam and Heidemann, 1974). These and the earlier results are shown in Figure 2. The cyanogen bromide peptide from which the sequences or compositions were derived is α2-CBl. It also includes the first three residues of the helical region.

It can be seen that the first 20 residues in the α1(I) chain of the seven species studied are highly conserved. Substitutions (or gaps) occur at only six positions, and they are all conservative. The α2 chain sequences show a much greater variability. However, the similarity among them and to the α1 chain sequences, as well as a common function, allows the conclusion that they are homologous.

The lysine residue at position 9, present in all α chains studied, is normally converted to the aldehyde allysine, which is a precursor of cross-links. It is known that in some cases it is first hydroxylated (Miller *et al.*, 1969; Barnes *et al.*, 1974) and is then presumably converted to hydroxy-allysine, which is also a precursor of cross-links (Chapter 8).

The sequences of the nonhelical N-terminal regions of collagens

```
                    1              5                 10                      15
α1(I):   Chick     pGlu-Met-Ser-Tyr-Gly-Tyr-Asp-Glu-Lys-Ser-Ala-Gly-    Val-Ala-Val-Pro-Gly-Pro-Met-
         Rat       pGlu-Met-Ser-Tyr-Gly-Tyr-Asp-Glu-Lys-Ser-Ala-Gly-    Val-Ser-Val-Pro-Gly-Pro-Met-
         Rabbit    pGlu-Met-Ser-Tyr-Gly-Tyr-Asp-Glu-Lys-Ser-Ala-Gly-    Val,Ser,Val,Pro,Gly,Pro-Met-
         Calf      pGlu-Leu-Ser-Tyr-Gly-Tyr-Asp-Glu-Lys-Ser-Thr-Gly-    Ile-Ser-Val-Pro-Gly-Pro-Met-
         Pig       pGlu,Leu,Ser,Tyr,Gly,Ile,Asp,Glu,Lys,Ser,Ala,Gly,Gly,Ile,Ser,Val,Pro,Gly,Pro-Met-
         Baboon    pGlu,Leu,Ser,Tyr,Gly,Tyr,Asp,Glu,Lys,Ser,Thr,Gly,Gly,Ile,Ser,Val,Pro,Gly,Pro-Met-
         Human     pGlu,Leu,Ser,Tyr,Gly,Tyr,Asp,Glu-Lys-Ser-Thr-Gly-Gly-Ile,Ser,Val,Pro,Gly,Pro-Met-

α2:      Chick     pGlu-        Tyr-Asp-    Pro-Ser-Lys-    Ala-Asp-    Phe-Gly-     Pro-Gly-Pro-Met-
         Rat       pGlu-        Tyr-Ser-    Asp-    Lys-Gly-Val-Ser-    Ala-Gly-     Pro-Gly-Pro-Met-
         Rabbit    pGlu-        Phe-        Asx-Gly-Lys,Gly,Gly-                     Pro,Gly,Pro-Met-
         Calf      pGlu-        Phe-        Asp-Ala-Lys-Gly-Gly-            Gly-     Pro-Gly-Pro-Met-
         Pig       pGlu,        Tyr,        Asp,Gly,Lys,Gly,Val,        Ala,Gly,     Pro,Gly,Pro-Met-
         Baboon    pGlu,        Tyr,        Asp,Gly,Lys,Gly,Val,        Leu,Gly,     Pro,Gly,Pro-Met-
         Human     pGlu,        Tyr,        Asp,Gly,Lys,Gly,Val,        Leu,Gly,     Pro,Gly,Pro-Met-
```

FIGURE 2. Amino acid sequences at the N-terminal nonhelical ends of the $\alpha 1(I)$ and $\alpha 2$ chains from several species. When the sequence is not known a likely sequence based on homology is shown; residues located in this way are separated by commas. The numbering system is based on rat (or calf) $\alpha 1(I)$ with gaps being inserted to maintain alignment. Residues 16–18 begin the helical region (Table 1). Residue 9 (Lys) may be hydroxylated and is normally converted to the aldehyde allysine (or hydroxyallysine), a precursor of cross-links. Residues 7 and 8 probably fall in SLS band 0 (Figure 6) and native fibril band I (Figure 7).

```
-Gly-Pro-Pro-Ser-Gly-Gly-Tyr-Asp-Leu-Ser-Phe-Leu-Pro-Gln-
1025                1030                1035
 Pro-Pro-Gln-Gln-Glx-Lys-Ala-His-Asp-Gly-Gly-Arg-Tyr-Tyr
    1040                1045                1050
```

FIGURE 3. The amino acid sequence at the C-terminal nonhelical end of calf αl(I). Residue 1044 (Lys), like residue 9 (Figure 2), is a precursor of cross-links. Residues 1032 and 1047 + 1050 probably fall in SLS bands 57 and 58 (Figure 6), respectively, and in native fibril bands VI and VII (Figure 7), respectively.

other than type I have not been determined. However, from the composition of the N-terminal cyanogen bromide peptide of α1(II), a dipeptide, it can be determined that the N-terminal sequence is pGlu-Met-, as in α1(I). The next peptide contains a hydroxylysine, which is consistent with a role for this region similar to the same region in α1(I) and α2 (Miller *et al.*, 1973). It is important to determine whether this is the case for other collagens since it is not at all certain that aldehyde-derived cross-links are common to all collagens.

The C-terminal nonhelical region in α chains avoided detection for some time because it is easily lost by proteolysis during isolation (Stark *et al.*, 1971). However, the use of proper precautions has allowed it to be isolated and its sequence to be determined (Rauterberg *et al.*, 1972*b*). The C-terminal sequence for calf collagen α1(I) is shown in Figure 3. The presence of lysine-containing and aldehyde-containing forms of the C-terminal chymotryptide indicate that this region, like the N-terminal region, is a cross-linking site. A similar region in rat collagen α1(I) (Stoltz *et al.*, 1972) and chick collagen α1(II) (Miller *et al.*, 1973) has been reported although its sequence has not been determined. Similar C-terminal regions for other α chain types have not yet been demonstrated but are likely to be present. A review by Rauterberg (1973) discusses the data in more detail.

Other than the fact that these terminal sequences cannot have the triple-chain collagen helix, the sequences themselves reveal little or nothing about structure and function. The regions are rich in amino acids with large hydrophobic side chains and contain charged amino acids, consistent with a role in some type of specific interaction perhaps associated with fibril formation (see Section V-B).

B. *Helical Regions*

Sequences from the helical portion of the α1(I) chains of type I collagen from rat, calf, and chick, the α1(II) chain of cartilage collagen

TABLE 1
Amino Acid Sequences in the Helical Region of Collagen Chains[a]

Residue	α1(I) rat	α1(I) calf	α1(I) chick	α1(II) steer	α2 rat	α2 calf	SLS band[b]	Fibril band[b]
18	Pro	Pro	Pro	Val	Pro	Pro		
19	Met	Met	Met	Met	Met	Met		
21	Pro	Pro	Pro	Pro	Leu	Leu		
22	Ser	Ser	Ala	Met	Met	Met		
24	Pro	Pro	Pro	Pro	Pro	Pro		
25	Arg	Arg	Arg	Arg	Arg	Arg	0A	II
27	Leu	Leu	Leu	Pro	Pro	Pro		
28	Hyp	Hyp	Hyp	Hyp	Hyp	Hyp		
30	Pro	Pro	Pro	Pro	Ala	Ala		
31	Hyp	Hyp	Hyp	Ala	Val	Ser		
33	Ala	Ala	Ala	Ala	Ala	Ala		
34	Hyp	Hyp	Hyp	Hyp	Hyp	Hyp		
36	Pro	Pro	Pro	Pro	Pro	Pro		
37	Gln	Gln	Gln	Gln	Gln	Gln		
39	Phe	Phe	Phe	Phe	Phe	Phe		
40	Gln	Gln	Gln	Gln	Gln	Gln		
42	Pro	Pro	Pro	Asn	Pro	Pro		
43	Hyp	Hyp	Hyp	Hyp	Ala	Hyp		
45	Glu	Glu	Glu	Glu	Glu	Glu	1	III
46	Hyp	Hyp	Hyp	Hyp	Hyp	Hyp		
48	Glu	Glu	Glu	Glu	Glu	Glu	1	III
49	Hyp	Hyp	Hyp	Hyp	Hyp	Hyp		
51	Ala	Ala	Ala	Val	Gln	Gln		
52	Ser	Ser	Ser	Ser	Hyp	Thr		
54	Pro	Pro	Pro	Pro	Pro	Pro		
55	Met	Met	Met	Met	Ala	Ala		
57	Pro	Pro	Pro	Pro	Pro	Ala		
58	Arg	Arg	Arg	Arg	Arg	Arg	2	III
60	Pro	Pro	Pro	Pro	Ala	Pro		
61	Hyp	Hyp	Ala	Hyp	Hyp	Hyp		
63	Pro	Pro	Pro	Pro	Pro	Pro		
64	Hyp	Hyp	Hyp	Hyp	Hyp	Hyp		
66	Lys	Lys	Lys	Lys		Lys	3	IV
67	Asn	Asn	Asn	Hyp		Ala		
69	Asp	Asp	Asp	Asp		Glu	3	IV
70	Asp	Asp	Asp	Asp		Asp	3	IV
72	Glu	Glu	Glu	Glu		His	3	IV
73	Ala	Ala	Ala	Ala		Hyp		
75	Lys	Lys	Lys	Lys		Lys	4	IV
76	Pro	Pro	Pro	Hyp		Pro		
78	Arg	Arg	Arg	Lys		Arg	4	IV
79	Hyp	Hyp	Hyp	Ser		Hyp		
81	Gln	Glu	Glu	Glu		Glu		

TABLE 1—*Continued*

Residue	α1(I) rat	α1(I) calf	α1(I) chick	α1(II) steer	α2 rat	α2 calf	SLS band[b]	Fibril band[b]
82	Arg	Arg	Arg	Arg		Arg	4	IV
84	Pro	Pro	Pro	Pro		Val		
85	Hyp	Hyp	Hyp	Hyp		Pro		
87	Pro	Pro	Pro	Pro		Pro		
88	Gln	Gln	Gln	Gln		Gln		
90	Ala	Ala	Ala	Ala		Ala		
91	Arg	Arg	Arg	Arg		Arg	4A	V
93	Leu	Leu	Leu	Phe		Phe		
94	Hyp	Hyp	Hyp	Hyp		Hyp		
96	Thr	Thr	Thr	Thr		Thr		
97	Ala	Ala	Ala	Hyp		Hyp		
99	Leu	Leu	Leu	Leu		Leu		
100	Hyp	Hyp	Hyp	Hyp		Hyp		
102	Met	Met	Met	Val		Phe		
103	Hyl#	Hyl	Hyl#	Hyl#		Hyl	5	VI
105	His	His	His	His		Ile		
106	Arg	Arg	Arg	Arg		Arg	5	VI
108	Phe	Phe	Phe	Tyr		His		
109	Ser	Ser	Ser	Hyp		Asn		
111	Leu	Leu	Leu	Leu		Leu		
112	Asp	Asp	Asp	Asp		Asp	5A	VII
114	Ala	Ala	Ala	Ala		Leu		
115	Lys	Lys	Lys	Hyl#		Thr	5A	VII
117	Asn	Asp	Glu	Glu		Gln	5A	VII
118	Thr	Ala	Hyp	Ala		Hyp		
120	Pro	Pro	Pro	Ala		Ala		
121	Ala	Ala	Ala	Hyp		Hyp		
123	Pro	Pro	Pro	Val		Val		
124	Lys	Lys	Lys	Hyl#		Hyl	6	VIII
126	Glu	Glu	Glu	Glu		Glu	6	VIII
127	Hyp	Hyp	Hyp	Ser		Hyp		
129	Ser	Ser	Ser	Ser		Ala		
130	Hyp	Hyp	Hyp	Hyp		Hyp		
132	Glx	Glu	Glu	Glx*		Glu	6	VIII
133	Asx	Asn	Asn	Asx*		Asn		
135	Ala	Ala	Ala	Ser*		Thr		
136	Hyp	Hyp	Hyp	Hyp*		Hyp		
138	Gln	Gln	Gln	Pro		Gln		
139	Met	Met	Met	Met		Hyl		
141	Pro			Pro		Ala		
142	Arg			Arg		Arg	7	IX
144	Leu			Leu		Leu		
145	Hyp			Hyp		Hyp		
147	Glu			Glu		Glu	7	IX

TABLE 1—*Continued*

Residue	α1(I) rat	α1(I) calf	α1(I) chick	α1(II) steer	α2 rat	α2 calf	SLS band[b]	Fibril band[b]
148	Arg			Arg		Arg	7	IX
150	Arg			Arg		Arg	7	IX
151	Hyp			Thr		Val		
153	Pro			Pro		Ala		
154	Hyp			Ala		Hyp		
156	Ser			Ala		Pro		
157	Ala			Ala		Ala		
159	Ala			Ala		Ala		
160	Arg			Arg		Arg	8	X
162	Asp			Asn		Ser	8	X
163	Asp			Asp		Asp	8	X
165	Ala			Gln		Ser		
166	Val			Hyp		Val		
168	Ala			Pro		Pro		
169	Ala			Ala		Val		
171	Pro			Pro		Pro		
172	Hyp			Hyp		Ala		
174	Pro			Pro		Pro		
175	Thr			Val		Ile		
177	Pro			Pro		Ser		
178	Thr			Ala		Ala		
180	Pro					Pro		
181	Hyp					Hyp		
183	Phe					Phe		
184	Hyp					Hyp		
186	Ala					Ala		
187	Ala					Hyp		
189	Ala					Pro		
190	Lys					Hyl	9	XI
192	Glu					Glu	9	XI
193	Ala					Leu		
195	Pro					Pro		
196	Gln					Val		
198	Ala					Asn		
199	Arg					Hyp	10	XI
201	Ser					Pro		
202	Glu					Ala	10	XI
204	Pro					Pro		
205	Gln					Ala		
207	Val					Pro		
208	Arg					Arg	10	XI
210	Glu					Glu	10	XI
211	Hyp					Val		
213	Pro					Leu		

TABLE 1—*Continued*

Residue	α1(I) rat	α1(I) calf	α1(I) chick	α1(II) steer	α2 rat	α2 calf	SLS band[b]	Fibril band[b]
214	Hyp					Hyp		
216	Pro					Leu		
217	Ala					Ser		
219	Ala					Pro		
220	Ala					Val		
222	Pro					Pro		
223	Ala					Hyp		
225	Asn					Asn		
226	Hyp					Ala		
228	Ala					Pro		
229	Asp					Asn	11	I
231	Gln					Leu		
232	Hyp					Hyp		
234	Ala					Ala		
235	Lys					Hyl	11	I
237	Ala					Ala		
238	Asn					Ala		
240	Ala					Leu		
241	Hyp					Hyp		
243	Ile					Val		
244	Ala					Ala		
246	Ala					Ala		
247	Hyp					Hyp		
249	Phe					Leu		
250	Hyp					Hyp		
252	Ala					Pro		
253	Arg					Arg	12	II
255	Pro					Ile		
256	Ser					Hyp		
258	Pro					Pro		
259	Gln					Val		
261	Pro					Ala		
262	Ser					Ala		
264	Ala					Ala		
265	Hyp					Thr		
267	Pro					Ala		
268	Lys					Arg	13	III
270	Asn					Leu		
271	Ser					Val		
273	Glu					Glu	13	III
274	Hyp					Hyp		
276	Ala					Pro		
277	Hyp					Ala		
279	Asn					Ser		

TABLE 1—*Continued*

Residue	α1(I) rat	α1(I) calf	α1(I) chick	α1(II) steer	α2 rat	α2 calf	SLS band[b]	Fibril band[b]
280	Lys					Hyl	14	III
282	Asp					Glu	14	III
283	Thr					Ser		
285	Ala					Asn		
286	Lys					Lys	14	III
288	Glu					Glu	14	III
289	Hyp					Hyp		
291	Pro					Ala		
292	Ala					Val		
294	Val					Gln		
295	Gln					Hyp		
297	Pro					Pro		
298	Hyp					Hyp		
300	Pro					Pro		
301	Ala					Ser		
303	Glu					Glu	15	IV
304	Glu					Glu	15	IV
306	Lys					Lys	15	IV
307	Arg					Arg	15	IV
309	Ala					Ser		
310	Arg					Thr	15	IV
312	Glu					Glu	15	IV
313	Hyp					Ile		
315	Pro					Pro		
316	Ser					Ala		
318	Leu					Pro		
319	Hyp					Hyp		
321	Pro					Pro		
322	Hyp					Hyp		
324	Glu					Leu	16	V
325	Arg					Arg	16	V
327	Gly					Asn		
328	Hyp					Hyp		
330	Ser					Ser		
331	Arg					Arg	16A	VI
333	Phe					Leu		
334	Hyp					Hyp		
336	Ala					Ala		
337	Asp					Asp	16A	VI
339	Val					Arg		
340	Ala					Ala		
342	Pro					Val		
343	Lys					Met	17	VII
345	Pro				Pro	Pro		

TABLE 1—*Continued*

Residue	α1(I) rat	α1(I) calf	α1(I) chick	α1(II) steer	α2 rat	α2 calf	SLS band[b]	Fibril band[b]
346	Ala				Hyp	Ala		
348	Glu				Asn	Ser	17	VII
349	Arg				Arg	Arg	17	VII
351	Ser				Thr	Thr		
352	Hyp				Ser	Ala		
354	Pro				Pro	Pro		
355	Ala				Ala	Ala		
357	Pro				Val	Val		
358	Lys				Arg	Arg	18	VIII
360	Ser				Pro*	Pro		
361	Hyp				Asx*	Asn		
363	Glu				Asx*	Asp	18	VIII
364	Ala				Ala*	Ser		
366	Arg				Arg*	Arg	18	VIII
367	Hyp				Hyp*	Hyp		
369	Glu				Glx*	Glu	18	VIII
370	Ala				Hyp*	Hyp		
372	Leu				Leu*	Leu		
373	Hyp				Met	Met		
375	Ala				Pro	Pro		
376	Lys				Arg	Arg	19	IX
378	Leu				Leu	Phe		
379	Thr				Hyp	Hyp		
381	Ser				Ser	Ser		
382	Hyp				Hyp	Hyp		
384	Ser				Asn	Asn		
385	Hyp				Val	Ile		
387	Pro				Pro	Pro		
388	Asp				Ala	Ala	20	IX
390	Lys				Lys	Lys	20	IX
391	Thr				Glu	Glu		
393	Pro				Pro	Pro		
394	Hyp				Val	Val		
396	Pro				Leu	Leu		
397	Ala				Hyp	Hyp		
399	Glx				Ile	Ile		
400	Asx				Asp	Asp	21	X
402	Arg				Arg	Arg	21	X
403	Hyp				Hyp	Hyp		
405	Pro				Pro	Pro		
406	Ala				Ile	Ile		
408	Pro				Pro	Pro		
409	Hyp				Ala	Ala		
411	Ala				Pro			

TABLE 1—*Continued*

Residue	α1(I) rat	α1(I) calf	α1(I) chick	α1(II) steer	α2 rat	α2 calf	SLS band[b]	Fibril band[b]
412	Arg				Arg		22	X
414	Gln				Glx			
415	Ala				Ala			
417	Val				Ala			
418	Met				Ile			
420	Phe	Phe	Phe	Phe	Phe			
421	Hyp	Hyp	Hyp	Hyp	Hyp			
423	Pro	Pro	Pro	Pro				
424	Lys	Lys	Lys	Hyl#			23	XI
426	Thr	Ala	Ala	Ala				
427	Ala	Ala	Ala	Asn				
429	Glu	Glu	Glu	Glu			23	XI
430	Hyp	Hyp	Hyp	Hyp				
432	Lys	Lys	Lys	Lys			23A	XI
433	Ala	Ala	Hyp	Ala				
435	Glu	Glu	Glu	Glu			23A	XI
436	Arg	Arg	Arg	Hyl#			23A	XI
438	Val	Val	Ala	Leu				
439	Hyp	Hyp	Hyp	Hyp				
441	Pro	Pro	Pro	Ala				
442	Hyp	Hyp	Hyp	Hyp				
444	Ala	Ala	Ala	Leu				
445	Val	Val	Val	Arg				
447	Pro	Pro	Ala	Leu				
448	Ala	Ala	Ala	Hyp				
450	Lys	Lys	Lys	Lys			24	XII
451	Asp	Asp	Asp	Asp			24	XII
453	Glu	Glu	Glu	Glu			24	XII
454	Ala	Ala	Ala	Thr				
456	Ala	Ala	Ala	Ala				
457	Gln	Gln	Gln	Ala				
459	Ala	Pro	Pro	Pro				
460	Hyp	Hyp	Pro	Hyp				
462	Pro	Pro	Pro	Pro				
463	Ala	Ala	Thr	Ala				
465	Pro	Pro	Pro	Pro				
466	Ala	Ala	Ala	Ala				
468	Glu	Glu	Glu	Glu			25	I
469	Arg	Arg	Arg	Arg			25	I
471	Glu	Glu	Glu	Glu			25	I
472	Gln	Gln	Gln	Gln				
474	Pro	Pro	Pro	Ala				
475	Ala	Ala	Ala	Hyp				
477	Ser	Ser	Ala	Pro				

TABLE 1—*Continued*

Residue	α1(I) rat	α1(I) calf	α1(I) chick	α1(II) steer	α2 rat	α2 calf	SLS band[b]	Fibril band[b]
478	Hyp	Hyp	Hyp	Ser				
480	Phe	Phe	Phe	Phe				
481	Gln	Gln	Gln	Gln				
483	Leu	Leu	Leu	Leu				
484	Hyp	Hyp	Hyp	Hyp				
486	Pro	Pro	Pro	Pro				
487	Ala	Ala	Ala	Hyp				
489	Pro	Pro	Pro	Pro				
490	Hyp	Hyp	Hyp	Hyp				
492	Glu	Glu	Glu	Glu			26	II
493	Ala	Ala	Ala	Hyp				
495	Lys	Lys	Lys				26	II
496	Hyp	Hyp	Hyp					
498	Glx	Glu	Glu				26	II
499	Glx	Gln	Gln					
501	Val	Val	Val					
502	Hyp	Hyp	Hyp					
504	Asp*	Asp	Asn				26A	III
505	Leu*	Leu	Ala					
507	Ala*	Ala	Ala					
508	Hyp*	Hyp	Hyp					
510	Pro	Pro	Pro					
511	Ser	Ser	Ala					
513	Ala	Ala	Ala					
514	Arg	Arg	Arg				27	III
516	Glu	Glu	Glu				27	III
517	Arg	Arg	Arg				27	III
519	Phe	Phe	Phe					
520	Hyp	Hyp	Hyp					
522	Glu	Glu	Glu				28	III
523	Arg	Arg	Arg				28	III
525	Val	Val	Val					
526	Gln	Glu	Gln				28	III
528	Pro	Pro	Pro					
529	Hyp	Hyp	Hyp					
531	Pro	Pro	Pro					
532	Ala	Ala	Gln					
534	Pro	Pro	Pro					
535	Arg	Arg	Arg				29	IV
537	Asn	Ala	Ala					
538	Asn	Asn	Asn					
540	Ala	Ala	Ala					
541	Hyp	Hyp	Hyp					
543	Asx	Asn	Asn					

TABLE 1—*Continued*

Residue	α1(I) rat	α1(I) calf	α1(I) chick	α1(II) steer	α2 rat	α2 calf	SLS band[b]	Fibril band[b]
544	Asx	Asp	Asp				30	IV
546	Ala	Ala	Ala					
547	Lys	Lys	Lys				30	IV
549	Asp	Asp	Asp				30	IV
550	Thr	Ala	Ala					
552	Ala	Ala	Ala					
553	Hyp	Hyp	Hyp					
555	Ala	Ala	Ala					
556	Hyp	Hyp	Hyp					
558	Ser	Ser	Asn					
559	Gln	Gln	Glu					
561	Ala	Ala	Pro					
562	Hyp	Hyp	Hyp					
564	Leu	Leu	Leu					
565	Glx	Gln	Glu					
567	Met	Met	Met					
568		Hyp	Hyp					
570		Glu	Glu				31	VI
571		Arg	Arg				31	VI
573		Ala	Ala					
574		Ala	Ala					
576		Leu	Leu					
577		Hyp	Hyp					
579		Pro	Ala					
580		Lys	Lys				32	VII
582		Asp	Asp				32	VII
583		Arg	Arg				32	VII
585		Asp	Asp				32	VII
586		Ala	Hyp					
588		Pro	Pro					
589		Lys	Lys				33	VIII
591		Ala	Ala					
592		Asp	Asp				33	VIII
594		Ala	Ala					
595		Pro	Pro					
597		Lys	Lys				33	VIII
598		Asp	Asp				33	VIII
600		Val	Leu					
601		Arg	Arg				33	VIII
603		Leu	Leu					
604		Thr	Thr					
606		Pro	Pro					
607		Ile	Ile					
609		Pro	Pro					

TABLE 1—*Continued*

Residue	α1(I) rat	α1(I) calf	α1(I) chick	α1(II) steer	α2 rat	α2 calf	SLS band[b]	Fibril band[b]
610		Hyp	Hyp					
612		Pro	Pro					
613		Ala	Ala					
615		Ala	Ala					
616		Hyp	Hyp					
618		Asp	Asp				34	IX
619		Lys	Lys				34	IX
621		Glu	Glu				34	IX
622		Ala	Ala					
624		Pro	Pro					
625		Ser	Ala					
627		Pro	Pro					
628		Ala	Ala					
630		Thr	Thr					
631		Arg	Arg				35	X
633		Ala	Ala					
634		Hyp	Hyp					
636		Asp	Asp				35	X
637		Arg	Arg				35	X
639		Glu	Glu				35	X
640		Hyp	Hyp					
642		Pro	Pro					
643		Hyp	Hyp					
645		Pro	Pro					
646		Ala	Ala					
648		Phe	Phe					
649		Ala	Ala					
651		Pro	Pro					
652		Hyp	Hyp					
654		Ala	Ala					
655		Asp	Asp				36	XI
657		Gln	Gln					
658		Hyp	Hyp					
660		Ala	Ala					
661		Lys	Lys				36	XI
663		Glu	Glu				36	XI
664		Hyp	Thr					
666		Asp	Asp				36	XI
667		Ala	Ala					
669		Ala	Ala					
670		Lys	Lys				36	XI
672		Asp	Asp				36	XI
673		Ala	Ala					
675		Pro	Pro					

TABLE 1—*Continued*

Residue	$\alpha 1$(I) rat	$\alpha 1$(I) calf	$\alpha 1$(I) chick	$\alpha 1$(II) steer	$\alpha 2$ rat	$\alpha 2$ calf	SLS band[b]	Fibril band[b]
676		Hyp	Hyp					
678		Pro	Pro					
679		Ala	Ala					
681		Pro	Pro					
682		Ala	Thr					
684		Pro	Ala					
685		Hyp	Hyp					
687		Pro	Pro					
688		Ile	Ala					
690		Asn	Glx					
691		Val	Val					
693		Ala	Ala					
694		Hyp	Hyp					
696		Pro	Pro					
697		Hyl	Hyl				37	I
699		Ala	Ala					
700		Arg	Arg				37	I
702		Ser	Ser					
703		Ala	Ala					
705		Pro	Pro					
706		Hyp	Hyp					
708		Ala	Ala					
709		Thr	Thr		Thr			
711		Phe	Phe		Phe			
712		Hyp	Hyp		Hyp			
714		Ala	Ala		Ala			
715		Ala	Ala		Ala			
717		Arg	Arg		Arg		38	II
718		Val	Val		Thr			
720		Pro	Pro		Pro			
721		Hyp	Hyp		Hyp			
723		Pro	Pro		Pro			
724		Ser	Ser		Ser			
726		Asn	Asn		Ile			
727		Ala	Ile		Thr			
729		Pro	Leu		Pro			
730		Hyp	Hyp		Hyp			
732		Pro	Pro		Pro			
733		Hyp	Hyp		Hyp			
735		Pro	Pro		Ala			
736		Ala	Ala		Ala			
738		Lys	Lys*		Lys		39	III
739		Glu	Glx*		Glu		39	III
741		Ser	Ser*		Ile			

TABLE 1—*Continued*

Residue	α1(I) rat	α1(I) calf	α1(I) chick	α1(II) steer	α2 rat	α2 calf	SLS band[b]	Fibril band[b]
742		Lys	Lys*		Lys		39	III
744		Pro	Pro		Pro			
745		Arg	Arg		Arg		39	III
747		Glu	Glu		Asp		39	III
748		Thr	Thr		Gln			
750		Pro	Pro		Pro			
751		Ala	Ala		Val			
753		Arg	Arg		Arg		40	III
754		Hyp	Hyp					
756		Glu	Glu				40	III
757		Val	Hyp					
759		Pro	Pro					
760		Hyp	Ala					
762		Pro	Pro					
763		Hyp	Hyp					
765		Pro	Pro					
766		Ala	Hyp					
768		Glu	Glu				41	IV
769		Lys	Lys				41	IV
771		Ala	Ala					
772		Hyp	Hyp					
774		Ala	Ala					
775		Asp	Asp				41	IV
777		Pro	Pro					
778		Ala	Ile					
780		Ala	Ala					
781		Hyp	Hyp					
783		Thr	Thr					
784		Pro	Pro					
786		Pro	Pro					
787		Gln	Gln					
789		Ile	Ile					
790		Ala	Ala					
792		Gln	Gln					
793		Arg	Arg				42	V
795		Val	Val					
796		Val	Val					
798		Leu	Leu					
799		Hyp	Hyp					
801		Gln	Gln					
802		Arg	Arg				43	VI
804		Glu	Glu				43	VI
805		Arg	Arg				43	VI
807		Phe	Phe					

TABLE 1—*Continued*

Residue	α1(I) rat	α1(I) calf	α1(I) chick	α1(II) steer	α2 rat	α2 calf	SLS band[b]	Fibril band[b]
808		Hyp	Hyp					
810		Leu	Leu					
811		Hyp	Hyp					
813		Pro	Pro					
814		Ser	Ser					
816		Glu	Glu				44	VII
817		Hyp	Hyp					
819		Lys	Lys				44	VII
820		Gln	Gln					
822		Pro	Pro					
823		Ser	Ser					
825		Ala	Ala					
826		Ser	Ser					
828		Glu	Glu				45	VIII
829		Arg	Arg				45	VIII
831		Pro	Pro					
832		Hyp	Hyp					
834		Pro	Pro					
835		Met	Met					
837		Pro	Pro					
838		Hyp	Hyp					
840		Leu	Leu					
841		Ala	Ala					
843		Pro	Pro					
844		Hyp	Hyp					
846		Glu	Glu				46	IX
847		Ser	Ala					
849		Arg	Arg				46	IX
850		Glu	Glu				46	IX
852		Ala	Ala					
853		Hyp	Hyp					
855		Ala	Ala					
856		Glu	Glu				47	IX
858		Ser	Ala					
859		Hyp	Hyp					
861		Arg	Arg				47	X
862		Asp	Asp				47	X
864		Ser	Ala					
865		Hyp	Ala					
867		Ala	Pro					
868		Lys	Lys				48	X
870		Asp	Asp				48	X
871		Arg	Arg				48	X
873		Glu	Glu				48	X

TABLE 1—*Continued*

Residue	α1(I) rat	α1(I) calf	α1(I) chick	α1(II) steer	α2 rat	α2 calf	SLS band[b]	Fibril band[b]
874		Thr	Thr					
876		Pro	Pro					
877		Ala	Ala					
879		Pro*	Pro					
880		Hyp*	Hyp					
882		Ala*	Ala					
883		Hyp*	Hyp					
885		Ala*	Ala					
886		Hyp*	Hyp					
888		Ala*	Ala					
889		Hyp*	Pro					
891		Pro	Pro					
892		Val	Val					
894		Pro	Pro					
895		Ala	Ala					
897		Lys	Lys				49	XI
898		Ser	Asn					
900		Asp	Asp				49	XI
901		Arg	Arg				49	XI
903		Glu	Glu				49	XI
904		Thr	Thr					
906		Pro	Pro					
907		Ala	Ala					
909		Pro	Pro					
910		Ile	Ala					
912		Pro	Pro					
913		Val	Hyp					
915		Pro	Pro					
916		Ala	Ala					
918		Ala	Ala					
919		Arg	Arg				50	XII
921		Pro	Pro					
922		Ala	Ala					
924		Pro	Pro					
925		Gln	Gln					
927		Pro	Pro					
928		Arg	Arg				51	I
930		Asx	Asp				51	I
931		Hyl	Hyl				51	I
933		Glx	Glu				51	I
934		Thr	Thr					
936		Glx	Glu				51	I
937		Glx	Gln					
939		Asx	Asp				52	I

TABLE 1—*Continued*

Residue	α1(I) rat	α1(I) calf	α1(I) chick	α1(II) steer	α2 rat	α2 calf	SLS band[b]	Fibril band[b]
940		Arg	Arg				52	I
942		Ile	Met					
943		Hyl					52	I
945		His						
946		Arg					52	I
948		Phe						
949		Ser						
951		Leu						
952		Gln						
954		Pro						
955		Hyp						
957		Pro						
958		Hyp						
960		Ser						
961		Hyp						
963		Glu					53	II
964		Gln						
966		Pro						
967		Ser						
969		Ala						
970		Ser						
972		Pro						
973		Ala						
975		Pro						
976		Arg					54	III
978		Pro						
979		Hyp						
981		Ser						
982		Ala						
984		Ser						
985		Hyp						
987		Lys					55	III
988		Asp					55	
990		Leu						
991		Asn						
993		Leu						
994		Hyp						
996		Pro						
997		Ile						
999		3 Hyp						
1000		Hyp						
1002		Pro						
1003		Arg					56	IV
1005		Arg					56	IV

TABLE 1—*Continued*

Residue	α1(I) rat	α1(I) calf	α1(I) chick	α1(II) steer	α2 rat	α2 calf	SLS band[b]	Fibril band[b]
1006		Thr						
1008		Asp					56	IV
1009		Ala						
1011		Pro						
1012		Ala						
1014		Pro						
1015		Hyp						
1017		Pro						
1018		Hyp						
1020		Pro						
1021		Hyp						
1023		Pro						
1024		Hyp						
1026		Pro						
1027		Pro						

[a] Glycine occurs in every third position beginning with residue 17 but is omitted to save space. Residues which have not been chemically ordered are followed by *. Hydroxylysine residues (Hyl) which are known to carry carbohydrate are followed by # References to these data are given in Table 2.

[b] The SLS and native fibril band designations, to which the basic and acidic residues in α1(I) are assigned, follow the nomenclature of Bruns and Gross (1973, 1974) with the addition of several SLS bands (designated A).

from steer, and the α2 chains of rat and calf collagen are assembled in Table 1. References from which the data were taken appear in Table 2. The numbering system is based on the rat (or calf) α1(I) chain which is the chain that has been most extensively studied. Residue 1 is the N-terminal pyroglutamic acid at the beginning of the nonhelical region (Figure 2). Gaps [or an extra residue for baboon and human α1(I)] are inserted where required to maintain homologous matching. So far this is necessary only in the nonhelical N-terminal regions.

A few short stretches of sequence from chains not included in Table 1 are available. These include: (1) guinea pig α2, residues 374–394, which are identical to rat α2 except for Ala at 385 and Met at 394; guinea pig α2, residues 114–132, which are identical to calf α2 except for Ala at 120; guinea pig α2, residues 23–43, which are identical to rat α2; rabbit α2, residues 23–49, which are identical to rat α2 except for Ala at 31 (Clark, Fietzek and Bornstein, 1976); (2) human, rabbit, pig (Fietzek *et al.*, 1974*b*), and chick α2 (Highberger *et al.*, 1971), residues 344–373, which are

TABLE 2
Literature References to Sequences in the Helical Region of Collagen Chains

Chain	Inclusive residues	Reference
Rat $\alpha1(I)$	20–55	Bornstein, 1967
	56–102	Butler and Ponds, 1971
	103–139	Butler, 1970
	140–238	Balian *et al.*, 1971
	239–418	Balian *et al.*, 1972
	419–567	Butler *et al.*, 1974*a*
Calf $\alpha1(I)$	20–139	Fietzek and Kuhn, 1975
	419–567	Fietzek *et al.*, 1972*a*
	568–835	Fietzek *et al.*, 1973
	836–948	Wendt *et al.*, 1972
	949–1027	Fietzek *et al.*, 1972*b*
Chick $\alpha1(I)$	17–55	Kang and Gross, 1970
	56–139	Kang *et al.*, 1975
	416–567	Dixit *et al.*, 1975*a*
	568–835	Highberger *et al.*, 1975
	836–942	Dixit *et al.*, 1975*b*
Steer $\alpha1(I)$	17–178	Butler *et al.*, 1976
	419–493	Butler *et al.*, 1974*b*
Rat $\alpha2$	20–64	Fietzek *et al.*, 1972*c*
	344–373	Highberger *et al.*, 1971
	374–421	Fietzek and Kuhn, 1974
	709–753	Fietzek and Kuhn, 1973
Calf $\alpha2$	23–343	Fietzek and Rexrodt, 1975
	344–373	Fietzek *et al.*, 1974*b*
	374–421	Fietzek and Kuhn, 1974

identical to calf and rat $\alpha2$ except at 346, 348, 352, 358, and 364 where conservative substitutions occur; (3) calf aorta $\alpha1(III)$, residues 140–165, 419–453, and 568–598, which show a close homology to other $\alpha1$ chains (Fietzek and Rauterberg, 1975).

C. *Hydroxylated Residues*

4-Hydroxyproline is only found preceding glycine in the helical region of the α chains of vertebrate collagens. In sequence studies on

$\alpha 1$(I)-CB2, it was noted that hydroxylation of proline preceding glycine is not always complete and may be different at a given site for the same collagen in different tissues (Bornstein, 1967). Partial hydroxylation has also been found in other parts of the $\alpha 1$(I) chain; some residues (76, 595, and 784, Table 1) apparently escape hydroxylation completely. Note, however, that residue 76 in steer $\alpha 1$(II) is hydroxylated even though the sequence around it is very similar to the same region in $\alpha 1$(I). A basis for understanding this variability now exists with the demonstration that hydroxylation of proline preceding glycine is necessary for molecular stability (Chapter 2) and that hydroxylation occurs only on random-coil α chains (Chapter 5). Apparently hydroxylation proceeds until sufficient hydroxyproline is formed for the chains to make a stable triple helix and then stops. The amount of hydroxylation required in general is somewhat less than the amount possible. Of course this concept does not explain the function (if any) of variable hydroxylation, nor does it imply that hydroxylation is the only controlling factor in helix formation.

It should be remembered that variability is found only at some sites. The overall degree of hydroxylation of the proline in a given collagen is constant, and must be constant within narrow limits, under normal and many abnormal conditions (see Chapter 5).

3-Hydroxyproline occurs to only a small extent in type I and II collagens. The one residue located in the sequence [999 in calf $\alpha 1$(I), Table 1] follows glycine and precedes 4-hydroxyproline. Studies on type IV collagen, which contains large amounts of 3-hydroxyproline, suggest that this may always be true (Gryder *et al.*, 1974). Nothing is known about the formation or function of this amino acid. The amount present in type I collagen from different species and tissues varies from 0 to 4 residues/α chain (Piez *et al.*, 1963).

Unlike 4-hydroxyproline, the hydroxylysine content of a collagen is highly variable from tissue to tissue and is a function of developmental stage or pathology (Piez and Likins, 1960; Miller *et al.*, 1967; Butler, 1973; Barnes, 1973; Barnes *et al.*, 1974; Cintron, 1974). This variability and its relationship to cross-linking is discussed elsewhere in this book (Chapters 4 and 8). It need only be noted here that only a few lysines appear to be fully hydroxylated in type I collagen of skin (residues 103, 697, 931, and 943, Table 1), although a number of other lysines are hydroxylated to a minor extent (Butler, 1968). Type II collagen, however, presents a different picture. Every lysine that preceeds glycine, in the sequence so far determined, is hydroxylated (Table 1).

It has already been noted (Section IV-A) that the N-terminal cross-link-precursor lysine, residue 9 (Figure 2), may be hydroxylated. The sequence around this residue is of course very different from that around other hydroxylysines, raising the possibility of different methods of biological regulation as well as different functions.

D. Carbohydrate

The evidence that the major and perhaps sole carbohydrate constituents in type I collagen are galactosylhydroxylysine and glucosylgalactosylhydroxylysine has been reviewed (Traub and Piez, 1971; Gallop *et al.*, 1972). Hydroxylysine residues that are known to be glycosylated in this fashion are indicated in Table 1. Analysis of the CNBr peptides from rat skin α1(I) indicates the presence of only one attachment site, residue 103. However, this may not be true for α1(I) from other tissues and other species. The α2 chain from rat skin contains two major sites of glycosylation, both located in α2-CB4 (Aguilar *et al.*, 1973), which contains residues 23–343.

Since the degree of hydroxylation of lysine in type I collagen varies, it is not unexpected that the amount of carbohydrate will vary. However, the amount and the ratio of mono- to diglycosylated residues also vary independently (Pinnell *et al.*, 1971; Barnes, 1973; Cintron, 1974).

Type II collagen contains about 10% carbohydrate, also as galactosylhydroxylysine and glucosylgalactosylhydroxylysine. The hydroxylysines so far identified in the sequence (residues 103, 115, 424, and 436, Table 1) are all fully glycosylated. A similar situation may hold for type IV collagen. In addition, type IV collagen has a heteropolysaccharide associated with it which, however, appears to be attached either to a noncollagenous protein or to a noncollagenous part of the type IV collagen molecule (Hudson and Spiro, 1972*b*; Kefalides, 1973).

E. Complement Component, C1q

Clq has a molecular weight of 410,000 and has six subunits, each containing three chains; about half of each chain appears to be collagen-like (Reid *et al.*, 1972). Sequence studies of one of the several types of chain confirm the collagen nature of this protein (Reid, 1974). As shown in Figure 4, there is a region of 78 residues consisting of typical collagen triplets Gly-X-Y where Y is sometimes hydroxyproline or hydroxylysine and X is sometimes proline. This region is preceded and followed by nontriplet regions. Presumably the triplet portion is present in the native

```
Ala-Pro-Asp-Gly-Lys-Hyl-Gly-Glx-Ala-Gly-Arg-Hyp-Gly-Arg-Arg-Gly-Arg-Hyp-
 1               5                  10                  15
Gly-Leu-Hyl-Gly-Glx-Glx-Gly-Glx-Hyp-Gly-Ala-Hyp-Gly-Ile-Arg-Gly-Thr-Ile-
    20                  25                  30                  35
Gly-Glx-Leu-Gly-Asx-Glx-Gly-Glx-Hyl-Gly-Pro-Ser-Gly-Asn-Pro-Gly-Lys-Val-
            40                  45                  50
Gly-Tyr-Hyp-Gly-Pro-Ser-Gly-Pro-Leu-Gly-Ala-Arg-Gly-Ile-Hyl-Gly-Ile-Hyl-
 55                 60                  65                  70
Gly-Thr-xxx-Gly-Pro-Ser-Gly-Asn-Ile-Lys-Glx-Gly-Asp-Gln-Pro-Arg-Pro-Ala-
        75                  80                  85                  90
Phe-Ser-Ala-Ile-Arg-
            95
```

FIGURE 4. The amino acid sequence of a portion of the A chain from human Clq showing the triplet collagen region (residues 4–81). Some aspects of the sequence are tentative. xxx is a unidentified residue (see Reid, 1974).

Clq as a triple-chain collagen helix. Its calculated length would be about 220 Å, assuming 2.86 Å per residue as in collagen.

Since it seems unlikely that collagen triplet sequences could have arisen independently, the Clq sequence was compared to the $\alpha 1(I)$ sequence by computer techniques (Barker and Dayhoff, 1972) to look for a similarily that might be ascribed to homology. Although the 78 collagen-like residues (4–81, Figure 5) from Clq were found to be considerably more similar to residues 131–208 than to any other region in rat $\alpha 1(I)$, comparison to random triplet sequences indicates that the statistical significance is too low to support a common derivation (Piez, unpublished). Of course, this does not prove separate origins since independent evolution could have erased most of the similarity.

It is certainly significant that the collagen molecular structure occurs in a protein that in other respects is not like interstitial collagen. Perhaps the rigid rodlike collagen structure will turn out to be useful in a variety of structural applications.

V. *Analysis of Sequences*

A. *Amino Acid Distribution*

The complete amino acid sequence of an α chain from a single species has not yet been determined. However, the data from rat, calf, and

TABLE 3
Distribution of Amino Acids Among Positions in the Collagen Triplet Gly-X-Y[a]

	Position 1	Position 2	Position 3	Total
4-Hydroxyproline		1[b]	113	114
Aspartic acid		16	15	31
Asparagine		7	5	12
Threonine		3	13	16
Serine		17	18	35
Glutamic acid		41	6	47
Glutamine		8	19	27
Proline		116	3	119
Glycine	337	1		338
Alanine		60	61	121
Valine		9	8	17
Methionine		2	5	7
Isoleucine		3	4	7
Leucine		18	1	19
Phenylalanine		12		12
Hydroxylysine			4	4
Lysine		12	20	32
Histidine		2		2
Arginine		9	42	51
Totals	337	337	337	1011

[a] Using a composite sequence from rat, calf and chick $\alpha 1(\mathrm{I})$; see text.
[b] 3-Hydroxyproline.

chick $\alpha 1(\mathrm{I})$ can be combined to form a complete sequence for the helical region as follows: residues 17–418, rat $\alpha 1(\mathrm{I})$; residues 419–1027, calf $\alpha 1(\mathrm{I})$; residues 132, 133, 930, 933, 936, 937, and 939 [which are Glx or Asx in rat or calf $\alpha 1(\mathrm{I})$], chick $\alpha 1(\mathrm{I})$; residues 399 and 400 [Glx and Asx in rat $\alpha 1(\mathrm{I})$] are assumed to be Gln and Asp, respectively, based on the most probable interpretation of the chemical data (Balian *et al.*, 1972). This composite sequence is useful for discussions of amino acid distribution and has been used to show a relationship between the distributions of charged amino acids and of large hydrophobic amino acids and molecular packing (Hulmes *et al.*, 1973; see Chapter 3). Species differences are few and would not affect general conclusions that can be made.

The distribution of amino acids among positions in the triplet Gly-X-Y in this composite sequence is shown in Table 3. Glycine, of course, always occupies position 1, as required by the triple-helical structure (Chapter 2), and it occurs elsewhere only once (in position 2, residue 327). The amino acids in positions 2 and 3 sometimes show a preference for one or the other position. Threonine prefers position 3 ($P < 10^{-2}$); glutamic acid prefers position 2 ($P < 10^{-7}$); leucine prefers position 2 ($P < 10^{-4}$); phenylalanine occurs only in position 2 ($P < 10^{-3}$); and arginine prefers position 3 ($P < 10^{-5}$). The reasons for these preferences are not clear. Serine and alanine (which are like threonine), aspartic acid (which is like glutamic acid), valine, methionine and isoleucine (which are like leucine and phenylalanine), and lysine (which is like arginine) are all equally distributed or nearly so. Salem and Traub (1975) have shown that some of the unequal distributions are consistent with intramolecular interactions that could stabilize the molecular structure.

Other analyses of the sequence data have produced a few other findings of interest (C.J. MacLean and K.A. Piez, unpublished results). These can be summarized as follows: (1) Although some triplets occur more (or less) frequently than expected, the differences are not large enough to be obviously significant or they result from the preference of certain amino acids for position 2 or 3 of the triplet. (2) Except for the five successive triplets of Gly-Pro-Hyp (or Pro) at the C-terminal end of the helical region, proline and hydroxyproline are somewhat more uniformly distributed than expected. This can perhaps be attributed to their role in stabilizing the helix (Chapter 2) and a need for this stabilization to be uniformly distributed. (3) Positively charged residues tend to be near negatively charged residues more often than expected, a finding that is probably important in considerations of molecular packing (Hulmes *et al.*, 1973). Specifically, parallel and aligned molecules, which would normally repulse one another because of opposing like charges, may attract one another. This feature is consistent with and supports the microfibril packing model discussed in Chapter 3 which permits parallel and aligned molecules in adjacent microfibrils. (4) There is no evidence for long-range internal homology. That is, the long collagen molecule probably did not arise by repetition of large DNA segments. Of course, multiple repetition of a gene coding for a primordial triplet such as Gly-Pro-Ala may well have given rise to the first collagen and lengthening could occur by repetition of triplets.

B. *Helical Cross-Link Sites*

Although analysis of the composite $\alpha 1(I)$ sequence does not give evidence for internal homology, it does show two remarkably similar regions, residues 101–111 and 941–951, a fact that was noted first by Wendt *et al.* (1972). They are Gly-Met-Hyl-Gly-His-Arg-Gly-Phe-Ser-Gly-Leu and the identical sequence except for Ile in place of Met (a conserva-

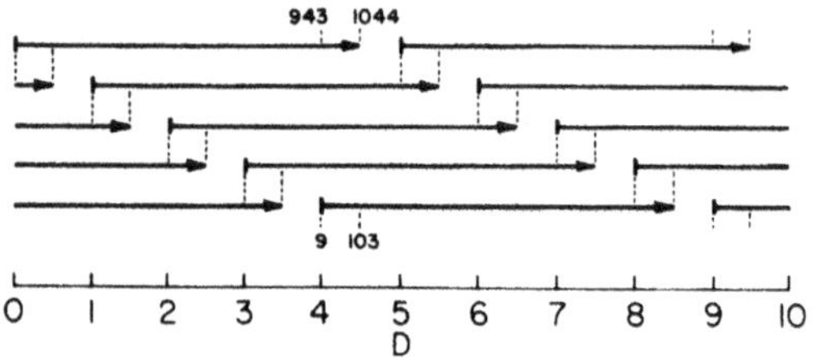

Figure 5. Petruska–Hodge packing of collagen molecules (arrows) showing likely positions for intermolecular cross-links (dashed lines). These positions predict the involvement of residue pairs 9–943 and 103–1044 in intermolecular cross-links (see text). The scale shows the native repeat, D = 668 Å = 234 residues.

tive substitution). It it highly unlikely that this similarity could have arisen by chance. The two histidines are the only ones in the helical region, and hydroxylysine and phenylalanine are rare amino acids. It might be expected, therefore, that these regions have a critical function. It is possible, as explained below, that they are helical cross-link sites.

It is known that the rodlike collagen molecules in the native fibril are staggered by multiples of 234 residues, a value referred to as D (Chapter 3), and that cross-links originate from lysine residues at positions 9 and 1044 in the nonhelical ends of the molecule. If it can be assumed that the nonhelical ends are extended with a residue spacing similar to the spacing in the helix, it can be readily calculated what part of the sequence residues 9 and 1044 would contact on adjacent molecules staggered by D, $2D$, $3D$, and $4D$. Referring to Figure 5, it can be seen that, since the distance between residues 9 and 1044 is 1035 residues or $4.42D$ (1035/234), residue 1044 will contact adjacent molecules $0.42D$ beyond integral values of D from residue 9. Therefore, residue 1044 will be near residues 107, 341, 575, and 809 [$(234 \times 0.42) + 9 + n234$, where $n = 0, 1, 2$, and 3]. Residue 9 will contact adjacent molecules at integral values of D beyond residue 9. That is, residue 9 will be near residues 243, 477, 711, and 945 ($234 + 9 + n234$, where $n = 0, 1, 2$, and 3). Examining the sequences around these positions (Table 1), it can be seen that two are particularly interesting. These are the two similar regions, 101–111 and 941–951, noted above. Both contain a hydroxylysine residue (residues 103 and 943), which is required for a cross-link, and both contain a histidine, which has also been implicated in cross-linking (Chapter 8).

A further reason for considering these regions to be significant is that both possible cross-links, 9–943 and 103–1035, would stabilize a $4D$ stagger between adjacent molecules (Figure 4). Independent electron optical (Zimmerman *et al.*, 1970), chemical (Kang, 1972; Miller *et al.*, 1973; Eyre and Glimcher, 1973; Dixit and Bensusan, 1973), and structural evidence (Chapter 3) support the idea that cross-links in these locations are important in the native collagen fibril. However, direct chemical evidence that residues 103 and 943 are involved is still lacking.

That the two putative helical cross-link regions are nearly identical and are rich in large hydrophobic residues implies that specific noncovalent interactions between them and the N- and C-terminal nonhelical cross-link regions occur prior to covalent cross-linking, perhaps as a necessary first step associated with fibril formation.

C. *Comparative Aspects*

The sequence data in Table 1 and Figure 2 have not been analyzed to determine evolutionary relationships or any other quantitative measures of species or tissue differences. Such analyses should probably await additional data.

A qualitative conclusion that can be made is that the α chains, whether from different species or tissues, are very similar, even omitting glycine from consideration. Where differences occur, they usually involve conservative substitutions. Since molecular structure only requires glycine in every third position and a certain amount of proline and hydroxyproline, the invariance of the other residues suggests that they are involved in specific intermolecular interactions that produce fibril structure and that this structure is very nearly the same for types I, II, and III collagen.

It might be expected that some insight into the significance of each amino acid to structure might be gained from examining the frequency with which one amino acid is substituted by another. From the data in Table 1 it can be calculated that, relative to their frequency in the composite $\alpha1$(I) sequence previously described: (1) Lysine and arginine are rarely substituted; when they are, one nearly always replaces the other. (2) Aspartic acid, glutamic acid, and phenylalanine are infrequently substituted; when they are, the first two usually replace one another. (3) Glutamine, proline, and alanine show a moderate degree of substitution; when they are, the last two usually replace one another. (4) Serine, methionine, and leucine show a relatively high frequency of substitution. (5) Asparagine, threonine, valine, and isoleucine are the most frequently substituted. There is insufficient glycine (in positions 2 or 3) and histidine to make a comparison. Tyrosine and cystine are absent.

These results suggest that the charged amino acids are the most critical and that the large hydrophobic amino acids, except for phenylalanine, are much less critical to structure. However, there is evidence that both groups of amino acids may be involved in molecular interactions (Hulmes *et al.*, 1973). Apparently simple explanations cannot be expected.

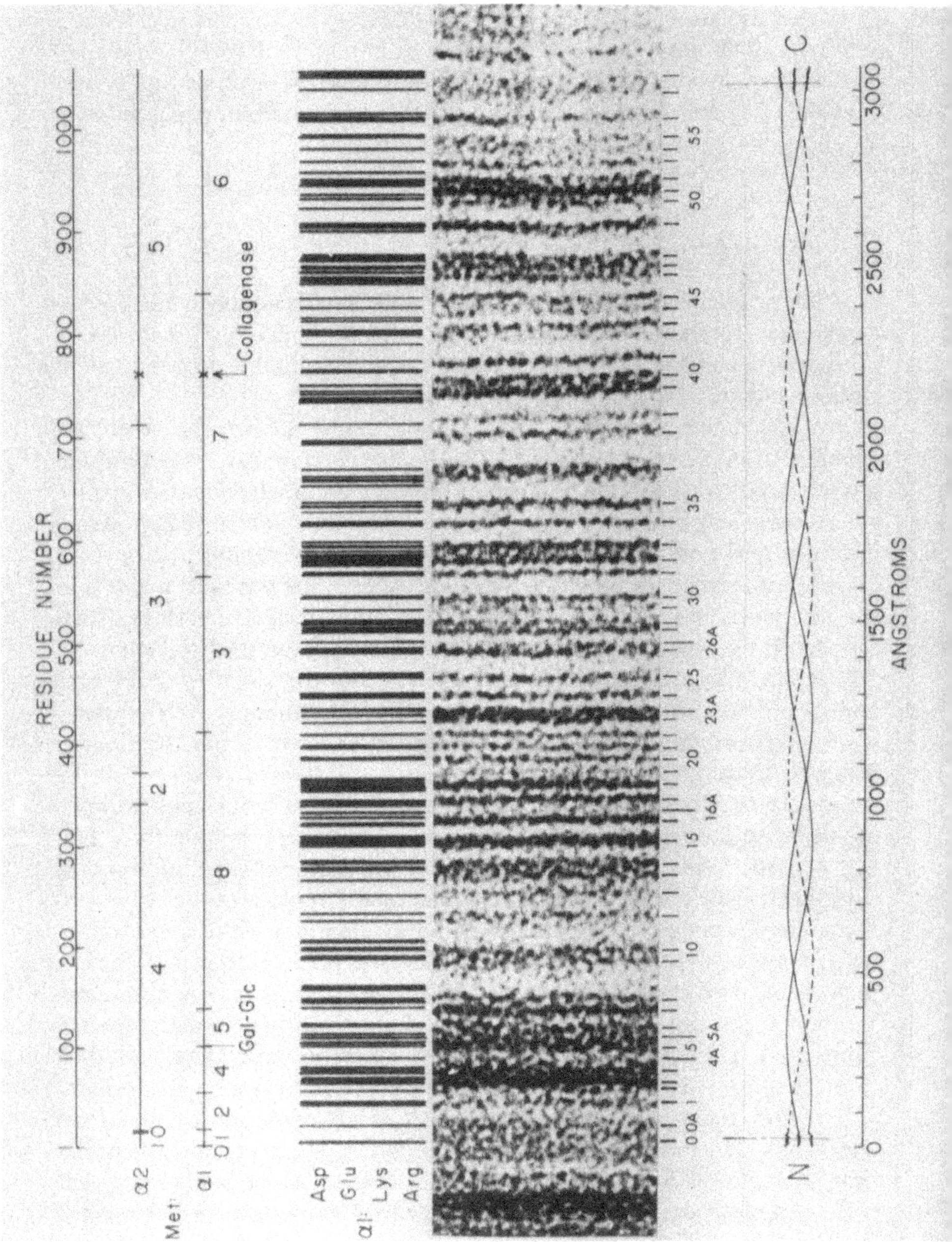
RESIDUE NUMBER
100 200 300 400 500 600 700 800 900 1000
Met: α2
α1
Collagenase
Gal-Glc
α1: Asp Glu Lys Arg
N
C
0 500 1000 1500 2000 2500 3000
ANGSTROMS

FIGURE 6. An electron micrograph of collagen in the SLS form (center) showing the relationship of the band pattern to the positions of aspartic acid (Asp), glutamic acid (Glu), lysine (Lys), arginine (Arg) in the amino acid sequence of the $\alpha 1(I)$ chain. The schematic collagen molecule (below) shows the relationship of the helical body and nonhelical ends to the SLS pattern. The distribution of methionine (Met) in rat $\alpha 1(I)$ and $\alpha 2$ (above, and see Figure 1) is shown. The bands are numbered according to the nomenclature suggested by Bruns and Gross (1973); several additional bands have been designated (0A, 4A, 5A, 16A, 23A, and 26A). The electron micrograph shows other SLS aggregates abutted to the central SLS aggregate to form N-N and C-C junctions, accounting for the planes of symmetry near the N- and C-terminal ends. There is probably a slight overlap of the ends. The C-terminal nonhelical end is often missing in preparations like this, which may explain the absence of a strong band 58. The electron micrograph was provided by K. Kühn.

Comparison of $\alpha 1(\mathrm{II})$ with $\alpha 1(\mathrm{I})$ has suggested to Butler *et al.* (1974*b*) that there are alternate variable and invariable regions along the chains. These could be related to different requirements for specificity in different parts of the collagen molecule.

VI. Electron Optical Information

A. The SLS Aggregate

The collagen molecule under certain *in vitro* conditions can form an aggregate that consists of a bundle of parallel molecules with their ends in register. This is referred to as the segment-long-spacing or SLS aggregate (see Bruns and Gross, 1973). Positively stained preparations give a characteristic band pattern in the electron microscope (Figure 6). The bands arise from clustering of charged residues which retain the heavy metal stain and are therefore electron dense. What one sees, in effect, is a radial projection from the bundle of molecules of the positions of charged amino acid chains along the molecule. Since the α chains are parallel and in register (or nearly so) in the molecule and the residue spacing is constant (Chapter 2), this is a form of primary structure information.

The value of the SLS aggregate is that it provides an easily accessible "fingerprint." Types I and II collagen (Trelstad *et al.*, 1970; Stark *et al.*, 1972) and renatured $\alpha 1(\mathrm{I})$ or $\alpha 2$ (Tkocz and Kuhn, 1969), which give $[\alpha 1(\mathrm{I})]_3$ and $(\alpha 2)_3$, from various species give very similar SLS band patterns showing the close relationship. Collagen fragments can be renatured and made into short SLS aggregates which allows the position in the molecule from which they arose to be identified by matching the peptide band pattern to the whole collagen SLS pattern. It is likely that type III collagen will form an SLS aggregate similar to types I and II. Type IV collagen, however, may not, since its molecular structure is very different (Olsen *et al.*, 1973).

It has been noted by several investigators that the positions of charged amino acid side chains in peptides from collagen show an excellent correlation with the SLS bands in the region of the molecule from which the peptide was obtained (see references in Table 2). This is shown in Figure 6 for the whole molecule. By comparison of the sequence to the pattern it is possible to assign every charged amino acid to a band (Chapman, 1974). These assignments are shown in Table 1 and Figures 2 and 3 (legends).

In making these assignments and examining SLS patterns, it appears that some of the bands designated by Bruns and Gross (1973) are doublets (5, 23, and 26) and that several very light bands were not given numbers (between 0 and 1, 4 and 5, and 16 and 17). These are designated 0A, 4A, 5A, 16A, 23A, and 26A (Figure 6 and Table 1). These features are uncertain in SLS aggregates, but they are confirmed by the sequence. The assignments made here differ only slightly from those made by Chapman (1974).

B. The Native Fibril

Electron micrographs of positively stained native collagen fibrils also have a characteristic band pattern that provides primary structure information. Since collagen molecules are staggered in a regular way, each repeat of the *D* period (668 Å or 234 residues) contains the contents of whole collagen molecules (Chapter 3). This can be seen by looking across a repeat in the schematic packing structure shown in Figure 5. The result is that each native fibril band can be viewed as arising from the superposition of SLS bands spaced 234 residues apart.

This fact has been utilized to reconstruct the native fibril band pattern from the SLS band pattern. This was done by Hodge and Schmitt (1960) by moving a sheet of photographic paper parallel to the molecular axis under a negative of the SLS band pattern and exposing at intervals equivalent to the native fibril period, and by Kuhn and Zimmer (1961) by similarly superimposing a densitometer trace of the SLS pattern on itself. Considerable trial and error must have been required since the *D* period was not well defined, but the resulting reconstructions matched actual micrographs and densitometer tracings of native fibrils quite well.

This can now be done with greater precision starting with a plot of the positions of the acidic and basic amino acids (Figure 6) rather than the SLS pattern. Repeated photographic exposures are made at intervals equivalent to 234 amino acid residues (Doyle *et al.*, 1974). The result (Figure 7) shows a very close match to the actual band pattern, particularly when an optically averaged pattern is used (Bruns and Gross, 1974). All of the bands, even the very light ones, can be seen, and the intensities are similar. The only differences are that, in the reconstructed pattern, band III shows a splitting and the triad of VI, VII, and VIII are closer together.

A more sophisticated reconstruction from the amino acid sequence was done by Chapman and Hardcastle (1974) utilizing computer methods. They tested several possible staggers and found that a stagger of 232–233 residues gave the best fit. This agrees well with the value of 234

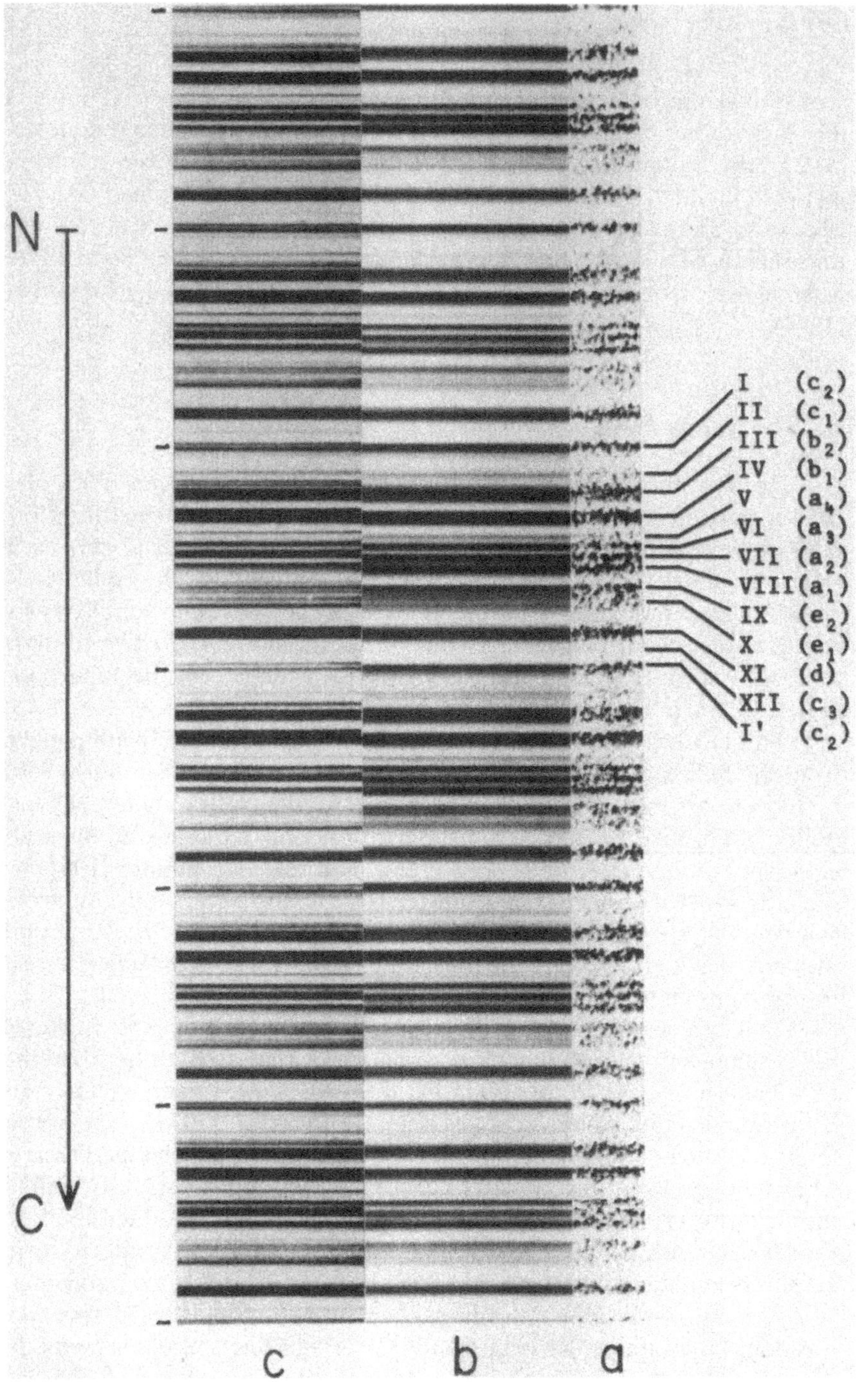

N
C
I (c2)
II (c1)
III (b2)
IV (b1)
V (a4)
VI (a3)
VII (a2)
VIII (a1)
IX (e2)
X (e1)
XI (d)
XII (c3)
I' (c2)
c
b
a

determined by an independent approach (Hulmes *et al.*, 1973; see Chapter 3).

Since the SLS and native fibril bands can be related to one another and the SLS pattern can be related to the sequence, the charged amino acids in $\alpha 1$(I) can be assigned to native fibril bands. These assignments are shown in Table 1 and Figures 2 and 3 (legends).

C. The Symmetrically Banded Fibril

Under some conditions that are not well defined, collagen precipitates from solution to form banded fibrils with the same period as native fibrils but with a symmetrical band pattern. This pattern has been explained as arising from microfibrils containing the native asymmetrical band pattern which are antiparallel and staggered; reconstruction of the band pattern from the proposed model shows a close match to the actual band pattern (Doyle *et al.*, 1974). In the native fibril, the microfibrils are parallel and aligned (see Chapter 3).

VII. Procollagen

The discovery that collagen is synthesized as a precursor, procollagen, that is larger than the final molecule opens a new area to primary structure studies. As reviewed in Chapter 6, the pro-α chains are longer at the N-terminal end and at the C-terminal end by several hundred amino acids. One or more specific proteases converts procollagen to collagen. The extensions are not collagen-like in the sense that they are accessible to proteolytic enzymes and contain little or no hydroxyproline and less proline and glycine than collagen. Nothing has yet been reported about their sequence or structure.

FIGURE 7. A comparison of the band pattern observed by electron microscopy of a positively stained native collagen fibril (a and b) with the band pattern reconstructed photographically from the positions of charged amino acids in the amino acids sequence (c) assuming a stagger of 234 residues between molecules. (b) is an optical average of (a) obtained by moving the photographic paper perpendicular to the fibril axis while printing (a). The Roman numerals show the band nonmenclature suggested by Bruns and Gross (1974); the older designations are given in parentheses. The native fibril repeat is marked by short dashes (left). The relative position and direction of a collagen molecule, in axial projection, appears at the left. The electron micrograph (a and b) was provided by R. Bruns and J. Gross.

ACKNOWLEDGMENT

I am indebted to Drs. Adams, Bruns, Bornstein, Butler, Clark, Fietzek, Gross, Highberger, Kang, Kuhn, Miller, and Reid for providing information prior to publication. I also thank Mrs. L. Bradley, Miss T. Giblin, and Mrs. M. Munsterteiger for their patient help in assembling data and preparing the manuscript.

References

Aguilar, J. H., Jacobs, H. G., Butler, W. T., and Cunningham, L. W., 1973, The distribution of carbohydrate groups in rat skin collagen, *J. Biol. Chem.* **248:**5106.

Allam, S. S., and Heidemann, E., 1974, Isolation, characterization and comparative studies of the N-terminal peptides from soluble pig skin collagen, *FEBS Lett.* **39:**187.

Balian, G., Click, E. M., and Bornstein, P., 1971, Structure of rat skin collagen α1-CB8. Amino acid sequence of the hydroxylamine-produced fragment HA1, *Biochemistry* **10:**4470.

Balian, G., Click, E. M., Hermodson, M. A., and Bornstein, P., 1972, Structure of rat skin collagen α1-CB8. Amino acid sequence of the hydroxylamine-produced fragment HA2, *Biochemistry* **11:**3798.

Barker, W. C., and Dayhoff, M. O., 1972, Detecting distant relationships: Computer methods and results, *in: Atlas of Protein Sequence and Structure* (M. O. Dayhoff, ed.), pp. 101–110, National Biomedical Research Foundation, Washington, D.C.

Barnes, M. J., 1973, Biochemistry of collagens from mineralized tissues, *in: Hard Tissue Growth, Repair and Remineralization,* pp. 247–261, ASP (Elsevier Excerpta Medica North-Holland), Amsterdam.

Barnes, M. J., Constable, B. J., Morton, L. F., and Royce, P. M., 1974, Age-related variations in hydroxylation of lysine and proline in collagen, *Biochem. J.* **139:**461.

Becker, U., and Timpl, R., 1972, Cyanogen bromide peptides of the rabbit collagen α1-chain, *FEBS Lett.* **27:**85.

Bornstein, P., 1967, Comparative sequence studies of rat skin and tendon collagen. I. Evidence for incomplete hydroxylation of individual prolyl residues in the normal proteins, *Biochemistry* **6:**3082.

Bruns, R. R., and Gross, J., 1973, Band pattern of the segment-long-spacing form of collagen. Its use in the analysis of primary structure, *Biochemistry* **12:**808.

Bruns, R. R., and Gross, J., 1974, High-resolution analysis of the modified quarter-stagger model of the collagen fibril, *Biopolymers* **13:**931.

Butler, W. T., 1968, Partial hydroxylation of certain lysines in collagen, *Science* **161:**796.

Butler, W. T., 1970, Chemical studies on the cyanogen bromide peptides of rat skin collagen. The covalent structure of α1-CB5, the major hexose-containing cyanogen bromide peptide of α1, *Biochemistry* **9:**44.

Butler, W. T., 1973, Concerning the high level of hydroxylysine in dentin collagen, *Alabama J. Med. Sci.* **10:**103.

Butler, W. T., and Ponds, S. L., 1971, Chemical studies on the cyanogen bromide peptides of rat skin collagen. Amino acid sequence of α1-CB4, *Biochemistry* **10:**2076.

Butler, W. T., Underwood, S. P., and Finch, J. E., Jr., 1974*a*, Chemical studies on the cyanogen bromide peptides of rat skin collagen. Amino acid sequence of α1-CB3, *Biochemistry* **13:**2946.

Butler, W. T., Miller, E. J., Finch, J. E., Jr., and Inagami, T., 1974*b*, Homologous regions of collagen α1(I) and α1(II) chains: Apparent clustering of variable and invariant amino acid residues, *Biochem. Biophys. Res. Commun.* **57:**190.

Butler, W. T., Miller, E. J., and Finch, J. E., Jr., 1976, The covalent structure of cartilage collagen. Amino acid sequence of the NH_2-terminal portion of the α1(II) chain, *Biochemistry* (in press).

Byers, P. H., McKenney, K. H., Lichtenstein, J. R., and Martin, G. R., 1974, Preparation of type III procollagen and collagen from rat skin, *Biochemistry* **13:**5243.

Chapman, J. A., 1974, The staining pattern of collagen fibrils. I. An analysis of electron micrographs, *Conn. Tissue Res.* **2:**137.

Chapman, J. A., and Hardcastle, R. A., 1974, The staining pattern of collagen fibrils II. A comparison with patterns computer-generated from the amino acid sequence, *Conn. Tissue Res.* **2:**151.

Chung, E., and Miller, E. J., 1974, Collagen polymorphism: Characterization of molecules with the chain composition $[\alpha 1(III)]_3$ in human tissues, *Science* **183:**1200.

Chung, E., Keele, E. M., and Miller, E. J., 1974, Isolation and characterization of the cyanogen bromide peptides from the α1(III) chain of human collagen, *Biochemistry* **13:**3459.

Cintron, C., 1974, Hydroxylysine glycosides in the collagen of normal and scarred rabbit corneas, *Biochem. Biophys. Res. Commun.* **60:**288.

Clark, C. C., and Bornstein, P., 1972, Cyanogen bromide cleavage of guinea pig skin collagen. Isolation and characterization of peptides from the α1 and α2 chains, *Biochemistry* **11:**1468.

Clark, C. C., Fietzek, P. P., and Bornstein, P., 1975, The order of cyanogen bromide peptides and location of carbohydrate in the α2 chain of guinea pig skin collagen, *Eur. J. Biochem.* **56:**327.

Dayhoff, M. O., 1972, *Atlas of Protein Sequence and Structure,* pp. D296–D301, National Biomedical Research Foundation, Washington, D.C.

Dixit, S. N., and Bensusan, H. B., 1973, The isolation of crosslinked peptides of collagen involving α1-CB6, *Biochem. Biophys. Res. Commun.* **52:**1.

Dixit, S. N., Kang, A. H., and Gross, J., 1975*a*, Covalent structure of collagen: Amino acid sequence of α1-CB3 of chick skin collagen, *Biochemistry* **14:**1929.

Dixit, S. N., Seyer, J. M., Oransky, A. D., Corbett, C., Kang, A. H., and Gross, J., 1975*b*, Covalent structure of collagen: Amino acid sequence of α1-CB6A of chick skin collagen, *Biochemistry* **14:**1933.

Doyle, B. B., Hulmes, D. J. S., Miller, A., Parry, D. A. D., Piez, K. A., and Woodhead-Galloway, J., 1974, Axially projected collagen structures, *Proc. R. Soc. B* **187:**37.

Epstein, E. H., Jr., 1974, $[\alpha 1(III)]_3$ Human skin collagen. Release by pepsin digestion and preponderance in fetal life, *J. Biol. Chem.* **49:**3225.

Eyre, D. R., and Glimcher, M. J., 1973, Collagen cross-linking. Isolation of cross-linked peptides from collagen of chicken bone, *Biochem. J.* **135:**393.

Fietzek, P. P., and Kuhn, K., 1973, The covalent structure of collagen: Amino acid sequence of the N-terminal region of α2-CB5 from rat skin collagen, *FEBS Lett.* **36:**289.

Fietzek, P. P., and Kuhn, K., 1974, The covalent structure of collagen: Amino acid sequence of the N-terminal region of α2-CB3 from calf skin collagen, *Z. Physiol. Chem.* **335:**647.

Fietzek, P. P., and Kuhn, K., 1975, The covalent structure of collagen: Amino acid sequence of α1-CB2, 4 and 5 from calf skin collagen, *Eur. J. Biochem.* **52:**77.

Fietzek, P. P., and Rauterberg, J., 1975, Cyanogen bromide peptides of type III collagen: First sequence analysis demonstrates homology with type I collagen, *FEBS Lett.* **49:**365.

Fietzek, P. P., and Rexrodt, F. W., 1975, The covalent structure of collagen. The amino acid sequence of α2-CB4 from calf skin collagen, *Eur. J. Biochem.* **59:**113.

Fietzek, P. P., Munch, M., Breitkreutz, D., and Kuhn, K., 1970, Isolation and characterization of the cyanogen bromide peptides from the α2 chain of calf skin collagen, *FEBS Lett.* **9:**229.

Fietzek, P. P., Wendt, P., Kell, I., and Kuhn, K., 1972*a*, The covalent structure of collagen: Amino acid sequence of α1-CB3 from calf skin collagen, *FEBS Lett.* **26:**74.

Fietzek, P. P., Rexrodt, F. W., Wendt, P., Stark, M., and Kuhn, K., 1972*b*, The covalent structure of collagen. Amino-acid sequence of peptide α1-CB6-C2, *Eur. J. Biochem.* **30:**163.

Fietzek, P. P., Kell, I., and Kuhn, K., 1972*c*, The covalent structure of collagen. Amino acid sequence of the N-terminal region of α2-CB4 from calf and rat skin collagen, *FEBS Lett.* **26:**66.

Fietzek, P. P., Rexrodt, F. W., Hopper, K. E., and Kuhn, K., 1973, The covalent structure of collagen. 2. The amino-acid sequence of α1-CB7 from calf skin collagen, *Eur. J. Biochem.* **38:**396.

Fietzek, P. P., Breitkreutz, D., and Kuhn, K., 1974*a*, Amino acid sequence of the amino-terminal region of calf skin collagen, *Biochim. Biophys. Acta* **365:**305.

Fietzek, P. P., Furthmayr, H., and Kuhn, K., 1974*b*, Comparative sequence studies on α2-CB2 from calf, human, rabbit, and pig-skin collagen, *Eur. J. Biochem.* **47:**257.

Gallop, P. M., Blumenfield, O. O., and Seifter, S., 1972, Structure and metabolism of connective tissue proteins, *Ann. Rev. Biochem.* **41:**617.

Gryder, R. M., Lamon, M., and Adams, E., 1974, Sequence position of 3-hydroxyproline in basement membrane collagen: Isolation of 3-hydroxyprolyl-4-hydroxyproline from swine kidney, *J. Biol. Chem.* **250:**2470.

Heinrich, W., Lange, P. M., Stirtz, T., Iancu, C., and Heidemann, E., 1971, Isolation and characterization of the large cyanogen bromide peptides from the α1- and α2-chains of pig skin collagen, *FEBS Lett.* **16:**63.

Highberger, J. H., Kang, A. H., and Gross, J., 1971, Comparative studies on the amino acid sequence of the α2-CB2 peptides from chick and rat skin collagens, *Biochemistry* **10:**610.

Highberger, J. H., Corbett, C., Kang, A., and Gross, J., 1975, The amino acid sequence of chick skin collagen α1-CB7, *Biochemistry* **14:**2872.

Hodge, A. J., and Schmitt, F. O., 1960, The charge profile of the tropocollagen macromolecule and the packing arrangement in native-type collagen fibrils, *Proc. Natl. Acad. Sci. U.S.A.* **46:**186.

Hudson, B. G., and Spiro, R. G., 1972*a*, Studies on the native and reduced alkylated renal glomerular basement membrane. Solubility, subunit size, and reaction with cyanogen bromide, *J. Biol. Chem.* **247:**4229.

Hudson, B. G., and Spiro, R. G., 1972*b*, Fractionation of glycoprotein components of the reduced alkylated renal glomerular basement membrane, *J. Biol. Chem.* **247:**4239.

Hulmes, D. J. S., Miller, A., Parry, D. A. D., Piez, K. A., and Woodhead-Galloway, J., 1973, Analysis of the primary structure of collagen for the origins of molecular packing, *J. Mol. Biol.* **79:**137.

Kang, A. H., 1972, Studies on the location of intermolecular cross-links in collagen. Isolation of a CNBr peptide containing δ-hydroxylysinonorleucine, *Biochemistry* **11**:1828.

Kang, A. H., and Gross, J., 1970, Amino acid sequence of cyanogen bromide peptides from the amino-terminal region of chick skin collagen, *Biochemistry* **9**:796.

Kang, A. H., Dixit, S. N., Corbett, C., and Gross, J., 1975, The covalent structure of collagen: Amino acid sequence of α1-CB5 glycopeptide and α1-CB4 from chick skin collagen, *J. Biol. Chem.* **250**:7428.

Kefalides, N. A., 1973, Structure and biosynthesis of basement membranes, *Int. Rev. Conn. Tissue Res.* **6**:63.

Kefalides, N. A., Tomichek, E., and Alper, R., 1974, Selective cleavage of basement membrane proteins at the carboxyl peptide linkages of methionine and at the amino peptide linkages of cysteine residues, *in: Proc. XXIInd Colloquium on Protides of Biological Fluids,* Brugge, Belgium, Pergamon Press (in press).

Kuhn, K., and Zimmer, E., 1961, Eigenschaften des Tropokollagen-Molekuls und deren Bedeutung für die Fibrillenbildung, *Z. Naturforsch.* **16B**:648.

Miller, E. J., 1971*a*, Isolation and characterization of a collagen from chick cartilage containing three identical α chains, *Biochemistry* **10**:1652.

Miller, E. J., 1971*b*, Isolation and characterization of the cyanogen bromide peptides from the α1(II) chain of chick cartilage collagen, *Biochemistry* **10**:3030.

Miller, E. J., 1972, Structural studies on cartilage collagen employing limited cleavage and solubilization with pepsin, *Biochemistry* **11**:4903.

Miller, E. J., 1973, A review of biochemical studies on the genetically distinct collagens of the skeletal system, *Clin. Orthop. Relat. Res.* **92**:260.

Miller, E. J., and Lunde, L. G., 1973, Isolation and characterization of the cyanogen bromide peptides from the α1(II) chain of bovine and human cartilage collagen, *Biochemistry* **12**:3153.

Miller, E. J., and Matukas, V. J., 1969, Chick cartilage collagen: A new type of α1 chain not present in bone or skin of the species, *Proc. Natl. Acad. Sci. U.S.A.* **64**:1264.

Miller, E. J., Martin, G. R., Piez, K. A., and Powers, M. J., 1967, Characterization of chick bone collagen and compositional changes associated with maturation, *J. Biol. Chem.* **242**:5481.

Miller, E. J., Lane, J. M., and Piez, K. A., 1969, Isolation and characterization of the peptides derived from the α1 chain of chick bone collagen after cyanogen bromide cleavage, *Biochemistry* **8**:30.

Miller, E. J., Epstein, E. H., Jr., and Piez, K. A., 1971, Identification of three genetically distinct collagens by cyanogen bromide cleavage of insoluble human skin and cartilage collagen, *Biochem. Biophys. Res. Commun.* **42**:1024.

Miller, E. J., Woodall, D. L., and Vail, M. S., 1973, Biosynthesis of cartilage collagen. Use of pulse labeling to order the cyanogen bromide peptides in the α1(II) chain, *J. Biol. Chem.* **248**:1666.

Olsen, B. R., Alper, R., and Kefalides, N. A., 1973, Structural characterization of a soluble fraction from lens-capsule basement membrane, *Eur. J. Biochem.* **38**:220.

Piez, K. A., 1967, Soluble collagen and the components resulting from it denaturation, *in: Treatise on Collagen, Vol. 1, Chemistry of Collagen* (G. N. Ramachandran, ed.), pp. 201–248, Academic Press, New York.

Piez, K. A., and Likins, R. C., 1960, The nature of collagen. II. Vertebrate collagens, *in: Calcification in Biological Systems,* pp. 411–420, American Association for the Advancement of Science, Washington, D.C.

Piez, K. A., Eigner, E. A., and Lewis, M. S., 1963, The chromatographic separation and amino acid composition of the subunits of several collagens, *Biochemistry* **2:**58.

Pinnell, S. R., Fox, R., and Krane, S. M., 1971, Human collagens: Differences in glycosylated hydroxylysines in skin and bone, *Biochim. Biophys. Acta.* **229:**119.

Rauterberg, J., 1973, The C-terminal non-helical portion of the collagen molecule, *Clin. Orthop. Relat. Res.* **97:**196.

Rauterberg, J., Timpl, R., and Furthmayr, H., 1972*a*, Structural characterization of N-terminal antigenic determinants in calf and human collagen, *Eur. J. Biochem.* **27:**231.

Rauterberg, J., Fietzek, P., Rexrodt, F., Becker, U., Stark, M., and Kuhn, K., 1972*b*, The amino acid sequence of the carboxyterminal nonhelical cross link region of the $\alpha 1$ chain of calf skin collagen, *FEBS Lett.* **21:**75.

Reid, K., 1974, A collagen-like amino acid sequence in a polypeptide chain of human Clq (a subcomponent of the first component of complement), *Biochem. J.* **141:**189.

Reid, K., Lowe, D., and Porter, R. R., 1972, Isolation and characterization of Clq, a subcomponent of the first component of complement, from human and rabbit sera, *Biochem. J.* **130:**749.

Salem, G., and Traub, W., 1975, Conformational implications of amino acid sequence regularities in collagen, *FEBS Lett.* **15:**94.

Stark, M., Rauterberg, J., and Kuhn, K., 1971, Evidence for a non-helical region at the carboxyl terminus of the collagen molecule, *FEBS Lett.* **13:**101.

Stark, M., Miller, E. J., and Kuhn, K., 1972, Comparative electron microscope studies on the collagen extracted from cartilage, bone and skin, *Eur. J. Biochem.* **27:**192.

Stoltz, M., Timpl, R., and Kuhn, K., 1972, Non-helical regions in rat collagen $\alpha 1$ chain, *FEBS Lett.* **26:**61.

Strawich, E., and Nimni, M. E., 1971, Properties of a collagen molecule containing three identical components extracted from bovine articular cartilage, *Biochemistry* **10:**3905.

Timpl, R., Furthmayr, H., Hahn, E., Becker, U., and Stoltz, M., 1973, Immunochemistry of collagen. Model properties of a natural protein antigen, *Behring Inst. Mitt.* **53:**66.

Tkocz, C., and Kuhn, K., 1969, The formation of triple-helical collagen molecules from $\alpha 1$ or $\alpha 2$ polypeptide chains, *Eur. J. Biochem.* **7:**454.

Traub, W., and Piez, K. A., 1971, The chemistry and structure of collagen, *Adv. Prot. Chem.* **25:**243.

Trelstad, R. L., 1974, Human aorta collagens: Evidence for three distinct species, *Biochem. Biophys. Res. Commun.* **57:**717.

Trelstad, R. L., Kang, A. H., Igarashi, S., and Gross, J., 1970, Isolation of two distinct collagens from chick cartilage, *Biochemistry* **9:**4993.

Trelstad, R. L., Kang, A. H., Toole, B. P., and Gross, J., 1972, Collagen heterogeneity. High resolution separation of native $[\alpha 1(\mathrm{I})]_2\alpha 2$ and $[\alpha 1(\mathrm{II})]_3$ and their component α chains, *J. Biol. Chem.* **247:**6469.

Volpin, D., and Veis, A., 1971, Isolation and characterization of the cyanogen bromide peptides from the $\alpha 1$ and $\alpha 2$ chains of acid-soluble bovine skin collagen, *Biochemistry* **10:**1751.

Volpin, D., and Veis, A., 1973, Cyanogen bromide peptides from insoluble skin and dentin bovine collagens, *Biochemistry* **12:**1452.

Wendt, P., von der Mark, K., Rexrodt, F., and Kuhn, K., 1972, The covalent structure of collagen. The amino-acid sequence of the 112-residues amino-terminal part of peptide α1-CB6 from calf-skin collagen, *Eur. J. Biochem.* **30:**169.

Zimmerman, B. K., Pikkarainen, J., Fietzek, P. P., and Kuhn, K., 1970, Cross-linkages in collagen. Demonstration of three different intermolecular bonds, *Eur. J. Biochem.* **16:**217.

2
Molecular Structure

G. N. RAMACHANDRAN AND C. RAMAKRISHNAN

I. Outline of the Structure

It is nowadays well known that the molecular structure of collagen is based on three intertwining helical polypeptide chains. A detailed description of this along with a brief summary of the related structures of polypeptide chains having amino acid residues commonly occurring in collagen, is contained in the chapter by Ramachandran (1967) in the *Treatise on Collagen,* Volume 1. Therefore, we shall only give a general account of the early evidences which led to the postulation of the triple-chain structure. However, the geometry and properties of the structure will be discussed in some detail, with special reference to the two modifications of the basic three-chain protofibril, namely the so-called one-bonded and two-bonded structures. These two structures differ only in the number of interchain hydrogen bonds (namely one and two, respectively) for a unit composed of three peptide units in the sequence (Gly-X-Y)$_n$ where X and Y may be any one of the amino acid residues. In fact, after the appearance of the *Treatise on Collagen,* the molecular structures of many interesting polymers related to collagen, such as (Gly-Pro-Pro)$_n$ have been worked out, which are quite relevant to the postulation of the correct structure for collagen. These, together with an account of the primary structure and of cross-links and fibril structure are contained in a review by Traub and Piez (1971). Because of the two reviews mentioned above, namely Ramachandran (1967) and Traub and Piez (1971), no attempt will be made in this

G. N. RAMACHANDRAN · Molecular Biophysics Unit, Indian Institute of Science, Bangalore 560 012, India C. RAMAKRISHNAN · Molecular Biophysics Unit, Indian Institute of Science, Bangalore 560 012, India.

chapter to make the early references complete. Only the significant aspects regarding the triple-helical structure which are contained in these two reviews would be mentioned here. On the other hand, an attempt will be made to make the chapter as self-contained as possible, so that it is not necessary to study the original references in order to get a basic picture of the collagen structure at the molecular level, as it is understood at present. For those who wish to have a general account of the collagen structure without going into details, the report of the 1967 John Arthur Wilson Memorial Lecture delivered to the American Leather Chemists' Association, would be a good introduction to the subject (Ramachandran, 1968).

As is implied by the title to this Section I, some of the general considerations regarding the building up of a molecular structure from theory will be included here. Also, a brief account of the structure of collagen, as it is understood today, is included in the later subsections, so that those who do not wish to have more detailed information need not read beyond this section. So also, the style of presentation adopted in this section will not be rigorous, but rather relaxed.

A. General Considerations Regarding Peptide Units

1. Amino Acid Residues and Peptide Units

Although the chemist takes the amino acid residue —NH—$C^{\alpha}(H^{\alpha}R)$—CO— as the building block of a protein chain or a polypeptide chain, the understanding of the conformation of such a chain is facilitated by taking the set of atoms in the backbone between one α-carbon atom and the next α-carbon atom, namely —$C_1^{\alpha}(H_1^{\alpha}R_1)$—$C_1O_1$—$N_2H_2$—$C_2^{\alpha}(H_2^{\alpha}R_2)$— as the structural unit. In the latter, the set of atoms starting from C_1^{α} and ending with C_2^{α} along the backbone of the chain has a very rigid structure, in that these atoms form essentially a planar structure as shown in Figure 1. The idea of a planar peptide unit goes back to the early times of peptide chemistry in the 1950s (Pauling, 1952, see also Pauling, 1960). However, it had been understood all the time that slight deviations from planarity can occur and that there may be deviations of individual atoms from the least-squares plane passing through all the atoms of the peptide unit. [For an account of the theory of the peptide unit and its conformation, see the reviews by Ramachandran and Sasisekharan (1968) and Scheraga (1968).] Therefore, in building the collagen structure a slight degree of nonplanarity may also be permissible, if the structure is made better, or more stable, by introducing nonplanar distortions.

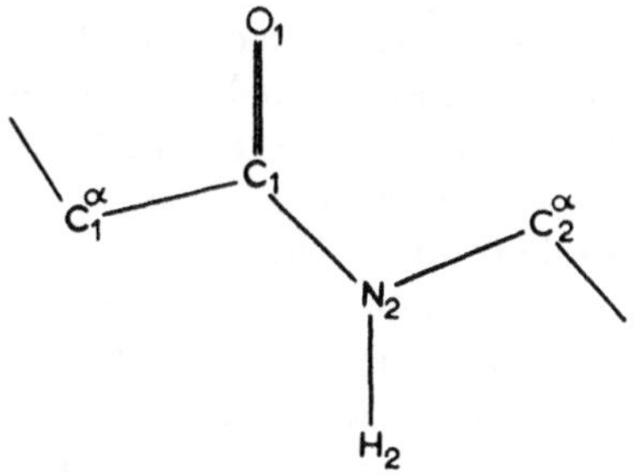

FIGURE 1. The planar peptide unit, in which the set of atoms C_1^α, C_1, O_1, N_1, H_2, and C_2^α are all coplanar. However, small deviations from strict planarity are permissible.

In the nomenclature adopted for the various atoms in a peptide chain as shown, for instance, in Figure 1, we follow the recommendations of the IUPAC-IUB Commission on Biochemical Nomenclature (1970). In this, the atoms belonging to the same amino acid residue all have the same index, while the peptide unit j consists of the atoms from C_j^α to the next C_{j+1}^α. If the nature of the amino acid residues j and $j + 1$ require to be stated more specifically, we shall call the peptide unit linking C_j^α (say Gly) and C_{j+1}^α (say Pro) as the peptide unit Gly-Pro.

2. *Cis and Trans Peptide Units*

Although the planar nature of the peptide unit gives it a unique conformation, it can occur in two modifications which are given the names *cis* and *trans*. These are shown in Figure 2. The only difference between the two is a rotation of 180° about the bond C—N. However, it is an observed fact that invariably only *trans* peptide units occur in protein structures. This is an aspect which has still not been fully understood and

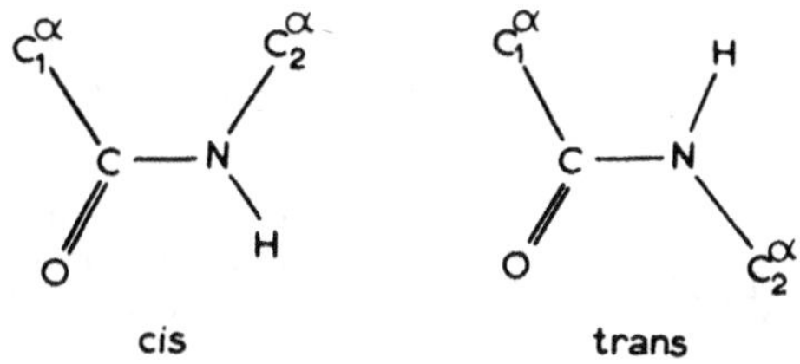

FIGURE 2. *Cis* and *trans* peptide units. The C=O and N-H bonds are on the same side of the central bond C-N in the former case, and on opposite sides in the latter case. Except in the case of X-Pro, *cis* peptide units are not observed, the *trans* conformation being universally valid normally.

explained, but which is of fundamental importance from the practical point of view, namely in building protein structures. This fact was not clear in the 50s and early 60s, and attempts had been made in those days to build structures for some of the fibrous proteins using both *cis* and *trans* peptide units. However, it is known nowadays that, if the residue corresponding to the α-carbon atom 2 is not proline or hydroxyproline, the peptide unit 1 linking C_1^α to C_2^α is always in the *trans* conformation. If the residue 2 is Pro, say, and the peptide unit 1 is of the form Gly-Pro or X-Pro, where X is any general amino acid with a β-carbon atom, then it is quite possible for the peptide unit 1 to be either *cis* or *trans*. In the above example, X can also be Pro (or Hyp), so that the sequence Pro-Pro for a peptide unit can also render it likely to have the *cis* conformation, in addition to the *trans*. In the case of such peptide units with Pro as the second residue, the free-energy difference between the two modifications seems to be negligible, and either one or the other may occur in order to suit the conditions required in building up that particular structure.

3. *Hydrogen Bonds*

When a hydrogen atom is linked by a covalent bond to an electronegative atom (such as N or O), it can form a second weak bond with another electronegative atom. In such a case, the direction of the covalent bond from the first (donor) atom to the proton will be pointing approximately in the direction of the second (acceptor) electronegative atom. In such a case, it is said that there is a hydrogen bonding between the donor and acceptor atoms, A_1 and A_2 (Figure 3). The distance between the atoms A_1 and A_2 is called the hydrogen bond length (we shall use the symbol l for this purpose) and the angle between the directions $A_1 \cdots A_2$ and A_1—H may be called the hydrogen bond angle (θ). The common types of hydrogen bonds of interest to proteins, in particular collagen, are NH$\cdots$O, OH$\cdots$O, and NH$\cdots$N. Further details regarding this type of bonding and its occurrence may be obtained from the review of Ramachandran and Sasisekharan (1968).

The energy of stabilization of a hydrogen bond is of the order of 3–5 kcal/mole as against the value of 0.3 to 1.0 kcal/mole for the nonbonded interactions between the various bonded atoms, which are the ones responsible for the packing of the molecule in polypeptide or protein

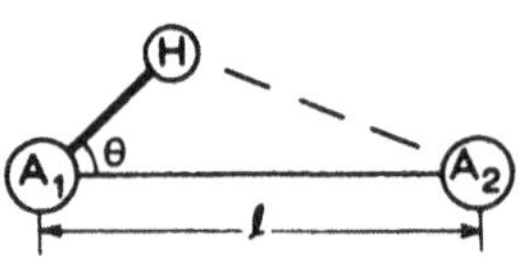

FIGURE 3. The donor A_1, the hydrogen atom H, and the acceptor atom A_2, in a hydrogen bond. The hydrogen bond length l and angle θ are also shown in the figure.

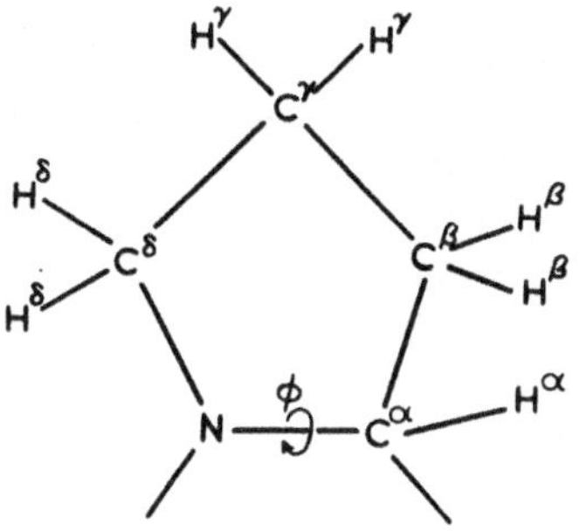

FIGURE 4. The side chain atoms in a prolyl residue. Because C^α is linked to N via covalent bonds in a 5-membered ring, the backbone dihedral angle ϕ about the bond N-C^α can only take values close to $-60°$.

structures. Since the hydrogen-bond energy is thus of the order of 5–10 times that of the nonbonded energy, the occurrence of hydrogen bonds is extremely important for the stabilization of a molecular structure. We shall see below how hydrogen bonds play a very important role in the stability of the collagen structure.

4. *Role of Imino Acid Residues*

As is well known, all the amino acid residues, other than proline and hydroxyproline,* contain a free NH group. Therefore, it is possible to have hydrogen bonds with the NH being a donor for forming such a hydrogen bond. On the other hand, if the residue is an imino acid residue, such as proline or hydroxyproline, the nitrogen is linked to the carbon atom via a 5-membered ring as shown in Figure 4. It is not possible to have the nitrogen of an imino acid residue take part in a hydrogen-bond formation. On the other hand, there is another factor that is relevant in the case of proline or hydroxyproline residues—namely that the 5-membered ring being rigid, there is no freedom of rotation about the bond N—C^α. As a consequence, the structure in that local region is more rigid than it would be if a Pro or Hyp residue did not occur. This may lead both to an advantage, as well as a disadvantage. If the necessary orientation at the local position for building the structure is what is

* Standard notations, such as Gly, Pro, Hyp, etc., will be used wherever necessary, for amino and imino acid residues.

required by the prolyl residue, then the structure would be stabilized by such a residue. This, in fact, occurs in collagen, as will be seen later. In some other circumstances, for example, the case of the α-helix which requires the NH hydrogen bond for its stability, the introduction of a proline residue not only removes a hydrogen bond but also makes the local region overcrowded, so that the helix is disrupted.

5. *Phi–Psi Plots*

In Section I-A-1, we saw that the peptide units have a rigid planar structure. Therefore, in a long peptide chain the only degrees of freedom are the rotations about the two bonds N—C^{α} and C^{α}—C meeting at the C^{α} atom. These two dihedral angles of rotation are given the names ϕ and ψ. They are extremely important in the description of peptide conformation and, therefore, we shall give a little more detail about these angles.

Using the standard definition of a dihedral angle (IUPAC-IUB Commission on Biochemical Nomenclature, 1970), which requires the specification of four atoms linked together and in which the rotational angle is about the bond joining the middle two atoms, the angles ϕ and ψ may be defined by the full description $\phi(C_1—N_2—C_2^{\alpha}—C_2)$ and $\psi(N_2—C_2^{\alpha}—C_2—N_3)$, as shown in Figure 5. These dihedral angles and their application to the description of protein conformation are described in good detail in the review by Ramachandran and Sasisekharan (1968)*. Here we shall show what is known as the contact map in which the allowed conformations of (ϕ,ψ) are shown (Figure 6). In this figure, the two types of regions, namely those bounded by contacts of normal limits and contacts of extreme limits, are as described in that review. If attention is directed to the region enclosed by the curve denoting the extent for extreme limits, then we might say that any allowed conformation for a protein chain at a local region, defined by the dihedral angles (ϕ_j, ψ_j) at atom C_j^{α} must be contained within this boundary. Mostly, they are found within the region bounded by normal contact limits. In Figure 6 are also

* Note, however, that the earlier literature including the review by Ramachandran and Sasisekharan (1968) assumes that the dihedral angle θ(A—B—C—D) is zero when A and D are *trans* to each other as far as the peptide backbone is concerned, while the latest conventions (IUPAC-IUB Commission on Biochemical Nomenclature, 1970) assumes $\theta = 0$ when A and D are *cis* to each other for all dihedral angles, both in the backbone and the side chain. Therefore, $\theta_{old} = \theta_{new} \pm 180°$, for backbone dihedral angles, a simple conversion rule that may be applied when comparing the literature using the old conventions and the new conventions. In the (ϕ,ψ) map shown in Figure 6, the origin is at the bottom left-hand corner according to the old rules and at the center according to the new rules. Following the new rules, the ranges of ϕ and ψ are taken to be from $-180°$ to $+180°$.

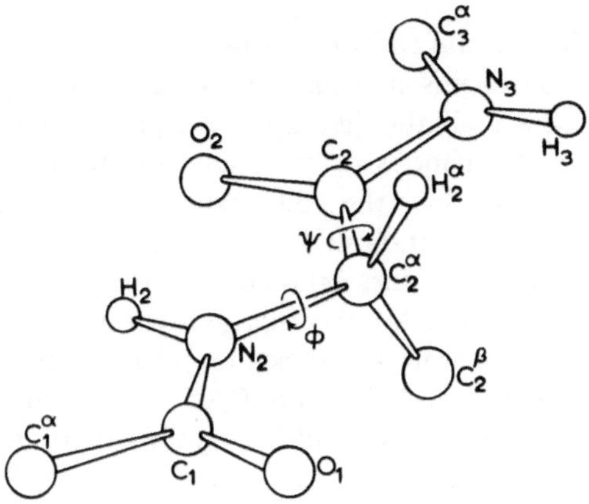

FIGURE 5. The dihedral angles ϕ and ψ for two linked peptide units. The orientation shown corresponds to the fully stretched conformation for which $\phi = 180°$, $\psi = 180°$.

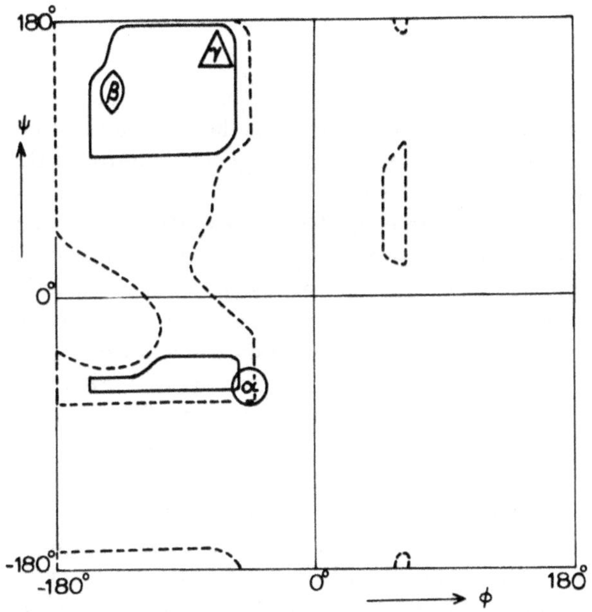

FIGURE 6. The contact map in the (ϕ, ψ) plane, according to the new conventions of IUPAC-IUB (1970). The regions enclosed by the continuous lines are fully allowed by the normal contact limits, while those enclosed by dashed lines are allowed only by the so-called extreme limits, in which the normal conditions are relaxed somewhat.

shown the conformations observed for the regular structures of polypeptide chains in which the chain takes up a helical conformation and in which all the (ϕ_j, ψ_j) are equal or nearly equal. The three particular conformations to be remembered are the right-handed α-helical conformation (α_P), the fully extended β-structure (β), and the collagen type structure, indicated by a triangle (in view of its approximate threefold symmetry) and denoted by the symbol γ. As will be seen from this, the value of ϕ for the collagen-type structure is close to $-60°$ and a simple examination of a skeletal model will show that under these circumstances, the bonds N_2H_2 and $C_2^\alpha C_2^\beta$ will lie in the same plane, so that the 5-membered ring can be readily closed and form the structure of the side chain of proline or hydroxyproline. This peculiar feature has great importance for the collagen structure, as will be seen later.

B. Amino Acid Composition of Collagen

The amino acid composition of collagen was known even in the 1940s. These early studies, as well as later investigations, showed a remarkable degree of uniformity in the general nature of the amino acid composition of collagen from a variety of sources. Considering the fractional number of residues in the polypeptide chain, it was well known that glycine forms very close to one third the total number of residues in the chain, in all samples that have been studied, including not only mammalian collagen, but also amphibian and avian collagens (Eastoe, 1967). Another characteristic and significant feature of the amino acid composition of collagen is that proline plus hydroxyproline form approximately 25% of the amino acid residues in the collagen chain. As already mentioned, these two imino acid residues have a peculiar feature, namely that the C^β atom is linked to the peptide nitrogen by the side chain, forming a 5-membered ring. The consequence of this is that the peptide unit linking the previous amino acid residue with the imino acid residue (Pro and Hyp) has relatively little freedom of rotation about the N—C^α bond. This imposes a considerable rigidity in the collagen chain, and it is interesting to note that the restriction imposed on the orientation of the peptide unit about the N—C^α bond in Pro and Hyp is just what is required to stabilize the collagen structure (note $\phi \simeq -60°$ for the conformation γ in Figure 6). Considering the other amino acid residues, alanine is approximately 10% and the polar side chains (Arg, Lys, Asp,

Glu), form 20% of the amino acid residues. Although these do not play a very significant role in the formation and structure of the triple helix, they are relatively important in the inter-triple-chain linkages which lead to fibril formation (see Chapter 1 and Chapter 3).

C. *Outline of the Molecular Structure in Relation to Amino Acid Composition*

1. *Glycine at Every Third Position*

Although the fact that glycyl residues form one-third the total number of residues in the polypeptide chain of collagen was known for quite some time, it was only in 1954 that a structure, which demanded this occurrence of glycine to the extent of 33%, was worked out (Ramachandran and Kartha, 1954). In this structure, it was definitely postulated that *every third residue in the primary sequence is glycine* and that the positions occupied by glycine residues cannot be replaced by any other residue containing a β-carbon atom in its side chain. The authors also assumed that all the peptide units which occur in this structure had the *trans* conformation for the peptide unit, unlike some of the earlier structures. The projection of the structure* down the central axis of the triple chain is shown in Figure 7a. Essentially, it consists of three polypeptide chains, each of which takes the shape of a threefold helix having a left-handed twist (symmetry 3_2 according to crystallographers), with a pitch of 9 Å containing three residues. The three helical chains are arranged about a central axis so that the three chains (marked A, B, and C in Figure 7) are also related to one another by a threefold screw axis, in this case also having a left-handed twist (symmetry 3_2).

* There are several ways of representing a complicated molecular structure on the plane of the paper. One is to make a perspective drawing which is similar to what it would look if a model of it were made and observed by the eye. However, in such a structure, the details are not clearly marked and it is difficult to appreciate the angle between various bonds, the distances between the atoms, and so on, purely by looking at the drawing. On the other hand, it is common in crystallography to give the molecular structure in the form of projections along a suitable direction on the plane of the paper kept normal to this direction. This diagram is what one would see if a model were illuminated by a parallel light and the shadow is observed on a screen. The third dimension of the model is indicated in the projection drawing by numbers, giving the distance from the plane of projection to the atom concerned. Using the information given by such a projection, it is quite easy to calculate the distances between the atoms, the angles between atom pairs, and so on. We shall follow this practice in our chapter.

The three chains are held together by interchain hydrogen bonds (indicated for instance between B and C in Figure 7a). The atoms, other than the backbone, which are connected to the α-carbon atoms are also marked for chain A in Figure 7b. It will be seen from this that while the bonds connecting the β-carbon atoms to C_2^α and C_3^α are pointing away from the center of the triple helix, the two hydrogen atoms attached to C_1^α (which form part of a glycine residue) are located near the center of the triple helix, and are pointing towards the center. There is no space for C^β to occur in this central region, and the three chains have to be separated appreciably from one another if C^β should occur at position 1. If this were done, the possible hydrogen bonds linking the chains A to B, B to C, and C to A cannot be formed as the relevant bond length becomes too large for good stability. Therefore, we may conclude that there cannot be a nonglycyl residue with its side chain located at C_1^α in any of the chains. Hence the residue at every C_1^α atom or the symmetry-related atoms C_4^α, C_7^α, etc., cannot be anything other than glycine. Thus, purely from consideration of the packing of collagen chain, we can conclude that *glycine must occur at every third position in the chain in all its ordered regions* in which the triple helix occurs.

It will be seen from Figure 7b that the symmetry which relates the three chains, defined by a rotation of $-120°$ about, and a translation of 3 Å along, the central axis of the triple helix, will transform chain A to chain B, chain B to chain C, and chain C back into chain A. It can also be noted from Figure 7b that there is no difficulty in fixing a prolyl side chain either at position 2 or at position 3. What is more interesting is the fact that the dihedral angle ϕ for proline, which has only a restricted range, is found to be perfectly satisfactory in the model as built from the above consideration. It is true, however, that the positions 2 and 3 need not always have an imino acid residue, Pro or Hyp, but whenever they occur, they can readily be fitted into the triple-chain structure at these positions without distorting the main chains. All these will be discussed in detail later. Here we are introducing these factors to indicate that experimental evidence from the primary-structure studies are fully satisfied by the properties of the triple helix. In fact, recently, the complete amino acid sequence of one of the three chains which occurs in the collagen triple helix has been completed (see Chapter 1 for details). These studies show clearly that in the chain of collagen which consists of more than 1000 residues, all except the first few and the last few residues have this repeating sequence, namely -Gly-X-Y-, and also that Pro and Hyp occur only in the positions 2 and 3. One more property that has been observed in this primary sequence is that a Hyp occurs only in the third position

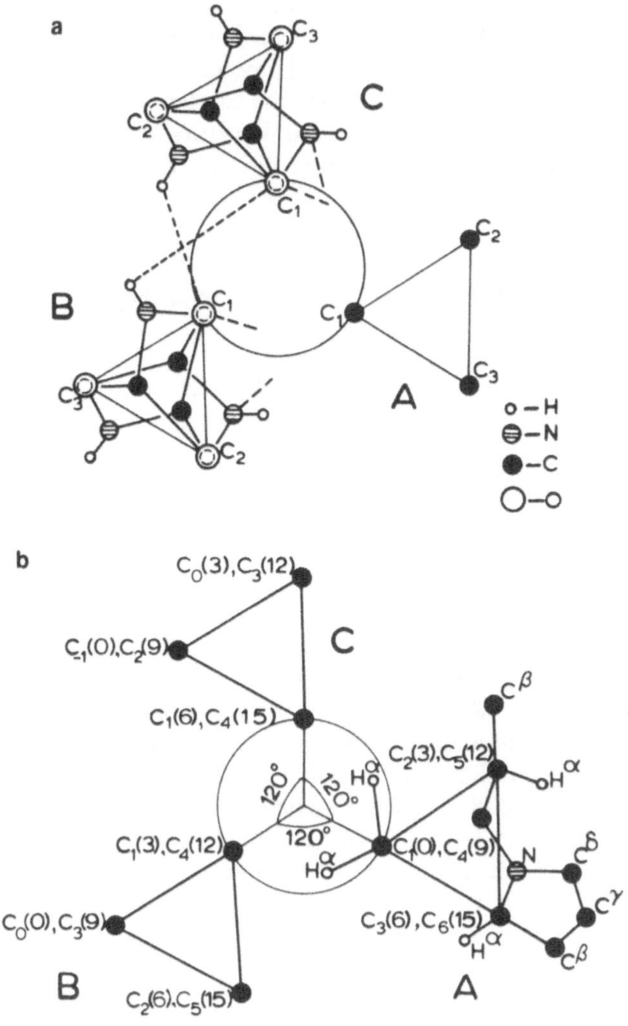

FIGURE 7. Projection of the Ramachandran–Kartha (1954) triple-helical structure of collagen, with an exact threefold screw-axis symmetry. (a) This shows the hydrogen bonds postulated, between chains B and C. (b) This indicates the angles and heights above the basal plane of the α-carbon atoms in the three chains. Note that there is only space for two H^{α}-atoms attached to C_1 on the inside of the triple helix, but not for a C^{β}-atom, while a C^{β} can be attached to C_2 and C_3 and their symmetry-related atoms.

and never in the second position, while Pro can occur occasionally in the third position also, although mostly it is found only in the second position. This peculiar feature, namely that Hyp occurs only in the third position, has an important part to play in the understanding of the hydrogen-bonding scheme in the collagen structure and the way in which the hydrogen bonds tend to stabilize the structure. This is discussed in Section II.

2. *Agreement with X-Ray Diffraction Pattern and the "Coiled-Coil" Structure*

The above simple triple-helical model does not, however, satisfy one particular fact observed from X-ray diffraction studies. This is that, whereas the number of units per turn (n) of the helix as obtained from the model is exactly 3, the same helical parameter n as deduced from the X-ray pattern turns out to be 3.3, with the rise per unit (h) being 2.9 Å. In order to be in accord with the X-ray evidence regarding these helical parameters of the structure, the simplified structure mentioned above has to be modified, and this was first done by Ramachandran and Kartha (1955). We shall not go into the details of how this was arrived at. They are available in a simple form elsewhere (Ramachandran, 1968). This modified structure, which is a coiled coil, is shown in outline in Figure 8, where each peptide unit is indicated by a line. It is interesting to note that this structure, in addition to satisfying the X-ray evidence, still possesses the requirement of the primary structure, namely that the residues at positions 1, 4, 7, etc., in each chain can only be glycyl, as they occur on the inside of the triple helix and do not have space for a C^{β} atom attached to C^{α}. Also, the other two positions, 2 or 3, and related ones can easily accommodate imino acid residues in addition to other residues containing β-carbon atoms.

This so-called "coiled-coil structure" was put forward by Ramachandran and Kartha (1955). In this structure, two hydrogen bonds for every three residues, connecting one chain with its neighbors, were assumed to occur. It was pointed out by Rich and Crick (1955) that it is more likely that only one hydrogen bond for every three residues would occur. Although the precise coordinates were not given by either Ramachandran and Kartha (1955) or by Rich and Crick (1955), they were worked out in later years and are available in the papers by Ramachandran *et al.* (1962) and Rich and Crick (1961). We shall not give the details of these two structures (known in the literature as the "two-bonded" and the "one-

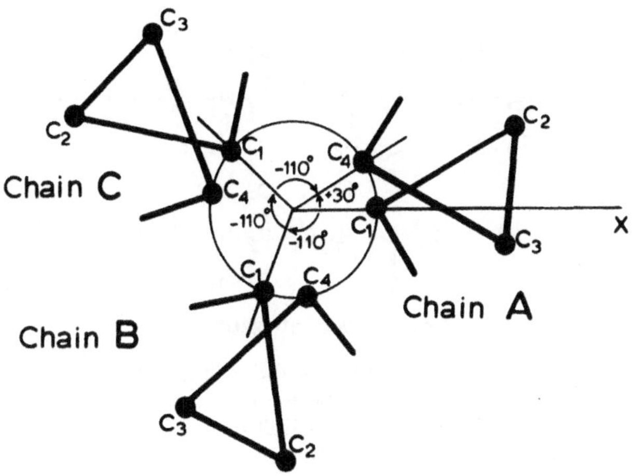

FIGURE 8. The coiled-coiled structure of the collagen backbone atoms, which is the correct basis of its molecular packing, as revealed by X-ray data. While the atom C_4 comes over C_1 in Figure 7, it is twisted by an angle of about 30° about the central axis (marked +30°) in this figure. As a consequence, the atoms C_1 of chains A and B, etc., are not related by a twist of −120°, as in Figure 7, but by one of −110°, leading to approximately 3.3 units per turn.

bonded" structures), since a reconciliation of the two suggestions has taken place recently (Ramachandran and Chandrasekaran, 1968). The main point of difference between the Madras group and the Cambridge group was this: that the former preferred a two-bonded structure, while the latter preferred a one-bonded structure (Figure 9a), believing that the two-bonded structure was too closely packed to be stable. There has been no controversy at all regarding the essential nature of the arrangement of the three helices or the degree of their supercoiling. Critical evidence to distinguish between these two structures has not been forthcoming until very recently, and what was available previously was not fully discriminatory between the two possibilities.

However, a very recent and carefully carried out experimental study by Yee, Englander, and von Hippel (1974) has conclusively proved that "native collagen has a two-bonded structure." What Yee *et al.* determined was, in effect, the number of NH groups per tripeptide which exchange freely with tritium when the sample was put in tritiated water. They found that approximately 1.7 NH groups per three residues exchange slowly with the medium, indicating that definitely more than one NH group is

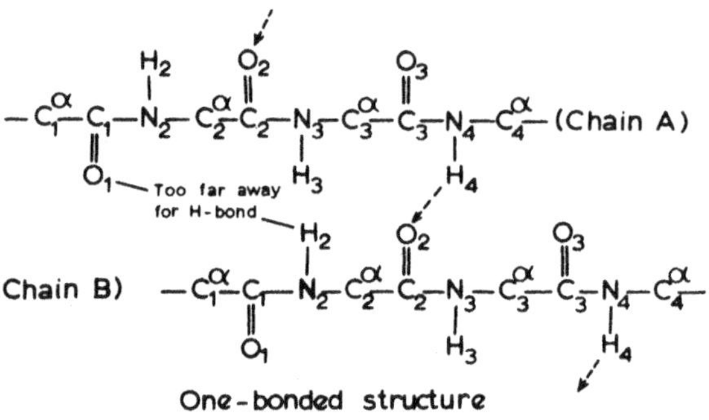

Figure 9. (a) The nature of the hydrogen bond which occurs between two neighboring chains A and B in the collagen triple helix, in the so-called "one-bonded" structure. The structure is really three-dimensional, and the one-dimensional diagram with parallel chains is only schematic.

involved in hydrogen bonding. Allowing for the occurrence of proline and hydroxyproline in positions 2 and 3, which do not have free NH groups, the number 1.7 instead of 2 for the number of hydrogen-bonded amino groups per triplet of residues is satisfactory.

In the meanwhile Ramachandran and Chandrasekaran (1968) had proposed an alternative two-bonded structure which reconciles the ideas

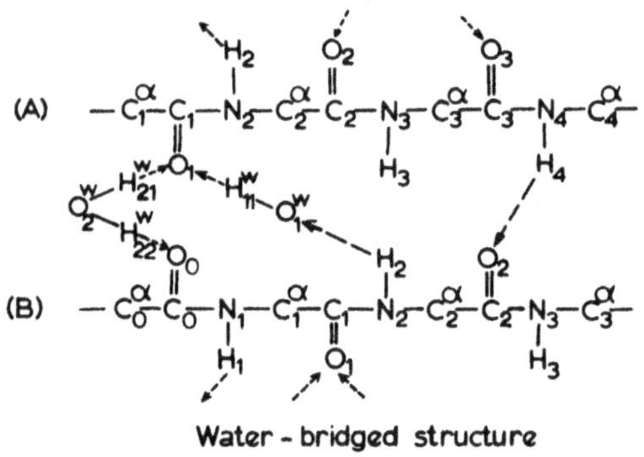

Figure 9. (b) The hydrogen bonds in the two-bonded water-bridged structure. Here, the hydrogen bond $N_4H_4(A)\cdots O_2(B)$ of the one-bonded structure is retained, while $N_2H_2(B)$ is hydrogen-bonded to a water oxygen O_1^w, which is linked to O_1 of chain A. In addition, a second water molecule $O_2^w(H_{21}^w, H_{22}^w)$ links O_1 of chain A to O_0 of chain B.

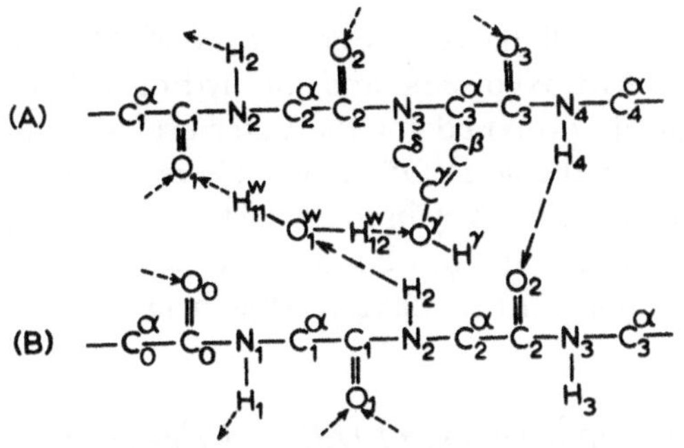

FIGURE 9. (c) This diagram illustrates how the second hydrogen attached to O_1^w can be hydrogen bonded to O^γ of chain A, belonging to a hydroxyproline side chain.

of the two different structures proposed earlier—namely, those having one and two hydrogen bonds per three residues. In this new structure, which we may call "the water-bridged structure" one of the interchain hydrogen bonds is between the backbone atoms of neighboring chains, while the other one is a hydrogen bond between the NH of the first peptide unit and a water molecule, the water molecule then hydrogen-bonding with a C=O of the neighboring chain. The essential outline of this two-bonded water-bridge structure is shown in Figure 9b, in which the two types of hydrogen bonds are shown between the chains A and B. In their report, Ramachandran and Chandrasekaran (1968) also found that this arrangement of the three helical chains could also have one more hydrogen bond via a second water molecule, between chains A and B, which is also shown in Figure 9b. This second water molecule donates both the protons for forming hydrogen bonds, and the two hydrogens of the water molecule are linked to C=O groups in chains A and B. In order to distinguish between the two types of water molecules, the former one, which provides an intermediary hydrogen bond linking N_2H_2 of chain B to O_1 of chain A, is called water molecule 1 (H_{11}^w—O_1^w—H_{12}^w); the second molecule which links the two chains is designated as the water molecule 2 (H_{21}^w—O_2^w—H_{22}^w). The observations of Yee *et al.* (1974) are fully in agreement with the postulation of the second interchain hydrogen bond being via a water bridge. A full account of the water-bridged structure and the evidences in favor of it will be discussed in Section II. Incidentally, the water bridge shown in Figure 9b can also form another hydrogen bond with the oxygen in the hydroxyl group of hydroxyproline, if the Hyp residue occurs at position 3 in the structure. This is shown schematically in

Figure 9c, where it will be seen that the bond O_1^w—H_{12}^w is pointing towards O^γ of the hydroxyproline and the hydrogen bond is possible if the position originally proposed for the water-bridge structure shown in Figure 9b is taken for the atom O^w. Thus, it is reasonable to assume that hydroxyproline has a role to play in the stability of the collagen triple helix by being involved in a hydrogen bond which links the O^γ of this rigid chain with the backbone of the collagen triple helix.

D. *Nature of the Conformation at Different Places in the Collagen Triple Helix*

In the description given above of the collagen structure, we had shown that some of the locations in the chain can only be occupied by glycyl residues and also that certain locations are preferred by the imino acid residues. Thus, the side chains at positions C_1, C_4, C_7, etc., can only correspond to the hydrogen of a glycyl residue, no larger residue being possible. On the other hand the positions 2 and 3 and their related positions 5, 6, 8, 9, etc., can accommodate a variety of side chains containing a β-carbon atom. But, what is interesting is the fact that the backbone conformation of the chain takes such a shape in the triple helix that the rigid 5-membered ring of proline or hydroxyproline can occur at either positions 2 or 3. By observation, while glycine is always found at every third position (see Chapter 1), proline and hydroxyproline are seen to occupy only approximately 40% of locations 2 and 3. In view of these, the basic triple-chain structure and its supercoiling may be expected to be the same as that observed with the other triple-chain helices in related polypeptides such as $(\text{Gly-Pro-Hyp})_n$ or $(\text{Gly-Ala-Pro})_n$. However, the details of the interchain hydrogen bonding may not be the same for different types of sequences that occur in the local regions of the triple helix. Thus, if proline occurs in position 2, it is not possible to have either a direct hydrogen bond or a water-bridged hydrogen bond, with N_2 as donor, since this nitrogen is involved in the 5-membered ring, with no proton attached to it. Similarly, while the hydroxyproline O^γ can readily receive a hydrogen bond from O_1^w, this may not be observed if a Hyp residue does not occur in position 3, but some other residue occurs there. Thus it should be borne in mind that, in the actual structure, the nature of the hydrogen bonding in different regions of the triple-helical protofibril may be different from one another, since there is no strictly repeating sequence of bonds and hydrogen bonds in the native collagen structure. This is, of course, to be expected since the primary sequence of collagen is not repetitive, except for the occurrence of glycine at every third position for a major portion of its length. In order to investigate the properties of different types of hydrogen bonds, it is therefore necessary to make studies on related polypeptides having sequences similar to those found in collagen, but in which the sequence repeats for every third residue. This will be discussed in Section III, dealing with polypeptides related to collagen.

II. Molecular Structure in Relation to Amino-Acid Sequence

A. Basic Structure

As was mentioned in Section I-C-2, the basic geometry of the triple-helical structure is describable by Figure 8. The three individual chains in Figure 8 are marked A, B, and C, and they are related to each other by the following properties, which are given in terms of the cylindrical polar coordinates, r, ϕ, z, of the representative atoms in the three chains. The z axis of the coordinate system is taken to be perpendicular to the plane of the paper passing through the center of the triple helix, with the x axis (corresponding to $\phi = 0°$) coinciding with the line joining the center to C_1 of the A chain. In terms of these, the relations are given by

$$\begin{array}{lll} C_1(A): r = r_0 \text{ Å} & \phi = 0° & z = 0.0 \text{ Å} \\ C_1(B): r = r_0 \text{ Å} & \phi = -110° & z = 2.9 \text{ Å} \\ C_1(C): r = r_0 \text{ Å} & \phi = -220° & z = 5.8 \text{ Å} \\ C_4(A): r = r_0 \text{ Å} & \phi = -330° \ (+30°) & z = 8.7 \text{ Å} \end{array} \quad (1)$$

where r_0 is the distance from the central axis to any one of the atoms at positions 1, 4, etc. It will be noticed that there is a helical repetition of atoms in a single chain for every three residues. For example, the atom C_4 can be obtained from the atom C_1 by a rotation of +30° about the z axis, followed by a translation of 8.7 Å along the same axis. Furthermore, the three chains themselves are related by a left-handed screw symmetry, which may be designated by the operations $\Delta\phi = -110°$ and $\Delta z = 2.9$ Å. The three chains A, B, and C are identical in their conformation, or shape, and one of them (A) is related to the other two by the relations

$$\begin{array}{llll} A \rightarrow B: r_{jB} = r_{jA} & \phi_{jB} = \phi_{jA} - 110° & z_{jB} = z_{jA} + 2.9 \text{ Å} & (2) \\ A \rightarrow C: r_{jC} = r_{jA} & \phi_{jC} = \phi_{jA} - 220° & z_{jC} = z_{jA} + 5.8 \text{ Å} & \end{array}$$

In Eq. (2), r_j, ϕ_j, and z_j are the cylindrical polar coordinates of any atom in the chain concerned, and the above relations are applicable not only to the α-carbon atoms but also to all the other atoms in the backbone of the chain.

It will be noticed that the data given by Eqs. (1) and (2) provide information regarding both the supercoiling of one chain as well as the relationships between the three supercoiled helices with one another. This relationship is given pictorially in Figure 8. Essentially, the different structures discussed in the literature have the same pattern, with only minor changes in the values of the repeat distance along the axis of the helix and the unit twist. On the other hand, as already mentioned, there have been two suggestions regarding the number of stabilizing interchain H bonds that the structure can have: (1) the structure can only accommodate one hydrogen bond between neighboring helices per triplet of residues and (2) two direct backbone–backbone hydrogen bonds can exist

TABLE 1
Coordinates of the Backbone Atoms of Three Residues in the One-Bonded Structure of Collagen[a,b,c]

Atom	r(Å)	ϕ(°)	z(Å)
N_1	2.0	−32	−1.0
H_1	1.8	−62	−1.2
C_1^{α}	1.4	0	0.0
C_1	2.4	−2	1.1
O_1	3.1	−22	1.5
N_2	3.0	22	1.6
C_2^{β}	4.9	34	2.8
C_2^{γ}	4.7	47	1.8
C_2^{δ}	3.2	50	1.6
C_2^{α}	4.0	18	2.7
C_2	3.2	12	4.0
O_2	2.0	22	4.2
N_3	4.0	1	4.8
C_3^{β}	5.0	−16	6.6
C_3^{γ}	6.0	−13	5.4
C_3^{δ}	5.4	−1	4.6
O_3^{γ}	6.1	−25	4.7
C_3^{α}	3.6	−9	6.0
C_3	3.1	8	7.1
O_3	3.8	23	7.3

[a] The sequence -Gly-Pro-Hyp- is assumed to occur in all the three chains.
[b] All hydrogens are omitted, except the amino hydrogens.
[c] The values are rounded off to 0.1 Å and 1°.

TABLE 2
Coordinates of the Atoms of Three Residues in the Two-Bonded Structure of Collagen[a,b]

Atom	r(Å)	ϕ(°)	z(Å)
N_1	2.0	−28	−0.9
H_1	2.3	−53	−0.7
C_1^{α}	1.2	0	0.0
H_{11}^{α}	1.2	43	−0.6
H_{12}^{α}	0.7	−55	0.4
C_1	2.2	10	1.1
O_1	3.2	−5	1.2
N_2	2.5	39	1.6
H_2	2.2	62	1.4
C_2^{α}	3.5	37	2.7
H_2^{α}	4.4	31	2.4
C_2^{β}	4.2	56	3.1
C_2	2.9	23	3.9
O_2	1.7	13	4.0
N_3	3.8	16	4.7
H_3	4.8	21	4.8
C_3^{α}	3.5	3	5.9
H_3^{α}	3.4	−12	5.6
C_3^{β}	4.9	3	6.7
C_3	2.6	19	6.8
O_3	2.7	45	6.4

[a] All atoms in the backbone and atoms in the side chain up to C^{β} are included.
[b] All values are rounded off to 0.1 Å and 1°.

between neighboring residues per triplet. Using the parameters unit twist (t = −110° and unit height (h) = 2.9 Å, these two structures may be described by the coordinates listed in Tables 1 and 2. The main structural difference between the two is that the one-bonded structure has only one hydrogen bond of the type $N_4H_4(A)\cdots O_2(B)$, while the two-bonded structure contains two hydrogen bonds, namely $N_2H_2(B)\cdots O_2(A)$ and $N_4H_4(A)\cdots O_3(B)$. [The diagram showing the projection of these two structures is not given here, but suitable diagrams are available in an earlier article by Ramachandran (1967). These are not given here because they are only historically important, and they have been superceded by later developments, which have reconciled the differing assumptions made by the two groups of workers.]

If the coordinates listed in Tables 1 and 2 are examined, it will be seen that the distance from the axis (r) of the atom C_1 is 1.4 Å in the case of the one-bonded structure, while it is as small as 1.2 Å for the two-bonded structure. Thus, as a whole, the three chains are packed much more closely together in the two-bonded structure than in the one-bonded structure. It has been generally felt by workers interested in molecular structure of biological compounds that perhaps the two-bonded structure is too closely packed to have good stability. Although calculations of energies using the usual potential functions, as adopted in the authors' laboratory for biopolymer conformational calculations, indicated that the two-bonded structure has a slightly lower energy than the one-bonded structure (Ramachandran and Venkatachalam, 1966), the argument cannot be taken to be rigorously valid because it depends to a large extent on the accuracy of the potential functions employed for this purpose. As mentioned in Section I, it has now been possible to obtain a reconciliation of the one-bonded structure and the two-bonded structure by having one hydrogen bond directly between the two neighboring polypeptide chains and one other hydrogen bond via a water molecule. Since the stabilizing energy of a hydrogen bond is much larger than the energies of non-bonded interactions, it is likely that the energy of this water-bridged structure would be lower than both the one-bonded and the two-bonded structures, and would, therefore, be a preferred configuration of the collagen triple helix. In view of this, we shall discuss in detail in this section only the water-bridged structure and its various properties.

B. The Water-Bridged Structure

The water-bridged structure was proposed by Ramachandran and Chandrasekaran (1968). In this structure, for which the coordinates are given in Table 3, there are three hydrogen bonds per triplet of residues. In two of these, the NH groups donate a proton each and these are (see Figure 9b):

H bond 1:	direct	$N_4H_4(A)\cdots O_2(B)$
H bond 2:	via water	$N_2H_2(B)\cdots O_1^w—H_{11}^w\cdots O_1(A)$

One of these is a direct interchain backbone hydrogen bond and the other is from the backbone NH group N_2H_2, of chain B, to the oxygen of water 1, which then donates one of its hydrogens to O_1 of chain A. Ramachandran and Chandrasekaran (1968) found that this structure (shown in Figure 10) can also accommodate one more water molecule, which has hydrogen bonds, respectively, with $O_1(A)$, $O_0(B)$:

H-bond 3:	from water	$O_1(A)\cdots H_{21}^w—O_2^w—H_{22}^w\cdots O_0(B)$

TABLE 3
Coordinates of the Atoms in the Three Residues of the Water-Bridged One-Bonded Structure (Only H^{α} and C^{β} in Positions 2 and 3)[a,b]

Atom	r(Å)	ϕ(°)	z(Å)
N_1	2.0	−32	−1.0
H_1	1.8	−62	−1.2
C_1^{α}	1.4	0	0.0
H_{11}^{α}	0.4	−21	0.4
H_{12}^{α}	1.8	34	−0.5
C_1	2.4	−2	1.1
O_1	3.1	−22	1.5
N_2	3.0	22	1.6
H_2	3.0	40	1.3
C_2^{α}	4.0	18	2.7
H_2^{α}	4.7	8	2.4
C_2^{β}	4.9	34	2.8
C_2	3.2	12	4.0
O_2	2.0	22	4.2
N_3	4.0	1	4.8
H_3	5.0	−1	4.7
C_3^{α}	3.6	−9	6.0
H_3^{α}	3.0	−25	5.8
C_3^{β}	5.0	−16	6.6
C_3	3.1	8	7.1
O_3	3.8	23	7.3
O_1^{W}	3.8	−42	3.8
H_{11}^{W}	3.5	−36	2.9
H_{12}^{W}	4.6	−34	4.1
O_2^{W}	3.5	−72	0.9
H_{21}^{W}	2.9	−58	1.0
H_{22}^{W}	3.1	−88	1.0

[a] All backbone atoms, atoms attached to the C^{α}'s and the water molecules are included.
[b] All values are rounded off to 0.1 Å and 1°.

It may be mentioned that in the early days when the two-bonded and one-bonded structures were rivals in the field, experimental studies regarding the number of hydrogen bonds per triplet of residues clearly favored the two-bonded structure (Harrington, 1964). In spite of this, there was a feeling among the theoretical workers in the field that the two-bonded structure of Ramachandran *et al.* (1962) was probably not the right structure, for reasons mentioned above. On the other hand, the one-bonded structure, whose hydrogen-bonding scheme is shown in Figure 9a,

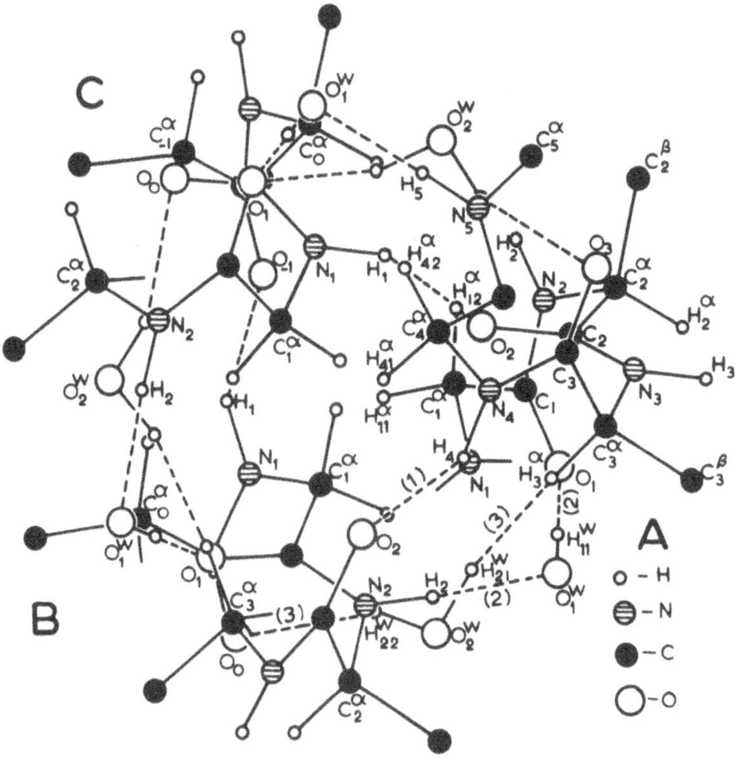

FIGURE 10. The water-bridged structure of collagen. The covalent bonds are shown by solid lines and the hydrogen bonds by broken lines. Note the direct interchain hydrogen bond $N_4H_4(A)\cdots O_2(B)$ and the second one between chains A and B of the type $N_2H_2(B)\cdots O_1^wH_{11}^w\cdots O_1(A)$.

has the defect that N_2H_2 (of the B chain) and O_1 (of the A chain) are too far away, although they are pointing toward each other in the structure. The newly proposed water-bridged structure got rid of this defect of the one-bonded structure (namely that it does not contain the maximum number of hydrogen bonds inside the triple chain that are possible). It has done this by having a water molecule O_1^w bridging $N_2H_2(B)$ and $O_1(A)$.

Thus, by forming hydrogen bonds from two of the amino groups per triplet, one directly to the neighboring chain carbonyl group and another via water, the experimental measurement of the number of NH protons which slowly exchange with deuterium, or tritium, under suitable conditions could be explained, while at the same time having the relaxed dimensions of the original one-bonded structure. In fact, a recent detailed experimental study made by Yee *et al.* (1974) has again substantiated the occurrence of two hydrogen bonds involving the amino groups in residues at positions 1 and 2, respectively. This structure is shown in Figure 10. It will be readily seen from this figure that the amino group N_3H_3 is pointing away from the center of the triple helix and cannot take part in any hydrogen bonding between neighboring chains in the triple helix. In fact, Yee *et al.* (1974) mention that their results appear to exclude one-bonded models for collagen, but that they cannot distinguish between the earlier two-bonded model, which contained two direct peptide-to-peptide hydrogen bonds per triplet, and the more recent proposal involving one direct bond and one cross-bridged via a water molecule. However, the present authors feel that the water-bridged two-bonded structure is definitely superior to the original two-bonded structure, for reasons mentioned earlier. In fact, there is one more reason in favor of this structure, which will be discussed in the next section, namely, that the same water molecule which cross-bridges the two peptide chains also can be involved in the formation of a hydrogen bond with the O^{γ} of a hydroxyl residue which occurs in position 3. The nature of the hydrogen bond in this case is shown schematically in Figure 9c.

C. *Hydrogen Bonding of Hyp Hydroxyl Group*

In fact, Figure 9c indicates an important role that the hydroxyl group of hydroxyproline in position 3 of the collagen peptide chain can play in stabilizing its triple-helical structure. It will be seen from this figure that the amino group at N_2 of chain B donates a proton to the water molecule which in turn donates one proton to O_1 of chain A and another proton to O^{γ} of the hydroxyproline side chain at position 3 in chain A. Figure 9c is only schematic, but its full structure in projection is shown in Figure 11 involving in particular the bridging produced by the water molecule both between the peptide backbones and between a backbone atom and the Hyp side-chain atom. The diagram shown in Figure 11 is slightly different from that given by Ramachandran *et al.* (1973), in which they proposed the crucial role for hydroxyproline in the structure of collagen, because it stabilizes the structure by forming additional hydrogen bonds. In the diagram given in the paper by Ramachandran *et al.* (1973), the OH group of Hyp was considered as donating a proton to the hydrogen bond O^{γ}—$H^{\gamma}\cdots O_1^{w}$. However, as pointed out by Ramachandran *et al.* (1975), a very simple rearrangement of the scheme of hydrogen bonding serves the double purpose of the hydroxyproline hydroxyl group taking part not only in receiving a hydrogen bond from the bridging water molecule, but

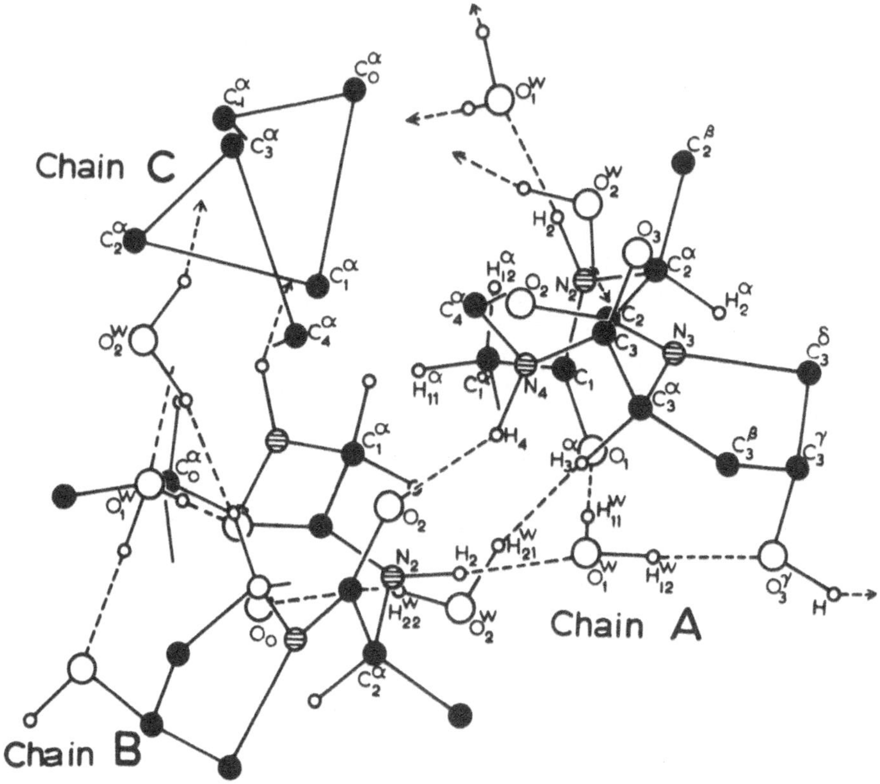

FIGURE 11. The water-bridged structure with a hydrogen bond from the water molecule to O^γ of the hydroxyproline side chain. Note that the group $O^\gamma H^\gamma$ can donate a proton to form a hydrogen bond between two neighboring triple-helical protofibrils (shown by an arrow).

also in donating another hydrogen bond $O^\gamma—H^\gamma \cdots O$ between one triple-helical chain and a neighboring triple-helical chain. Thus Hyp serves the purpose not only of stabilizing a triple-chain protofibril, but also in providing a cross-link (only of the hydrogen-bonding type and not a covalent bond) between one protofibril and the next. (See Bansal *et al.*, 1975b). Simple calculations show that the center-to-center distance between two neighboring protofibrils is about 12 Å when this hydrogen bond is a direct bond between O^γ of one protofibril and a carbonyl oxygen of the neighboring protofibril. This agrees very well with the minimum distance between triple-helical axes observed for ordered structures of collagen fibers when their X-ray pattern is taken at low humidity. However, when the humidity is increased, this distance increases and probably more water molecules come between the protofibrils and separate them from one another by forming a layer of water molecules in between.

Other Evidences for the Water-Bridged Structure

We shall not discuss separately the evidences in favor of the water-bridged structure both with and without Hyp because, when Hyp does not occur in the local region at position 3, the water molecule would still serve to form the bridge as shown in Figure 9b and in detail in Figure 10. In such a case, one of the protons of the water molecule, which bridges the two chains, is pointing outward, and molecules from the solvent medium could conceivably come and disturb this water molecule. Hence, the structure is not expected to be particularly stable without Hyp also occurring in the third position. As will be mentioned in the next section, in which peptides having triplets of repeating sequences are discussed, the stability of the polymer (Gly-Ala-Pro)$_n$ is not as good as that of (Gly-Ala-Hyp)$_n$ in forming a triple-chain structure. As will be readily seen from Figure 9c and Figure 11, the latter, namely (Gly-Ala-Hyp)$_n$, can readily form all the hydrogen bonds indicated in Figure 9c, while one of these is distinctly absent in (Gly-Ala-Pro)$_n$. What is more, the bridging water is also subject to attack from the water medium in that polymer. Hence there will be an overall lack of stability in the case of (Gly-Ala-Pro)$_n$. This is mentioned here because it has a close relationship with the hydrogen-bonded structure as observed in native collagen.

In addition to the above water molecules, Figures 9b and 9c also show the possibility of a second water molecule O_2^w bridging the neighboring chains A and B by donating a proton each to O_1 of chain A and O_0 of chain B. This means that in the regions of the collagen structure, in which the second residue is not Pro, there can be two water molecules firmly bound to the protofibril. This fact is also supported by an analysis of the anisotropies measured with proton and deuterium resonances in conjunction with water absorption data of collagen (Berendsen, 1972). Berendsen found that *two* water molecules per *three* amino acid residues are firmly bound to the protofibril, which is in agreement with the model proposed by Ramachandran and Chandrasekaran (1968) and described above.

D. *Evidence for the Role of Hydroxyproline in Stabilizing Collagen*

As discussed in the previous section, the most satisfactory structure of collagen in which the hydroxyproline residue plays an important part is the one in which it stabilizes the structure by forming a hydrogen bond with a water molecule, which is then firmly linked to *three* groups in the neighborhood. Taking the water oxygen O_1^w of Figure 9c, these are (1) the amino group N_2H_2 of chain B, (2) the carbonyl group $C_1{=}O_1$ of chain A, and (3) the γ-hydroxyl group $O^\gamma H^\gamma$ of the Hyp side chain in chain A. This would mean that the hydroxylation of proline in position 3 would play a fundamental part in stabilizing the collagen structure. It is well known that in collagen, hydroxyproline occurs only in position 3, and that a special enzyme, proline hydroxylase, exists in the metabolic system that adds the hydroxyl group to the C^γ of proline occurring in position 3. The primary sequence of amino acid residues given in Chapter 1 does not contain even one example of Hyp in position 2, although a small number of prolines in position 3 may not be (fully) hydroxylated. In recent years, the action of proline hydroxylase has been studied on synthetic analogs having the repeating sequences -Gly-X-Y- (Hutton *et al.*, 1968; Kivirikko *et al.*, 1969, 1972; Kikuchi *et al.*, 1969; for more references see Chapters 5 and 11). Invariably, the hydroxylation occurs only on proline at position Y, i.e., the third residue starting with Gly as first in the triplet sequence. It is very satisfying to note that the molecular structure proposed for collagen also can have a hydrogen bond to the Hyp hydroxyl group only if it occurs in the third position. Further, it must also occur as a *trans* hydroxyprolyl residue, which is found to be the case in native collagen almost exclusively.

It is interesting to note that the theoretical picture for the role of hydroxyproline as leading to greater stability of the collagen triple helix has been supported by experimental evidences. One such evidence for the role of hydroxyproline in stabilizing the triple helix of collagen has been obtained by Berg and Prockop (1973), who found that the nonhydroxylated form of collagen, which was extracted from embryonic tendon cells, had a value for the melting temperature (T_m) which was 15° lower than the T_m of a hydroxylated form of collagen from the same source. We shall not discuss this experiment more in detail since it is discussed in Chapter 5. It is only necessary to point out that the modified protocollagen was also shown to consist of polypeptides having the same size as the $\alpha 1$ and $\alpha 2$ chains of normal collagen and that they exhibited a thermal transition in optical rotation studies similar to collagen. Therefore, the two forms of collagen differed only in the fact that there were many more Hyp residues in position 3 in the normal collagen than in the modified unhydroxylated collagen.

A similar conclusion that "hydroxyproline stabilizes the triple helix of chicken tendon collagen" has been arrived at by Jiminez *et al.* (1973). In this study, the thermal stability of unhydroxylated collagen relative to hydroxylated collagen was investigated using pepsin digestion at various temperatures as an enzymatic probe of conformation. The results clearly indicate that the unhydroxylated molecules have a denaturation temperature between 20° and 25°C, while the hydroxylated molecules are stable at least up to 35°C. This study also thus confirms the fact that the presence of hydroxyproline in the collagen triple helix contributes significantly to its thermal stability.

III. Structures of Synthetic Polypeptides Related to Collagen

Synthetic polypeptides have been found to be very good models of fibrous protein structures. For example, the well-known α-helix, which is the basic structure for the KMEF class of fibrous proteins, has been found to occur in fibers prepared from poly-γ-benzyl-L-glutamate and poly-L-alanine (Bamford *et al.*, 1956; Brown and Trotter, 1956). Several other polypeptides have been shown to have the α structure in solution. Similarly, some sequential polypeptides containing alternating glycyl and alanyl residues have proved useful as models for the study of the β structure as it occurs in the silk proteins (Fraser *et al.*, 1965). In the same way, studies on polypeptides resembling collagen in amino acid sequence could prove to be useful models for the study of the properties of the collagen structure. An excellent review has been published by Traub and Piez (1971), in which the structure in the fibrous form of a number of sequential polypeptides have been reported and analyzed. Another extensive and useful review is by Kobayashi and Isemura (1972). In view of these, and in view of the fact that work reported later than these reviews is quite small, but for the example of (Gly-Pro-Hyp)$_n$, we shall only give a brief summary of the main results and discuss some of the recent studies on the subject. The discussion will also be essentially restricted to the solid (fibrous) state, and solution studies will not be considered in detail.

As mentioned in Section II, collagen has glycine as every third residue in its ordered region and, therefore, its sequence can be described by -Gly-X_1-Y_1-Gly-X_2-Y_2-Gly-. In this, X_1, Y_1, X_2, Y_2, etc., are any amino acid residues. However, in view of the fact that imino acid residues Pro and Hyp occur quite frequently in the positions X and Y in the chain, we shall consider only the following sets of combinations in which increasing amounts of the imino acid residues are present.

We shall be considering sequential polypeptides having a repeat of 3 residues with the sequence (Gly-X-Y)$_n$. In this, both X and Y can be amino acid residues (indicated by the symbol A, including glycine) or an imino acid residue (indicated by the symbol I). In that case, the synthetic models for collagen can be divided into four classes:

1. Those with X = Y = A, that is $(\text{Gly-A-A})_n$.
2. Those with an I in position X and an A in position Y, namely$(\text{Gly-I-A})_n$.
3. Those in which X = A and Y = I, namely $(\text{Gly-A-I})_n$.
4. Those which both X and Y are imino-acid residues, namely$(\text{Gly-I-I})_n$.

An examination of the collagen structure indicates that all these possibilities could occur in the triple-chain structure. As mentioned in Section II, the glycine residues occur on the inside of the triple helix and both the positions X and Y, namely positions 2 and 3 following glycine, can be occupied by an amino acid or an imino acid residue. However, imino acid residues occurring in these positions make the structure more rigid because the dihedral angle ϕ (which is close to $-60°$) cannot have an appreciable variation when the 5-membered ring of the imino acid side chain links the α-carbon atom with the previous nitrogen. In view of this, one would expect that, of the four types mentioned, the type $(\text{Gly-I-I})_n$ would readily form triple helices while $(\text{Gly-I-A})_n$ and $(\text{Gly-A-I})_n$ would be less stable in such an arrangement and $(\text{Gly-A-A})_n$ may only form the structure under suitable conditions, but in general, may not form such a structure.

A. *Homopolypeptides*

We will be considering briefly in this section homopolypeptides related to collagen such as polyglycine, poly-L-proline, and poly-L-hydroxyproline. In all these cases, one of the modifications of the crystalline structure of the homopolypeptide is closely related to the collagen triple helix—if not in triple-helix formation—at least in the dihedral angles (ϕ,ψ) being close to those observed in the collagen structure.

1. *Polyglycine*

The structural form of polyglycine which resembles collagen is polyglycine II, which was first described by Crick and Rich (1955). In this structure, the polypeptide helices have 3 residues per turn and the unit height is 3.1 Å. The chains are packed in a hexagonal array linked by

NH···O hydrogen bonds. A modification of this structure has been proposed by Ramachandran *et al.* (1966) in which there is an additional CH···O hydrogen bond between the chains. The existence of the CH···O hydrogen bond is supported by the infrared study of Krimm *et al.* (1967). A possible structure of polyglycine II with direct and inverted chains has been given by Ramachandran *et al.* (1967). One interesting feature in this structure is that the CH···O hydrogen bonds can be formed only between like chains and thus not all the α-carbon atoms have their CH groups hydrogen-bonded. This has also been confirmed by the infrared studies of Krimm *et al.* (1967).

2. *Poly-L-proline*

As in the case of polyglycine, two modifications of poly-L-proline are known, namely poly-L-proline I and poly-L-proline II. Of these, form I has been shown to have all the peptide units in the *cis* conformation (Traub and Shmueli, 1963). On the other hand, the peptide units are all in the *trans* conformation in poly-L-proline II (Cowan and McGavin, 1955). This structure was later revised by Sasisekharan (1959*a*). In this revised structure, it has been shown that CH···O hydrogen bonds occur between a $C^{\gamma}H_2$ group of one chain and a carbonyl oxygen of another chain. This structure has $n = 3$ and $h = 3.12$ Å, values which are close to those of polyglycine II. The chains are also packed in a hexagonal lattice. Arnott and Dover (1968), in their least-squares refinement, considered the possibility of a random orientation (up or down of the chain). However, their refinement favored only the form in which the chains are all running in the same direction. In solution, poly-L-proline I and poly-L-proline II are interconvertible under suitable conditions (solvents). Details are available from the review of Traub and Piez (1971).

3. *Poly-L-hydroxyproline*

Sasisekharan (1959*b*) observed two forms of this polymer, each giving different X-ray patterns. He determined the crystal structure of one of these forms, which is similar to collagen and to polyglycine and poly-L-proline II. As in the latter polymers, in poly-L-hydroxyproline the chains have a threefold symmetry with a unit height of 3.05 Å, the chains being packed in a hexagonal lattice with OH···O hydrogen bonds between $O^{\gamma}H^{\gamma}$ and a carbonyl oxygen of the backbone of a neighboring chain.

Thus, we note that the three homopolypeptides, polyglycine, polyproline, and polyhydroxyproline, tend to take up a helical structure with a threefold symmetry and a unit height close to 3.1 Å. However, they do not show the supercoiling observed in collagen. If this supercoiling is taken into account and a value of 30° for three residues is taken for the second coiling, then the repeat comes down to a value close to 2.9 Å, as observed in collagen. Hence, it can be assumed that the basic chain structure of these three polypeptides is essentially the same as that observed in collagen and indicated as γ in Figure 6.

B. *Polytripeptides*

We have mentioned in the last section that four types of polytripeptides are particularly relevant for our study. Indicating always the polypeptides with Gly as the starting residue, these four are: (1) (Gly-A_2-A_3)$_n$, (2) (Gly-I_2-A_3)$_n$, (3) (Gly-A_2-I_3)$_n$, and (4) (Gly-I_2-I_3)$_n$. Table V of the review by Traub and Piez (1971) gives the results in a summarized form. Because of this, we shall not dwell in detail on these, but shall give only a summary and the salient features of the structures for the sake of completeness. Also, we shall focus our attention on the structure of these polypeptides in their solid form, that is in the form of fibers and only mention in passing their behavior in solution and the structures taken up by them in solution.

1. *Polytripeptides with Repeating Sequence of the Type Gly-A_2-A_3*

It is known from sequence studies on collagen that some regions of the chain, rich in polar groups, do not have an imino acid either in position 2 or 3, so that the sequence in these local regions will be of the type Gly-A_2-A_3. In view of this, polytripeptides of the form (Gly-A_2-A_3)$_n$ are of interest in relation to collagen. The structure of a few polytripeptides with sequences of this type are known: (Gly-Gly-Ala)$_n$ (Andries *et al.*, 1971); (Gly-Ala-Ala)$_n$ (Doyle *et al.*, 1970); [Gly-Gly-Glu (OEt)]$_n$ and [Gly-Ala-Glu (OEt)]$_n$ (Anderson *et al.*, 1970; Andries and Walton, 1971). It is interesting that almost all of these do not exhibit either a triple helical structure or even a chain conformation of the polyglycine II type. They invariably exhibit the β structure. The only exception is the polymer (Gly-Gly-Ala)$_n$ which shows two forms, one having a superfolded cross-β-structure and the other corresponding to the polyglycine II structure with a unit height per residue of $h = 3.14$ Å.

It is therefore of interest to note that the triple-helical structure that occurs in collagen, which presumably also exists for regions having the local sequence -Gly-A_2-A_3-, has been produced by the effect of the other sequences having the triple-helical structure and carrying regions also having these sequences along with them. It is interesting that, in water solution also, the polypeptide $(\text{Gly-Ala-Ala})_n$ is not collagen-like but has a random coil structure, with indications of some α or β structures occurring in it (Doyle *et al.*, 1970). Presumably the other polymers mentioned above also would not indicate any collagen-like structure in solution.

2. *Polytripeptides with Repeating Sequence Gly-I_2-A_3*

In this type of compound the middle residue at the second position can be a proline or hydroxyproline. Traub and Piez (Table V, 1971) list a series of compounds of this type. In most of them, I = Pro and A_3 is one of the amino acid residues Ala, ϵ-tosyl-Lys, Lys HCl, Phe, Ser, and Tyr. All these compounds exhibit a triple-helical structure in the solid state. It is interesting that $(\text{Gly-Pro-Ala})_n$ in solution also exhibits a collagen-like structure in water and in polyhydric alcohols (Brown *et al.*, 1972). Here again, it is reasonable to expect that most of the polymers with Pro at the second position and one of the common amino acid residues in the third position would also exhibit collagen-like features in solution. However, it has been reported by Kitaoka *et al.* (1958) that $(\text{Gly-Pro-Leu})_n$ is not collagen-like in the solid state. This requires confirmation.

3. *Polytripeptides with Repeating Sequence Gly-A_2-I_3*

In this type of compound the third position is occupied by an imino acid. In this case also, several examples are listed by Traub and Piez (1971) regarding their fibrous structure. All of them take up either the triple helical or the polyproline II structure. They include examples of $(\text{Gly-Ala-Pro})_n$, $(\text{Gly-Ala-Hyp})_n$, $(\text{Gly-Ser-Pro})_n$, and $(\text{Gly-Ser-Hyp})_n$. On examination, it is seen that of the two types of compounds in which either Pro or Hyp is in position 3, the ones that have Hyp in position 3 readily form the triple helix. However, the ones having Pro in the third position sometimes form the polyproline II structure and sometimes exhibit the collagen-like triple helical structure. From this, it can be concluded that $(\text{Gly-A}_2\text{-Hyp})_n$ is more stable in a collagen-like structure than $(\text{Gly-A}_2\text{-Pro})_n$. This agrees with the ideas mentioned in Section II regarding the hydrogen-bond formation in the triple-helical structure by the hydroxyl group of hydroxyproline. In solution, $(\text{Gly-Ala-Pro})_n$ has been investigated (Doyle *et al.*, 1971) and apparently does not exhibit an ordered structure in water, but shows a collagen-like conformation in trifluoroethanol and aqueous ethylene glycol.

An interesting observation has been made by Sutoh and Noda (1974) that the -Gly-Ala-Pro- sequence takes up the collagen-like structure even in solution in water if it is made as a sandwich with -Gly-Pro-Pro- sequences on either side. They studied the sequence (Pro-Pro-Gly)$_n$(Ala-Pro-Gly)$_m$(Pro-Pro-Gly)$_n$ with the material very carefully synthesized by the solid-phase synthesis technique. It is interesting that even when $m = 5$ and $n = 5$, i.e., when 5 sequences of Gly-Ala-Pro are kept in between, indications of the ordered collagen-like conformation were obtained.

4. *Polytripeptides with Repeating Sequence Gly-I_2-I_3*

There are four possible polypeptides belonging to this group—(Gly-Pro-Pro)$_n$, (Gly-Pro-Hyp)$_n$, (Gly-Hyp-Pro)$_n$, and (Gly-Hyp-Hyp)$_n$. All these four polypeptides have been studied by X-ray diffraction, some by the Russian workers (Rogulenkova *et al.,* 1964; Andreeva *et al.,* 1967, 1970) and some by the Israeli workers (Yonath and Traub, 1969) and Japanese workers (Okuyama *et al.,* 1972). It is interesting that all the four types of polymers exhibit the triple-helical structure of collagen in their solid state. The oligomers (pentamers and decamers) of Gly-Pro-Pro and Gly-Pro-Hyp have been found to exhibit the collagen-like fold in solution (Sakakibara *et al*., 1968, 1973; Sutoh and Noda, 1974).

It is an interesting fact that (Gly-Pro-Pro)$_n$ exhibits a peculiar arrangement of a single-chain triple helix in which one third of the chain runs in a direction opposite to the other two thirds. This has been studied in solution by Engel (1967) and is very similar to what occurs in the case of *Ascaris* collagen where denaturation, or removal of the regular structure, does not lead to a lowering of the molecular weight by a factor of 3 (McBride and Harrington, 1967). A possible model of a single-chain triple helical structure for (Gly-Pro-Pro)$_n$ has been given by Ramachandran *et al.* (1968). Recent morphological evidence on lamellar single crystals of poly(Gly-Pro-Pro) by Andries and Walton (1970) supports this type of arrangement for this polymer.

C. *Polyhexapeptides*

So far, only the Israeli group (Segal *et al*., 1969) seems to have studied in the solid state repeating hexapeptide sequences as models of the collagen structure. Remembering the occurrence of glycine at every third position in collagen, the suitable examples studied by these workers were (Gly-Pro-Ala-Gly-Pro-Pro)$_n$, (Gly-Ala-Pro-Gly-Pro-Pro)$_n$, (Gly-Ala-Ala-Gly-

Pro-Pro)$_n$ and (Gly-Ala-Pro-Gly-Pro-Ala)$_n$. They found that all four polymers show an X-ray diffraction pattern essentially similar to that given by collagen, indicating that they would have the same basic triple-helical structure as in collagen.

Of the four polypeptides mentioned above, the example of (Gly-Ala-Ala-Gly-Pro-Pro)$_n$ is particularly interesting. In this case, although (Gly-Ala-Ala)$_n$ by itself forms only a β structure, it takes up the collagen fold when it is put in association with a tripeptide of the sequence (Gly-Pro-Pro) on either side. Similarly, (Gly-Ala-Pro)$_n$ does not always readily form a collagen-like structure; but -Gly-Ala-Pro- in association with -Gly-Pro-Pro- in a hexapeptide sequence readily forms the collagen structure. These studies clearly indicate that if a polypeptide chain has a variety of sequences, with Pro and Hyp occurring fairly frequently in its sequence (all the while having Gly at every third position), then such a chain could readily associate to form a triple-helical structure. This is probably what happens in native collagen. The excellent work done by the Israeli group and by the Russian group on synthetic polypeptides have made it quite possible to understand why collagen exhibits a triple-helical structure throughout its length in spite of the fact that in several places it has only the sequence -Gly-A_1-A_2-. The sandwich oligomers prepared by the Japanese workers also prove that the more actively helix-promoting regions may tend to fill up the deficiency in helix promotion exhibited by some other regions of the chain.

D. *The Relative Stability of (Gly-Pro-Pro)$_n$ and (Gly-Pro-Hyp)$_n$*

In Section II it was mentioned that the collagen molecule with unhydroxylated prolines in position 3 is much less stable than the normal collagen molecule with Hyp in position 3. A reasonable cause for this difference was pointed out to be the possibility of there being hydrogen bonds stabilizing the structure which involved the $O^{\gamma}H^{\gamma}$ group of the hydroxyproline side chain. This naturally raises the question of whether a similar difference would be found between the polymers of the tripeptides Gly-Pro-Pro and Gly-Pro-Hyp. However, it will be noticed that the water-bridging linkage cannot occur in either of these two polypeptides because Pro occurs in position 2 and there is no free NH in the second residue in the triplet to donate a proton for the hydrogen bond to the water through which it is linked to the next residue. However, it is very interesting that a recent work by Sakakibara *et al.* (1973) showed that the melting temperature of (Pro-Hyp-Gly)$_{10}$ is about 35° higher than the melting temperature of (Pro-Pro-Gly)$_{10}$. The two molecules are exactly the same chain length so any differences arising from differences in molecular weight are avoided.

Although, at first sight, it would appear that the observations on the above two oligomers would indicate a support of the general theory that hydroxyproline leads to greater stability than proline in position 3 in collagen, the fact that a hydrogen bond through water would be formed in the water-bridged structure proposed by Ramachandran *et al.* (1973) makes it difficult to understand the phenomenon in this case. However, a very interesting suggestion has been made by Berg *et al.* (1973) that a new type of hydrogen-bonded structure can occur in the triple helix of (Gly-Pro-Hyp)$_n$, in which there are two interchain hydrogen bonds, one the standard $N_4H_4 \cdots O_2$, and the second one a direct interchain hydrogen bond between the hydroxyl group of hydroxyproline in one chain and the carbonyl of glycine in the adjacent chain. For such a hydrogen bond to be possible, it is necessary that the peptide unit occurring between glycine and proline should have the *cis* conformation for each triplet. Berg *et al.* have verified the reasonableness of this linkage by means of model building and rough preliminary calculations. Recent work done in our laboratory by Bansal *et al.* (1975*a*) has confirmed the above proposal made by Berg *et al.* (1973). The unit height *h* is as low as 2.75 Å in this model, as contrasted with 2.9 Å for collagen, and also the unit twist is 113° as contrasted with 110° for collagen. It is found that the hydroxyproline side chain can be involved in a direct hydrogen bond with the carbonyl of a neighboring chain only if the *cis* peptide unit is introduced between residues 1 and 2 in the triplet. The hydrogen-bonding scheme is shown in

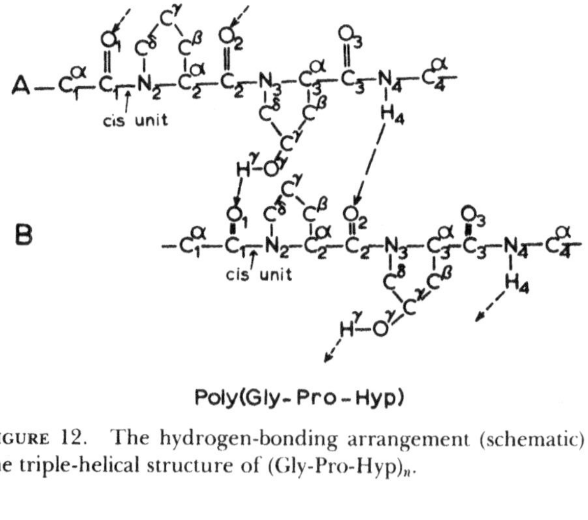

FIGURE 12. The hydrogen-bonding arrangement (schematic) in the triple-helical structure of (Gly-Pro-Hyp)$_n$.

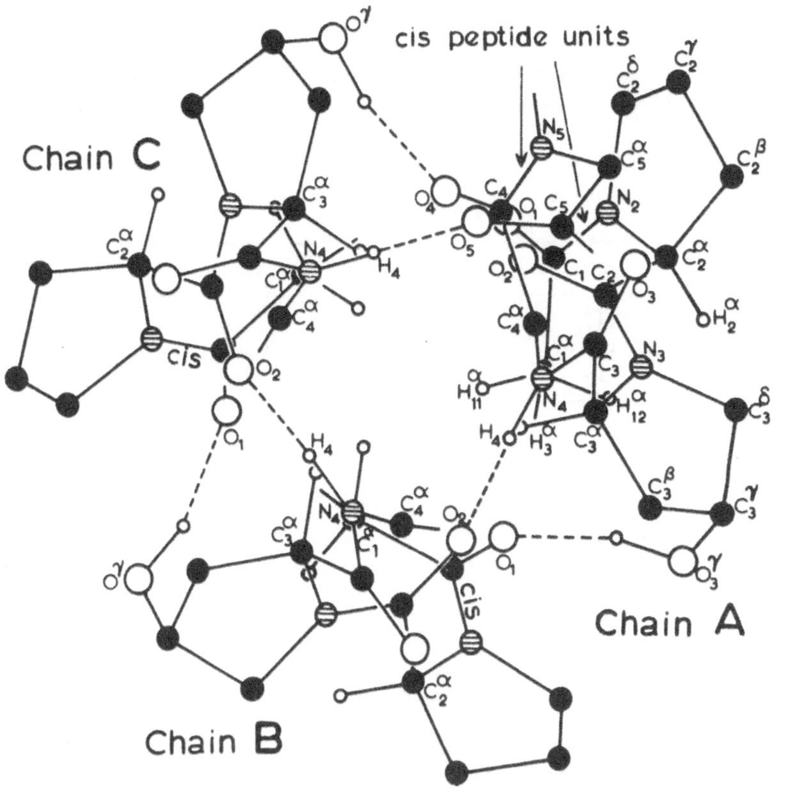

FIGURE 13. Projection of the triple-helical structure of (Gly-Pro-Hyp)$_n$. Note the interchain hydrogen bond $N_4H_4(A)\cdots O_2(B)$ and the one from the side chain $O_3^{\gamma}H_3^{\gamma}(A)\cdots O_1(B)$ between the chains A and B.

Figure 12, analogous to Figure 9 for the all-*trans* structure. The projection of this structure is given in Figure 13, which is sufficiently accurate to indicate that the model is possible and is not violated by any unreasonable short contacts. Table 4 gives the dihedral angles in the backbone of the chain as calculated in our theoretical model and as originally proposed by Berg *et al.* (1973) from their model-building studies. It will be seen that the values of the two do not differ from each other by more than 30° for any dihedral angle, indicating that the features of the model as proposed by Berg *et al.* (1973) from model building of the chains are all correct and reasonable.

It is a very interesting question whether this unusual structure with a *cis* peptide bond and a side chain–backbone hydrogen bond involving the Hyp residue would be observed to occur in collagen. As is clear from the

TABLE 4
Dihedral Angles and Other Parameters Related to the Triple-Helical Structure of Poly(Gly-Pro-Hyp) with a cis Peptide Unit (Unit Twist = −113.0°; Unit Height = 2.80 Å)

	Our calculations				Model of Berg *et al.* (1973)	
Atom	$\tau(°)$	$\phi(°)$	$\psi(°)$	$\omega(°)$[a]	$\phi(°)$	$\psi(°)$
C_1^α	115	51	−143	0	40	−120
C_2^α	112	−71	152	180	−80	180
C_3^α	113	−80	173	180	−80	180

Hydrogen-Bond Parameters

Nature of hydrogen bond	l(Å)	$\theta(°)$
N_4H_4 (chain A)···O_2(chain B)	2.83	23.8
$O_3^\gamma H_3^\gamma$ (chain A)···O_1(chain B)	2.86	23.9

[a] Same in our calculations and in the model of Berg *et al.* (1973).

above discussion, the occurrence of this type of structure requires that the sequence be -Gly-Pro-Hyp- in all three chains and that it should occur at least for a distance involving 2 if not 3 triplets. There are very few regions, if any, in the collagen primary structure in which the sequence -Gly-Pro-Hyp- is repeated for 2 successive triplets. It is also not yet known whether, at the same height or position in the three chains, the sequence -Gly-Pro-Hyp- occurs together. In the absence of positive information regarding the above factors, it should be assumed that this unusual structure with a *cis* peptide bond probably does not occur in the collagen triple helix. However, both local sequences -Gly-Pro-Pro- and -Gly-Pro-Hyp- have a great tendency to form the collagen triple helix because both Pro and Hyp are imino acids with the 5-membered side chain restricting the ϕ rotation to very close to the value required for the triple helix. Therefore, we should rather imagine that sequences of this type in collagen tend to stabilize its triple helical structure in the normal way by having all its residues as *trans* and having a conformation close to that required for building the triple helix.

E. *Hybrid Formation between Collagen and Synthetic Polypeptides*

We have seen above that synthetic polypeptides such as (Gly-Pro-Pro)$_n$ and (Gly-Pro-Ala)$_n$ are good examples of polymers which take up a triple-helical structure, similar to collagen, in their fibrous state. The question, therefore, arises whether these polypeptides would form collagen-like fibers in a hybrid form along with the $\alpha 1$ chain of natural collagen. Such an experiment has been done by Heidemann *et al.* (1973), who have shown that the $\alpha 1$ chain of calf-skin collagen can form hybrids in the presence of the synthetic polypeptides (Gly-Pro-Pro)$_n$ and (Gly-Pro-Ala)$_n$, and to a very much smaller extent, (Gly-Pro-Ser)$_n$. They showed that the hybridization reaction increases with increasing amounts of the synthetic polypeptide and also with increasing molecular weight of the synthetic material. The hybrids could readily be identified by the amino acid composition of the fibrous material that is found.

This type of experiment could be used for finding out which type of tripeptides more readily form the collagen-type triple helix and is mentioned here because it has good potentialities for future application.

Acknowledgments

This review was mostly written at Bangalore, but the final preparation was done at Chicago. We are greatly indebted to Miss Manju Bansal for her assistance and Dr. V. S. Ananthanarayanan for reading through the manuscript and for useful suggestions. We wish to thank Professor Robert Haselkorn of Chicago for providing the necessary facilities for completing the work in Chicago. Thanks are also due to Mr. T. K. Raveendran for typing and Mr. H. T. Doreyawar for the figures.

We wish to acknowledge the assistance received from the National Institutes of Health, U.S. Public Health Service in the form of grants AM-15964 in Bangalore and AM-11493 in Chicago.

References

Anderson, J. M., Rippon, W. B., and Walton, A. G., 1970, Model tripeptides for collagen, *Biochem. Biophys. Res. Commun.* **39:**802.

Andreeva, N. S., Esipova, N. G., Millionova, M. I., Rogulenkova, V. N., and Shibnev, V. A., 1967, Polypeptides with regular sequences of amino acids as the models of collagen structure, *in: Conformation of Biopolymers* (G. N. Ramachandran, ed.), Vol. 2, pp. 469–481, Academic Press, New York.

Andreeva, N. S., Esipova, N. G., Millionova, M. I., Rogulenkova, V. N., Tumanyan, V. G., and Shibnev, V. A., 1970, Synthetic regular polytripeptides and proteins of collagen class, *Biofizika* **15:**198.

Andries, J. C., and Walton, A. G., 1970, Morphological evidence for antiparallel peptide chains in poly glycyl prolyl proline, *J. Mol. Biol.* **54:**579.

Andries, J. C., and Walton, A. G., 1971, The morphology of poly(Gly-Ala-Glu(OEt)), *J. Mol. Biol.* **56:**515.

Andries, J. C., Anderson, J. M., and Walton, A. G., 1971, Morphological and structural studies of poly(Gly-Gly-Ala), *Biopolymers* **10:**1049.

Arnott, S., and Dover, S. D., 1968, The structure of poly-L-proline II, *Acta Cryst.* **B24:**599.

Bamford, C. H., Elliott, A., and Hanby, W. E., 1956, *Synthetic Polypeptides*, Ch. VII, Academic Press, New York.

Bansal, M., Ramakrishnan, C., and Ramachandran, G. N., 1975*a*, A triple-helical model for (Gly-Pro-Hyp)^n with *cis* peptide units, *Biopolymers* **14:**2457.

Bansal, M., Ramakrishnan, C., and Ramachandran, G. N., 1975*b*, Stabilization of the collagen structure by hydroxyproline residues, *Proc. Ind. Acad. Sci.* **A 82:**152.

Berendsen, H. J. C., 1972, Interaction of water and proteins, *in; Enzymes—Structure and Function* (FEBS Proc. 8th Meeting), Vol. 29, pp. 19–27, North-Holland, Amsterdam.

Berg, R. A., and Prockop, D. J., 1973, The thermal transition of a non-hydroxylated form of collagen. Evidence for a role for hydroxyproline in stabilizing the triple-helix of collagen, *Biochem. Biophys. Res. Commun.* **52:**115.

Berg, R. A., Kishida, Y., Kobayashi, Y., Inouye, K., Tonelli, A. E., Sakakibara, S., and Prockop, D. J., 1973, A model for the triple-helical structure of (Pro-Hyp-Gly)$_{10}$ involving a *cis* peptide bond and inter-chain hydrogen-bonding to the hydroxyl group of hydroxyproline, *Biochim. Biophys. Acta* **328:**553.

Brown, L., and Trotter, I. F., 1956, X-ray studies of poly-L-alanine, *Trans. Faraday Soc.* **52:**537.

Brown, F. R., III, diCorato, A., Lorenzi, G. P., and Blout, E. R., 1972, Synthesis and structural studies of two collagen analogues: Poly(L-prolyl-L-seryl-glycyl) and poly(L-prolyl-L-alanyl-glycyl), *J. Mol. Biol.* **63:**85.

Cowan, P. M., and McGavin, S., 1955, Structure of poly-L-proline, *Nature (London)* **176:**501.

Crick, F. H. C., and Rich, A., 1955, Structure of polyglycine II, *Nature (London)* **176:**780.

Doyle, B. B., Traub, W., Lorenzi, G. P., Brown, F. R., III, and Blout, E. R., 1970, Synthesis and structural investigations of poly(L-alanyl-L-alanyl-glycine), *J. Mol. Biol.* **51:**47.

Doyle, B. B., Traub, W., Lorenzi, G. P., and Blout, E. R., 1971, Conformational investigations on the polypeptide and oligopeptides with the repeating sequence L-alanyl-L-prolyl glycine, *Biochemistry* **10:**3052.

Eastoe, J. E., 1967, Composition of collagen and allied proteins, *in: Treatise on Collagen* (G. N. Ramachandran, ed.), Vol. 1, pp. 1–72, Academic Press, New York.

Engel, J., 1967, Conformational transitions of poly-L-proline and poly(L-prolyl-glycyl-L-proline), *in: Conformation of Biopolymers* (G. N. Ramachandran, ed.), Vol. 2, pp. 483–497, Academic Press, New York.

Fraser, R. D. B., MacRae, T. P., Stewart, F. H. C., and Suzuki, E., 1965, Poly-L-alanyl glycine, *J. Mol. Biol.* **11:**706.

Harrington, W. F., 1964, On the arrangement of the hydrogen bonds in the structure of collagen, *J. Mol. Biol.* **9:**613.

Heidemann, E. R., Harrap, B. S., and Schiele, H. D., 1973, Hybrid formation between collagen and synthetic polypeptides, *Biochemistry* **12**:2958.

Hutton, J. J., Marglin, A., Witkop, B., Kurtz, J., Berger, A., and Udenfriend, S., 1968, Synthetic polypeptides as substrates and inhibitors of collagen proline hydroxylase, *Arch. Biochem. Biophys.* **125**:779.

IUPAC-IUB Commission on Biochemical Nomenclature, 1970, *J. Mol. Biol.* **52**:1.

Jiminez, S., Harsch, M., and Rosenbloom, J., 1973, Hydroxyproline stabilizes the triple helix of chicken tendon collagen, *Biochem. Biophys. Res. Commun.* **52**:106.

Kikuchi, Y., Fujimoto, D., and Taniya, N., 1969, The enzyme hydroxylation of protocollagen models, *Biochem. J.* **115**:569.

Kitaoka, H., Sakakibara, S., and Tani, H., 1958, Synthesis of poly(L-prolyl-L-leucylglycyl). An attempted synthesis of model collagen, *Bull. Chem. Soc. Jpn.* **31**:802.

Kivirikko, K. I., Prockop, D. J., Lorenzi, G. P., and Blout, E. R., 1969, Oligopeptides with the sequences Ala-Pro-Gly and Gly-Pro-Gly as substrates or inhibitors for protocollagen proline hydroxylase, *J. Biol. Chem.* **244**:2755.

Kivirikko, K. I., Kishida, Y., Sakakibara, S., and Prockop, D. J., 1972, Hydroxylation of $(X\text{-Pro-Gly})_n$ by protocollagen proline hydroxylase, *Biochim. Biophys. Acta* **271**:347.

Kobayashi, Y., and Isemura, T., 1972, Polypeptides related to collagen and its triple helical structure, *Progr. Polym. Sci. Jpn.* **3**:315.

Krimm, S., Kuroiwa, K., and Rebane, T., 1967, Infrared studies of C—H···O═C hydrogen bonding in polyglycine II, *in: Conformation of Biopolymers* (G. N. Ramachandran, ed.), Vol. 2, pp. 439–447, Academic Press, New York.

McBride, E. W., and Harrington, W. F., 1967, Helix-coil transition in collagen. Evidence for a single-stranded triple helix, *Biochemistry* **6**:1499.

Okuyama, K., Tanaka, N., Ashida, T., Kakudo, M., Sakakibara, S., and Kishida, Y., 1972, An X-ray study of the synthetic polypeptide $(\text{Pro-Pro-Gly})_{10}$, *J. Mol. Biol.* **72**:571.

Pauling, L., 1952, The planarity of the amide group in polypeptides, *J. Am. Chem. Soc.* **74**:3964.

Pauling, L., 1960, *The Nature of the Chemical Bond,* p. 281, Cornell University Press, Ithaca, New York.

Ramachandran, G. N., 1967, Structure of collagen at the molecular level, *in: Treatise on Collagen* (G. N. Ramachandran, ed.), Vol. 1, pp. 103–183, Academic Press, New York.

Ramachandran, G. N., 1968, Molecular architecture of collagen, *J. Am. Leather Chem. Assoc.* **63**:160.

Ramachandran, G. N., and Chandrasekaran, R., 1968, Interchain hydrogen bonds via bound water molecules in the collagen triple helix, *Biopolymers* **6**:1649.

Ramachandran, G. N., and Kartha, G., 1954, Structure of collagen, *Nature (London)* **174**:269.

Ramachandran, G. N., and Kartha, G., 1955, Structure of collagen, *Nature (London)* **176**:593.

Ramachandran, G. N., and Sasisekharan, V., 1968, Conformation of polypeptides and proteins, *Adv. Protein Chem.* **23**:283.

Ramachandran, G. N., and Venkatachalam, C. M., 1966, The stability of the two-bonded collagen triple helix, *Biochim. Biophys. Acta* **120**:457.

Ramachandran, G. N., Sasisekharan, V., and Thathachari, Y. T., 1962, Structure of collagen at the molecular level, *in: Collagen* (N. Ramanathan, ed.), pp. 81–116, Interscience Publishers, New York.

Ramachandran, G. N., Sasisekharan, V., and Ramakrishnan, C., 1966, Molecular structure of polyglycine II, *Biochim. Biophys. Acta* **112**:168.

Ramachandran, G. N., Ramakrishnan, C., and Venkatachalam, C. M., 1967, Structure of polyglycine II with direct and inverted chains, *in; Conformation of Biopolymers* (G. N. Ramachandran, ed.), Vol. 2, pp. 429–438, Academic Press, New York.

Ramachandran, G. N., Doyle, B. B., and Blout, E. R., 1968, Single-chain triple helical structure, *Biopolymers* **6:**1771.

Ramachandran, G. N., Bansal, M., and Bhatnagar, R. S., 1973, A hypothesis on the role of hydroxyproline in stabilizing collagen structure, *Biochim. Biophys. Acta* **322:**166.

Ramachandran, G. N., Bansal, M., and Ramakrishnan, C., 1975, Hydroxyproline stabilises both intra-fibrillar structure as well as inter-protofibrillar linkages in collagen, *Curr. Sci.* **44:**1.

Rich, A., and Crick, F. H. C., 1955, The structure of collagen, *Nature (London)* **176:**915.

Rich, A., and Crick, F. H. C., 1961, The molecular structure of collagen, *J. Mol. Biol.* **3:**483.

Rogulenkova, V. N., Millionova, M. I., and Andreeva, N. S., 1964, On the close structural similarity between poly-Gly-L-Pro-L-Hypro and collagen, *J. Mol. Biol.* **9:**253.

Sakakibara, S., Kishida, Y., Kikuchi, Y., Sakai, R., and Kakiuchi, K., 1968, Synthesis of poly-(L-prolyl-L-prolyl-glycyl) of defined molecular weights, *Bull. Chem. Soc. Jpn.* **41:**1273.

Sakakibara, S., Inouye, K. I., Shudo, K., Kishida, Y., Kobayashi, Y., and Prockop, D. J., 1973, Synthesis of (Pro-Hyp-Gly)$_n$ of defined molecular weights. Evidence for the stabilization of collagen triple helix by hydroxyproline, *Biochim. Biophys. Acta* **303:**198.

Sasisekharan, V., 1959*a*, Structure of poly-L-proline II, *Acta Cryst.* **12:**897.

Sasisekharan, V., 1959*b*, Structure of poly-L-hydroxyproline A, *Acta Cryst.* **12:**903.

Scheraga, H. A., 1968, Calculations of conformations of polypeptides, *Adv. Phys. Org. Chem.* **6:**103.

Segal, D. M., Traub, W., and Yonath, A., 1969, Polymers of tripeptides as collagen models. VIII. X-ray studies of four polyhexapeptides, *J. Mol. Biol.* **43:**519.

Sutoh, K., and Noda, H., 1974, Conformational change of the triple helical structure. III. Stabilizing forces in the triple helix, *Biopolymers* **13:**2461.

Traub, W., and Piez, K. A., 1971, The chemistry and structure of collagen, *Adv. Protein Chem.* **25:**243.

Traub, W., and Shmueli, U., 1963, Structure of Poly-L-proline I, *in: Aspects of Protein Structure* (G. N. Ramachandran, ed.) pp. 81–92, Academic Press, New York.

Yee, R. Y., Englander, S. W., and von Hippel, P. M., 1974, Native collagen has a two-bonded structure, *J. Mol. Biol.* **83:**1.

Yonath, A., and Traub, W., 1969, Polymers of tripeptides as collagen models. IV. Structure analysis of poly(L-prolyl-glycyl-L-proline), *J. Mol. Biol.* **43:**461.

3
Molecular Packing in Collagen Fibrils

A. MILLER

I. Introduction

In this chapter I will discuss the three-dimensional arrangement of collagen molecules in the fibrils of connective tissue. This is a problem of quite general significance since the nature of the molecular packing in similar systems such as the fibrils of muscles and other biological fibers is still unknown. Once the sequence and symmetry of the individual collagen molecules are known, it is natural to proceed to inquire how these molecules aggregate to form the functional units of connective tissue. As Crick (1966) recognized, "the superlattice of collagen is a neglected problem and it is time somebody took it up again."

The "superlattice of collagen" is, of course, the molecular arrangement in the fibrils. The issues then become, What are the appropriate experimental techniques for determining the molecular arrangement in biological fibers?, Are there any discernible principles of arrangement or assembly of long molecules into fibers analogous to those noted by Caspar and Klug (1962) for the arrangement of globular subunits in virus particles?, and Is it possible to recognize any relationships between the molecular arrangement and the biological function of the fibers?

At present the first question is still the main concern, but even at this early stage the results of studies on molecular arrangement have suggested tentative answers to the second and third questions.

A. MILLER · Laboratory of Molecular Biophysics, Zoology Department, Oxford University, Oxford, England. Present address: European Molecular Biology Laboratory, Grenoble, France.

Here I will describe the broad range of methods that has proved necessary to provide the information required for a model of molecular arrangement. Electron microscopy by itself is not sufficient to resolve the details of how the long molecules pack nor are the electron micrographs always immediately relevant to the native structures. To electron microscopy we have added X-ray diffraction, neutron diffraction, electron diffraction, and optical diffraction from electron micrographs. The results obtained by such techniques have then been supplemented by knowledge of the molecular structure and its amino acid sequence as well as the positions of covalent intermolecular bonds. In view of the lack of a direct method for the determination of molecular arrangement, the strategy is to use the information from a given approach to place restrictions on the possible models for the molecular arrangement and then, on conflating these restrictions, to indicate a unique model.

The extent to which we have a unique model for the molecular arrangement of collagen in tendon is the principal subject discussed here. However, we shall also note some wider implications of the present results, in particular the recognition of the origins of the molecular specificity which is the basis of self-assembly of long molecules into fibers and the fact that in collagen it may be possible to recognize the essential features of each level in the structural hierarchy from molecule to tissue.

The characteristic collagen fibril is observed in the electron microscope to have a regular banded appearance (Hall *et al.*, 1942; Schmitt *et al.*, 1942; Wolpers, 1943). The period of the bands is 668 Å, as estimated by X-ray diffraction patterns from native rat-tail tendons (Bear, 1942; Miller and Parry, 1973). In tendons the fibrils run parallel to each other and are usually 1000–3000 Å in diameter. In other tissues such as skin or cartilage the fibrils have a similar appearance to those in tendon but they are frequently narrower and are not parallel to each other but interwoven in a complex network. There is also a greater proportion of noncollagen material in these tissues compared with tendon. The detailed appearance of the repeating unit of length 670 Å (D) in the band pattern, depends on how the fibrils have been stained. Negative staining divides the period into two roughly equal bands, one darkly stained and one lightly stained (Tromans *et al.*, 1963; Olsen, 1963; Hodge and Petruska, 1963). Positive staining results in about 12 narrow darkly stained bands which produce a polarized D period (Gross and Schmitt, 1948; Nemetschck *et al.*, 1955).

The collagen molecule is known to be of length 2990 Å (L) (Gross *et al.*, 1954; Boedeker and Doty, 1956; Hall, 1956; Hodge and Petruska, 1963). Thus L is approximately equal to $4.5D$. Our question is: How are the molecules of length L arranged in three dimensions to give a fibril with an axial period of D? It is obvious that with the information we have mentioned so far, there is a large number of possible solutions to this problem.

As mentioned above, we shall now show how certain kinds of evidence reduce the number of possible solutions. First we shall show that there is a well-established solution for the axial arrangement of the molecules. This may be regarded as a one-dimensional solution to the structure. Clearly any three-dimensional solution must be consistent with the one-dimensional solution, and we shall then go on to discuss the types of three-dimensional arrangements that must be considered.

II. The Collagen Molecule

The primary sequence and secondary structure of the collagen molecule have been discussed in the previous chapters so here we shall only summarize the points essential to our argument.

The molecule has a rodlike shape of about 3000 × 15 Å. It consists of three similar polypeptide chains termed α chains, wound round each other to form a three-strand rope. Each α chain is some 1050 residues long and, apart from in the telopeptides (of 16 and 25 residues at the N- and C-terminal ends respectively of one type of α chain), the amino acid glycine occurs as every third residue for the whole length of the molecule (see summary in Hulmes *et al.*, 1973). High-angle X-ray diffraction patterns reveal a helix of pitch 9.5 Å with subunits spaced axially by 2.86 Å. Molecular model building indicated how the three chains could be arranged on such a helix. Each α chain is itself a helix with approximately three residues per turn and when the chains are wound round each other, the glycine residues are arranged so that they form the core of the three-strand rope. The three chains are related by the helix of pitch 9.5 Å which implies that one α chain is related to another by an azimuthal rotation of 108° and an axial translation of 2.86 Å. The axis of an individual α chain follows a helical path of pitch about 87 Å. The accuracy of measurement of the 9.5-Å and 2.86-Å layer lines indicate that the pitch of this major helix followed by the axis of a single α chain could be within the range 76–114 Å, that is, 27–40 amino acid residues.

As we have said, a helix with parameters P = 9.5 Å and h = 2.86 Å relates to α chains. It also relates the nth residue to the $(n + 3)$th residue within a single α chain. The exact arrangement of a triplet of three adjacent amino acids in a single α chain is still being debated and must be determined by chemical evidence, model building, and comparison of calculated and observed X-ray diffraction patterns.

The helix with parameters $P = 9.5$ Å and $h = 2.86$ Å is close to having 10 residues in three turns. Such an integral helix has 10-fold symmetry in projection down the helix axis. Furthermore, when the -(Gly-X_n-Y_n)- residues are plotted on such a helix, in projection down the helix axis the X_n and Y_{n+1} residues lie close to the same azimuthal positions. Hence all of the amino acid residues, other than those in the glycine core, will lie close to 10 equally spaced edges on the circumference of the three-strand rope molecule.

III. One-Dimensional Arrangement

A. Electron Microscopy and Amino Acid Sequence

By a one-dimensional arrangement we mean the axial molecular arrangement or the structure as projected onto the fiber axis. A one-dimensional solution to the arrangement of collagen molecules in native fibrils was proposed by Schmitt *et al.* (1955). They suggested that the D (668 Å) repeat could be generated if molecules of length L were staggered axially by a distance D with respect to their neighbors. At that time it was thought that L was equal to $4D$ so this was called the "quarter-stagger" model and, because of the integral relation between L and D, the model involved the end-to-end contact of molecules. When the more accurate

FIGURE 1. Diagram of the collagen molecule. Each segment *a, b, c,* and *d* is 668 Å(D) long; segment e is about $0.5D$. The segments are similar in their distribution of large apolar residues (see Hulmes *et al.*, 1973, and text) but are distinct in that they have different electron-density profiles when projected on to the molecular axis.

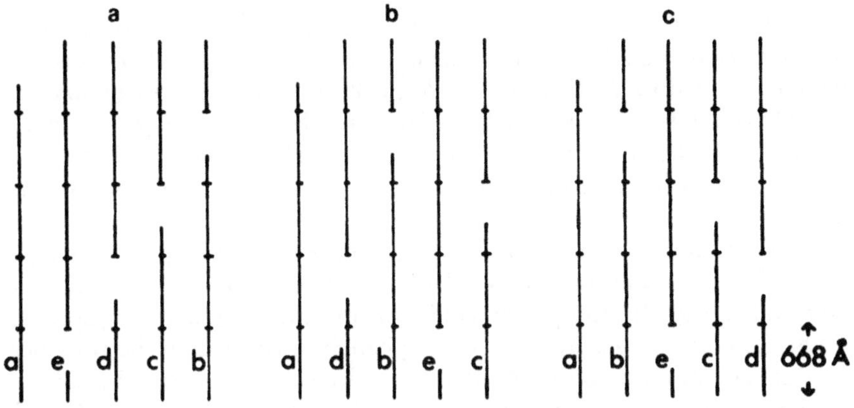

FIGURE 2. Some two-dimensional molecular arrangements which have a periodicity of D when projected on to the fibril axis. In each section of height D there is, in projection, a contribution from one of each of the segments *a, b, c, d,* and *e*. (From Doyle *et al.*, 1974*a*). (a) This arrangement was first illustrated by Hodge and Petruska (1963). (b) This and (a) are the only two possible regular arrangements which have a period of D when projected on to the fibril axis. (c) Illustration of an irregular molecular arrangement with a period of D when projected on to the fibril axis.

value of $L = 4{,}4D$ was determined by Hodge and Petruska (1963), they pointed out that a D stagger between neighboring molecules led to a structure with gaps of some 300 Å between the ends of molecules. In this solution, the D period contained an overlap region of $0.4D$ and a gap region of $0.6D$. This explained the one light and one dark band seen in electron micrographs of negatively stained fibrils.

The Hodge-Petruska model is shown in Figure 2a. As drawn, Figure 2 is, of course, a two-dimensional arrangement, but on the basis of the available data, it was not possible to distinguish this from other possible two-dimensional arrangements, one of which is shown in Figure 2b. We shall confine ourselves in this section to deducing the correct one-dimensional or axial molecular arrangement.

The collagen molecule of length L may be divided into four segments of length D which are labeled *a, b, c,* and *d* in Figure 1, and a fifth segment *e* of about $0.5D$. As we shall see later, the segments, although chemically similar to each other, are not identical. Formally any arrangement of molecules which has a projected axial period of D must contain equal numbers of each of the five segments in each D period (Doyle *et al.*, 1974*a*).

The simplest bundle of collagen molecules with a true axial period of D will contain five molecules in transverse section, each contributing a different segment to the *D* period as, for example, in the Hodge–Petruska solution (Figure 2). If, for convenience, we term such simple *D*-periodic bundles "unit fibrils," then the simplest collagen fibril with a true axial *D* period would be one in which the unit fibrils were arranged in axial register. Whether or not this is the molecular arrangement within a real fibril may be investigated by considering the detailed structure of a real *D* period.

Hodge and Schmitt (1960) obtained support for the quarter-stagger model from electron microscopy of polymorphic forms of collagen. They obtained collagen in the polymorphic form termed segment-long-spacing (SLS). SLS may be obtained by reprecipitation of collagen in the presence of a small concentration of ATP. The molecules line up side by side in parallel register to form a segment of fixed length equal to the molecular length *L*, and variable width. Electron micrographs of the SLS form contain 58 closely spaced dark bands under positive staining (Bruns and Gross, 1973) and it has now been confirmed that these correspond to the location of charged amino acid residues (von der Mark *et al.*, 1970; Doyle *et al.*, 1974*b*; Piez, this volume, Chapter 1, Figure 7). Hodge and Schmitt (1960) photographically superimposed the SLS band pattern upon itself staggered by integral values of the *D* period and synthesized a pattern of bands with the *D* repeat which correspond to the positively stained native band pattern from collagen. A similar reconstruction was made graphically by Kuhn and Zimmer (1961). In retrospect it is possible to see that these workers used the correct value of *D* in their reconstruction even though the precise molecular length was not known. Their work has been confirmed using our present knowledge of *L* and the amino acid sequence (Doyle *et al.*, 1974*b*; Chapman and Hardcastle, 1974; Chapman, 1974).

The fact that these optical syntheses were successful indicates that molecular segments *a, b, c, d,* and *e* are in register throughout the fibril.

It is important to establish that a collagen fibril is composed of in register simple unit fibrils because there are other arrangements which theoretically could produce fibrils with a true axial period of *D*. If unit fibrils were grouped into pairs such that within each pair the unit fibrils were related by a non-*D* axial translation, then each pair would have a true axial period of *D*, and if the pairs were arranged in axial register the resulting fibril would be *D* periodic. However, the projection of the amino acid sequence within a pair on to the fibril axis would be different from the in-register unit fibrils and thus the pattern of positively stained bands seen in electron micrographs of native collagen fibrils would also be different from that in the pattern reconstructed assuming 'unit fibrils' are in register. Hence we can conclude that non-*D* axial staggers between unit fibrils do not occur in native collagen fibrils; later (Section IV-B-3) we shall see that non-D axial staggers can occur between narrow D-periodic native-like fibrils in certain polymorphic forms of collagen.

Knowledge of the one-dimensional structure and the amino acid sequence of a complete α1 chain allows us to inquire about the origins of the molecular arrangement described above. The first point to test is whether the D repeat can be shown to arise from the primary sequence. This was done (Hulmes *et al.*, 1973) by translating the sequence of the 1011 amino acids of the triple-helical region past itself and scoring for favorable interactions between opposing amino acids. It was found that

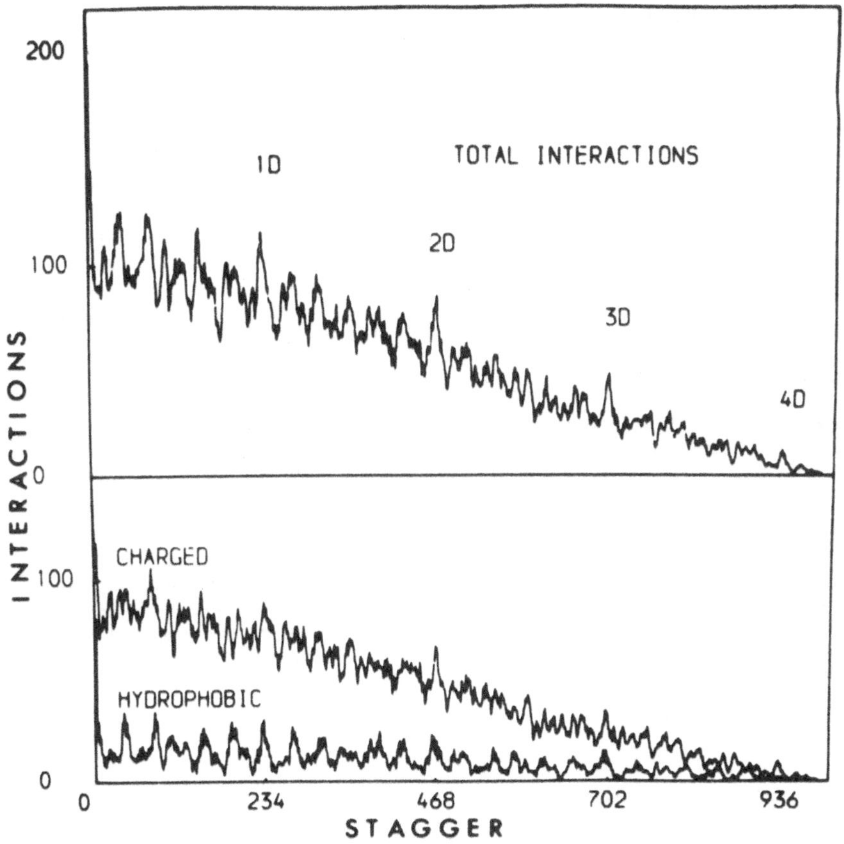

Figure 3. A complete plot of the number of large hydrophobic and charge interactions and their total between two collagen molecules (on the ordinate) as a function of the stagger between them. The stagger (on the abscissa) is measured in residues. The D intervals are marked. Note the periodicity and the symmetry in the curve of the hydrophobic residues alone (from Hulmes *et al.*, 1973).

interactions between amino acids of opposite charge and between large hydrophobic amino acids was maximal when the chains were staggered by 0*D*, 1*D*, 2*D*, 3*D*, and 4*D*, where *D* is 234 residues (Figure 3). Taking *D* as 670 Å as determined by the low-angle X-ray diffraction pattern from native rat-tail tendon, this gives an average axial separation between amino acid residues of 2.86 ± 0.02 Å, in good agreement with the value of 2.866 Å determined independently by high-angle X-ray diffraction (Miller, Parry, and Wray, unpublished results). Hence we may conclude that the *D* repeat in collagen fibrils has its origins in the amino acid sequence. Figure 3 shows that the self-assembly of collagen molecules to form native collagen fibrils is a result of the amino acid sequence.

The bottom curve in Figure 3 is an autocorrelation function of the distribution of large hydrophobic amino acids in the molecule akin to the Patterson function in X-ray crystallography. The fact that there are planes of symmetry in Figure 3 in the curve corresponding to interactions between hydrophobic residues such that peaks at x and $(D - x)$ are identical, implies a *D* period in the distribution of large hydrophobic residues in the collagen molecule. If in the collagen fibril there is a 1*D* axial translation between nearest-neighbor molecules, this would lead us to

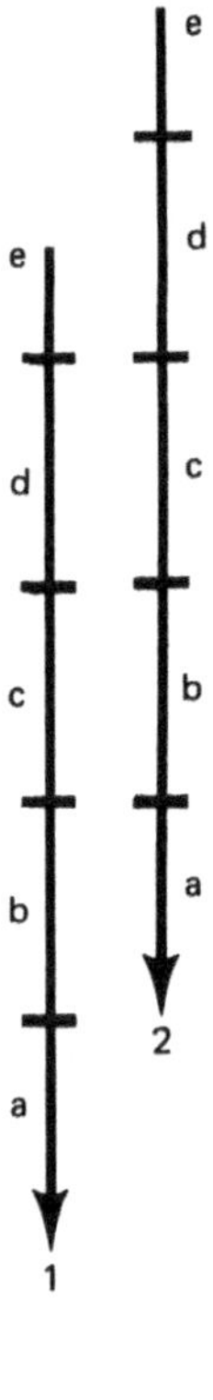

FIGURE 4. Two adjacent collagen molecules 1 and 2 staggered axially by *D*. The distribution of interacting apolar amino acid residues in a_2 complements that in b_1. Hence a_1, b_1, c_1, d_1, and e_1 are pseudoequivalent in their distribution of interacting apolar amino acids. If the number of interacting apolar residues is a large fraction of the total apolar residues in the molecule, then the molecule will show a pseudoperiodicity of *D* in the distribution of its apolar residues.

expect a pseudoperiod of *D* in the hydrophobic amino acid residues within one molecule. In Figure 4 two molecules labeled 1 and 2 are staggered by *D*. Segment a_2 complements b_1 and hence a_1 complements b_1. Similarly, b_1 complements c_1, c_1 complements d_1, and d_1 complements e_1. Molecule 1 thus has a pseudoperiod of *D* if the complementarity is of the like–like type as with hydrophobic residues, so the arrangement shown in Figure 2a is consistent with the interaction curve in Figure 3. The arrangement in Figure 2b also has a *D* repeat in projection and would also lead one to expect the molecules to have a pseudorepeat of *D* in the hydrophobic residues.

In summary, we have shown by the sequence–interaction curve that in one dimension the collagen molecule has a pseudorepeat, but not a true repeat, of *D*. The native fibril does, however, have a true *D* repeat in axial projection, and this is brought about by there being equal numbers of the five segments *a, b, c, d,* and *e* of the molecule being arranged in register within a slab of the fibril of height *D*.

B. X-Ray Diffraction Studies

The low-angle X-ray diffraction pattern from tendons contains a series of meridional reflections (Bear, 1942) which index on a period of 668 Å for native rat-tail tendon (Miller and Parry, 1973). The origin of these reflections is clearly the *D* periodicity in the collagen fibrils. Since the reflections are on the meridian of the diffraction pattern, they contain information about the structure of the collagen fibril as projected on to the fibril axis, that is, the one-dimensional structure discussed in this section. The intensities of the meridional reflections contain information about the one-dimensional variation in electron density within a single *D* period. If the phases of the reflections were also known, then the electron-density variation along the fibril axis could be computed (ignoring scaling factors) by evaluating

$$\rho(z) = \sum_h /F^h/\cos\,(2\pi\, hz + \alpha^h) \qquad (1)$$

where F^h is the amplitude of the *h*th order reflection and may be obtained from I^h where I^h is the intensity and α^h the phase of the *h*th order reflection. The phases of the reflections cannot be readily determined. Several attempts have been made to derive the one-dimensional electron-density variation along the fibril axis from the low-angle meridional X-ray reflections (Kaesburg and Shurman, 1953; Burge and Randall, 1955;

Tomlin and Worthington, 1956; Bear and Morgan, 1957; Ericson and Tomlin, 1959; Ellis and McGavin, 1970; Chandross and Bear, 1973). Since the amino acid sequence of the $\alpha 1$ chain and much of the $\alpha 2$ chain is now known, it is likely that a complete solution to the one-dimensional molecular arrangement in collagen will be possible in the near future, so we shall give a brief review of the situation to date. Two possible strategies exist. One is to determine the phases of the X-ray reflections by some technique such as the isomorphous replacement method, and thus evaluate $\rho(z)$ directly from Eq. (1). Any model for the one-dimensional molecular arrangement of collagen such as that discussed earlier in this section and derived from electron microscopy, may then be tested since it should have a calculated $\rho(z)$ identical to that evaluated by Eq. (1). Another approach is to calculate $\rho(z)$ for a proposed model and then predict the intensities of the X-ray reflections by Fourier inverting $\rho(z)$. The calculated and observed intensities may then be compared. This second approach does not use phase data and does not, in general, result in a unique model.

A prominent feature about the first few orders of the low-angle meridional reflections from wet tendon is that the odd orders are of strong intensity whereas the even orders are weak. This suggests that the electron-density variation in a D period is a step-function of length about $0.5D$. Kaesburg and Schurman (1953) and Tomlin and Worthington (1956) made a more detailed analysis and concluded that the D period contained a step-function of length $0.46D$. A step-function of length $(1 - 0.46)D$ would produce an identical set of reflection intensities and Bear and Morgan (1957) favored $0.54D$. Tomlin and Worthington (1956) also noted that the orders in the region of the 20th were relatively intense even from collagen in different conditions, and suggested the existence of an axial pseudoperiod of about 33 Å. Burge and Randall (1955) explored the relationship between the X-ray reflections and the set of fine bands observed in electron micrographs of positively stained fibrils. Ericson and Tomlin (1959) studied the effect of heavy-metal stains, in particular silver nitrate and iodine, on the X-ray reflections. They argued that only the step-function of $0.46D$ was consistent with the positive staining effects of the heavy metals. They also reported drastic changes in the X-ray reflection intensities from specimens dehydrated at 200°C and from those stained with phosphotungstic acid and warned of the difficulty in analyzing the positively stained banding pattern in electron micrographs in terms of electron density.

Ellis and McGavin (1970) studied the meridional X-ray diffraction pattern from elastoidin. They drew attention to the fact that the feature of high intensity around the 20th order was present in these patterns as well as in the patterns from tendons. However, the first few orders of the pattern from elastoidin did not have the strong–odd, weak–even feature observed with tendons. Ellis and McGavin (1970) concluded from the first observation that the detailed structure of elastoidin collagen was closely similar to that of tendon collagen; in particular they shared the same strong Fourier component corresponding to a pseudoperiod of 30 Å. However, the low-order differences indicated some low-resolution differences between the structures. They made the assumption, known to be invalid, that the electron-density profile was centrosymmetric and applied the method of isomorphous replacement to assign phases to the reflections and calculated $\rho(z)$ from Eq. (1) using about 35 orders of diffraction. The assumption of centrosymmetry means that the fine structure in the electron-density profile will contain errors. The $\rho(z)$ for collagen from tendon resembled a step-function of length $0.46D$, and the authors pointed out the consistency of this with the Hodge–Petruska model of overlaps (of about $0.4D$) and gaps (of about $0.6D$). Tomlin (1955) had suggested that the step function could result from alternating molecular overlap and gaps. The $\rho(z)$ for collagen in elastoidin differed in that it showed a large peak of electron density near the center of the gap region, but it was not easy to suggest an interpretation of this.

Chandross and Bear (1973) have constructed detailed models for the one-dimensional electron-density distribution and compared these with the intensities of the observed X-ray reflections. The models are used to calculate phases, and $\rho(z)$ curves are calculated using observed amplitudes and calculated phases as well as calculated amplitudes and calculated phases. A start was made from the earlier models of Bear and Morgan (1957) and the step-function and pattern of fine bands varied so as to give best agreement with the intensities of the first 15 orders of the meridional X-ray diffraction pattern. For wet collagen it was concluded that the step-function was $0.46D$ and the pattern of fine ripples superimposed on this was noted to agree well with the positions of the narrow positively stained bands observed in electron micrographs of collagen fibrils. A disadvantage of this method is that the use of a model to calculate phases inevitably results in a profile similar to the original model and is thus not a good test of the model.

There are various possible ways of determining the phases of the X-ray reflections directly so as to produce an electron-density curve. One method attempted by Ellis and McGavin (1970) is that of isomorphous replacement. Another is to measure the intensities of the reflections when the fiber is immersed in solutions of known and different electron densities. It has already been demonstrated (R. B. Jones and A. Miller, unpublished results) that X-ray diffraction patterns from rat-tail tendons immersed in sucrose solutions dramatically alter the intensities of the meridional reflections. Neutron diffraction patterns have also been obtained from native rat-tail tendon (White *et al.*, 1976). These show the first few orders of the low-angle meridional reflections, and the intensities are altered when the tendon is transferred from H_2O to D_2O which has a different scattering cross section. Some results from these attempts at phase determination seem likely in the near future.

A complete solution to the one-dimensional molecular arrangement in collagen has still to be achieved. As mentioned above, this would require a demonstration that the $\rho(z)$ derived from a sequence of amino acids, and possibly other molecules, was the same as a $\rho(z)$ derived from the X-ray reflections. Perhaps the work of Chandross and Bear comes close to this since the fine ripples of $\rho(z)$ correspond to the positively stained bands in electron micrographs which in turn represent the location of charged amino acids (von der Mark *et al.*, 1970; Doyle *et al.*, 1974*b*) which tend to contain a larger number of electrons than apolar amino acids. Another partial correlation comes from the observation (Doyle *et al.*, 1974*b*) that the Fourier transform of an axially projected, suitably staggered array of $\alpha 1$ chain amino acid sequences contains a maximum in the region corresponding to the 20th order of diffraction. A complete solution will require knowledge of the $\alpha 2$ chain amino acid sequence and the positions of other molecules such as sugars which are arranged with a D periodicity.

IV. Three-Dimensional Molecular Arrangement

In the previous section, we saw that the data are sufficient to establish the one-dimensional distribution of electron density in collagen, that is, the structure as projected on to the fiber axis. This obviously restricts the possible three-dimensional arrangements to those compatible with the one-dimensional solution. In this section we shall distinguish between features of the three-dimensional molecular arrangement which are well established and those which are still problematic. We shall call the fibril axis the z axis and assume for the present that the x and y axes lie in a plane at right angles to the z axis. Let us consider an x–y plane which is a transverse section through the fibril and denote the positions of collagen molecules in this plane by circles. If each circle is labeled with *a, b, c, d,* or *e* to indicate which segment of the collagen molecule lies in this particular plane, then we may restate our knowledge of the one-dimensional structure by saying that in this plane there must be equal numbers of each of the segments *a, b, c, d,* and *e*. The problem of the three-dimensional arrangement of the molecules in collagen then becomes one of determining how the labeled circles are arranged in the *x–y* plane.

We may get some idea of the broad nature of possible three-dimensional arrangements by posing three general questions.

1. Irrespective of the labels in the circles, are they randomly positioned in the *x–y* plane or are they positioned on a regular lattice?
2. Irrespective of the labels in the circles, are they grouped into clusters or are the individual circles in equivalent positions?
3. If a lattice and/or clustering is present, how are the labels *a, b, c, d,* or *e* distributed in the lattice and/or clusters?

A. *Relative Lateral Positions of Molecules*

The X-ray diffraction patterns from native rat-tail tendon (North *et al.*, 1954; Miller and Wray, 1971) firmly establish that in a transverse plane

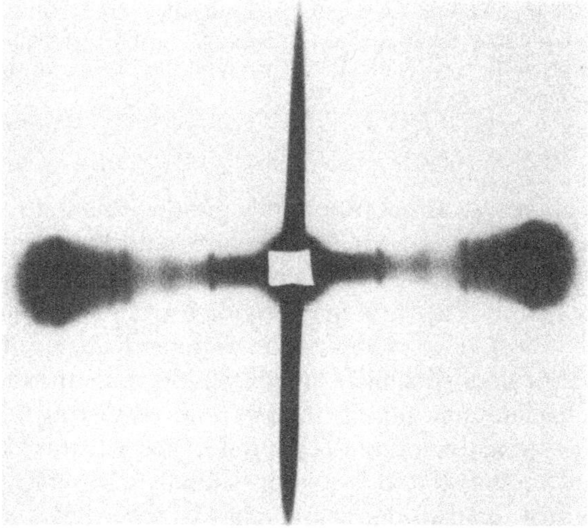

FIGURE 5. Medium-angle X-ray diffraction pattern from rat-tail tendon in the native state (from Miller and Wray, 1971). The first sharp reflection in the near-equatorial region is at a spacing corresponding to 38 Å. Note the other sharp reflections, the broad intensity maximum in the region corresponding to 13 Å in the lateral direction, and the fanning of the intensity in the near-equatorial region.

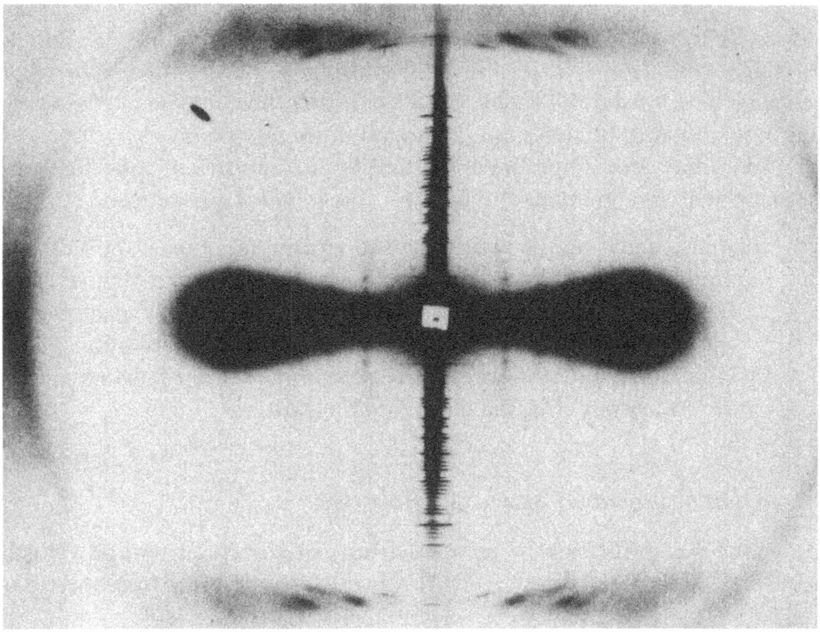

FIGURE 6. A longer exposure of Figure 5 (from Miller and Wray, 1971). Note the sampled 9.5-Å layer line, the row lines connecting the sharp reflections on the 9.5-Å layer line to those in the near-equatorial region, and the fanning of the intensity about the 9.5-Å layer line.

the collagen molecules are not randomly positioned but are arranged on a regular lattice. In the equatorial and near-equatorial region of the X-ray diffraction patterns (Figure 5) there are sharp reflections, the lateral breadth of which indicate that they originate from crystallites of coherent width about 1000 Å. This is close to the average diameter of a fibril in rat-tail tendon. Row lines of similar lateral breadth pass through these near-equatorial reflections and sample the layer line of spacing 9.5 Å (Figure 6). This layer line arises from the three-strand-rope structure of the collagen molecule, and the fact that it is split up laterally into reflections which are connected by row lines to the near-equatorial set means that the collagen molecules are arranged on a regular three-dimensional lattice.

The row lines in Figure 6 are observed to consist of reflections more closely spaced in the meridional direction than the meridional reflections. This is particularly obvious in the row line of lateral spacing 38 Å both close to the equator and where it intersects the 9.5 Å layer line. At present it is not possible to estimate the precise separation between these reflections, hence the true repeat of collagen along the fiber axis is still not known. It is, however, evident that a single fibril consists of collagen molecules arranged on a three-dimensional lattice.

The sharp near-equatorial reflections occur in the X-ray diffraction patterns from tendons enclosed in atmospheres with a range of humidities. However, dry tendons or tendons that have been fully swollen in water do not give the sharp reflections, which means that the lattice is fairly readily disordered or broken down. The effect of humidity on the structure suggested that water is incorporated into the three-dimensional lattice of the native fibrils. As we shall see later, electron micrographs of transverse sections of collagen have not yet allowed visualization of this regular lattice.

B. *Evidence Concerning Clustering of Molecules*

We shall now discuss the question of whether, in projection on to the x–y plane, the collagen molecules are in equivalent positions or grouped into clusters. Evidence relevant to this comes from electron microscopy and X-ray diffraction.

1. *Electron Microscopy*

In electron micrographs of longitudinal views of negatively stained collagen fibrils fine filaments have frequently been observed particularly in the "gap" region. Estimates of the diameter of these filaments have varied from 15 Å to 50 Å (Tromans *et al.*, 1963; Olsen, 1963), and with certain stains it is possible to see these filaments as continuous throughout the gap and overlap regions (Hosemann and Nemetschek, 1973). It is difficult to estimate the exact diameter of filaments from electron microscopy since the effects of stain on the appearance of the filament is not known, but some of these authors have concluded that their observations are good evidence that the filaments are several molecular diameters thick.

Electron micrographs have been obtained of collagen fibrils which are in a state of partial formation or dissolution (see, for example, Hosemann and Nemetschek, 1973). One side of the fibril is well formed and shows the characteristic D period; the other side is frayed into a bundle of long, fine filaments. Some electron micrographs show a continuous gradation from a well-formed to a completely frayed fibril, and these reveal that in the native fibril the filaments are packed together approximately parallel to the fibril axis. Hosemann and Nemetschek (1973) estimate that they are 35 Å across, but once more it is difficult to estimate the exact diameter of these filaments.

Filaments estimated to have similar diameter have been observed in electron micrographs of more fully dispersed collagen, sometimes termed polymeric collagen (Chapman and Steven, 1966; Steven, 1970; Veis *et al.*, 1970).

Investigations of mineral nucleation in collagen have also provided information about the lateral packing of molecules (Höhling *et al.*, 1974). Electron microscopy of the mineralizing front of rat incisor mantle dentine reveals bundles of needle-shaped crystals of hydroxyapatite. These needles are embedded in collagen fibrils and lie parallel to the fibril axis. Estimates of the lateral distance between these needles gave values of 30–60 Å with most values lying between 38 and 44 Å. Estimates of the diameter of the needles gave most values lying between 17 and 21 Å and 21 and 25 Å in two different electron micrographs. Measurements were also made of the lateral separation between hydroxyapatite needles in the mineralizing border of rat-tail bone. Most values of the lateral separation lay between 41 and 47 Å, while estimates of the diameters lay between 17 and 21 Å. Höhling *et al.* (1974) suggest that these observations are consistent with the hydroxyapatite needles occupying the spaces between 38-Å-diameter collagen filaments packed on a tetragonal lattice, as in the model of Miller and Parry (1973) (discussed in Section IV-C-1).

Lateral clustering of collagen molecules has also been investigated by electron microscopy of transverse sections through fibrils. Sections through tendons treated in the normal way for electron microscopy generally do not show any fine structure in projection onto the x–y plane. However, recently examples of some fine structure have been reported. In cornea, the collagen fibrils are smaller than in tendon (200–500 Å in diameter) and are separated by a nonfibrillar matrix. Smith and Frame (1969) observed that in transverse section the fibrils of cornea appeared to be made up of particles about 50 Å in diameter, packed fairly close together, but not on any recognizable lattice. Bouteille and Pease (1971) carried out similar studies on collagen fibrils from rabbit aorta. They used a new method of specimen preparation which involved inert dehydration of the collagen by gradual replacement of water by ethylene glycol. The tissue was embedded in polyhydroxymethacrylate and stained with uranyl acetate and lead nitrate. Electron micrographs of transverse sections through specimens treated in this way showed particles of about 50 Å diameter which had been swollen apart during preparation of the specimen. Höhling *et al.* (1974) have also studied transverse sections of the collagen fibrils of rat-tail tendon treated by the inert dehydration method of Bouteille and Pease (1971), embedded in polyhydroxymethacrylate, and stained with uranyl acetate and lead acetate. They observe similar dotlike nuclei and internuclei distances between 30 and 56 Å with most distances lying between 38 and 48 Å. Höhling *et al.* (1974) propose that the nuclei of stain lie in the spaces between clusters of collagen molecules rather than on the clusters themselves.

As in the case of longitudinal views, it is difficult to make a precise estimate of the diameter of the objects which give rise to the 50-Å clusters of stain. In transverse section this is particularly difficult because the particles are approximately circular and their apparent diameter is very sensitive to the position of the focus of the electron beam (Haydon, 1968, 1969; Thon, 1966*a,b*; Millward, 1970). Taken by itself, therefore, the electron microscopy is certainly suggestive that the collagen molecules cluster into filaments with diameter approximately 50 Å, but it is not conclusive. The existence of continuous filaments certainly implies that there must be places along their length where at least two molecules run side by side, since it is known that end-to-end contact of molecules does not occur in the fibrils. Zimmerman *et al.* (1970) have shown that 4*D* periodic strands can be isolated from collagen which has been treated briefly with pepsin, and although this material appears different from the so-called polymeric collagen of Veis and Steven described above, it is difficult, at present, to know precisely how the two types of filaments are related. The separate fine filaments do not necessarily mean that the molecules are geometrically clustered in the fibril, but may simply reflect a regular lateral distribution of weaker intermolecular bonds which are most readily broken by acid, stain, or swelling treatments. In intact fibrils the strands of stain seen in negative staining may be "gaps" in the structure which need only be one molecule wide as in the Hodge–Petruska model. The observation of filaments in intact fibrils is not by itself suggestive of clusters since any regular lattice would be likely to appear filamentous. Further information from the observation of filaments in intact fibrils depends on analysis of the lateral spacings recorded between the filaments.

2. *X-Ray Diffraction*

Another technique which may be used to investigate the possible existence of lateral clusters of molecules is X-ray diffraction. The X-ray diffraction patterns from rat-tail tendon have been used to develop a detailed model which involves lateral clustering of the molecules (Miller and Wray, 1971). The diameter of the collagen molecule is 13–15 Å. In the equatorial and near-equatorial region of the X-ray diffraction pattern there is a maximum in the intensity at an R value (R is the distance on the X-ray diffraction pattern parallel to the equator and is measured in $Å^{-1}$.

TABLE 1
Observed and Calculated Row Line Spacings for Rat-Tail Tendon Collagen h, k Indices[a]

Observed	Calculated	Cell 1	Cell 2	Cell 3	Intensity
38.00	38.47	10	11	20	ms
26.51E	27.20	11	20	22	w
24.61	24.33	—	21	31	w
18.90	19.24	20	22	40	ms
17.48E	17.20	21	31	42	ms
~15.7	15.39	—	—	34	vw
15.03	15.09	—	32	51	w
13.69	13.60	22	40	44	vs
13.28	13.19	—	41	53	vs
12.64E	12.82	30	33	60	vs
12.24	12.17	31	42	62	vw
11.41	11.47	—	—	63	w
10.31	10.10	—	52	73	w
9.64E	9.62	40	44	80	w
8.65	8.6	42	62	84	w
8.15	8.1	—	63	93	w
7.2	7.15	52	73	104	w
6.8	6.8	44	80	88	w
6.3	6.4	60	66	120	w

[a] Tetragonal unit cells of side are 38.47, $38.47\sqrt{2}$, and 38.47×2Å, respectively. Some of these spacings have already been reported by Miller and Wray (1971) and Wray (1972). Intensities are estimated as vs (very strong), ms (moderately strong), w (weak), and vw (very weak). Those reflections with equatorial components are marked E after the observed spacing.

Each R value corresponds to a d spacing (Å in real space) corresponding to 12–14 Å which is clearly related to the intermolecular distance. However, there are also sharp reflections between this region and the origin, that is, at R values corresponding to longer lateral spacings than the intermolecular distance. A list of the measured d spacings of the equatorial and near-equatorial reflections is given in Table 1. The innermost reflection corresponds to a spacing of 38.5 Å. The true equatorial reflections originate from the projection of the structure down the fiber axis on to an x–y plane, and some equatorial reflections (such as the one with a d spacing of 17.3 Å) have a d-spacing which is larger than the intermolecular distance. If the molecules were straight and aligned parallel to the fiber axis, the observation of equatorial reflections of relatively long spacing could very simply be explained by supposing that they were arranged in regular clusters. However, the X-ray diffraction patterns contain unambiguous evidence that the molecules are not parallel to the fibril axis but inclined at an angle of a few degrees. This may be deduced from the distribution of intensity about the equator and about the 9.5-Å layer line. The Z splitting of the near-equatorial reflections is linearly proportional to R so that the distribution of intensity in this region has a fanlike appearance. There are also intense reflections on the equator. A similar fanning occurs about the 9.5-Å layer line. This fanning, since it contains relatively sharp reflections, cannot be caused by either a short coherent length of the molecule or by relative misalignment of the fibrils or fibers in the X-ray beam; it could only arise from tilting of the molecules.

This unambiguous evidence from X-ray diffraction patterns that the molecules are tilted to the fiber axis means that the equatorial reflections corresponding to spacings such as 17.3 Å, which are longer than the intermolecular distance, cannot be taken as simple indication of lateral clustering. As will be shown later, the positions of the reflections and the molecular tilt can be readily reconciled with a cluster having a helical symmetry.

The effect of very dilute solutions of heavy-metal stain on the X-ray diffraction pattern is of some interest. Weak solutions of sodium silicotungstate had the effect of enhancing the intensity of the 38-Å row line relative to the intensity of the rest of the X-ray diffraction pattern. This was interpreted by Miller and Wray (1971) as suggesting that an object of lateral dimension 38 Å was being outlined by the stain. Similar observations were made with rat-tail tendons fixed with formaldehyde and stained with phosphotungstic acid at pH 2 (Hosemann, 1973; Hosemann and Nemetschek, 1973). This observation implies that within the crystal lattice, there is an important vector perpendicular to the fibril axis and of length 38 Å separating sites in the structure which absorb heavy metal stain.

3. Electron Microscopy of Obliquely Striated Collagen

A further source of data which suggests a specific lateral clustering of the molecules in collagen is electron microscopy of an obliquely striated polymorphic form of reprecipitated collagen.

It is known that collagen molecules may be reprecipitated in a variety of different polymorphic forms; which polymorph occurs is determined mainly by the conditions of precipitation and to a lesser extent by the source of the collagen. Polymorphs with oblique striations have been described (Schmitt, 1956; Kuhn *et al.*, 1964), and in one form of reprecipitated collagen from cartilage the basis of the oblique striations was clearly due to D-periodic, narrow subfibrils staggered axially with respect to each other by about 90 Å (Bruns *et al.*, 1973).

Cartilage collagen differs from skin and tendon collagen in a number of respects, chiefly in that the α chains of the triple helix of the collagen molecule are identical and the telopeptides are somewhat larger. Thus cartilage collagen may be represented by $3[\alpha 1(\mathrm{II})]$ as distinct from the $2[\alpha 1(\mathrm{I})]\ \alpha 2$ form of skin and tendon collagen (see, for example, Chapter 1). Despite this difference the cartilage collagen resembles that from skin and tendon in many ways. Cartilage collagen exists as D-periodic fibrils in cartilage, and in the SLS polymorph the patterns of positively stained bands from the two types of collagen are closely similar (see Doyle *et al.*, 1974*b*). Bruns *et al.* (1973) showed that when cartilage collagen was

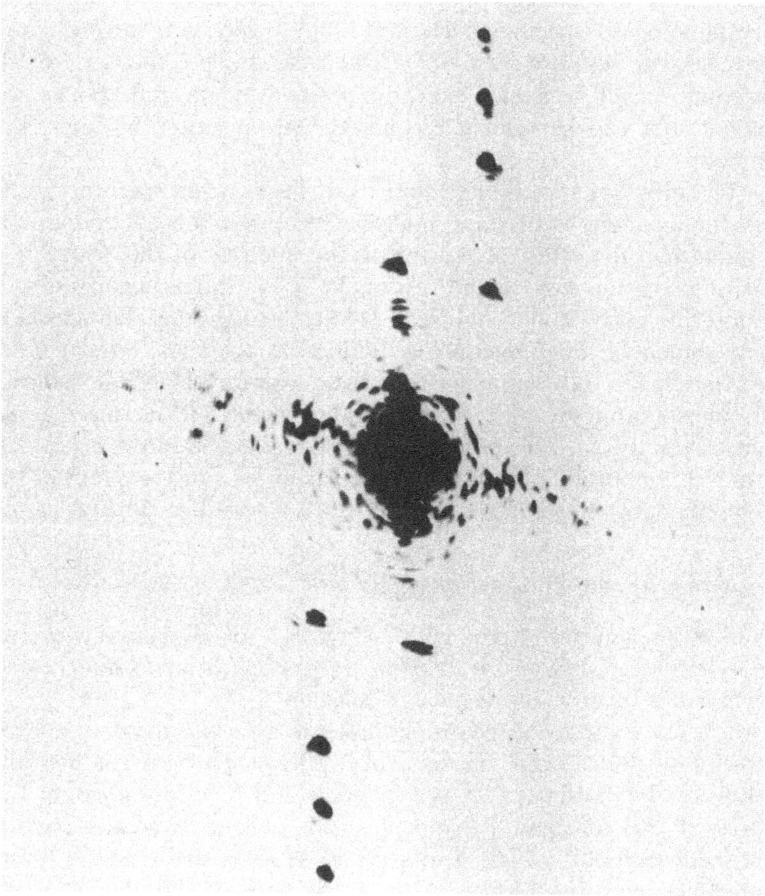

FIGURE 7. Optical diffraction pattern of Figure 8. This and the next five figures are from Doyle *et al.* (1974*a*).

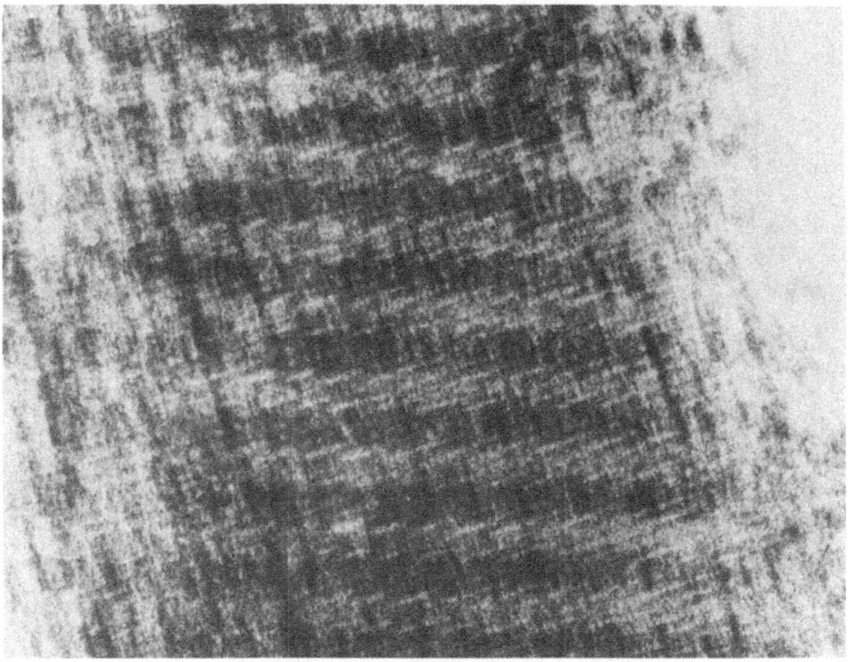

FIGURE 8. Electron micrographs of a negatively stained obliquely striated tactoid of collagen. The subfibril width is about 380 Å and subfibrils are staggered axially by about 120 Å with respect to their neighbors. (X99,000).

reprecipitated at low pH, a considerable amount occurred in the form of obliquely striated tactoids in which the striations were inclined at about 67° to the long axis of the tactoid. Electron micrographs of negatively stained obliquely striated tactoids showed that they were composed of narrow, cross-striated units, which the authors termed subfibrils, arranged approximately parallel to one another and to the longitudinal axis of the tactoid. Each subfibril was about 140 Å wide and had the same dark–light banding pattern as an electron micrograph of a negatively stained reconstituted skin collagen fibril. Neighboring subfibrils were shifted axially by about 90 Å.

This interpretation of the oblique striations as due to axially staggered D-periodic subfibrils is confirmed by optical diffraction (Doyle *et al.*, 1974*a*). Figure 7 shows an optical diffraction pattern of an electron micrograph of a tactoid of obliquely striated collagen. The optical diffraction pattern contains sharp reflections on several oblique row lines. The separation parallel to the meridian of neighboring reflections along a row line is $1/D$ due to the D periodicity of the subfibrils, and the row lines

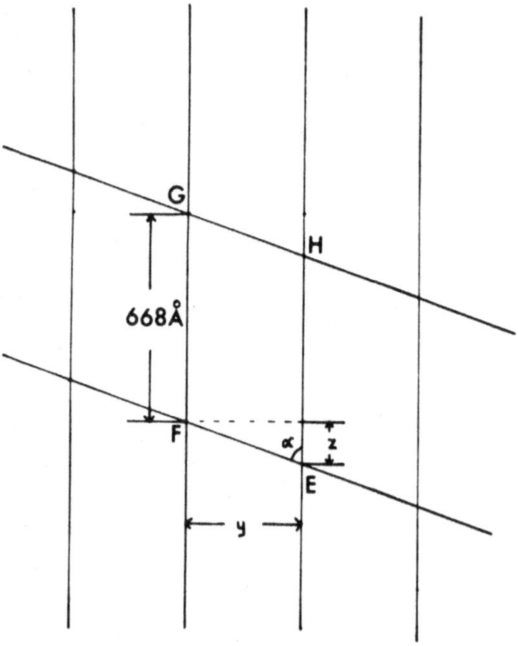

FIGURE 9. Diagram of the lattice on the tactoid of Figure 8. One unit cell EFGH is indicated.

themselves are perpendicular to the vector relating neighboring subfibrils. This is illustrated in Figures 8 and 9. Figure 9 is a diagram of the lattice in Figure 8. $EH = FG = D$ (668 Å). The diameter of the subfibrils is y, the axial shift between nearest neighbors z, and α, the angle of the oblique striations to the longitudinal axis of the tactoid. Figure 10 is a diagram of the diffraction pattern in Figure 7, and its relation to the tactoid lattice is shown. The observed reflections occur at the intersections of two sets of lines. The layer lines (parallel to the equator) are spaced apart by a distance $1/D$ parallel to the meridian. The oblique row lines make an angle α with the equator and are spaced apart by a distance $1/EF$. The zero-order row line intersects the origin of the diffraction pattern; the first-order row line makes an intercept on the meridian at $1/z$ and on the equator at $1/y$. Note that the meridian and equator are defined here as at right angles to each other. Their absolute orientation is arbitrary but the meridian is usually selected as closely parallel to the fiber axis. Here, the equator is defined as at right angles to the subfibril axis; there will only be *true* meridional reflections if z is an integral submultiple of D.

Further studies on obliquely striated reprecipitated cartilage collagen (Doyle *et al.*, 1974*a*) revealed that the subfibril width could vary but the axial shift between neighboring subfibrils is fairly constant. A tactoid with y

= 380 Å is shown in Figure 8. A rough idea of the subfibril width can be obtained readily be measuring α, the angle of the oblique striations to the longitudinal axis of the tactoid; more precise values of y and z may be obtained from optical diffraction patterns.

When the variability of subfibril width was recognized we realized that this obliquely striated polymorph might afford a favorable system to explore for the smallest diameter of filament which still displays the D period characteristic of collagen fibrils. It is obvious that the smaller the diameter of a D-periodic filament, the greater the restrictions on the possible three-dimensional molecular arrangements which could generate the D period. Initially this search was carried out by measuring α, the inclination of the oblique striations to the longitudinal axis of the tactoids since, given that z is relatively constant, $y \sim \sin \alpha$. The smallest value of α which we observed was about 23° (Doyle *et al.*, 1974*a*). This corresponds to a y of about 40 Å. However, in such steeply inclined oblique striations it is

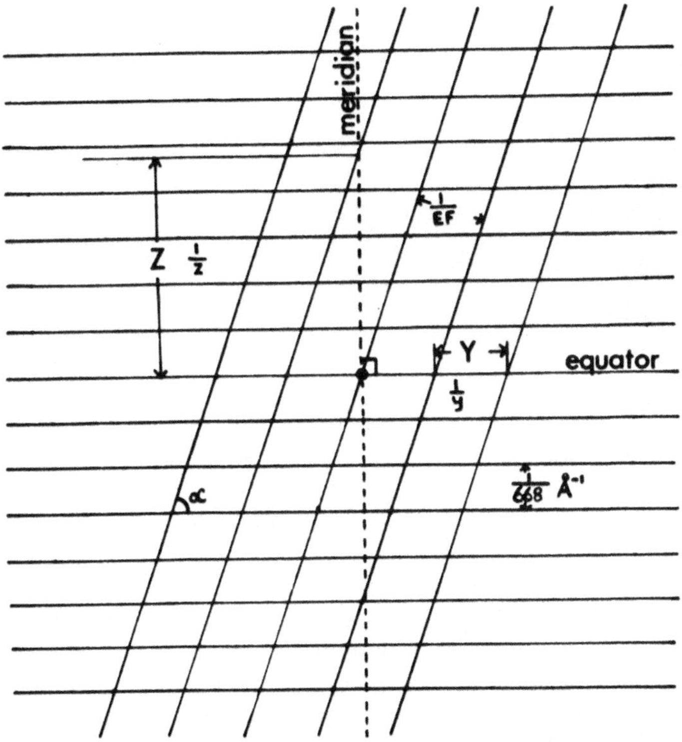

FIGURE 10. Diagram of the lattice reciprocal to that in Figure 9. This figure corresponds to optical diffraction patterns such as that in Figure 7.

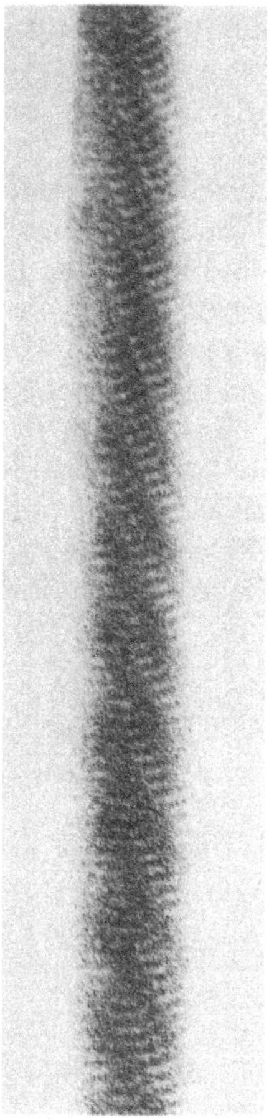

FIGURE 11. Electron micrograph of chick cartilage collagen reconstituted under the conditions described by Bruns *et al.* (1973) and negatively stained with uranyl acetate (X120,000).

difficult to visualize the individual subfibrils in the tactoid. That the origin of the steeply inclined striations ($\alpha \sim 23°$) was the same as the tactoids with less-steeply inclined striations (higher values of α) was confirmed by optical diffraction. Figure 11 shows an electron micrograph of obliquely striated reprecipitated cartilage collagen with $\alpha \sim 23°$. Figure 12 is an optical diffraction pattern of Figure 11. In Figure 12 the diffraction pattern consists of oblique rows of reflections separated by a distance of $1/D$ parallel to the meridian; this confirms that the tactoids contain D-periodic filaments. This corresponds to the fact that in Figure 12 a line parallel to the tactoid axis intercepts the oblique striations at axial intervals of D. In Figure 11 it is not easy to see the axial shifts of z relating the narrow D periodic filaments.

If the oblique striations in Figure 12 were straight with no z steps, the diffraction pattern could contain only a zero-order oblique row of reflections. In Figure 12 the existence of a first- (and weaker second-) order row of reflections indicates that the oblique striations in Figure 11 do have a periodicity along their length. The meridional and equatorial intercepts of the first-order oblique row of reflections indicates that the periodicity along the oblique striations of Figure 11 has components y = 40 Å and z = 97 Å. Hence the optical diffraction patterns enable us to conclude that these tactoids are composed of parallel, D-periodic filaments of width 40 Å staggered axially by 97 Å.

The above observations were on collagen from cartilage. However, obliquely striated tactoids have also been obtained from skin collagen (Kuhn *et al.*, 1964). These tactoids resembled the ones we obtained with cartilage collagen. An optical diffraction pattern of the electron micrograph of Kuhn *et al.* (1964) showed the same features as Figure 12, so this tactoid is also composed of parallel, stepped D-periodic filaments. Intercepts of the first-order row of reflections on the equator and meridian, respectively, enabled us to estimate y = 37 Å and z = 99 Å.

The conclusion which may be derived from the above discussion is that a filament 37–40 Å *wide* carries the D period when projected on to the filament axis. We are formally unable to go further than this since we do not know the complete three-dimensional structure of the obliquely

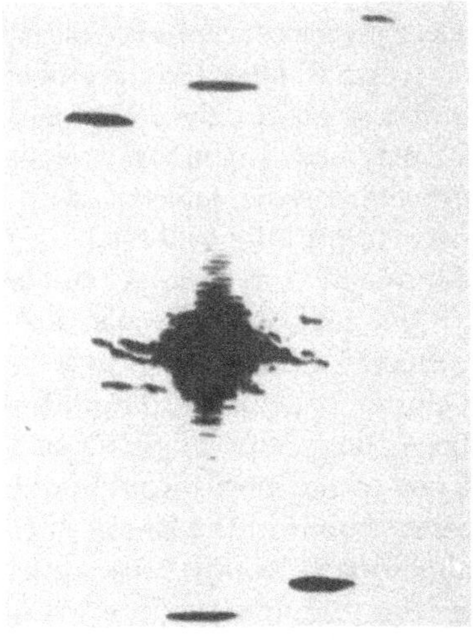

FIGURE 12. Optical diffraction pattern of the electron micrograph in Figure 11. By analogy with Figure 7 it may be deduced that the tactoid in Figure 11 is composed of 40-Å wide, D periodic subfibrils staggered axially by 97-Å with respect to their neighbors.

striated tactoids. If the *D* periodic filaments are cylindrical, then they would be 37–40 Å in diameter. While this may be a likely situation we cannot conclude it from these observations since we do not know the *thickness* of the *D*-periodic filament which is 37–40 Å wide. An estimate of the depth of penetration of stain into the tactoids would clearly limit the possible thickness. The stain obviously penetrates sufficiently to reveal the *D* periodicity in the subfibrils. However, they are strictly *D*-periodic and there is no sign of a 90–100-Å axial shift of *D*-periodic objects within the subfibrils. Hence, either the stain does not penetrate far and thus the subfibrils of period *D* are more or less cylindrical or the axial shift does not occur in the tactoids between subfibrils related by a lateral vector perpendicular to the tactoid surface. A further possible source of inaccuracy is that we have not tried to estimate the inclination of the tactoids to the electron beam during electron microscopy. It is unlikely that the angles involved will greatly alter the conclusions of this section.

The conclusion here is similar to that drawn from the X-ray diffraction data described in Section IV-B-2, but in this section it refers to collagen in a reprecipitated obliquely striated form; the conclusion from X-ray diffraction referred to the native tendons.

4. *Summary*

The X-ray diffraction patterns from stained tendons and the observation of filaments in electron micrographs of fibrils indicates that stain goes into the fibril with a regular lateral repeat of about 40 Å (the X-ray evidence gives 38 Å). These observations on intact fibrils do not necessarily imply clusters, but may be revealing the regular occurrence of "gaps" (and hence the *e* segments of the molecules) on a lateral lattice. These observations, therefore, support the well-established fact that the fibril is a crystal and further indicate 38 Å as an important lateral repeat distance.

When we ask if the collagen molecules are clustered laterally, the essential question is whether there exists a cluster such that intermolecular bonding within the cluster is geometrically different from that between clusters. The existence of long, fine filaments of about 40 Å diameter in electron micrographs of frayed fibrils would imply a regular distribution of weaker intermolecular bonds in the lateral direction. However, it could be argued that if the broken bonds were closely similar to the intact intrafilament bonds, the use of the term cluster, while semantically justifiable, would be unhelpful when used to describe the geometry of molecular arrangement in a fibril. It is also necessary to know the true axial periodicity of the fine filaments, D, $4D$, or some other, in order to establish the molecular arrangement within them.

The only strong evidence for discrete entities with the D period is that from the obliquely striated tactoids. The 40-Å-wide subfibrils are obviously clusters, and only the uncertainty about their thickness prevents a cogent demonstration of a unique structure. This, of course, demonstrates the ability of collagen molecules to assemble into narrow D periodic filaments rather than the existence of such filaments in native fibrils. Miller and Wray (1971) suggested that X-ray diffraction evidence of a structural rearrangement from the native to a polymorphic, orthomorphic molecular arrangement, with the preservation of the 38-Å row line, could imply the discrete nature of the 38-Å-wide bundles.

We conclude that there is unambiguous evidence that the collagen molecules are arranged on a regular lattice in which there is an important lateral vector of length 40 Å. We further conclude that there is suggestive, although inconclusive, evidence that the molecules are grouped laterally into clusters of width, probably of diameter 40 Å, and that these clusters are very long, parallel to the fibril axis, and have a projected axial period of D.

C. *The Three-Dimensional Lattice*

Subsections A and B above show that the collagen molecules are definitely arranged in a regular lattice in the fibrils and that 38 Å is an important lattice vector. This might be related to an interfilament distance or an intersheet distance. In this section we shall discuss the attempts which have been made to index the sharp equatorial and near-equatorial reflections in the X-ray diffraction patterns and the implications of these for molecular packing. Obviously the correct indexing scheme will give the correct three-dimensional lattice in the fibrils and settle the uncertainties with which we concluded the last section.

The most recent set of spacings for the equatorial and near-equatorial reflections in the X-ray diffraction pattern from rat-tail tendon has been published by Miller and Parry (1973) (see Table 1). I am not aware at present of any other set of observed spacings [apart from those of Hosemann and Nemetchek (1973) discussed below] nor have we in this laboratory so far obtained sufficient extra data to warrant modification of Table 1.

This list differs from that of North *et al.* (1954) mainly in that more

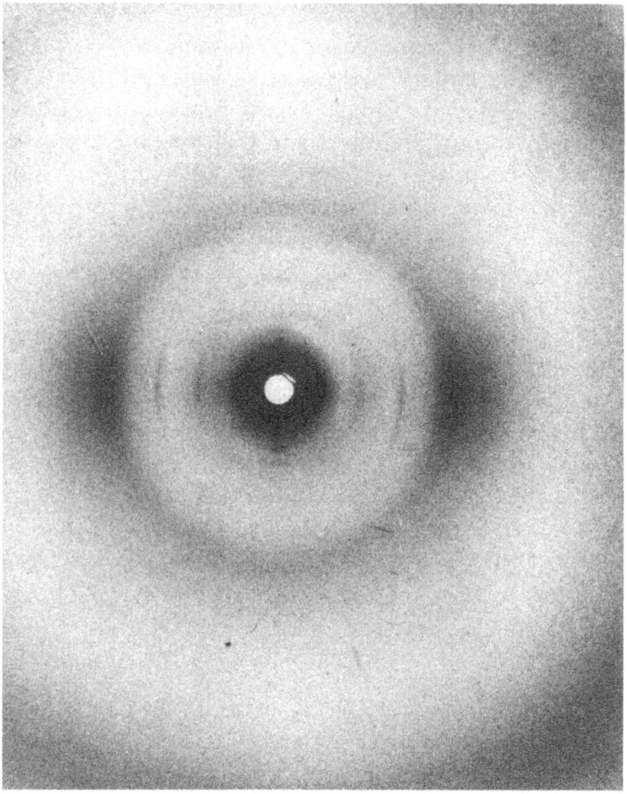

FIGURE 13. Medium-angle X-ray diffraction pattern from dried toad sartorius muscle. Note sharp arc-shaped reflections in the near-equatorial region. These index as orders of a 48-Å spacing and are probably due to recrystallized lipid bilayers.

reflections were observed and some closely spaced reflections were resolved. The reflection at spacing about 49 Å can be removed by treating the tendon with an ethanol–water mixture which leaves the rest of the pattern unchanged. A set of reflections indexing on 49 Å has been observed in α-keratin and attributed to lipid (Fraser *et al.*, 1964). Our results confirm this interpretation of the 49-Å reflection from tendon. Similar reflections attributed to lipid are frequently visible in diffraction patterns from muscle (Miller, unpublished results; see also Figure 13) and preparations of membranes (Pollard *et al.*, 1973). It is presumed that phospholipid possibly from a membrane source may readily recrystallize in a bilayer form similar to lecithin (Levine and Wilkins, 1971). This explanation means that the reflection at 49 Å from tendon must be excluded from consideration.

Two main indexing schemes have been suggested, one based on a tetragonal lattice of side 38 Å, the other on a hexagonal lattice of side 15 Å.

1. *Tetragonal Lattices*

Examination of the spacings of the reflections in Table 1 shows that they are close to the positions predicted by a tetragonal unit cell of side 38 Å. The (1,0), (1,1), (2,0), (2,1), (3,0), and (3,1) positions are all accounted for and cell 1 in Table 1 shows that some two thirds of the observed reflections are predicted by this cell. There are additional reflections but, between the (3,1) position and the origin, these are all close to the positions predicted by the tetragonal cell, leading Miller and Parry (1973) to conclude that the true cell must be one which resembles the tetragonal cell of 38 × 38 Å very closely. Various cells were tried which involved slight deformation of the sides or angles of the tetragonal cell. Cells in which the sides were of different length seemed improbable in view of the single (1,0) reflection at a spacing of 38 Å. Cells with equal sides but an angle close to, but different from, 90° seemed attractive since this could lead to a single (1,0) reflection and splitting of the (1,1) reflection into two components (around 26 Å) as is observed. However, the "split (1,1)" reflections are not displaced about the true tetragonal (1,1) position as the above distortion would predict. The 26.5-Å reflection is very close to the true (1,1) one, the 24.6-Å reflection at a higher R value. The modification to a simple tetragonal cell suggested by Miller and Parry (1973) stemmed from the observation that the true equatorial reflections could all be indexed on the simple cell; hence in projection onto an x–y plane the simple cell was adequate, but it is not adequate for the whole three-dimensional structure. A simple way of modifying the cell without affecting the projected structure is to move neighboring cells in an axial direction with respect to each other, and in cell 2 of Miller and Parry (1973) nearest-neighbor cells are related to each other by axial shift of $z = \frac{1}{2}$. This accounted for all but two reflections, the 11.4 Å being a particularly intense exception. Fraser *et al.* (1974) showed that all the reflections could be accounted for by a tetragonal cell of side 2 × 38 Å and space-group $P4_1$ or $P4_3$. In this solution the near-fit obtained by the simpler cells is a natural consequence of the symmetries of the simple cells. We shall discuss this in more detail in Section V and show how the collagen molecules may be accommodated in it.

An alternative indexing scheme based on a near-tetragonal cell has been proposed by Nemetschek and Hosemann (see Hosemann, 1973). In this scheme the sides of the tetragonal cell are made slightly different but the angle is maintained at 90°. This interpretation runs into difficulties since the intense sharp reflections at spacings corresponding to 13.6 and 12.6 Å are too close to the origin to index as the (3,0) and (0,3) of the fundamental unit cell. Hosemann (1973) accounts for this by proposing a disordered array of molecules within the cells which he claims will move the (3,0) and (0,3) reflections closer to the origin. He supports this contention by optical diffraction patterns. There are, however, some points against this interpretation. First the unit cell predicts more reflections than are observed by us (Table 1), although North *et al.* (1954) did report some of the reflections required by this unit cell. Second, if the lattice (as distinct from unit-cell contents) is disordered, then this will lead to broadening of the reflections in a direction parallel to the equator, especially with increased distances from the meridian. Such a broadening is evident in the optical diffraction patterns shown by Hosemann (1973) and suggests that the proposed model does involve a disordered lattice. However, the observed breadth of the reflections in the X-ray diffraction pattern in a direction parallel to the equator does not increase with increasing R to any measurable extent (Miller and Wray, 1971). Even the reflection at a spacing of 9.6 Å still corresponds to a coherent "crystallite" of 1000 Å across. If there is no lattice disorder in the Hosemann model, but only disorder of the molecules within the cell, then the underlying broad intensity maximum around 13 Å may be shifted closer to the origin, but the positions of the Bragg reflections [the (3,0) and (0,3)] will not be altered by such a disorder.

The Hosemann–Nemetschek model may be thought of as based on either sheets of collagen molecules or microfibrils (i.e., clusters now termed "octafibrils" by the authors). The undistorted unit cell of about 38 × 38 Å may be considered as orthogonally intersecting pairs of sheets of collagen molecules. One feature of the model as presented by Hosemann (1973) is that each unit cell contains 8 molecular diameters and thus a single unit cell could not itself have the D period since this requires $5n$ molecular diameters where n is an integer. Furthermore, the octafibril of this model does not have the advantage of helical symmetry. The X-ray patterns of Miller and Wray (1971) indicate that if the Hosemann–Nemetschek indexing is correct, then their cell must itself be D periodic. If it contained 5 molecules, the octafibril would be the microfibril of Smith (1968). If the cell contains 10 molecules, then the packing density in the overlap region will be greater than that in crystals of (Gly-Pro-Pro)$_{10}$ (see Sakakibara *et al.*, 1972).

Recently Hosemann *et al.* (1974) have proposed that the 8 disordered units in a single unit cell are not single molecular diameters but correspond to sections through strands comprised of molecules linked by a 4*D* axial stagger. Hence in the overlap region this model does have 10 molecular diameters and in the gap region 8 molecular diameters. The unit cell now has the required *D* period, and it contains the equivalent of two microfibrils of the type suggested by Smith (1968) and discussed in Section V. This model still has the advantage that it may be adapted to a sheet structure or a microfibril (octafibril) structure. Presumably a helical octafibril could be produced by the repeated *D*-staggering of dimers of in-register collagen molecules.

2. *Hexagonal Lattices*

If rodlike molecules of cylindrical symmetry are close-packed parallel to each other, then they would, in transverse section, lie on a hexagonal lattice. It has generally been supposed that the collagen molecules are packed on a hexagonal lattice in fibrils, although no direct evidence has been produced for this mode of packing. Some indications that collagen molecules could be packed on a hexagonal lattice have been presented by Katz and Li (1972, 1973, 1974). This is based on experimental estimates of the intermolecular volume within a collagen fibril. A hexagonal lattice has also been proposed by Kuhn (1969) and Macfarlane (1971).

There has been little systematic analysis of the possible hexagonal lattices and their predicted diffraction patterns. If it is assumed that in a hexagonally packed array at least one line of molecules will be related by the Hodge–Petruska axial stagger (*D* and 4*D*), then there are three kinds of lattices possible. These are listed below.

			Fit with X-ray pattern
1. Axial	staggers—2 (*D* and 4*D*)	(e.g., Kuhn, 1969)	Bad
	1 (0*D*)		
2.	2 (*D* and 4*D*)	(e.g., Macfarlane, 1971)	Good
	1 (2*D* and 3*D*)	(e.g., Katz and Li, 1973)	
3.	1 (*D* and 4*D*)		Good
	2 (2*D* and 3*D*)		

If the molecules are tilted to the fiber axis or, if parallel to the fiber axis, are in different azimuthal orientations, these lattices will yield unit cells containing more than one molecular diameter when projected down

TABLE 2
Observed and Calculated d Spacings for Monoclinic Cell[a]

Index	Calculated spacing (Å)	Observed spacing (Å)
(1,0)	38	38.5
		26.5
(0,1) (−1,1) (1,1)	24.8	24.6
(−2,1)	18.4	18.9
(2,0)	19.0	
		17.5
		15.7
		15.0
(2,1) (−3,1) (−1,2)	13.2	13.7
		13.3
		12.6
(3,0)	12.6	12.24
(0,2) (−2,2)	12.4	11.41
(1,2) (−3,2)	10.9	10.31
(2,2) (−4,2)	9.2	9.64
(4,0)	9.5	

[a] Monoclinic cell $a = 2.642x$; $b = 1.732x$; $\gamma = 109°$

the fiber axis. Case 1 above produces one type of cell; cases 2 and 3 produce a second type of dimension $a = \sqrt{7}x$, $b = \sqrt{3}x$, $\gamma = 109°$ (x = intermolecular distance) (Macfarlane, 1971). The reflections predicted by the first type of cell do not resemble those observed in the X-ray diffraction pattern, but those predicted from the second cell are listed in Table 2 and may be seen to be quite close to some of the observed reflections. The disagreement with the observed reflections is more serious than that of a simple 38 × 38 Å tetragonal cell (see Table 1). In particular the hexagonal cell does not predict splitting of the 26/24 reflections nor the 13.6/13.2-Å pair, neither does it predict the intense reflections at 17.3- and 11.4-Å spacings. This model may be modified by assuming that the lateral intermolecular spacing within a sheet of molecules may be different from that between sheets. Thus the exact ratio between a and b which the hexagonal array demands may be relaxed, as may the value of γ.

Of the three types of hexagonal lattice listed above we note that type 1 contains molecules with the set of axial staggers ($0D$, $1D$, and $4D$) which have, at present, been observed, but does not produce a unit cell which agrees with the X-ray pattern. Types 2 and 3 are in reasonable agreement with the X-ray pattern but do not contain the observed set of axial staggers.

As indicated above, one way of producing such a hexagonal cell in projection down the fiber axis would be if the molecules were in different azimuthal orientations. If we assume again that the intermolecular contact within a sheet is particularly important, then we see that the azimuthal rotations should lead to a repeat after 5 molecules in a Hodge–Petruska sheet. Thus the molecule should have interacting edges ($2\pi/10$) apart in azimuth. In the description of the symmetry of the collagen molecule (Section II) we noted that the molecule may indeed have just such a set of equally spaced edges. The molecular symmetry cannot be used simply to deduce molecular packing however. α-Helices crystallize on lattices with different symmetries and (Pro-Pro-Gly)$_{10}$ has been shown to crystallize in a tetragonal space group. Furthermore this same molecular symmetry can be used to support other types of molecular packing (Segrest and Cunningham, 1971) (Section V).

Much of the available experimental data relevant to a model for the molecular arrangement in collagen fibrils has now been surveyed in this article. The next step is to see to what extent this data points to a unique model. It will quickly become clear that several models have been proposed recently, and so I will center the rest of the discussion round what I consider, at present, to be the model which is consistent with the widest range of data. This model is described in Section V, and, while it is not possible to claim unambiguous proof of the model, it receives strong support from several sources. It will be termed the heuristic model. The intention is that by taking this model as an origin for a hypothetical–deductive approach, the epithet may be justified.

V. Heuristic Model

The most compact form of the minimum structural unit with the D period when projected on to the fiber axis is the five-stranded microfibril suggested by Smith (1968). Smith started from the Hodge–Petruska arrangement shown in Figure 1. He pointed out that a minimum Hodge–Petruska set, five molecules wide, could be wrapped around the surface of a cylinder so that its opposite parallel edges came in contact. The resulting cylindrical structure, termed a microfibril, has the appeal that within it the collagen molecules are related as subunits on a helix and are thus in equivalent positions. Since the molecular diameter is in the range 13–15 Å, the diameter of such a microfibril would be 35–41 Å. Smith supported his suggestion by reference to the observation of particles of about this diameter in electron-micrograph transverse sections of collagen fibrils in cornea (Smith and Frame, 1969), although such observations must be interpreted with caution (Haydon, 1968, 1969; Millward, 1970). The existence of such microfibrils is further supported by the rest of the data from electron microscopy and X-ray diffraction described here in Section IV-B. The heuristic model is, therefore, based on the microfibril.

If the molecules on the Smith microfibril are represented by points, the microfibril may be plotted as a radial projection (Figure 14). The Smith microfibril has a fivefold screw axis (5_1 or 5_4). The points lying on the 5_1 axis may be connected by a single line to give a single-strand helix.

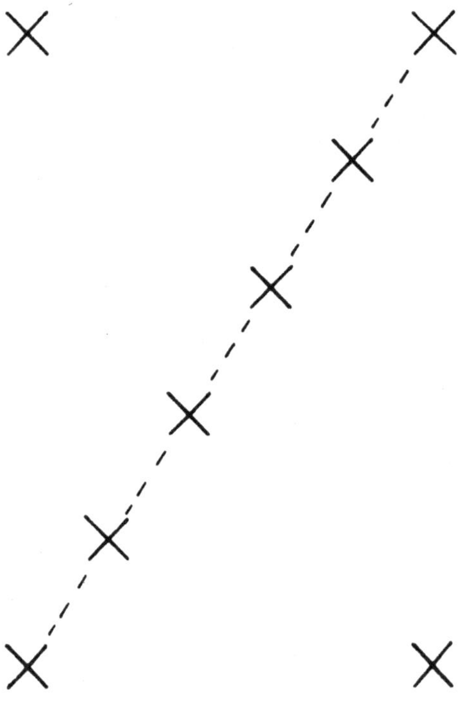

FIGURE 14. The crosses represent a radial projection of a 5_1 helix. A radial projection is formed by projecting the helix outwards along radii on to a cylindrical surface. The cylinder is then cut along a line parallel to the helix axis and opened out to give a flat sheet. The broken line traces the single-strand helical path which connects the crosses. In the Smith (1968) modification of the Hodge–Petruska (1963) net, the 5_1 helix connects molecules staggered axially by $1D$.

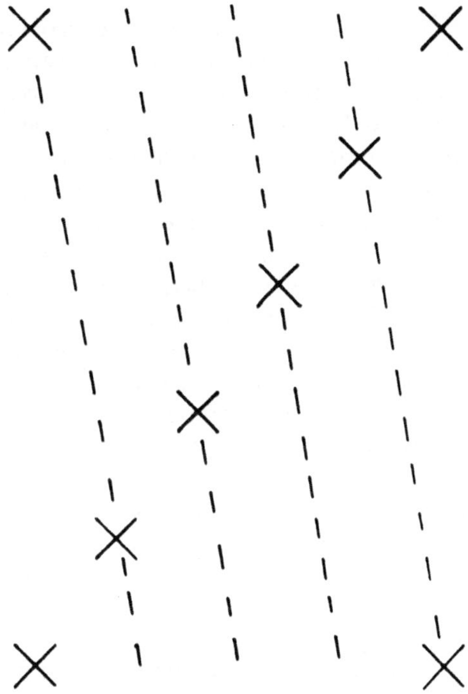

FIGURE 15. As in Figure 14, a radial projection of a 5_1 helix of crosses. The dotted lines show the four-start helix by which the 5_1 helix may alternatively be described. In the Smith (1968) modification of a Hodge–Petruska (1963) net, the four-start helix contacts molecules linked by an axial stagger of $4D$.

Alternatively these points may be joined by a set of four lines (Figure 15) to produce a four-strand helix. The physical meaning of a strand in this case would be molecules linked in a $4D$ stagger with a molecular overlap of about $0.5D$. If the points in Figure 14 are connected by five sets of vertical lines, the microfibril may be described as a five-strand cable where each strand is coincident with a molecular axis. There are three alternative ways of describing the same 5_1 helix—one-stranded, four-stranded, and five-stranded. I shall term it a five-stranded microfibril to emphasize the feature of five molecular segments *a, b, c, d,* and *e* (see Section III-A) which occur in register within a *D* period.

The next relevant factor is that collagen fibrils are three-dimensional crystals (see Section IV-A). This means that the microfibrils must be regularly positioned with respect to each other. It is known that the molecular segments *a, b, c, d,* and *e* are in register throughout a fibril (Section III-A), so it remains to determine their arrangement in the x–y plane (Section IV-A). This depends on the indexing of the equatorial and near-equatorial reflections in the X-ray diffraction pattern. In Section IV-C reasons were given for preferring a tetragonal cell of side 2×38.5 Å and space group $P4_3$. Within this cell are four microfibrils each with true axial repeat of 4*D* and a projected axial repeat of *D* in which the molecules are related by a quasifourfold screw axis. Microfibrils are placed around 4_3 axes. This model, in addition to accounting for the positions of the observed X-ray reflections, also explains why many of the reflections can be accounted for by smaller tetragonal unit cells of sides 38 Å (space group $P4_1$) and 54 Å (space group $I4_1$), respectively (Fraser *et al.*, 1974). In this model adjacent microfibrils are related by a twofold axis, and thus there is a 0*D* axial stagger at all intermicrofibrillar contacts. Zimmerman *et al.* (1970) have described electron-microscope evidence for a 0*D* axial stagger between molecules in collagen fibrils.

The space group required the microfibril to have a projected axial period of *D* but a true axial period of 4*D*. This puts restrictions on the possible molecular arrangements within the microfibril (see Doyle *et al.*, 1974*a*) and, in fact, there are only two possible regular molecular arrangements within a five-strand microfibril. One of these (that of Smith, 1968) has intermolecular axial shifts of *D* and 4*D*; the other of 2*D* and 3*D* (Figure 2). The chemical evidence for covalent links between molecules with a 4*D* axial shift (Kang, 1972) supports the Smith model if the covalent linking is between adjacent molecules on the microfibril periphery. Zimmerman *et al.* (1970) also describe electron-microscope evidence for the existence of 1*D* and 4*D* staggered molecules in collagen fibrils.

The next factor to be incorporated into the model is the inclination of the molecules to the fibril axis. In terms of tetragonally packed microfibrils this might be achieved by inclination of straight microfibrils. However, the space group $P4_3$ suggests that some molecular inclination involves coiling of the molecules round the microfibril axis so that the 5_1 symmetry in the microfibril with straight molecules is converted into a fourfold symmetry. There are many degrees of tilt which would bring about such a symmetry conversion, so we assume that this mechanism accounts for all of the tilt and use the extent of the fanning of the intensity in the near-equatorial and 9.5-Å layer line of the X-ray diffraction pattern to define the degree of molecular coiling. This results in an estimate of a value of about 700 Å for the pitch of the coil of the collagen molecular axis.

The supercoiling may explain the puzzling discovery of a periodicity in the large hydrophobic residues within the *D* period (Figure 3). This shorter period was 2*D*/11. If we imagine a five-stranded microfibril supercoiled with a pitch of about 700 Å, the line of contact of this microfibril with a neighboring microfibril would intersect the five supercoiled strands at axial intervals of 700/5 Å which, while not precisely 2*D*/11, is of that order. The period of 2*D*/11 would ensure a gently inclined edge of apolar residues on the outside of the microfibril, possibly associated with microfibril packing. That is, of course, highly speculative at present.

The heuristic model may be readily summarized. The collagen molecules are arranged in long five-stranded microfibrils with 1*D* and 4*D* axial intermolecular staggers, where D = 668 Å. These microfibrils are supercoiled with a pitch of about 700 Å and arranged face to face throughout the fibril on a tetragonal lattice of side 38.5 Å.

This model is consistent with a wide range of diverse observations from electron microscopy, X-ray diffraction, and the location of intermolecular chemical cross-links. It has the attraction that the helical microfibril is defined by a single intermolecular stagger, that of 1*D*. This automatically produces the 4*D* stagger within the microfibril, and the *D*-period of the whole fibril is produced by an in-register (0*D*) interaction between microfibrils.

It is important to stress that available evidence does not permit discrimination of the detailed molecular structure within the microfibril but only of the symmetry and molecular topology described above. It is not possible to decide the exact relation between molecules in the gap or overlap regions nor to decide whether molecular packing differs between these two regions. The supercoil of the molecules may not be regular. Since the coherent length of the triple helical crystallites is only about 150 Å (Miller and Wray, 1971; Hosemann *et al.*, 1974) it is possible that the arrangement is more like the segmented rope structure once suggested for α-proteins (Fraser and MacRae, 1961) in which short (150 Å) lengths of straight collagen molecules are tilted about the microfibril axis. It is still not possible to decide the "hand" of the helix of molecular ends or of the helix followed by the coiled molecular axis.

Before embarking on criticisms of the heuristic model, I will mention a topic which at present is rather equivocal in its use for discriminating between models but which may yet provide some deciding considerations. The topic is the symmetry of the collagen molecule. Attention has been drawn to this by Segrest and Cunningham (1971, 1973). The point made by these authors is, first, that two parallel cylinders will touch along an edge parallel to the cylinder axis and that a helical α chain will contribute to this edge at intervals of the pitch of the helix, and, second, that the two identical $\alpha 1$ chains in the collagen molecule are related to each other by an azimuthal rotation of 108°. Hence if one $\alpha 1$ chain provides an interacting edge, the other $\alpha 1$ chain will provide an identical edge 108° around the molecule. Now 108° is the internal angle of a pentagon so this could relate the molecular symmetry to the symmetry of molecular packing in the overlap region of the microfibril (Segrest and Cunningham, 1973).

However, as these authors pointed out earlier (1971), the existence of such edges on a molecule could equally indicate a pleated-sheet arrangement of molecules.

It may also be important to note (Section II) that the X_n and Y_{n+1} amino acid residues lie at approximately the same azimuthal positions. Thus the nonglycine amino acids are confined to 10 equally spaced edges around the surface of the molecule, emphasizing the natural 10-fold symmetry of the molecule. As indicated in Section I, this would be the appropriate molecular symmetry if neighboring molecules in a sheet were rotated by $2\pi/5$ so as to form an array which repeated after 5 molecules.

Sakakibara *et al.* (1972) and Okuyama *et al.* (1972) have reported X-ray diffraction studies on crystals of $(\text{Gly-Pro-Pro})_{10}$. The crystals have a near-tetragonal lattice of side about 26.5 Å with four triple-helical molecules in the unit cell of space group $P2_12_12_1$. The distance between molecules is about 13.6 Å. The molecules have a dominant axial period of 2.86 Å. The unit cell has a true axial period of 102.6 Å with every fifth a pseudorepeat. The diffraction pattern is interpreted in terms of an α-chain helix pitch of 60 Å. If this is correct, the symmetry of $(\text{Gly-Pro-Pro})_{10}$ is significantly different from that of the collagen molecule (Section 2).

VI. *Criticisms of the Heuristic Model and Discussion of Other Models*

In this section, I will discuss some criticisms of the heuristic model and try to indicate what sort of future experiments might be critical in testing, refining, modifying, or replacing the model. In so doing consideration will be given to the advantages of other proposed models in the hope that comparisons may lead to a more generally accepted model.

The most serious drawback of the heuristic model is that, so far, it has not been possible to visualize the tetragonal lattice of microfibrils in electron micrographs of thin transverse sections of collagen fibrils. What we have to go on are reports of dotlike particles of stain separated by about 40 Å but not arranged in any detectable order. Longitudinal views of fibrils in electron micrographs certainly supplement this in that they show filaments of a similar spacing, but this lack of direct evidence for the nature of the lateral packing of the molecules is the main experimental *lacuna* at present, and it obviously affects any proposed model. A tetragonal lattice of rather larger dimensions has been observed by Knight and Hunt (1974) in transverse sections through the egg case of the dogfish. This is known from high-angle X-ray diffraction patterns to contain collagen and so is of great value in demonstrating that this sort of lattice can occur in collagenous structures.

A. Fibril Density

An outstanding difference between the two tetragonal models and the hexagonal model discussed in Section IV-C lies in the densities of molecular packing. Knowledge of the density of the collagen fibrils should, therefore, allow discrimination between the different models, and this approach is explicitly proposed by Katz and Li (1972, 1973, 1974). For comparison the volume occupied by a tripeptide in the various models is listed in Table 3 and, in addition, the volume occupied by a tripeptide in the (Gly-Pro-Pro)$_n$ structures that have been studied by X-ray diffraction. For the purpose of this calculation I have based the estimate for the Macfarlane hexagonal model on an interplanar distance of 12.6 Å, and the Katz and Li model on 13 Å. The effects of slightly different values of the interplanar distance can be readily estimated. Three methods are available for estimating the density of collagen fibrils. These are compari-

TABLE 3
Volume per Tripeptide for Proposed Collagen Models

	Volume for tripeptide (Å^3)		
	Overlap	Gap	Average
Collagen			
Heuristic model			
(38 × 38 Å)	840	1040	940
Tetragonal			
Hosemann and Nemetschek, 1973			
(38 × 35 Å)	380	480	430
Hexagonal models			
Katz and Li, 1973	570	710	640
Macfarlane, 1971	525	655	610
Polypeptides			
(Gly-Pro-Pro)$_{10}$			
(Sakakibara *et al.*, 1972)			505
(Gly-Pro-Pro)$_n$ dry films			
(Traub and Yonath, 1966)			390

son with molecular packing in crystals of $(Gly\text{-}Pro\text{-}Pro)_{10}$, direct measurement of density of collagen fibrils, and estimates of the intermolecular volume within a fibril.

First we note that the volume per tripeptide in $(Gly\text{-}Pro\text{-}Pro)_{10}$ is 505 Å^3 (Sakakibara *et al.*, 1972). If a similar density of packing is assumed to occur in collagen fibrils and allowance is made for the different side-chain volumes in collagen compared with $(Gly\text{-}Pro\text{-}Pro)_{10}$, this would provide a value of 615 Å as an average volume per tripeptide in collagen. The average side-chain volume in collagen is about 10% greater than in $(Gly\text{-}Pro\text{-}Pro)_{10}$, and a factor of 4.5/5.0 is required to allow for the gaps in collagen. This estimate of 615 Å for the volume per tripeptide in collagen is likely to be a minimum figure because of the compact nature of the proline side chains and because no allowance is made for the presence of other components of the collagen fibril such as sugar or water molecules.

The density of dry collagen has been obtained as 1.41 g/cm^3 (Pomeroy and Mitton, 1951) and that of wet tendon with the fibrils still not swollen apart as 1.16 g/cm^3 (Rougvie and Bear, 1953; Bear, 1956). This indicates that the percentage volume occupied by collagen within a fibril is 39%, assuming that all the swelling is taken up by the fibrils. The average exclusion volume of an amino acid residue in collagen is 110 Å^3. This means that the average volume per tripeptide is $3 \times 110 \times 100/39$ Å^3, i.e., 850 Å^3.

Finally, Katz and Li (1972, 1973) have carried out experiments designed to estimate the intermolecular volume in collagen fibrils. This is based on the use of radioactive probe molecules. If these molecules enter only the interfibrillar space, then they may be used to estimate the interfibrillar volume. By subtracting this from an estimate of the total intermolecular volume provided by the total water content of the fiber, a value of 1.14 ml/g of collagen is obtained for the intermolecular volume within the fibrils. The estimates were made on reconstituted collagen, and the authors point out that this intermolecular volume within a fibril is consistent with a hexagonally packed array of collagen molecules but is too small for a tetragonally packed array of five-stranded microfibrils.

It is evident that some of these estimates of fibril density are closest to that predicted by the hexagonal model for native tendon. The density value for the Hosemann–Nemetschek tetragonal model is greater than that of $(Gly\text{-}Pro\text{-}Pro)_{10}$ crystals and it is greater than the estimates from the other two methods. Hence, unless some very special packing occurs in collagen, this model seems unlikely. The heuristic model, on the other hand, is rather loosely packed with a density less than some of the estimated values but in good agreement with the measured values. No allowance has been made in the calculations for the effect of other molecules such as disaccharides, which are known to be covalently linked to collagen, or for the effect of the nontriple-helical telopeptides.

B. X-Ray Diffraction Patterns

As pointed out in Section IV-C, the medium-angle X-ray diffraction patterns from native collagen contain a set of sharp reflections in the near-equatorial region. The correct indexing of these reflections will lead to the lateral unit cell in which the collagen molecules are packed. However, in addition to the heuristic model, an alternative tetragonal cell has been proposed by Nemetschek and Hosemann (1973), and here we have shown that the observed reflections also occur close to the positions predicted by a hexagonal arrangement of molecules.

The positions of the true equatorial reflections in the X-ray diffraction pattern are of some importance since they contain information about the structure as projected down the fiber axis. These are observed at spacings corresponding to 26.5, 17.5, 12.6, and 9.6 Å (Wray, 1972; Miller and Parry, 1973). In the heuristic model the microfibril is a coiled-coiled-coil. Thus the near-equatorial region is made up of closely spaced layer lines, the Z values of which are determined by the pitch of the axis of the coiled triple-chain molecule. If these layer lines are sampled by row lines, the precise intensity at the intersection of the layer line and row line will depend on the orientation of the microfibril with respect to the crystal lattice vectors. This is difficult to predict in general and makes the coiled microfibril model difficult to test at this point. Models which account for the molecular tilt in terms of straight-tilted molecules make the prediction of the positions of reflections simpler. In this case the true equatorial reflections will lie on a straight line through the lattice, passing through the origin and perpendicular to the meridian. It has not yet been shown that the equatorial reflections do lie on such a line in any of the lattices proposed. Nemetschek and Hosemann (1973) propose such a line which, of course, defines the direction of molecular tilt with respect to the lattice. However, this is subject to the criticisms we made in section IV-C-1. Furthermore, some of the equatorial reflections, but not others, become off-equatorial with small changes in the condition of the tendons.

There is one feature of the X-ray diffraction pattern which at first sight might appear to favor straight-tilted rather than coiled molecules, but in order to explain this we must first describe yet another important point about the structure of native rat-tail tendon. This is based on the observation (Miller and Wray, 1971) that the row lines are inclined to the meridian by about 3°. The "meridional" reflections, therefore, lie along lines inclined to the true meridian by 2–3° (Nemetschek and Hosemann, 1973). Split meridional reflections can occur if fibers occur in two orientations at a small number of degrees to each other. This may be seen in low-angle X-ray diffraction patterns from fibers which have not been stretched sufficiently to remove the crimp described by Diamant *et al.* (1972). However, the tilted row lines observed by Miller and Wray (1971) could not have been produced by this effect. The fibers had been stretched so as to remove the crimp, and the inner and outer arms of the row line are quite different in appearance. Therefore, the tilted row lines must arise from a nonmonoclinic lattice. The vectors connecting equivalent sites across the fibril are, therefore, not precisely at right angles to the fibril axis. This is particularly clear in Figure 1b of Miller and Wray (1971). The inner arm of the row line samples the 9.5-Å layer line closer to the meridian and at a lower value of Z than the outer arm. This sort of effect could occur if straight-tilted molecules were inclined so that the molecular transform was tilted in a direction opposite to that tilt of the row lines.

In spite of this observation, further features of the X-ray pattern led Miller and Wray (1971) to conclude that the explanation was not as simple as this and to favor the coiled coil. However, the observation described above may indicate something like a segmented rope structure. Under certain conditions the tendon gives an X-ray diffraction pattern with row lines parallel to the meridian, but still with fanning of the intensity in the near-equatorial region. The lattice is, therefore, at least monoclinic, and the molecules are still tilted in this so-called orthomorphic state (Miller and Wray, 1971).

The explanation for the tilted row lines is still not clear, but since there does exist a three-dimensional crystal lattice, it seems most likely that this lattice will be sheared.

The effect of heavy-metal stain on the X-ray diffraction is commonly to enhance the intensity of the 38-Å row line (Miller and Wray, 1971; Hosemann and Nemetschek, 1973). Therefore, no matter what the lateral arrangement of collagen molecules is, 38 Å is the length of an important vector in it. This is consistent with the observation of a 40-Å-wide, D-periodic filament by electron microscopy (Doyle *et al.*, 1974*a*) and fits in well with both the heuristic model and that of Hosemann and Nemetschek. In the case of a hexagonal model it would be puzzling why the other period in the monoclinic cell ($\sqrt{3} \times 12.6$ Å) is not more frequently represented in the X-ray diffraction patterns from metal-stained specimens. If a tetragonal solution is incorrect, the dominance of the 38-Å row line would tend to suggest a molecular arrangement in which sheets of molecules were important and the intermolecular packing within the sheet was different from that between the sheets. It may be recalled that it is necessary to depart from a precise hexagonal array to explain the positions of the sharp reflections in the X-ray diffraction pattern (Section IV-C-2). The simplest relaxation of the exact hexagonal array would be to preserve equivalent relationships between molecules within a sheet and then allow the sheets to pack with different intermolecular contacts to those within the sheet. That it is the 38-Å row line which is dominant means that the sheet is not a linear array of molecules in close contact but is more likely to be a zig-zag sheet rather like the one Segrest and Cunningham (1971) showed to be a possible arrangement of helices with 10 residues in three turns (Section V). In this interpretation the crystallinity of a sheet is good, but the order between sheets is not so well developed. The D-periodic narrow filaments observed by Doyle *et al.* (1974*a*) could then consist of a raft, five molecules wide, packed in a zig-zag fashion.

When tendons are treated more extensively with solutions of heavy-metal stain, the effect on the X-ray diffraction is dramatic (Wray, 1972; Nemetschek and Hosemann, 1973). The series of sharp row lines disappear and are replaced by an intense column of intensity parallel to the meridian at an R value corresponding to about 115 $Å^{-1}$. The row line at 38 Å even under moderate heavy-metal staining remained sharp in a direction parallel to the equator, indicating that the fibril lattice was preserved. However, the intensity at an R value of 115 $Å^{-1}$ is not confined to a sharp row line but exists as layer line streaks. This means that the fibril lattice has broken down and been replaced by structures of width 100–200 Å.

Nemetschek and Hosemann (1973) and Hosemann *et al.* (1974) have drawn attention to the variation in intensity along the columns at $R = 115$ $Å^{-1}$ in a direction parallel to the meridian. Intensity maxima occur close to spacings that index as orders of $D/5$ or 135 Å. These authors point out that this intensity distribution results from the distribution of the heavy-metal stain along the fibril axis and refer to the narrow transverse bands seen in electron micrographs of positively stained fibrils. This is essentially the same topic as that discussed in Section III-B namely, the one-dimensional structure of collagen fibrils and, indeed, the intensity distribution along the 115-Å column closely follows the meridional intensities from a similarly stained material. Nemetschek and Hosemann (1973) and Hosemann *et al.* (1974) suggest that the 135-Å pseudoperiod may be related to the coherent scattering length of the triple-helical collagen molecule which is in the range 100–200 Å (Miller and Wray, 1971; Nemetschek and Hosemann, 1973). Assuming that the heavy-metal stain enters amorphous regions in the molecule, it is concluded that the fibril consists of segments spaced axially 135 Å apart. These segments consist of crystalline regions of about 115 Å and the amorphous regions of about 20 Å into which the heavy-metal stain penetrates.

This is an attractive idea, and the short, coherent length of the collagen molecule must be taken into account in the complete one-dimensional solution. The regions of the molecule taking up heavy metals under conditions of positive staining cannot be all amorphous since these regions are too close together (von der Mark, 1970; Doyle *et al.*, 1974*b*) and the optical diffraction pattern of the positively stained bands shows enhancement of the sixth rather than the fifth order of the *D* period (Doyle *et al.*, 1974*b*). However, the effect of negative staining is not completely clear. The limited coherent length could be related to the length of segments in a segmented rope.

C. Electron Microscopy

There is, unfortunately, no evidence from electron microscopy concerning the crystalline lattice in the collagen fibrils. Observations of the dotlike particles of stain have been reported from thin transverse sections through fibrils, but these particles are not evidently on a regular lattice. Perhaps further analysis of the filamentous structures seen in longitudinal views of fibrils is the most promising means of investigating the nature of the lateral molecular packing, but it will be recalled that X-ray diffraction studies showed that the regularity of the lattice was highly sensitive to the water content. Knight and Hunt (1974) discovered a well-developed tetragonal lattice in the collagen from the egg case of dogfish. Since the longitudinal period (370 Å) and banding pattern are different from those in tendon, it is not clear how the two molecular packing arrangements are related. However, this discovery shows that molecules with the collagen conformation can pack in a tetragonal cell. Höhling *et al.* (1974) and Katz and Li (1974) find evidence from electron microscopy of mineralized tissue consistent with tetragonally and hexagonally packed models, respectively.

A different type of observation concerning the molecular packing in collagen fibrils is the twisted or helical appearance that, under certain conditions, fibrils show in the electron microscope. This has been reported recently by Bouteille and Pease (1971) who studied inertly dehydrated, unfixed collagen from rabbit aorta, and by Rayns (1974) from examination of collagen from freeze-fractured glycerinated beef heart. In these studies the whole fibril appears to be a multistrand helix of pitch 700–800 Å. Rayns (1974) found the helix was consistently right-handed and composed of filaments about 70 Å in diameter. It is difficult at present to relate these observations to the molecular packing in native tendon where the collagen molecules do not make an angle greater than 4° with the fibril axis as estimated by X-ray diffraction. The helix could be noncollagenous material on the fibril surface, or it could be due to a structural rearrangement which occurs during preparation for electron microscopy or, less likely, it might represent a different type of molecular packing of collagen than in tendon.

Some, but not all, of the low-angle meridional reflections from native rat-tail tendon are streaked out parallel to the equator. The occurrence of these streaks is not obviously related to the intensity of the corresponding meridional reflection, and they do not occur in the diffraction patterns from collagen treated with ethanol–ether mixtures; the meridional reflections are unaltered after such treatment (Miller and Wray, unpublished results). This suggests that noncollagenous material is arranged axially with the D period but in a structure of fairly narrow width as indicated by the streaks. The streaks could be caused by a layer of noncollagenous material arranged with an axial period of D on the surface of the collagen fibrils.

VII. *Conclusions*

The last few years have witnessed a rapid increase in the number of studies on the three-dimensional arrangement of molecules in collagen fibrils. Considerable progress has been made and well-established findings can be summarized.

1. A collagen fibril is a single crystal and thus the molecules are arranged on a regular three-dimensional lattice.
2. The collagen molecules are not parallel to the fibril axis but tilted by about 4°.
3. In native tendon at least one lattice vector is not precisely at right angles to the fibril axis but inclined, at about 3°, to a plane perpendicular to the fibril axis.
4. A vector of length 38 Å is important in the lateral arrangement of molecules, i.e., the structure, in the plane at (90 − 3)° to the fibril axis.
5. There is chemical evidence that collagen molecules, shifted axially by $4D$, are covalently linked in the fibril.
6. There is chemical and electron-microscopic evidence that in the fibril molecules are related by axial shifts of $0D$, $1D$, and $4D$.

7. When projected on to the fibril axis (more precisely along a line at 87° to the fibril axis), the molecular arrangement has a periodicity of *D* (670 Å).

The complete three-dimensional molecular arrangement has not yet been determined. Three main models have been proposed, and in this chapter my aim has been to point up the differences between them in order to assist in the discovery of a more firmly established model. A heuristic model has been described which is preferred here since it appears to be consistent with a wide range of types of evidence. This model consists of five-strand microfibrils, supercoiled and packed face to face on a tetragonal lattice. Some possible difficulties with absent low-angle near-equatorial reflections in the X-ray diffraction pattern and a rather low density of molecular packing are criticisms of this model at present. A second model based on a closely similar near-tetragonal cell has been proposed, but this cell contains an octofibril rather than a five-strand microfibril. Criticisms of this model are that it has a high density of molecular packing, that the crystallographic justification for the special type of disorder required is still not explicit, and that so far it has not been related to the triple-helical molecular conformation as have the other models. A third model is based on a hexagonal array of collagen molecules. The specific proposals which have been made would require modification to fit the X-ray diffraction patterns from either native or heavy-metal-stained tendons and to fit the observed axial intermolecular shifts. It is hoped that dialectic based on these three models will result in a fruitful synthesis.

The demonstration that the origins of the axial intermolecular stagger lies in the amino acid sequence of the collagen molecule provides an understanding of the self-assembly of the fibril and gives a strong hint of the basis of molecular specificity in other systems such as muscle. This success opens the way for analysis of the origins of the various polymorphic forms of collagen (Doyle *et al.*, 1975*a*) and encourages attempts to extend this approach to three dimensions. It now seems likely that in the near future we shall have, for collagen, an understanding of the physical and chemical interactions involved at each level from molecule to tissue. Apart from its considerable academic interest, this would be of immense value in clarifying our understanding of the connective-tissue diseases.

Note Added in Proof

Some relevant work has been done in the last year or so. The amino acid sequence of collagen has been analyzed further, and it appears that there may be some significance in the fact that the oppositely ionizable amino acids tend to occur on the molecule in pairs separated by no more than 2 amino acid residues (Doyle *et al.*, 1975*b*). Interaction curves were calculated on the assumption that these pairs could interact as dipoles and this produced a dominant maximum at an intermolecular stagger of $1D$, but not at the other integral multiples of D. Piez and Torchia (1975) agreed that the ion pairs were significant for intermolecular interactions, and suggested a specific structural mechanism for the interaction. It is clear that further progress on the analysis of the amino acid sequence will involve three-dimensional molecular packing. That this pairing or non-pairing is structurally significant is further supported by studies on the polymorphic forms of collagen (Doyle *et al.*, 1975*a*). The FLS series has been analyzed and shown to be explicable as different combinations of only four kinds of intermolecular overlap. These four overlaps also are the positions at which the unpaired positively ionizable amino acids are closest to each other in the interacting molecules. The muccopolysaccharides which precipitate collagen in the FLS forms are long, flexible, negatively charged molecules, and it seems likely that they act by bridging the unpaired positively charged amino acids in collagen.

Some progress has been made toward the complete solution of the one-dimensional structure described in Section IIIB. The neutron diffraction pattern shows a set of intense meridional reflections (White *et al.* 1976). When the intensities of the first 22 orders are measured from rat-tail tendon in D_2O, the observed values agree well with the values calculated by Fourier-inverting the molecular scattering profile of the Hodge–Petruska one-dimensional structure based on the known amino acid sequence of the collagen $\alpha 1$ chain. The clustering of the polar and apolar amino acids into periods of $2D/11$, $D/6$, and $D/5$ dominates the intensities of the neutron reflections D_2O. The difficulties of obtaining a good fit with the X-ray reflections (see Doyle *et al.*, 1974*a*) were paralleled by similar difficulties with the neutron diffraction pattern for collagen in H_2O and can now be recognized as due to the large contribution of the telopeptides to the latter patterns (Doyle *et al.*, 1976; Hulmes, 1975). Thus a complete understanding of the one-dimensional structure to amino acid resolution is now available.

Further models have been proposed for the molecular packing. Woodhead-Galloway *et al.* (1975) accept the tetragonal indexing with slight modification but suggest that two-stranded microfibrils (Burge, 1965) are supercoiled and arranged in the tetragonal cell. It has the disadvantage that the density of molecular packing suggested for native collagen (which contains a considerable amount of water) exceeds that of dry fibers of $(\text{Gly-Pro-Pro})_n$ and of crystals of $(\text{Gly-Pro-Pro})_{10}$. The symmetry of this model is not defined although the microfibrils are said to have a longitudinal repeat of $5D$. It is proposed that the microfibrils are supercoiled in opposite directions and close-packed; in the explanation of the equatorial X-ray reflections no account is taken of the fact that microfibrils coiled with different hands will be different in projection along the microfibril axis.

Veis and Yuan (1975) have proposed a model closely similar to that of Fraser *et al.* (1974) in that it is based on the five-strand microfibril packed in a tetragonal lattice of side 38 Å, with a superlattice due to the microfibril orientations. This model goes further since it contains details of the relation between the structures of gap and overlap regions. No doubt it will be possible to test such models when relevant experimental observations become available.

A new method has recently been developed for the digital processing of fiber diffraction patterns (Fraser *et al.*, 1976). This involves the use of a recording microdensitometer to scan an X-ray diffraction photograph and to store the intensity data as an array in a computer mass storage memory. The observed intensities can then be corrected on a point-by-point basis and the intensity array on the photograph converted to an intensity array in undistorted reciprocal space. This is, of course, the kind of correction made routinely to X-ray crystallographic data. In the case of fiber diffraction data, where the intensity distribution can be continuous in some parts but confined to Bragg reflections in other parts, the corrections are more complex. This new approach is an attempt to put the fiber diffraction data on a secure quantitative basis.

The method of Fraser *et al.* (1976) has been applied by Fraser, Macrae, and Miller (unpublished) to the X-ray diffraction patterns from collagen (such as those in Figures 5 and 6). An interesting preliminary result concerns the distribution of sharp reflections on the 9.5 $Å^{-1}$-layer line in the X-ray diffraction pattern from native collagen. When converted to undistorted reciprocal space, these reflections lie in R-positions corresponding to a tetragonal cell 38 × 38 Å sheared in the z direction. The splitting of the row lines indicates that this shearing preserves the equivalence of the equal sides (38 Å) of the tetragonal cell. This is particularly obvious in the case of the 38 Å row line, which splits into a single pair of row lines in the rotation diffraction pattern. This rotation pattern is produced because the collagen "crystals" are the fibrils which, while parallel to each other, do not appear to have any preferred azimuthal orientation.

ACKNOWLEDGMENTS

It is a pleasure to acknowledge that this chapter has been greatly influenced by discussion with my colleagues in this laboratory, Barbara

Doyle, Bruce Fraser, David Hukins, David Hulmes, David Parry, Karl Piez, Stephen White, John Woodhead-Galloway, and John Wray.

References

Bear, R. S., 1942, Long X-ray diffraction spacings of collagen, *J. Am. Chem. Soc.* **64**:727.

Bear, R. S., 1952, The structure of collagen fibrils, *Adv. Protein Chem.* **7**:69.

Bear, R. S., 1956, The structure of collagen molecules and fibrils, *J. Biophys. Biochem. Cytol.* **2**:363.

Bear, R. S., and Morgan, R. S., 1957, *in: Connective Tissue* (R. E. Turnbridge, ed.), p. 321, Blackwell, Oxford.

Boedetker, H., and Doty, P., 1956, The native and denatured states of soluble collagen, *J. Am. Chem. Soc.* **78**:4267.

Bouteille, M., and Pease, D. C., 1971, The tridimensional structure of native collagenous fibrils, their proteinaceous filaments, *J. Ultrastruct. Res.* **35**:314.

Bruns, R. R., and Gross, J., 1973, *Biochemistry* **12**:808.

Bruns, R. R., Trelstad, R. L., and Gross, J., 1973, Cartilage collagen: A staggered substructure in reconstituted fibrils, *Science* **181**:269.

Burge, R. E., 1965, *Structure and Function of Connective and Skeletal Tissue* (S. Fitton Jackson *et al.*, eds.), p. 2, Butterworth, London.

Burge, R. E., and Randall, J. T., 1955, The equivalence of electron microscopic and X-ray observations on collagen fibres, *Proc. R. Soc. A* **233**:1.

Caspar, D. L. D., and Klug, A., 1962, Physical principles in the construction of regular viruses, *Cold Spring Harbor Symp. Quant. Biol.* **27**:1.

Chandross, R. J., and Bear, R. S., 1973, Improved profiles of electron density distribution along collagen fibrils, *Biophys. J.* **13**:1030.

Chapman, J. A., 1974, *Connect. Tissue Res.* **2**:137.

Chapman, J. A., and Hardcastle, 1974, *Connect. Tissue Res.* **2**:151–159.

Chapman, J. A., and Steven, F. S., 1966, *in: Biochimie et Physiologie du Tissu Conjontif* (P. Compte, ed.), p. 65, Lyon.

Crick, F. H. C., 1966, *in: Principles of Biomolecular Organisation* ed. (G. E. W. Wolstenholm and M. O'Connor, eds.), pp. 131–132, Churchill, London.

Diamant, J., Keller, A., Baer, E., Litt, M., and Arridge, R. G. C., 1972, Collagen; Ultrastructure and its relation to mechanical properties as a function of ageing, *Proc. R. Soc. B* **180**:293.

Doyle, B. B., Hulmes, D. J. S., Miller, A., Parry, D. A. D., Piez, K. A., and Woodhead-Galloway, J., 1974*a*, A D-periodic narrow filament in collagen, *Proc. R. Soc. B* **186**:67.

Doyle, B. B., Hulmes, D. J. S., Miller, A., Parry, D. A. D., Piez, K. A., and Woodhead-Galloway, J., 1974*b*, Axially projected collagen structures, *Proc. R. Soc. B* **187**:37.

Doyle, B. B., Hukins, D. W. L., Hulmes, D. J. S., Miller, A., and Woodhead-Galloway, J., 1975*a*, Collagen polymorphism: Its origins in the amino-acid sequence, *J. Mol. Biol.* **91**:79.

Doyle, B. B., Hukins, D. W. L., Hulmes, D. J. S., Miller, A., Rattew, C. J., and Woodhead-Galloway, J., 1975*b*, *Biochem. Biophys. Res. Comm.* **60**:858.

Doyle, B. B., Haas, J., Hulmes, D. J. S., Jenkins, G., Miller, A., Timmins, P., and White, J. W., 1976, *Proc. Brookhaven Symposium* (in press).

Ellis, D. O., and McGavin, S., 1970, The structure of collagen—an X-ray study, *J. Ultrastruct. Res.* **32**:191.

Ericson, L. G., and Tomlin, S. G., 1959, Further studies of low-angle X-ray diffraction patterns of collagen, *Proc. R. Soc. A* **252**:197.

Fraser, R. D. B., and MacRae, T. P., 1961, The α-configuration of fibrous proteins, *Nature* **189**:572.

Fraser, R. D. B., MacRae, T. P., Rodgers, G. E., and Filshie, B. K., 1963, Lipids in keratinised tissue, *J. Mol. Biol.* **7**:90.

Fraser, R. D. B., Miller, A., and Parry, D. A. D., 1974, Packing of microfibrils in collagen, *J. Mol. Biol.* **83**:281.

Fraser, R. D. B., Macrae, T. P., Miller, A., and Rowlands, R. J., 1976, Digital processing of fiber diffraction patterns. *J. Appl. Cryst.* **9**:81.

Gross, J., and Schmitt, F. O., 1948, Further progress in the electron microscopy of collagen. *J. Am. Leather Chem. Assoc.* **43**:658.

Gross, J., Highberger, J. H., and Schmidt, F. O., 1954, Collagen structures considered as states of aggregation of a kinetic unit. The tropocollagen particle. *Proc. Natl. Acad. Sci. U.S.A.* **40**:679.

Hall, C. E., 1956, Visualisation of individual macromolecules with the electron microscope, *Proc. Natl. Acad. Sci. U.S.A.* **42**:801.

Hall, C. E., Jakus, M. A., and Schmitt, F. O., 1942, Electron microscope observations of collagen, *J. Am. Chem. Soc.* **64**:1234.

Haydon, G. B., 1968, On the interpretation of high resolution electron micrographs of macromolecules, *J. Ultrastruct. Res.* **25**:349.

Haydon, G. B., 1969, Electron phase and amplitude images of stained biological thin sections, *J. Microsc.* **89**:73.

Hodge, A. J., and Petruska, J. A., 1963, Recent studies with the electron microscope on ordered aggregates of the tropocollagen molecule, *in: Aspects of Protein Structure* (G. N. Ramachandran, ed.), p. 289, Academic Press, London.

Hodge, A. J., and Schmitt, F. O., 1960, The charge profile of the tropocollagen macromolecule and packing arrangement in native type collagen fibrils, *Proc. Natl. Acad. Sci. U.S.A.* **46**:186.

Höhling, H. J., Ashton, B. A., and Köster, H. D., 1974, *Cell Tissue Res.* **148**:11.

Hosemann, R., 1973, Paracrystals in biopolymers and synthetic polypeptides, *Endeavour* **32**:99.

Hosemann, R., and Nemetschek, T., 1973, Reaktions Abäufe zwischen Phosphorwolframsäure und Kollagen, *Kolloid-Z.Z. Polym.* **351**:53.

Hosemann, R., Dreissig, W., and Nemetschek, T., 1974, Schachtelhalmstructure of the octafibrils in collagen, *J. Mol. Biol.* **83**:275.

Hulmes, D. J. S., 1975, D.Phil. Thesis, Oxford University.

Hulmes, D. J. S., Miller, A., Parry, D. A. D., Piez, K. A., and Woodhead-Galloway, J., 1973, Analysis of the primary structure of collagen for the origins of molecular packing, *J. Mol. Biol.* **79**:137.

Kaesberg, P., and Shurman, M., 1953, Further evidence concerning the periodic structure in collagen, *Biochem. Biophys. Acta* **11**:1.

Kang, A. H., 1972, Studies on the location of intermolecular cross-links in collagen, *Biochemistry* **11**:1828.

Katz, E. P., and Li, S. T., 1972, The molecular packing of collagen in mineralised and non-mineralised tissues, *Biochem. Biophys. Res. Commun.* **46**:1368.

Katz, E. P., and Li, S. T., 1973, The intermolecular space of reconstituted collagen fibrils, *J. Mol. Biol.* **73**:351.

Katz, E. P., and Li, S. T., 1974, Structure and function of bone collagen fibrils, *J. Mol. Biol.* **80**:1.

Knight, D. P., and Hunt, S., 1974, Fibril structure of collagen in egg capsule of dogfish, *Nature (London)* **249:**380.

Kuhn, K., 1969, The structure of collagen, *Essays Biochem.* **5:**59.

Kuhn, K., and Zimmer, E., 1961, Eigenschaften des Tropocollagen-Moleküls und deren Bedeutung für die Fibrillenbildung, *Z. Naturforsch.* **16:**648.

Kuhn, K., Fietzek, P., and Kuhn, J., 1966, The action of proteolytic enzymes on collagen, *Biochem. Z.* **344:**418.

Kuhn, K., Kuhn, J., and Schuppler, G., 1964, Kollagenfibrillen mit anormalen Querstreifungsmuster, *Naturwissenschaften* **51:**337.

Levine, Y. K., and Wilkins, M. H. F., 1971, Structure of oriented lipid bilayers, *Nature (London) New Biol.* **230:**69.

Macfarlane, E. F., 1971, Molecular packing structure of collagen, *Search* **2:**171.

Miller, A., and Parry, D. A. D., 1973, Structure and packing of microfibrils in collagen, *J. Mol. Biol.* **75:**441.

Miller, A., and Wray, J. S., 1971, Molecular packing in collagen, *Nature (London)* **230:**437.

Millward, G. R., 1970, The substructure of α-keratin microfibrils, *J. Ultrastruct. Res.* **31:**349.

Nemetschek, T., and Hosemann, R., 1973, A kink model of native collagen, *Kolloid-Z.Z. Polym.* **251:**1044.

Nemetschek, T., Grassmann, W., and Hofmann, U., 1955, Über die hochunterteilte Querstreifung des Kollagens, *Z. Naturforsch B.* **10:**61.

North, A. C. T., Cowan, P. M., and Randall, J. T., 1954, Structural units in collagen fibrils, *Nature* **174:**1142.

Okuyama, K., Tanaka, N., Ashida, T., Kakudo, M., Sakakibara, S., and Kishida, Y., 1972, An x-ray study of the synthetic polypeptide $(\text{Pro-Pro-Gly})_{10}$, *J. Mol. Biol.* **72:**571.

Olsen, B. R., 1963, Electron microscope studies on collagen. I: Native collagen fibrils. *Z. Zellforsch.* **59:**199; Electron microscope studies on collagen. II: Mechanism of linear polymerization of tropocollagen molecules, **59:**184.

Piez, K. A., and Torchia, D., 1975, *Nature* **258:**87.

Pollard, H., Miller, A., and Cox, C., 1973, Synaptic vesicles: Structure of chromaffin granule membranes, *J. Supramol. Struct.* **1:**295.

Pomeroy, C. D., and Mitton, R. J., 1951, The real densities of chrome and vegetable-tanned leathers, *J. Soc. Leather Trades Chem.* **35:**360.

Rayns, D. G., 1974, Collagen from frozen fractured glycerinated beef heart, *J. Ultrastruct. Res.* **48:**59.

Rougvie, M. A., and Bear, R. S., 1953, An X-ray diffraction investigation of swelling by collagen, *J. Am. Leather Chem. Assoc.* **48:**735.

Sakakibara, S., Kishida, Y., Okuyama, K., Tanaka, N., Ahsida, T., and Kakudo, M., 1972, Single crystals of $(\text{Pro-Pro-Gly})_{10}$, a synthetic polypeptide model of collagen, *J. Mol. Biol.* **65:**371.

Schmitt, F. O., 1956, Macromolecular interaction patterns in biological systems, *Proc. Am. Phil. Soc.* **100:**476.

Schmitt, F. O., Gross, J., and Highberger, J. H., 1955, Tropocollagen and the properties of fibrous collagen, *Exp. Cell. Res. (Suppl.)* **3:**326.

Schmitt, F. O., Hall, C. E., and Jakus, M. A., 1942, Electron microscope investigations of the structure of collagen, *J. Cell. Comp. Physiol.* **20:**11.

Segrest, J. P., and Cunningham, L. W., 1971, Molecular basis for fibrillar aggregation of tropocollagen, *Nature (London), New Biol.* **234:**26.

Segrest, J. P., and Cunningham, L. W., 1973, Unit fibril models derived from the molecular topography of collagen, *Biopolymers* **12:**825.

Smith, J. W., 1968, Molecular pattern in native collagen, *Nature* **219:**157.
Smith, J. W., and Frame, J., 1969, Observations of the collagen and protein–polysaccaride complex of rabbit corneal stroma, *J. Cell Sci.* **4:**421.
Steven, F. S., 1970, Isolation and characterisation of polymeric collagen from complex connective tissue, *in: Chemistry and Molecular Biology of the Intercellular Matrix* (E. A. Balazs, ed.), p. 43, Academic Press, New York.
Thon, F., 1966*a*, Zur Defokussierungsabhängigkeit des Phasenkontrastes bei der elektronenmikroskopischen Abbildung, *Z. Naturforsch.* **21:**476.
Thon, F., 1966*b*, Imaging properties of the electron microscope near the theoretical limit of resolution, *in: Electron Microscopy 1966* (R. Uycota, ed.), Vol. 1, p. 23, Maruzen, Tokyo.
Tomlin, S. G., 1955, The structure of collagen fibres, *Proc. Int. Wool Textile Res. Conf. B*, p. 187, Melbourne, Australia.
Tomlin, S. G., and Worthington, C. R., 1956, Low-angle X-ray diffraction patterns of collagen, *Proc. R. Soc. A* **235:**189.
Traub, W., and Yonath, A., 1966. Polymers of tripeptides as collagen models I. X-ray studies of poly(L-prolyl-glycyl-L-proline) and related polypeptides, *J. Mol. Biol.* **16:**404.
Tromans, W. J., Horne, R. W., Gresham, G. A., and Bailey, A. J., 1963, Electron microscope studies on the structure of collagen fibrils by negative staining, *Z. Zellforsch.* **58:**798.
Veis, A., and Yuan, L., 1975. *Biopolymers* **14:**895.
Veis, A., Bhatnagar, R. S., Shuttleworth, C. A., and Mussell, S., 1970, The solubilization of mature, polymeric collagen fibrils by lyotropic relaxation, *Biochim. Biophys. Acta.* **200:**97.
von der Mark, K., Wendt, P., Rexrodt, F., and Kuhn, K., 1970, Direct evidence for a correlation between amino-acid sequence and cross striation pattern of collagen, *FEBS Lett.* **11:**105.
White, J. W., Miller, A., and Ibel, K., 1976, Neutron diffraction by collagen. *J. Chem. Soc. Faraday Trans. II* **72:**435.
Wolpers, C., 1943, Kollagenquerstreifung und Grundsubstanz. Klin. Wochenschr. **22:**624.
Woodhead-Galloway, J., Hukins, D. W. L., and Wray, J. S., 1975, *Biochem. Biophys. Res. Comm.* **64:**1237.
Wray, J. S., 1972, D. Phil. thesis, Oxford University.
Yonath, A., and Traub, W., 1969, Polymers of tripeptides as collagen models IV. Structure analysis of poly(L-prolyl-glycyl-L-proline), *J. Mol. Biol.* **43:**461.
Zimmerman, B. K., Pikkarainen, J., Fietzek, P. P., and Kuhn, K., 1970, Cross-linkages in collagen, *Eur. J. Biochem.* **16:**217.

4
Cross-Linking

MARVIN LAWRENCE TANZER

I. *Introduction*

The explosion of information concerning collagen biosynthesis, structure, and function has been amply documented by many recent review articles which have become increasingly specialized in their scope (Gallop *et al.*, 1972; Grant and Prockop, 1972; Traub and Piez, 1971; Davison, 1973). The rapid developments in this field can be attributed both to the generous support to science in the past 20 years, primarily by the United States government, and to the establishment of many active laboratories by young investigators who were trained during the era of rapid expansion of the health sciences. This amplification of collagen research is now yielding a number of important insights into the biology of the protein, and the phenomenon of cross-linking constitutes one of the areas in which substantial progress has been made.

Historically, cross-linking within proteins has largely been a consequence of the formation of disulfide bonds between specific cysteine residues. It is generally thought that correct folding of a newly synthesized protein places the reactive sulfhydryl groups in suitable proximity for disulfide formation to ensue. This scheme is subject to some reservations because multichain molecules such as insulin do not behave in this manner although the single-chain biosynthetic precursor, proinsulin, does follow the general rule. Disulfide cross-links have not been of importance in collagen biochemistry because most vertebrate collagens do not contain cysteine. However, the picture is now changing because disulfide bonds are present in *Ascaris* collagen, in basement membrane collagens, in type

MARVIN LAWRENCE TANZER · Department of Biochemistry, University of Connecticut Health Center, Farmington, Connecticut 06032.

III collagen, and in the procollagen precursors of many vertebrate collagens. In each of these cases, the disulfides have been shown to be important in the structural integrity of the protein.

The other major interest in protein cross-linking has been the introduction of synthetic cross-links into specific sites of proteins. The most versatile method is to use bifunctional reagents, containing two reactive groups, which will form bridges between two amino acid side chains. The general classes of such bifunctional reagents include *N*-substituted maleimides, alkyl and aryl halides, isocyanates, acylating compounds, imidoesters, and aldehydes (Wold, 1972; Slobin, 1972). The availability of quite selective agents such as the imidoesters, which are soluble in water and which react under mild conditions, has kindled interest in probing the three-dimensional relationships of proteins in complex structures such as membranes and ribosomes. This level of complexity is formally analogous to that encountered in the packing of collagen molecules into fibrils; probing the spatial relationships by means of specific cross-links is feasible in all of these instances. In the case of collagen, a high degree of specificity can be obtained. This specificity is a consequence of the facts that collagen *spontaneously* undergoes cross-linking in the fibrillar form and that the general chemical nature of the natural cross-links has been elucidated.

The cross-linking of collagen is conceptually similar to cysteine cross-linking in that intrinsic groups of the protein form bridges. The detection and isolation of the collagen cross-links has relied upon employing selective chemical reagents and, in a sense, this approach is analogous to the introduction of exogenous cross-links because, up to the present time, the major way in which the cross-linked amino acids have been isolated has required prior chemical treatment of the protein. Consequently, the ultimate form of these cross-links is not known, and one is limited to discussing the information obtained by the methods of chemical pretreatment. Such knowledge has, however, provided a basis for indirect studies in which presumptive "final" cross-links have been detected.

The major insight into the general nature of collagen cross-linking was provided by selective inhibitors, the lathyrogens. These compounds effectively prevent collagen and elastin cross-linking and, consequently, alter the tensile properties of those tissues which contain these two proteins. The story of lathyrism has been well reviewed in recent years (Levene, 1973; Tanzer, 1965) and suffice to say, the lathyritic phenomenon is now a classic example of specific inhibition of a biochemical pathway. Although the mechanistic details remain to be elucidated, one can summarize the phenomenon in the following way: collagen molecules serve as substrate for the enzyme, lysyl oxidase, which converts certain

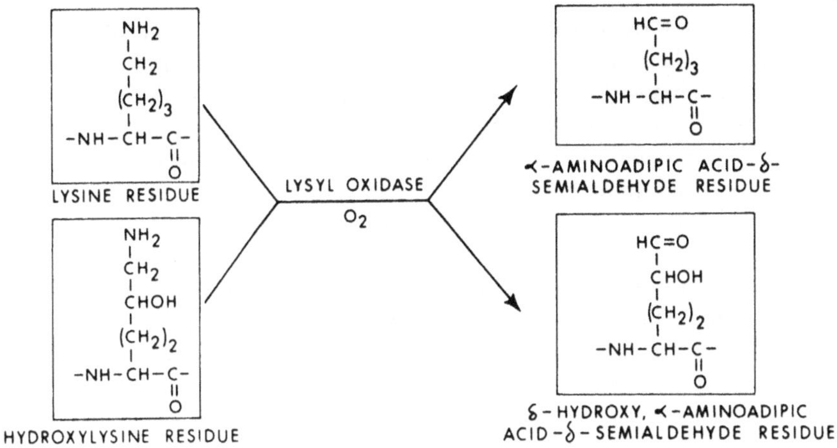

FIGURE 1

lysine and hydroxylysine ε-amino groups into aldehydes (Figure 1). (See Note Added in Proof at the end of this chapter.) The lathyrogens, such as β-aminopropionitrile, specifically inhibit lysyl oxidase by irreversibly combining with the enzyme (Narayanan *et al.*, 1972). Consequently, the collagen molecules do not undergo enzymatic modification, which is a prerequisite for cross-linking. In the normal situation, collagen molecules, which contain the aldehydic groups, self-assemble into fibers which then become cross-linked because of reactions that occur between the aldehydic groups and other amino acids of adjoining molecules. When lysyl oxidase is inhibited by β-aminopropionitrile, the unmodified collagen molecules still self-assemble, but cross-linking does not occur because aldehydes are not present in the collagen molecules.

The realization that endogenous aldehydes in collagen participate in cross-linking the natural fibers formed by the protein has stimulated interest in elucidating the chemical nature, specific locations, and biologic distribution of the cross-links. Considerable information has already accrued in these areas and will be the subject of the ensuing discussion.

II. *Chemistry of Cross-Links*

Once it became established that aldehydes were intrinsic components of purified collagen and that the absence of such aldehydes interfered with normal cross-linking, the next logical step was to show which chemical reactions occurred during cross-linking. The range of reactions was necessarily restricted because there are a limited number of amino acid side chains which could react with aldehydes and, *a priori*, it seemed most likely that ϵ-NH_2 groups were the best candidate. α-NH_2 groups could be eliminated as participants because most collagens contain amino-terminal pyroglutamate (Gallop *et al.*, 1972; Traub and Piez, 1971).

The experimental approach independently chosen by several investigators seeking to identify the cross-links was that of chemical reduction, using sodium borohydride (Bailey, 1968; Tanzer, 1967; Blumenfeld and Gallop, 1966). This reagent has the advantage of being quite selective and relatively mild in its reactivity and can be obtained in radioactive form. The latter property enabled one to label all reducible compounds with tritium, thereby providing a very convenient tag for subsequent isolation of the reduced substances. Unfortunately, under certain circumstances, some unexpected reactions do occur with sodium borohydride, but this was recognized and the problem identified (Paz *et al.*, 1969, 1970). The studies which have since evolved seem to be based on very specific and selective chemical reductions, but it is important to be alert for any unanticipated modifications which might occur.

The early studies of the effects of borohydride reduction demonstrated that both physical and chemical properties of tendons and reconstituted collagen fibers were markedly changed by exposure to sodium borohydride. The tensile strength of tendons was significantly increased by prior reduction with $NaBH_4$, and the reduction also made the tendons resistant to the disruptive effects of compounds such as cysteine and penicillamine (Bailey, 1968). Reconstituted collagen fibers of the native type became very insoluble when exposed to $NaBH_4$ and also underwent much greater thermal shrinkage than control fibers (Tanzer, 1968). Other studies showed that borohydride-treated skin yielded far less extractable collagen than normal skin and that the maximal tension which developed upon heating the treated skin was about threefold greater than the control skin (Balian *et al.*, 1969). The subunit composition of solutions of denatured collagen, which had been pretreated with $NaBH_4$ while in the native fibril form, showed a marked difference compared to both normal reconstituted fibrils and to denatured collagen solutions which had been pretreated with $NaBH_4$. The difference was that, compared to the controls, the treated native fibrils showed about a 30% diminution of monomer and dimer components and an equivalent increase in a heterogeneous population of larger-molecular-weight components (Tanzer, 1968). The specificity of the borohydride effect upon the reconstituted fibrils was related both to correct molecular packing in the fibrils and to the availability of aldehyde and ϵ-NH_2 groups.

Incorporation of tritium into collagen from $NaB[^3H]_4$ was shown to range from 0.5 to 9 moles, depending upon the origin of the particular collagen preparation (Tanzer, 1968; Paz *et al.*, 1969; Deshmukh and Nimni, 1971). Following hydrolysis of the protein, a number of radioactive components were detected, and the appearance of specific components could be related to: (1) the conditions of hydrolysis, especially acid *vs.* alkaline methods (Paz *et al.*, 1969; Davison *et al.*, 1972; Franzblau *et al.*, 1970); (2) the method of collagen extraction (Deshmukh and Nimni, 1969, 1971); (3) the physical state of the collagen during $NaB[^3H]_4$ reduction, especially fibrils *vs.* solution, and also the type of fibrils (Tanzer and Mechanic, 1968; Kang *et al.*, 1970); and (4) the denatured *vs.* native form of intact tendons (Bailey and Lister, 1968). Such studies enabled one to more closely focus upon those components which might be cross-links, rather than attempt to examine all of the radioactive substances.

Thus, the scene was set for isolating and characterizing the presumptive cross-links, and several laboratories have developed this area in the past six years. The methods used for isolation have largely been those of ion-exchange chromatography, progressing from columns of limited resolving power to more highly resolving systems such as those employed in contemporary amino acid analyzers. In several cases, the latter technique has provided detection of isomeric forms of the cross-links, particularly those which contain a hydroxylysyl moiety as a structural component. Characterization of the molecular structure of the isolated cross-links also relied extensively upon contemporary methods of organic analysis, most notably nuclear magnetic resonance spectroscopy and high- and low-resolution mass spectrometry. In certain instances, the more classical tools of elemental analysis, colorimetric analyses, electrophoresis, titration, osmometry, and periodate degradation have been employed. The results obtained by these latter methods have usually provided supporting evidence for the proposed structures. Finally, in the case of the less complex cross-links, proof of structure has been completed by organic synthesis of the postulated compound followed by the demonstration of identity of the natural and synthetic substances. Another important area of knowledge which also contributed to these studies, in a general way, was the prior elucidation of the cross-links of elastin (Franzblau, 1971; Partridge, 1969). The relationship between elastin and collagen cross-linking had previously been inferred from the fact that lathyrogens interfered with the cross-linking of both proteins.

The cross-links which have been isolated from collagen have all been characterized in their chemically reduced form, and the assumption is made that they exist in the protein in an unsaturated state, primarily as aldimine structures (Schiff bases). Some rearrangements of the aldimine configuration may also occur, as will be outlined later. To date, the only known aldehydes which give rise to the cross-links are α-aminoadipic acid δ-semialdehyde and δ-hydroxy, α-aminoadipic acid δ-semialdehyde (see Gallop *et al.,* 1972, for multiple references). Both of these compounds are presumably formed by the same enzyme, lysyl oxidase, although no direct evidence is available concerning this point.

Lysinonorleucine was detected and isolated from both naturally occurring and reconstituted collagen fibrils by several laboratories (Kang *et al.*, 1970; Tanzer and Mechanic, 1970; Bailey and Peach, 1971). It was found to be identical to synthetic lysinonorleucine, both chromatographically and by mass spectral analysis. It may arise as a consequence of the incomplete hydroxylation of lysine residues (Butler, 1968); periodate oxidation studies support this proposal (Tanzer and Mechanic, 1970). The aldimine form of lysinonorleucine has been detected in a number of collagenous tissues (Table 1), and it generally is less abundant than the other reducible cross-links.

Hydroxylysinonorleucine was the first reducible cross-link to be isolated and characterized from native collagen (Bailey and Peach, 1968). The structure was postulated from analysis of the mass spectrum of the trifluoroacetyl methyl ester derivative of the natural compound and proven by showing mass spectral and electrophoretic identity of the natural and synthetic compounds. Independent proof of the structure of hydroxylysinonorleucine, isolated from reconstituted collagen fibrils, was provided by Tanzer *et al.* (1970) who showed mass spectral identity of the natural and synthetic compounds, as their acetylated, permethylated derivatives. Evidence supporting the lactonization of hydroxylysinonorleucine was also presented. The aldimine form of hydroxylysinonorleucine is widely distributed in collagenous tissues and can potentially arise from two different routes: condensation of hydroxylysine and α-aminoadipic acid δ-semialdehyde or condensation of lysine and δ-hydroxy, α-aminoadipic acid δ-semialdehyde. Reduction with $NaB[^3H]_4$ would place the tritium in different locations in each instance, although the final product would be hydroxylysinonorleucine. Periodate oxidation should distinguish which route is predominant; in the case of calf-skin collagen it appears that hydroxylysine and α-aminoadipic acid δ-semialdehyde form the major route to the cross-link (Tanzer and Mechanic, 1970). The unreduced cross-link is dissociated by nonphysiologic pH conditions and by certain aminothiols (Bailey *et al.*, 1970). It also disappears upon thermal denaturation of collagen fibers (Bailey *et al.*, 1970). It is found in enzymic hydrolysates of $NaB[^3H]_4$ reduced calf-skin collagen, as are the other reducible cross-links (Bensusan, 1972; Housley *et al.*, 1974), indicating that they are not artifacts of acid or alkaline hydrolysis. Hydroxylysinonorleucine potentially may be present in collagen in its glycosylated forms (*O*-galactosyl and *O*-galactosylglucosyl derivatives), similar to dihydroxylysinonorleucine (see below).

Dihydroxylysinonorleucine is the most prominent cross-link in bone and dentin collagen. Its structure was postulated on the basis of the mass spectrum of the permethylated, isobutyloxycarbonyl derivative (Mechanic and Tanzer, 1970) and corroborated by the mass spectrum of the trifluoroacetyl methyl ester derivative (Mechanic *et al.*, 1971). Concurrently, Bailey and his associates isolated a compound from bone and dentin collagen which was originally termed "syndesinol" (Bailey *et al.*, 1969) while the unreduced form was termed "syndesine." The postulated structure was that of an aldol condensation product and was based upon elemental analysis and upon mass spectral analyses of the trifluoroacetyl methyl ester and ethyl ester derivatives. The postulated structure was

TABLE 1
Reduced Collagen Cross-Links—Structure and Distribution

Cross-link	Tissue; species	Reference
Lysinonorleucine 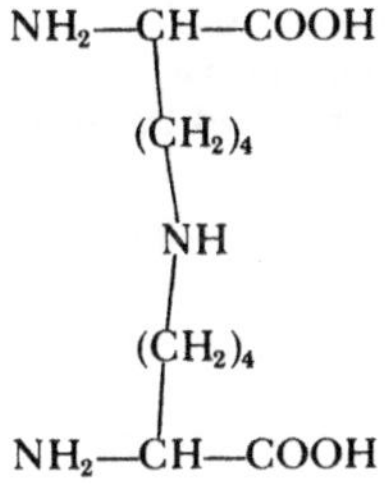	Tail tendon; rat and calf	Davison *et al.*, 1972
	Skin; chicken	Kang *et al.*, 1970
	Intervertebral disk and scapula; bovine	Bailey and Peach, 1971
	Skin; calf	Bensusan, 1972
	Skin and tendon; bovine	Mechanic and Tanzer, 1970
	Skin and nasal cartilage; calf	Tanzer and Mechanic, 1970
	Sclera and cornea; rabbit	Tanzer *et al.*, 1973*b*
	Basement membranes; sheep, lamb, and bovine	Tanzer and Kefalides, 1973
	Skin; rat	Deshmukh and Nimni, 1972
	Tendons; bovine	Shimokomaki *et al.*, 1972
	Tendon; rat	Cannon and Davison, 1973
	Fibroblasts, cell culture; mouse	Levene *et al.*, 1972
	Skin; human	Mechanic, 1972
	Sponge	Eyre and Glimcher, 1971
	Scales, swimbladder, skin; codfish	Bailey, 1970
	Elastoidin; dogfish	Bailey, 1970

TABLE 1—*Continued*

Cross-link	Tissue; species	Reference
Hydroxylysinonorleucine 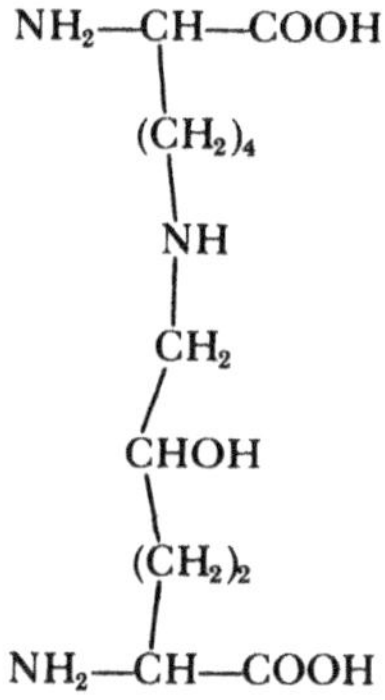	Tail tendon; rat and calf	Bailey and Peach, 1968; Davison *et al.*, 1972
	Skin; chicken	Franzblau *et al.*, 1970
	Tail tendon; rat	Franzblau *et al.*, 1970
	Intervertebral disk and scapula; bovine	Bailey and Peach, 1971
	Skin; calf	Bensusan, 1972; Tanzer *et al.*, 1970
	Tendons; rat, calf, chick embryo	Bailey *et al.*, 1970
	Skin and tendon; bovine	Mechanic and Tanzer, 1970
	Bone and dentin; fetal and adult, bovine	Mechanic *et al.*, 1971
	Bone and dentin; chick, human and bovine	Bailey *et al.*, 1969
	Sclera and cornea; rabbit	Tanzer *et al.*, 1973*b*
	Basement membranes; sheep, lamb and bovine	Tanzer and Kefalides, 1973
	Skin; rat	Deshmukh and Nimni, 1972
	Tail tendon; rat	Kang, 1972
	Skin; calf	DeLuque *et al.*, 1970
	Tendons; bovine	Shimokomaki *et al.*, 1972
	Tendon; rat	Cannon and Davison, 1973
	Tendon, cartilage, skin; calf	Bailey and Lapiere, 1973
	Fibroblasts; cell culture; mouse	Levene *et al.*, 1972
	Cartilage, bone, skin; human	Eyre and Glimcher, 1972
	Skin; human	Mechanic, 1972
	Bone; chicken	Mechanic *et al.*, 1972
	Skin; guinea pig	Forrest *et al.*, 1972
	Cornea; calf	Lian *et al.*, 1973
	Body wall; sea amenome	Bailey, 1971
	Skin, cartilage; bovine, avian, dogfish, lamprey, human	Bailey, 1971
	Cuticle; earthworm	Bailey, 1971
	Sponge	Eyre and Glimcher, 1971
	Body wall; sea urchin	Eyre and Glimcher, 1971
	Scales, swimbladder, skin; codfish	Bailey, 1970
	Elastoidin; dogfish	Bailey, 1970

TABLE 1—*Continued*

Cross-link	Tissue; species	Reference
Dihydroxylysinonorleucine	Tail tendon; rat and calf	Davison *et al.*, 1972
	Intervertebral disk and scapula; bovine	Bailey and Peach, 1971
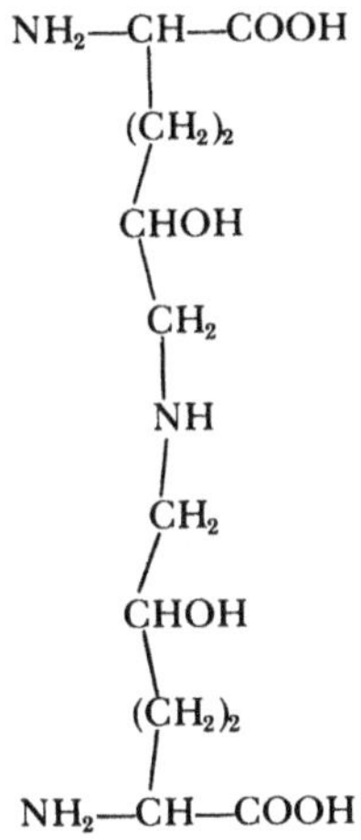	Tendons; rat, calf, chick embryo	Bailey *et al.*, 1970
	Skin; calf	Bensusan, 1972
	Skin and tendon; bovine	Mechanic and Tanzer, 1970
	Bone and dentin; fetal and adult bovine	Mechanic *et al.*, 1971
	Bone and dentin; chick, human and bovine	Bailey *et al.*, 1969
	Sternal cartilage; chicken	Miller and Robertson, 1973
	Bone, dentin and tendon; bovine and human	Davis, 1973
	Basement membranes; sheep, lamb, and bovine	Tanzer and Kefalides, 1973
	Bone; chicken and bovine	Eyre and Glimcher, 1973*a,c*
	Dentin; bovine	Kuboki *et al.*, 1973
	Tendons; bovine	Shimokomaki *et al.*, 1972
	Tendon; rat	Cannon and Davison, 1973
	Tendon, cartilage skin; calf	Bailey and Lapiere, 1973
	Skin, tendon, cartilage; bovine	Bailey and Shimokomaki, 1971
	Fibroblasts; cell culture; mouse	Levene *et al.*, 1972
	Cartilage, bone, skin; human	Eyre and Glimcher, 1972
	Skin; human	Mechanic, 1972
	Bone; chicken	Mechanic *et al.*, 1972
	Skin; guinea pig	Forrest *et al.*, 1972
	Skin, cartilage; bovine, avian, dogfish, lamprey, human, sea anemone	Bailey, 1971
	Body wall; sea anemone	Bailey, 1971
	Cuticle; earthworm	Bailey, 1971
	Sponge	Eyre and Glimcher, 1971
	Body wall; sea urchin, sea cucumber	Eyre and Glimcher, 1971
	Swimbladder, skin, scales; codfish	Bailey, 1970
	Elastoidin; dogfish	Bailey, 1970

TABLE 1—*Continued*

Cross-link	Tissue; species	Reference
Hydroxymerodesmosine 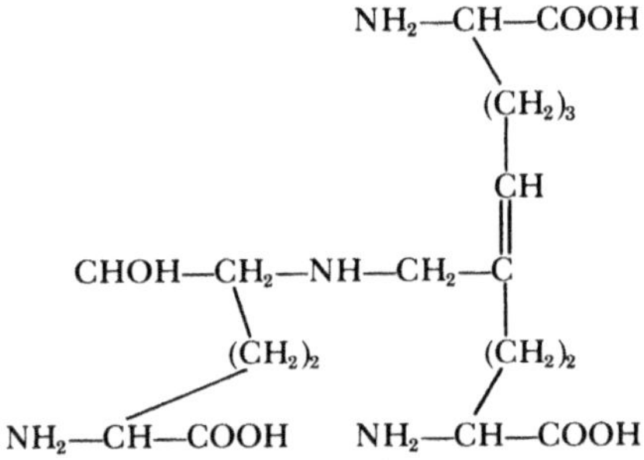	Skin; calf	Tanzer *et al.*, 1973*a*
	Basement membranes; sheep, lamb, and bovine	Tanzer and Kefalides, 1973
Aldol histidine 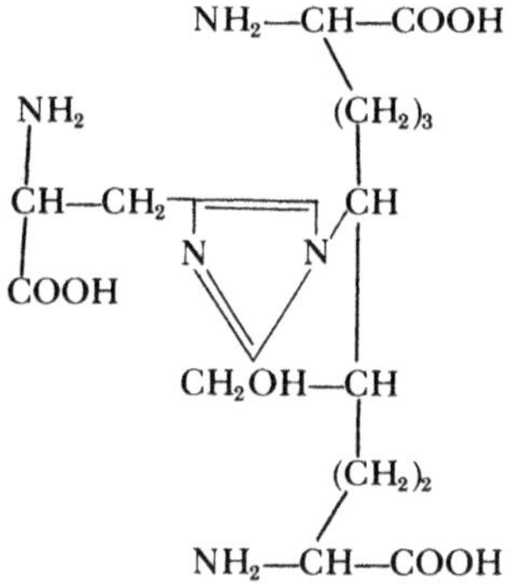	Skin; bovine	Fairweather *et al.*, 1972; Tanzer *et al.*, 1973*b*
	Basement membranes; sheep, lamb and bovine	Tanzer and Kefalides, 1973
Histidino-hydroxymerodesmosine 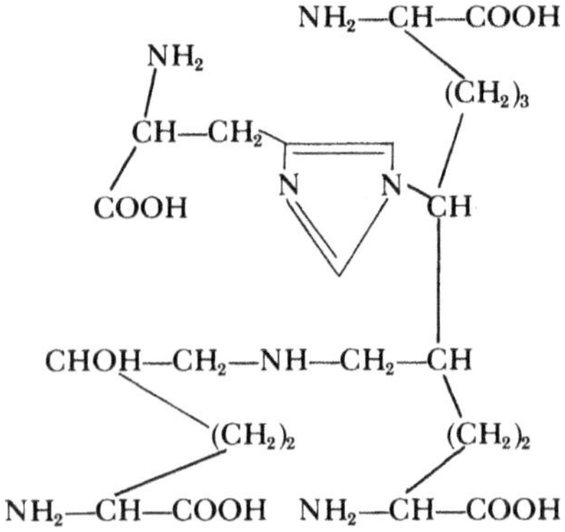	Skin; chicken	Franzblau *et al.*, 1970
	Tail tendon; rat	Franzblau *et al.*, 1970
	Tendons; rat, calf, chick embryo	Bailey *et al.*, 1970
	Skin; calf	Bensusan, 1972
	Skin; bovine	Tanzer *et al.*, 1973*b*
	Cornea and sclera; rabbit	Tanzer *et al.*, 1973*b*
	Skin; rat	Deshmukh and Nimni, 1972
	Tail tendon; rat	Kang, 1972
	Tendons; bovine	Shimokomaki *et al.*, 1972
	Tendon, cartilage, skin; calf	Bailey and Lapiere, 1973
	Tendon, cartilage, skin; bovine	Bailey and Shimokomaki, 1971
	Fibroblasts, cell culture; mouse	Levene *et al.*, 1972

TABLE 1—*Continued*

Cross-link	Tissue; species	Reference
Histidino-hydroxymerodes-mosine (continued)	Bone, skin; human	Eyre and Glimcher, 1972
	Skin; human	Mechanic, 1972
	Skin; guinea pig	Forrest *et al.*, 1972
	Cornea; calf	Lian *et al.*, 1973
	Skin, cartilage; bovine, avian, dogfish, lamprey	Bailey, 1971
	Swimbladder, skin, scales; codfish	Bailey, 1970
	Elastoidin; dogfish	Bailey, 1970

shown to be incorrect by these workers who subsequently identified the compound as dihydroxylysinonorleucine (Davis and Bailey, 1971). They completed the structural proof by synthesizing the cross-link and showed identity of the natural and synthetic substances. Unreduced dihydroxylysinonorleucine may be present *in situ* in at least two forms, the aldimine and the α-ketoamine (Fairweather, 1972; Tanzer, 1973), and indirect studies support this concept (Mechanic, 1974; Eyre and Glimcher, 1973*a*; Robins and Bailey, 1973*a*; Miller and Robertson, 1973). Other indirect studies suggest rearrangements involving cyclic intermediates (Davis, 1973). Recent evidence indicates that the unreduced dihydroxylysinonorleucine is primarily in the glycosylated form, containing *O*-galactosylglucose (Eyre and Glimcher, 1973*b*; Robins and Bailey, 1974). It has been shown that the unreduced compound *in situ* is more resistant to disruption by chemical agents and thermal denaturation than the other reducible cross-links (Bailey *et al.*, 1970). Perhaps glycosylation and/or the ability to undergo rearrangements plays a role in such stability. Natural reduction of this cross-link is thought to occur in bone collagen, as determined by isotopic dilution (Mechanic *et al.*, 1971), although there is disagreement about this point (Robins *et al.*, 1973).

Hydroxymerodesmosine was obtained from calf-skin collagen and subjected to mass spectral analysis as its trifluoroacetyl methyl ester derivative (Tanzer *et al.*, 1973*a*). The postulated structure was based upon comparison with the mass spectra of similar derivatives of merodesmosine, desmosine, hydroxylysinonorleucine, and dihydroxylysinonorleucine. Two alternative pathways for the formation of the unreduced form of hydroxymerodesmosine have been proposed.

Aldol histidine was isolated from insoluble cow-skin collagen and was characterized by NMR spectroscopy and mass spectrometry. The latter analysis was performed on both the trifluoroacetyl methyl ester derivative and the acetyl permethyl derivative (Fairweather *et al.*, 1972). Supporting evidence for the postulated structure was also provided by high-resolution mass spectrometry, ultraviolet spectrometry, and colorimetric methods. The physiologic significance of this compound is questioned by Bailey and co-workers (Robins and Bailey, 1973*b*) although aldol histidine is a major cross-link in both native and reconstituted fibrils of cow-skin collagen (Tanzer *et al.*, 1973*b*); it is not prominent in many other collagens which have been examined. This situation is analogous to other cross-links, especially dihydroxylysinonorleucine, which predominates in bone and dentin collagens.

Histidinohydroxymerodesmosine is present in most collagens and was shown to arise in reconstituted fibrils concurrent with the disappearance of an intramolecular-cross-link, aldol-condensation product (Franzblau *et al.*, 1970). It was termed "posthistidine" because of its elution location during ion-exchange chromatography. It was also called "fraction C" by Bailey (Bailey *et al.*, 1970) who detected it in insoluble, native collagens. Histidinohydroxymerodesmosine has been isolated from insoluble calf-skin collagen and was characterized by both physicochemical and chemical techniques (Tanzer *et al.*, 1973*b*; Hunt and Morris, 1973). The most informative data was obtained from PMR and [^{13}C]NMR studies coupled with high- and low-resolution mass spectrometry. This information provided the basis for postulating the structure of the cross-link, which is supported by the various chemical and degradative techniques. It was recognized years ago that unreduced histidinohydroxymerodesmosine was quite labile, particularly when collagen was exposed to solvents at acid pH (Kang *et al.*, 1970). This phenomenon has been confirmed (Davison *et al.*, 1972) and elaborated upon (Robins and Bailey, 1973*b*). The argument has been presented by the latter workers, that because the cross-link does not occur at nonphysiologic pH and its occurrence may be dependent upon nonprotonation of the imidazole ring of histidine, that it may be an artifact. By the same criteria, most of the other reducible cross-links would also be artifacts, as they also disrupt at nonphysiological pH levels (Davison *et al.*, 1972). Indeed, it has been shown that just above pH 8 and just below pH 7 the tensile properties of rat-tail tendon are markedly diminished (Bailey, 1968), apparently due to the lability of intermolecular cross-links. Unlike aldol histidine, histidinohydroxymerodesmosine is widely distributed in most collagenous tissues and is often prominent in such tissues. Several alternative pathways can account for the formation of this cross-link, and it is particularly notable that it can be considered to arise from three substances, each of which constitutes only a very small proportion of collagen, i.e., aldol condensation product, histidine, and hydroxylysine.

The implication of this analysis is that highly specific interactions must occur during collagen fibril formation to bring these reactive amino acids into correct proximity.

Finally, two classes of substances, the *N^{ϵ}-hexosyl lysines* and *N^{ϵ}-hexosyl hydroxylysines* have been isolated and characterized from cartilage collagen and bovine skin, respectively (Tanzer *et al.*, 1972; Robins and Bailey, 1972). The N^{ϵ}-hexosyl lysines have been fractionated into N^{ϵ}-glucitol lysine and N^{ϵ}-mannitol lysine. The origin and physiologic significance of all of these hexosyl-derived compounds is not clear, but some role may become apparent in the future.

The less-complex reduced cross-links have been synthesized by several different routes. Lysinonorleucine was formed by the alkylation of the ϵ-NH_2 group of lysine derivatives by brombutylhydantoin, followed by opening of the hydantoin ring and removal of the blocking groups (Franzblau *et al.*, 1969; Tanzer and Mechanic, 1970). Hydroxylysinonorleucine was obtained by a similar route (Tanzer *et al.*, 1970; Bailey and Peach, 1968), and more recently by a new procedure involving the 2-amino-5-hexenoic acid phenylhydantoin derivative (Davis and Bailey, 1972). This latter approach also served as a means for synthesizing dihydroxylysinonorleucine and the alcohols hydroxynorleucine and dihydroxynorleucine. It will be a much more formidable task to synthesize the larger cross-links, especially since a considerable number of isomeric forms are potentially possible.

A completely novel approach to the detection and chemical derivatization of the carbonyl-derived compounds in connective tissues has recently been described and has been successfully applied to collagen and elastin (Pereyra *et al.*, 1973, 1974). This method, a modified Strecker reaction employing cyanide and NH_3, has independently established the chemical structure of the carbonyl-derived compounds and has pointed the way to additional, unidentified substances in the connective tissue proteins (Pereyra *et al.*, 1974).

III. *Cross-Link Location*

Any consideration of intermolecular cross-linking in collagen immediately invokes the problem of molecular packing in collagenous tissues. This controversial and complex subject has given rise to a number of models of packing, none of which satisfy all of the available data. There are also a number of factors of recent vintage to consider, each of which could have a profound influence on the geometry of molecular associations. Such factors include the presence of procollagen in mature tissues (Veis *et al.*, 1972, 1973; Clark and Veis, 1972), the presence of two genetic types of collagen in a specific tissue (Miller and Matukas, 1974), the variation in carbohydrate content of various collagens (Kefalides, 1973), and the presence of other prosthetic groups in collagen such as phosphate and silicon (Veis and Perry, 1967; Schwarz, 1973). Although the techniques of X-ray diffraction and electron microscopy, as well as other physical methods (Katz and Li, 1973) can provide statistical information concerning spatial relationships, they are limited in resolving power, both inherently and by the specific complexities of collagen noted above. Ideally, of course, one would like to be able to locate every atom in space in the collagen polymer.

As noted in the introduction, one approach to locating adjacent proteins, e.g., in a ribosome or other assembly, is by the introduction of covalent cross-links, particularly if the conditions of reaction do not produce distortion. If the constituent proteins which become cross-linked have been previously defined in a biological and structural fashion, then some inferences can be made concerning possible spatial relationships and interactions. In the case of collagen, one is basically dealing with a repeating monomer unit (albeit subject to the complexities previously noted) whose primary structure is already well characterized. Thus, one can use that structural information in conjunction with the structures of the reducible cross-links in an effort to decipher the mode of packing in collagen fibers. A possible complication in this approach is that the reducible cross-links may form at multiple loci because the reactive aldehydes have been detected at many places on the collagen polypeptide backbone (Deshmukh and Nimni, 1971). This distribution may account for the suggestion that three general groups of cross-links occur in collagen, namely side-to-side bonds, head-to-tail bonds, and end-to-end bonds (Zimmermann *et al.*, 1970); the first two groups are thought to involve the reducible cross-links. The tactical approach taken by several laboratories has been to initially reduce the Schiff base cross-links (or, more recently, to omit reduction) and then to break up the polymeric collagen with specific reagents or enzymes (Kang, 1972; Volpin and Veis, 1971, 1973; Dixit and Bensusan, 1973; Miller, 1971; Eyre and Glimcher, 1973*a,c*; DeLuque *et al.*, 1970; Kuboki *et al.*, 1973). The task is then one of selecting, from the mixtures, specific peptides which contain an intermolecular cross-link which joins two or more polypeptide chains.

This approach has met with some success and a number of cross-linked peptides have been reported. Those peptides obtained following CNBr digestion have been characterized by comparison of their amino acid compositions with the known CNBr peptides isolated from collagen molecules. Thus, unreduced bovine-skin collagen and unreduced bovine dentin both contain a peptide whose composition can largely be accounted for by a cross-link joining the peptides α1-CB6 and α1-CB(0,1) (Volpin and Veis, 1971, 1973). The nature of the cross-link was not determined but it may be hydroxylysinonorleucine since a similar peptide containing this cross-link was isolated from borohydride-reduced rat-tail tendon collagen (Kang, 1972). In this latter study, many other partially purified cross-link-containing peptides were noted, indicating that intermolecular cross-links form at several different sites between adjacent molecules. Cross-linking of cartilage collagen, which contains primarily one type of α chain, was shown to involve peptides CB4 and CB9 (Miller, 1971). In a subsequent study, these workers showed that the cross-link was dihydroxylysinonorleucine and that the same peptide could be isolated from nonreduced cartilage collagen (Miller and Robertson, 1973). They provided evidence which suggested that the unreduced cross-link was stable because of a rearrangement of the Schiff base form, in agreement with an earlier postulate (Fairweather, 1972) and other evidence (Eyre and Glimcher, 1973*a*; Robins and Bailey, 1973*a*). Three peptides have been obtained from reduced bovine-skin collagen, two of which seem to contain α1-CB6 cross-linked (via an unknown cross-link) to α1-CB(0,1). In one case there is a 1:1 ratio of peptides and in the other case a 2:1 ratio (Dixit and Bensusan, 1973). The third peptide contains α1-CB6 and one other CB peptide whose identity is not certain. In similar fashion reduced chicken-bone collagen has yielded α1-CB6 as a partially purified peptide which contains dihydroxylysinonorleucine and hydroxylysinonorleucine, thought to be cross-linked to a small CB peptide (Eyre and Glimcher, 1973*c*). The same collagen preparation was also digested with bacterial collagenase and a highly purified peptide containing two polypeptide chains, cross-linked by dihydroxylysinonorleucine, was isolated. This peptide appeared to be identical with one previously characterized by the same investigators after its isolation from calf-bone collagen (Eyre and Glimcher, 1973*a*). The other interesting features of this peptide were that: (1) it could be obtained without prior reduction, probably due to a stable rearrangement of the Schiff base form; and (2) it contained glucose and galactose, probably as the *O*-galactosylglucosyl derivative of hydroxylysine.

Smaller cross-linked peptides have been isolated from protease digests of reconstituted calf-skin collagen and from bovine dentin (DeLuque *et al.*, 1970; Kuboki *et al.*, 1973). In the former case, three peptides containing hydroxylysinonorleucine were isolated and ranged in size from four to seven amino acid residues (counting the cross-link as one residue). In the latter case two peptides, containing the cross-link dihydroxylysinonorleucine, were isolated and contained 26 and 38 residues, respectively. Comparison of the composition of these peptides with the known CB peptides of bovine dentin allowed the authors to conclude that neither peptide contains the NH_2-terminal portion of dentin collagen.

As these various studies progress, it should eventually be possible to construct a map illustrating the packing arrangement of collagen molecules and the location of specific intermolecular cross-links. In addition, the variations imposed by tissue specificity and species specificity may become apparent.

IV. Cross-Link Biology

The considerations which are grouped in this area include the metabolic fate of the cross-links such as *in vivo* reduction or conversion to related compounds, the distribution of cross-links as a function of tissue type, animal species, and in relation to aging, and finally, the possible relationship of cross-linking to other physiologic and pathologic phenomena.

The question as to whether natural reduction of cross-links occurs is a controversial issue at present. The evidence which supports such reduction, in both reconstituted and native collagen fibrils, comes from two types of experiments. First, isotope dilution studies using $NaB[^2H]_4$ in combination with mass spectrometry showed that 25–50% of hydroxylysinonorleucine and dihydroxylysinonorleucine in bovine-bone collagen became reduced *in vivo* (Mechanic *et al.*, 1971). Second, in studies of reconstituted fibrils, in which the rat-skin collagen had been labeled with $[^{14}C]$lysine, four nonreducible cross-links developed progressively and one of these cross-links incorporated protons from the aqueous solvent (Deshmukh and Nimni, 1972). Two of the four cross-links, which did not incorporate protons, were shown to cochromatograph with lysinonorleucine and hydroxylysinonorleucine.

The evidence which indicates that natural reduction does not occur in native fibrils comes from three types of experiments. First, attempts to directly detect the reduced cross-links in large amounts of hydrolysate were unsuccessful (Robins *et al.*, 1973). Second, isotope dilution studies were done in which several tritiated cross-links were added to hydrolysates of bovine tendons and were reisolated; no significant change in specific activity was detected (Robins *et al.*, 1973). Third, the $NaB[^2H]_4$ isotope dilution experiment described above was repeated, and the hydroxylysinonorleucine obtained from bovine tendon was examined by mass spectrometry (Robins *et al.*, 1973). This latter experiment is impossible to evaluate because very selected portions of the spectrum are illustrated and even the peaks which derive from the presence of naturally occurring ^{13}C are omitted.

Thus, at this juncture one is left with conflicting data, and the solution to the question of *in vivo* reduction remains to be determined by future studies. Conceivably, the process of reduction may be species or tissue specific, accounting for the divergent results. It should also be noted that, in the case of the cross-links of elastin, especially lysinonorleucine, there is agreement about *in vivo* reduction (Franzblau, 1971; Partridge, 1969).

The distribution and relative abundance of the borohydride-reducible compounds has been studied in some detail. Most of the results are subject to the reservation that only qualitative data are given, i.e., chromatographic profiles showing relative peak areas; the total uptake of tritium and its concentration in the tissue is usually not considered. Another problem, that of destruction during hydrolysis, has recently been carefully evaluated (Davison *et al.*, 1972). These authors have also described the other limiting factors in quantitative interpretation of the elution diagrams.

The borohydride-reducible compounds are detectable in connective tissues from many sources (Table 1), including those from vertebrates and invertebrates and including basement membranes from vertebrates. With regard to aging, comparison of the elution diagrams of tissues of increasing age show some differences, but there is disagreement as to the interpretation of the results (Shimokomaki *et al.*, 1972; Cannon and Davison, 1973). The progressive insolubility of collagenous tissues in certain solvents, coincident with a progressive increase in polymeric components is often assumed to be indicative of progressive cross-linking. The relationship is clearly complex, involving such variables as total number of cross-links, cross-link location, change in chemical structure of cross-links, noncovalent interactions, progressive dehydration, and closer packing of molecules. Indirect evidence also suggests that formation of reducible cross-links may be impaired when the NH_2-terminal procollagen appendage remains intact (Bailey and Lapiere, 1973). As noted earlier, procollagen components are present in connective tissues (Veis *et al.*, 1972, 1973; Clark and Veis, 1972) and may affect both the packing mode and cross-linking pattern.

As the structures of the carbonyl-derived compounds have been elucidated, interest in comparing normal and abnormal collagenous tissues has gained momentum. Two hereditary disorders, Ehlers–Danlos syndrome and hydroxylysine-deficient collagen, have been shown to demonstrate abnormal cross-linking patterns in the skin of the affected individuals (Mechanic, 1972; Eyre and Glimcher, 1972). In both instances, there seems to be a relative deficiency of the normal borohydride-reducible compounds concomitant with the appearance of previously undetected substances.

In other instances, apparent quantitative changes in the relative abundance of reducible compounds has been reported. Experimental rickets affected the ratio of dihydroxylysinonorleucine to hydroxylysinonorleucine in chicken bones (Mechanic *et al.*, 1972). The cross-link pattern of dermal scars is different than the surrounding skin (Forrest *et al.*, 1972) and has been ascribed to differences in the molecular sites of cross-linking. Finally, it has been postulated that the increased solubility of the collagen formed by Marfan's syndrome cells in culture may reflect impaired cross-linking (Priest *et al.*, 1973).

V. Epilogue

The pathway leading from biosynthesis of procollagen to the final connective tissue architecture of collagen fibers is quite complex, involving a number of posttranslational modifications. At the outset, it is not certain whether the initial gene product is a single polypeptide chain, containing alternate collagen and noncollagen regions, or if individual pro-α chains are synthesized. In the latter case some mechanism must be invoked to provide for correct molecular composition and assembly. The modifications of proline hydroxylation, lysine hydroxylation, and glycosylation occur intracellularly on the growing procollagen which is then destined for export from the cell. Limited proteolysis of procollagen extensions may also occur intracellularly, but no direct evidence is available about this point. It is clear, however, that extracellular, limited proteolysis of the secreted procollagen does occur, as well as extracellular modification by lysyl oxidase. Both collagen molecules and procollagen molecules become incorporated into the growing fibers, and it appears that lysyl oxidase may continue to catalyze carbonyl-group formation along the individual molecules. The packing arrangement of the molecules provides the spatial orientation for cross-linking, and in part dictates the chemical nature of the resultant cross-links. These carbonyl adducts appear to convert into other compounds, either by reduction or by subsequent chemical reactions. The final collagen fiber, in most instances, lasts the life of the animal and may participate in the phenomenon of aging.

Multiple opportunities exist for the regulation or pathologic alteration of such a complex pathway. In addition, the occurrence of several distinct genetic types of collagen implies another stage of control, at the level of gene expression. Conceivably, all of the known regulatory processes which are well described in prokaryotes may be operative in the expression of collagen biosynthesis, postsynthetic modifications, fibril formation, and collagen turnover.

Note Added in Proof

In the 18 months since the original manuscript was completed, several important advances have been made. The enzyme lysyl oxidase has been purified to high specific activity and its properties studied (Siegel, 1974). This enzyme is active on both collagen and elastin, in their polymerized state, and the kinetics of activity indicate that fibrous collagen is a much more suitable substrate than solutions of collagen molecules. The isolated enzyme is active toward both lysyl and hydroxylysyl groups in collagen, although the basis for enzymatic selection of specific ϵ-NH_2 groups in the substrate is unknown; this point is of particular interest because the structures of fibrous collagen and the elastin polymer are quite different, and also the degree of lysine oxidation is much greater in elastin.

Inasmuch as the structures of the majority of the borohydride-reducible cross-links have been elucidated, interest has turned to the ultimate fate of these substances, especially since their content usually diminishes with age (Robins *et al.*, 1973; Fujii and Tanzer, 1974). However, in some tissues, this may not be the case (Davison and Patel, 1975), although there is some alteration in allysine distribution on the α chains with age. Another consideration which has been the focus of attention is the precise content of the different forms of the cross-links *in situ*, especially with regard to rearrangement configurations. Although all studies agree that Amadori types of rearrangements occur, there is considerable disagreement concerning the extent of this phenomenon (Mechanic, 1974; Mechanic *et al.*, 1974; Davis, 1973; Bailey *et al.*, 1974; Robins and Bailey, 1975). The varying results may reflect methodological factors which have significant effects upon the cross-linking and ultrastructure of collagen fibrils (Lian *et al.*, 1973). This point is further emphasized by thermal stability studies (Jackson *et al.*, 1974) and by derivation studies which were directed toward elucidating the ultimate fate of the reducible cross-links in mineralized collagens (Davis and Risen, 1974; Davis *et al.*, 1975).

Thus, it is problematic as to what influences the degree of rearrangement of a Schiff base cross-link in a protein and to what degree the previous treatment of the tissue influences the results. At any rate, one collagen cross-link has now been reported which occurs *in situ* and can be isolated directly from acid hydrolysates (Housley *et al.*, 1975). This substance, hydroxyaldol-histidine, is not reducible by reagents of the borohydride group and consequently may be viewed as a "stable" rather than an "intermediate" cross-link. Its structure is consistent with a condensation of allysine, hydroxyallysine, and histidine. The mechanistic details

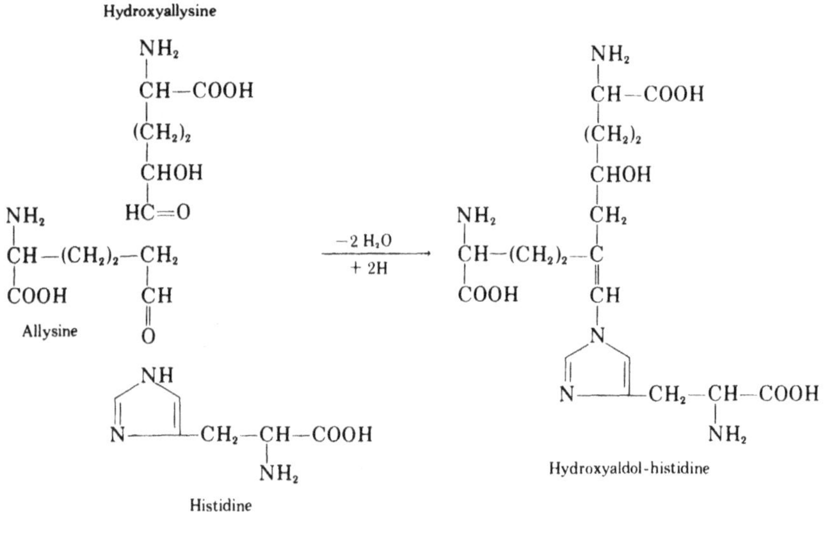

of such a condensation remain to be elucidated, including the implied *in situ* reduction, similar to lysinonorleucine formation (Gallop and Paz, 1975). Thus, as described in the main body of this chapter, histidine residues react with the endogenous aldehydes of collagen, forming unique cross-link structures which play an important role in the properties of the protein.

Isolation and characterization of cross-link-containing peptides has also advanced in recent years. Indeed, the new cross-link noted above was isolated from a pure peptide as well as from collagen itself. The cross-linked peptide contained three polypeptide chains, two of which were derived from a portion of the $\alpha 1(I)$ sequence (Becker *et al.*, 1975). Four other cross-linked peptides have also been characterized by these investigators, using both immunologic and biochemical criteria. The immunologic approach has been adopted by others but specific peptides were not isolated and identified (Chidlow *et al.,* 1974*a,b*). Another means of separating cross-linked peptides from non-cross-linked peptides depends upon the ability of the former to more readily adopt a helical conformation (Stimler and Tanzer, 1974); hydroxylapatite chromatography is used to distinguish between the native (cross-linked) and denatured (non-cross-linked) peptides.

The majority of the cross-linked peptides which have been isolated contain peptide segments from both the helical regions and NH_2- or COOH-terminal regions, as anticipated from the location of most of the collagen aldehydes within the quarter-stagger packing mode of collagen molecules in native fibrils. Recently, a cross-linked tripeptide, prolylhydroxylysinohydroxynorleucylvaline, was isolated from calf bone collagen (Fujii *et al.,* 1975) and its structure was consistent with a cross-link uniting two helical regions. This peptide, although small, could be attributed to very specific locations in the known primary structure of collagen α chains.

Interest in determining the relative abundance of the reducible cross-links in a variety of physiological and pathological states has continued. Thus, the study of bone cross-linking in rickets has been extended (Mechanic *et al.,* 1975) as well as studies of scleroderma and induced granulomas (Herbert *et al.,* 1974; Bailey *et al.,* 1973). Clearly, quantitative differences in cross-link content occur in these examples but it is difficult to attribute unique interpretations to the data. Patients with homocysteinuria may have disturbances of collagen cross-linking owing to interaction of homocysteine with collagen aldehydes (Kang and Trelstad, 1973).

References

Bailey, A. J., 1968, Intermediate labile intermolecular crosslinks in collagen fibres, *Biochem. Biophys. Acta* **160:**447.

Bailey, A. J., 1970, Comparative studies on the nature of the crosslinks in the collagen of various fish tissues, *Biochim. Biophys. Acta* **221:**652.

Bailey, A. J., 1971, Comparative studies on the nature of the cross-link stabilizing the collagen fibres of invertebrates, cyclostomes and elasmobranchs, *FEBS Lett.* **18:**154.

Bailey, A. J., and Lapiere, C. M., 1973, Effect of an additional peptide extension of the N-terminus of collagen from dermatosparactic calves on the cross-linking of the collagen fibres, *Eur. J. Biochem.* **34:**91.

Bailey, A. J., and Lister, D., 1968, Thermally labile cross-links in native collagen, *Nature (London)* **220:**280.

Bailey, A. J., and Peach, C. M., 1968, Isolation and structural identification of a labile intermolecular crosslink in collagen, *Biochem. Biophys. Res. Commun.* **33:**812.

Bailey, A. J., and Peach, C. M., 1971, The chemistry of the collagen cross-links. The absence of reduction of dehydrolysinonorleucine and dehydrohydroxylysinonorleucine *in vivo*, *Biochem. J.* **121**:257.

Bailey, A. J., and Shimokomaki, M. S., 1971, Age related changes in the reducible cross-links of collagen, *FEBS Lett.* **16**:86.

Bailey, A. J., Fowler, L. J., and Peach, C. M., 1969, Identification of two interchain crosslinks of bone and dentine collagen, *Biochem. Biophys. Res. Commun.* **35**:663.

Bailey, A. J., Peach, C. M., and Fowler, L. J., 1970, Chemistry of the collagen crosslinks. Isolation and characterization of two intermediate intermolecular crosslinks in collagen, *Biochem. J.* **117**:819.

Bailey, A. J., Bazin, S., and Delauney, A., 1973, Changes in the nature of the collagen during development and resorption of granulation tissue, *Biochim. Biophys. Acta* **328**:383.

Bailey, A. J., Robins, S. P., and Balian, G., 1974, Biological significance of the intermolecular cross-links of collagen, *Nature* **251**:105.

Balian, G. A., Bowes, J. H., and Cater, C. W., 1969, Stabilization of crosslinks in collagen by borohydride reduction, *Biochim. Biophys. Acta* **181**:331.

Becker, U., Furthmayr, H., and Timpl, R., 1975, Tryptic peptides from the cross-linking regions of insoluble calf skin collagen, *Hoppe-Seyler's Z. Physiol. Chem.* **356**:21.

Bensusan, H. B., 1972, Investigation of crosslinks in collagen following an enzymic hydrolysis, *Biochim. Biophys. Acta* **285**:447.

Blumenfeld, O. O., and Gallop, P. M., 1966, Amino aldehydes in collagen: The nature of a probable crosslink, *Proc. Natl. Acad. Sci. U.S.A.* **56**:1260.

Butler, W. T., 1968, Partial hydroxylation of certain lysines in collagen, *Science* **161**:796.

Cannon, D. J., and Davison, P. F., 1973, Cross-linking and aging in rat tendon collagen, *Exp. Gerontol.* **8**:51.

Chidlow, J. W., Bourne, F. J., and Bailey, A. J., 1974*a*, Production of hyperimmune serum against collagen and its use for the isolation of specific collagen peptides on immunosorbent columns, *FEBS Lett.* **41**:248.

Chidlow, J. W., Bourne, F. J., and Bailey, A. J., 1974*b*, Sheep antibodies to soluble rat collagen; Isolation of cross-linked peptides by affinity chromatography, *Immunology* **27**:665.

Clark, C. C., and Veis, A., 1972, High molecular weight α chains in acid-soluble collagen and their role in fibriliogenesis, *Biochemistry* **11**:494.

Davis, N. R., 1973, Stable crosslinks of collagen, *Biochem. Biophys. Res. Commun.* **54**:914.

Davis, N. R., and Bailey, A. J., 1971, Chemical synthesis of an intermolecular crosslink of collagen: A reevaluation of the structure of syndesine, *Biochem. Biophys. Res. Commun.* **45**:1416.

Davis, N. R., and Bailey, A. J., 1972, The chemistry of the collagen cross-links. A convenient synthesis of the reduction products of several collagen cross-links and cross-link precursors, *Biochem. J.* **129**:91.

Davis, N. R., and Risen, D. M., 1974, Mature collagen cross-links, *Biochem. Biophys. Res. Commun.* **61**:673.

Davis, N. R., Risen, D. M., and Pringle, G. D., 1975, Stable, nonreducible cross-links of mature collagen, *Biochemistry* **14**:2031.

Davison, P. F., 1973, Homeostasis in extracellular tissues: Insights from studies on collagen, *CRC Crit. Rev. Biochem.* **1**:201.

Davison, P. F., and Patel, A., 1975, Age related changes in aldehyde location on rat tail tendon collagen, *Biochem. Biophys. Res. Commun.* **65**:983.

Davison, P. F., Cannon, D. J., and Andersson, L. P., 1972, The effects of acetic acid on collagen cross-links, *Connect. Tissue Res.* **1**:205.

DeLuque, O., Mechanic, G., and Tanzer, M. L., 1970, Isolation of peptides containing the cross-link, hydroxylysinonorleucine, from reconstituted collagen fibrils, *Biochemistry* **9:**4987.

Deshmukh, K., and Nimni, M. E., 1969, Chemical changes associated with aging of collagen *in vivo* and *in vitro, Biochem. J.* **112:**397.

Deshmukh, K., and Nimni, M. E., 1971, Characterization of the aldehydes present on the cyanogen bromide peptides from mature rat skin collagen, *Biochemistry* **10:**1640.

Deshmukh, K., and Nimni, M. E., 1972, Identification of stable intermolecular crosslinks present in reconstituted native collagen fibres, *Biochem. Biophys. Res. Commun.* **46:**175.

Dixit, S. N., and Bensusan, H. B., 1973, The isolation of crosslinked peptides of collagen involving α1-CB6, *Biochem. Biophys. Res. Commun.* **52:**1.

Eyre, D. R., and Glimcher, M. J., 1971, Comparative biochemistry of collagen crosslinks: Reducible bonds in invertebrate collagens, *Biochim. Biophys. Acta* **243:**525.

Eyre, D. R., and Glimcher, M. J., 1972, Reducible crosslinks in hydroxylysine-deficient collagens of a heritable disorder of connective tissue, *Proc. Natl. Acad. Sci. U.S.A.* **69:**2594.

Eyre, D. R., and Glimcher, M. J., 1973*a*, Analysis of a crosslinked peptide from calf bone collagen: Evidence that hydroxylysyl glycoside participates in the crosslink, *Biochem. Biophys. Res. Commun.* **52:**663.

Eyre, D. R., and Glimcher, M. J., 1973*b*, Evidence for glycosylated cross-links in body-wall collagen of the sea cucumber, *Thyone briareus, Proc. Soc. Exp. Biol. Med.* **144:**400.

Eyre, D. R., and Glimcher, M. J., 1973*c*, Collagen cross-linking. Isolation of cross-linked peptides from collagen of chicken bone, *Biochem. J.* **135:**393.

Fairweather, R. B., 1972, personal communication.

Fairweather, R. B., Tanzer, M. L., and Gallop, P. M., 1972, Aldol-histidine, a new trifunctional collagen crosslink, *Biochem. Biophys. Res. Commun.* **48:**1311.

Forrest, L., Shuttleworth, A., Jackson, D. S., and Mechanic, G. L., 1972, A comparison between the reducible intermolecular crosslinks of the collagens of mature dermis and young dermal scar tissue of the guinea pig, *Biochem. Biophys. Res. Commun.* **46:**1776.

Franzblau, C., 1971, Elastin, *in: Comprehensive Biochemistry* (M. Florkin and E. H. Stoltz, eds.), Vol. 26, Part C, pp. 659–712, Elsevier, Amsterdam.

Franzblau, C., Faris, B., and Papaioannau, R., 1969, Lysinonorleucine. A new amino acid from hydrolysates of elastin, *Biochemistry* **8:**2833.

Franzblau, C., Kang, A. H., and Faris, B., 1970, *In vitro* formation of intermolecular crosslinks in chick skin collagen, *Biochem. Biophys. Res. Commun.* **40:**437.

Fujii, K., and Tanzer, M. L., 1974, Age related changes in the reducible cross-links of human tendon collagen, *FEBS Lett.* **43:**300.

Fujii, K., Corcoran, D., and Tanzer, M. L., 1975, Isolation and structure of a cross-linked tripeptide from calf bone collagen, *Biochemistry* **14:**4409.

Gallop, P. M., and Paz, M. A., 1975, Posttranslational protein modifications, with special attention to collagen and elastin, *Physiol. Rev.* **55:**418.

Gallop, P. M., Blumenfeld, O. O., and Seifter, S., 1972, Structure and metabolism of connective tissue proteins, *Annu. Rev. Biochem.* **41:**617.

Grant, M. E., and Prockop, D. J., 1972, Biosynthesis of collagen, *N. Engl. J. Med.* **286:**194.

Herbert, C. M., Lindberg, K. A., Jayson, M. I. V., and Bailey, A. J., 1974, Biosynthesis and maturation of skin collagen in scleroderma, and effect of D-penicillamine, *Lancet* **I:**187.

Housley, T. J., Tanzer, M. L., and Bensusan, H. B., 1974, Release of reducible collagen crosslinks by total enzymic hydrolysis, *Biochim. Biophys. Acta* **365:**405.

Housley, T. J., Tanzer, M. L., Henson, E., and Gallop, P. M., 1975, Collagen cross-linking: isolation of hydroxyaldol-histidine, a naturally-occurring cross-link, *Biochem. Biophys. Res. Commun.* **67:**824.

Hunt, E., and Morris, H. R., 1973, Collagen cross-links. A mass-spectrometric and ^{1}H- and ^{13}C-nuclear-magnetic-resonance study, *Biochem. J.* **135:**833.

Jackson, D. S., Ayad, S., and Mechanic, G., 1974, Effect of heat on some collagen cross-links, *Biochim. Biophys. Acta* **336:**100.

Kang, A. H., 1972, Studies on the location of intermolecular cross-links in collagen. Isolation of a CNBr peptide containing δ-hydroxylysinonorleucine, *Biochemistry* **11:**1828.

Kang, A. H., and Trelstad, R. L., 1973, A collagen defect in homocystinuria, *J. Clin. Invest.* **52:**2571.

Kang, A. H., Faris, B., and Franzblau, C., 1970, The *in vitro* formation of intermolecular crosslinks in chick skin collagen, *Biochem. Biophys. Res. Commun.* **39:**175.

Katz, E. P., and Li, S. T., 1973, The intermolecular space of reconstituted collagen fibrils, *J. Mol. Biol.* **73:**351.

Kefalides, N. A., 1973, Structure and biosynthesis of basement membranes, *in: International Review of Connective Tissue Research* (D. A. Hall and D. S. Jackson, eds.), Vol. 6, pp. 63–104, Academic Press, New York.

Kuboki, Y., Tanzer, M. L., and Mechanic, G. L., 1973, Isolation of polypeptides containing the intermolecular cross-link δ,δ'-dihydroxylysinonorleucine from dentin collagen, *Arch. Biochem. Biophys.* **158:**106.

Levene, C. I., 1973, Lathyrism, *in: Molecular Pathology of Connective Tissues* (R. Pérez-Tamayo and M. Rojkind, eds.), pp. 177–228, Marcel Dekker, New York.

Levene, C. I., Bates, C. J., and Bailey, A. J., 1972, Biosynthesis of collagen cross-links in cultured 3T6 fibroblasts; effect of lathyrogens and ascorbic acid, *Biochim. Biophys. Acta* **263:**574.

Lian, J. B., Morris, S., Faris, B., Albright, J., and Franzblau, C., 1973, The effects of acetic acid and pepsin on the cross-linkages and ultrastructure of corneal collagen, *Biochim. Biophys. Acta* **328:**193.

Mechanic, G., 1972, Crosslinking of collagen in a heritable disorder of connective tissue: Ehlers–Danlos syndrome, *Biochem. Biophys. Res. Commun.* **47:**267.

Mechanic, G. L., 1974, Collagen crosslinks: Direct evidence of a reducible stable form of the Schiff base Δ^6-dehydro-5,5'-dihydroxylysinonorleucine as 5-keto-5'-hydroxylysinonorleucine in bone collagen, *Biochem. Biophys. Res. Commun.* **56:**923.

Mechanic, G., and Tanzer, M. L., 1970, Biochemistry of collagen cross-linking. Isolation of a new cross-link, hydroxylysinohydroxynorleucine, and its reduced precursor, dihydroxynorleucine, from bovine tendon, *Biochem. Biophys. Res. Commun.* **41:**1597.

Mechanic, G., Gallop, P. M., and Tanzer, M. L., 1971, The nature of crosslinking in collagens from mineralized tissues, *Biochem. Biophys. Res. Commun.* **45:**644.

Mechanic, G., Toverud, S. U., and Ramp, W. K., 1972, Quantitative changes of bone collagen crosslinks and precursors in vitamin D deficiency, *Biochem. Biophys. Res. Commun.* **47:**760.

Mechanic, G. L., Kuboki, Y., Shimokawa, H., Nakamoto, K., Sasaki, S., and Kawanishi, Y., 1974, Collagen cross-links: direct quantitative determination of stable structural cross-links in bone and dentin collagens, *Biochem. Biophys. Res. Commun.* **60:**756.

Mechanic, G. L., Toverud, S. V., Ramp, W. K., and Gonnerman, W. A., 1975, The effect of vitamin D on the structural cross-links and maturation of chick bone collagen, *Biochim. Biophys. Acta* **393:**419.

Miller, E. J., 1971, Collagen cross-linking: Identification of two cyanogen bromide peptides containing sites of intermolecular cross-link formation in cartilage collagen, *Biochem. Biophys. Res. Commun.* **45**:444.

Miller, E. J., and Matukas, V. J., 1974, Biosynthesis of collagen, *Fed. Proc.* **33**:1197.

Miller, E. J., and Robertson, P. B., 1973, The stability of collagen cross-links when derived from hydroxylysyl residues, *Biochem. Biophys. Res. Commun.* **54**:432.

Narayanan, A. S., Siegel, R. C., and Martin, G. R., 1972, On the inhibition of lysyl oxidase by β-amino-propionitrile, *Biochem. Biophys. Res. Commun.* **46**:745.

Partridge, S. M., 1969, Elastin, biosynthesis and structure, *Gerontologia* **15**:85.

Paz, M. A., Lent, R. W., Faris, B., Franzblau, C., Blumenfeld, O. O., and Gallop, P. M., 1969, Aldehydes in native and denatured calf skin tropocollagen, *Biochem. Biophys. Res. Commun.* **34**:221.

Paz, M. A., Henson, E., Rombauer, R., Abrash, L., Blumenfeld, O. O., and Gallop, P. M., 1970, α-Amino alcohols as products of a reductive side reaction of denatured collagen with sodium borohydride, *Biochemistry* **9**:2123.

Pereyra, B., Paz, M. A., Gallop, P. M., and Blumenfeld, O. O., 1973, Systematic identification of aldehydes and aldehyde-derived cross-links in elastin by methods including a modified Strecker reaction, *Biochem. Biophys. Res. Commun.* **55**:96.

Pereyra, B., Blumenfeld, O. O., Paz, M. A., and Gallop, P. M., 1974, Maturation analysis in connective tissue proteins by ^{14}C-cyanide incorporation, *J. Biol. Chem.* **249**:2212.

Priest, R. E., Moinudoin, J. F., and Priest, J. H., 1973, Collagen of Marfan syndrome is abnormally soluble, *Nature* **245**:264.

Robins, S. P., and Bailey, A. J., 1972, Age-related changes in collagen: The identification of reducible lysine-carbohydrate condensation products, *Biochem. Biophys. Res. Commun.* **48**:76.

Robins, S. P., and Bailey, A. J., 1973*a*, Relative stabilities of the intermediate reducible crosslinks present in collagen fibres, *FEBS Lett.* **33**:167.

Robins, S. P., and Bailey, A. J., 1973*b*, The chemistry of the collagen cross-links. The characterization of fraction C, a possible artifact produced during the production of collagen fibres with borohydride, *Biochem. J.* **135**:657.

Robins, S. P., and Bailey, A. J., 1974, Isolation and characterization of glycosyl derivatives of the reducible cross-links in collagens, *FEBS Lett.* **38**:334.

Robins, S. P., and Bailey, A. J., 1975, The chemistry of the collagen crosslinks, the mechanism of stabilization of the reducible intermediate cross-links, *Biochem. J.* **149**:381.

Robins, S. P., Shimokomaki, M., and Bailey, A. J., 1973, The chemistry of the collagen cross-links. Age-related changes in the reducible components of intact bovine collagen fibres, *Biochem. J.* **131**:771.

Schwarz, K., 1973, A bound form of silicon in glycosaminoglycans and polyuronides, *Proc. Natl. Acad. Sci. U.S.A.* **70**:1608.

Shimokomaki, M., Elsden, D. F., and Bailey, A. J., 1972, Meat tenderness: Age related changes in bovine intramuscular collagen, *J. Food Sci.* **37**:892.

Siegel, R., 1974, Biosynthesis of collagen cross-links: Increased activity of purified lysyl oxidase with reconstituted collagen fibrils, *Proc. Natl. Acad. Sci. USA* **71**:4826.

Slobin, L. I., 1972, Use of bifunctional imidoesters in the study of ribosome topography, *J. Mol. Biol.* **64**:297.

Stimler, N., and Tanzer, M. L., 1974, Purification of large cross-linked peptides from insoluble calf bone and skin collagens by hydroxyapatite chromatography, *Biochim. Biophys. Acta* **365**:425.

Tanzer, M. L., 1965, Experimental lathyrism, *in: International Review of Connective Tissue Research* (D. A. Hall, ed.), Vol. 3, pp. 91–112, Academic Press, New York.

Tanzer, M. L., 1967, Collagen crosslinks: Stabilization by borohydride reduction, *Biochim. Biophyhys. Acta* **133**:584.

Tanzer, M. L., 1968, Intermolecular crosslinks in reconstituted collagen fibrils. Evidence for the nature of the covalent bonds, *J. Biol. Chem.* **243**:4045.

Tanzer, M. L., 1973, Cross-linking of collagen, *Science* **180**:561.

Tanzer, M. L., and Kefalides, N. A., 1973, Collagen crosslinks: Occurrence in basement membrane collagens, *Biochem. Biophys. Res. Commun.* **51**:775.

Tanzer, M. L., and Mechanic, G., 1968, Collagen reduction by sodium borohydride: Effects of reconstitution, maturation and lathyrism, *Biochem. Biophys. Res. Commun.* **32**:885.

Tanzer, M. L., and Mechanic, G., 1970, Isolation of lysinonorleucine from collagen, *Biochem. Biophys. Res. Commun.* **39**:183.

Tanzer, M. L., Mechanic, G., and Gallop, P. M., 1970, Isolation of hydroxylysinonorleucine and its lactone from reconstituted collagen fibrils, *Biochim. Biophys. Acta* **207**:548.

Tanzer, M. L., Fairweather, R., and Gallop, P. M., 1972, Collagen cross-links: Isolation of reduced N^{ϵ}-hexosylhydroxylysine from borohydride-reduced calf skin insoluble collagen, *Arch. Biochem. Biophys.* **151**:137.

Tanzer, M. L., Fairweather, R., and Gallop, P. M., 1973*a*, Isolation of the crosslink, hydroxymerodesmosine, from borohydride-reduced collagen, *Biochim. Biophys. Acta* **310**:130.

Tanzer, M. L., Housley, T., Berube, L., Fairweather, R., Franzblau, C., and Gallop, P. M., 1973*b*, Structure of two histidine-containing cross-links from collagen, *J. Biol. Chem.* **248**:393.

Traub, W., and Piez, K. A., 1971, The chemistry and structure of collagen, *Adv. Protein Chem.* **25**:243.

Veis, A., and Perry, A., 1967, The phosphoprotein of the dentin matrix, *Biochemistry* **6**:2409.

Veis, A., Anesey, J. R., Garvin, J. E., and Dimuzio, M. T., 1972, High molecular weight collagen: A long-lived intermediate in the biogenesis of collagen fibrils, *Biochem. Biophys. Res. Commun.* **48**:1404.

Veis, A., Anesey, J., Yuan, L., and Levy, S. J., 1973, Evidence for an amino-terminal extension in high-molecular-weight collagens from mature bovine skin, *Proc. Natl. Acad. Sci. U.S.A.* **70**:1464.

Volpin, D., and Veis, A., 1971, Differences between CNBr peptides of soluble and insoluble bovine collagens, *Biochem. Biophys. Res. Commun.* **44**:804.

Volpin, D., and Veis, A., 1973, Cyanogen bromide peptides from insoluble skin and dentin bovine collagens, *Biochemistry* **12**:1452.

Wold, F., 1972, Bifunctional reagents, *in: Methods in Enzymology* (C. H. W. Hirs and S. N. Timasheff, eds.), Vol. 25, pp. 623–651, Academic Press, New York.

Zimmermann, B. D., Pikkarainen, J., Fietzek, P. P., and Kühn, K., 1970, Cross-linkages in collagen. Demonstration of three different intermolecular bonds, *Eur. J. Biochem.* **16**:217.

5

Intracellular Steps in the Biosynthesis of Collagen

DARWIN J. PROCKOP, RICHARD A. BERG, KARI I. KIVIRIKKO, AND JOUNI UITTO

I. *Introduction*

The biosynthesis of collagen has many similarities to the biosynthesis of other proteins synthesized for "export," but collagen biosynthesis is distinguished by at least two prominent features: (1) the protein is first synthesized as a precursor form which fulfills several important functions, and (2) the biosynthesis involves several unusual posttranslational modifications which occur after assembly of amino acids into the three polypeptide chains of the molecule and which are essential for some of its critical structural features.

The precursor form of collagen is known as procollagen, and it differs from collagen in that it contains additional peptide extensions on the three polypeptide chains of the molecule. (For recent reviews on procollagen, see Schofield and Prockop, 1973; Martin *et al.*, 1974; Bornstein, 1974; Miller and Matukas, 1974). The additional peptide extensions make procollagen more soluble than collagen, and one of the functions of procollagen is to serve as a "transport" form which prevents

---

DARWIN J. PROCKOP, RICHARD A. BERG, KARI I. KIVIRIKKO and JOUNI UITTO · Department of Biochemistry, College of Medicine and Dentistry of New Jersey, Rutgers Medical School, Piscataway, New Jersey, 08854. Investigative work reported here was supported in part by N.I.H. research grants AM-16,516 and AM-16,186 from the U.S. Public Health Service. Present address for K.I.K. is Department of Medical Chemistry, University of Oulu, Oulu, Finland. Present address for J.U. is Division of Dermatology, Department of Medicine, Washington University School of Medicine, St. Louis, Missouri.

premature fiber formation during the biosynthetic process. Proteolytic conversion of procollagen to collagen occurs after the molecule is secreted from cells and the extracellular peptidase, or peptidases, which promote the conversion apparently help to determine when the protein will polymerize into normal fibers. However, a number of additional functions have been ascribed to procollagen, including the important suggestion that the additional peptide extensions on procollagen are essential for association of the three polypeptide chains of collagen and folding of these polypeptides into the triple-helical conformation of the native molecule.*

It is now generally recognized that the synthesis of most extracellular proteins involves enzymic reactions which occur after translation of the appropriate messenger RNAs, but collagen biosynthesis appears to involve more extensive posttranslational reactions than any other protein examined to date. In particular, the hydroxyproline, hydroxylysine, and glycosylated hydroxylysine found in collagen are all introduced into the molecule as the result of posttranslational modifications of prolyl and lysyl residues. In addition, the interchain disulfide bonds found among the peptide extensions of procollagen are also the result of posttranslational modifications which may be mediated by an enzyme or several enzymes. At least one of these posttranslational modifications, the hydroxylation of peptidyl proline to hydroxyproline, is essential for folding of the three polypeptide chains into the correct triple-helical structure. The importance of these posttranslational reactions in procollagen biosynthesis has been further emphasized by the demonstration that if one or more of the reactions is inhibited so that the protein does not fold into its correct structure, its secretion by cells is markedly altered so that either it is not secreted or it is secreted at a markedly reduced rate.

II. Transcription and Translation

A. Multiplicity of Genes for Collagen

The isolation of five different kinds of α chains from the collagens in different tissues clearly demonstrates that higher organisms must contain

* The possibility that the peptide extensions on procollagen may help to direct fiber formation has been suggested by Veis *et al.* (1972, 1973). It is also possible that the peptide extensions on procollagen may play a role in the binding of polysomes synthesizing procollagen to the membranes of the endoplasmic reticulum, much as has been suggested for small amino acid sequences at the amino-terminal ends of light chains of immunoglobulins (Milstein *et al.*, 1972) and of albumin (Judah *et al.*, 1973). For more complete discussions of the functions of procollagen see Schofield and Prockop (1973).

at least five different genes for collagen (see Chapter 1). As discussed below, variability in "posttranslational" modifications of the protein largely, but not completely, explains the variations among different collagens in their contents of hydroxyproline, hydroxylysine, and glycosylated hydroxylysine. The other differences in content and distribution of amino acids in the polypeptide chains reflect differences in the structural genes used for synthesis. Comparison of amino acid sequences from various regions of the α chains of type I and type II collagens indicated (Butler *et al.*, 1974) that some regions are more subject to genetic variability than others, but the significance of this observation is not yet apparent.

B. *Nature of the Initially Synthesized Polypeptide Chains*

Although a consensus has not been achieved (see below), the published reports from most laboratories agree that the collagenous polypeptides first synthesized are about 40% larger than the α chains of collagen because of peptide extensions on the ends of the chains* (Layman *et al.*, 1971; Jimenez *et al.*, 1971; 1973*a*; Bellamy and Bornstein, 1971; Dehm *et al.*, 1972; Uitto *et al.*, 1972b; Bornstein *et al.*, 1972; Ehrlich and Bornstein, 1972*a*; Goldberg *et al.*, 1972; Tsai and Green, 1972; Veis *et al.*, 1972; Grant *et al.*, 1972*a,b,* 1973; Dehm and Prockop, 1973; Jimenez *et al.*, 1973; Goldberg and Sherr, 1973; Sherr *et al.*, 1973; Clark *et al.*, 1973). The amino acid compositions of peptide extensions from type I (Uitto *et al.*, 1972*b*; Furthmayr *et al.*, 1972; von der Mark and Bornstein, 1973; Kohn *et al.*, 1974), type II (Uitto *et al.*, 1975*a*), and type III (Byers *et al.*, 1974) procollagen indicate that they differ in structure from collagen in that they contain less glycine and they contain little or no hydroxyproline or hydroxylysine. They also contain more acidic amino acids and more aromatic amino acids. Most importantly, they contain half-cystine residues which are not found in the triple-helical regions of type I and type II collagens. The amino acids in the extensions appear to be linked to the α chain portions by peptide bonds, and therefore it is generally assumed that the individual pro-α chains are assembled not as subunits but as continuous chains.

Several earlier hypotheses about the assembly of α chains or pro-α chains have now been discounted. For example, the suggestion that α

* By analogy with the α, β, γ chains of collagen, the individual polypeptide chains of procollagen are generally called pro-α chains and frequently the term pro-β chain is used to describe three such chains linked by interchain disulfide bonds. The terms pro-α and pro-γ chains are also used to refer to the polypeptides of protocollagen, the molecule which is identical to procollagen except that it contains no hydroxyproline or hydroxylysine and is correspondingly rich in proline and lysine.

chains were assembled from short peptides (Gallop *et al.*, 1967) has been disproven by sequencing studies on isolated α chains (Bornstein, 1970) and by Dintzis-type biosynthetic studies with type I and type II collagen (Vuust and Piez, 1971, 1972; Miller *et al.*, 1973).

Another earlier suggestion was that the three chains are first synthesized as a single, large polypeptide which is subsequently cleaved into three pro-α chains (Church *et al.*, 1971). This suggestion was attractive in part because it provided a mechanism whereby cells synthesizing type I collagen could efficiently produce two $\alpha 1$ chains for every $\alpha 2$ chain synthesized. The situation might therefore be analogous to the synthesis of the polio-virus proteins which apparently are first synthesized as a single polypeptide chain which is then cleaved to several separate proteins (Jacobson *et al.*, 1970). It might also be analogous to the synthesis of insulin which is first synthesized as a single polypeptide and then cleaved to two subunit peptides (Rubenstein and Steiner, 1970). However, evidence against a single polypeptide precursor for collagen has been developed by several different approaches. In one case, a Dintzis-type experiment on the synthesis of type I collagen demonstrated that synthesis of the $\alpha 1$ and $\alpha 2$ chains were initiated at about the same time, and that the initiation did not occur sequentially as one might expect if $\alpha 1$ and $\alpha 2$ chains were synthesized as components of a continuous polypeptide chain (Vuust and Piez, 1971, 1972). A second line of evidence against a single polypeptide precursor has been developed by the use of short-term pulse-label and chase experiments with tendon or cartilage cells incubated in suspension (Schofield *et al.*, 1974*a,b;* Harwood *et al.*, 1973; Uitto and Prockop, 1974*b*). With pulse-labeling periods ranging from less than one third the time required to synthesize a pro-α chain (see below) to pulse-labeling periods many times the synthesis time, no evidence was found for the presence of a polypeptide chain larger than a pro-α chain. A third line of evidence against the single polypeptide chain precursor has come from examining the size of polysomes which are synthesizing collagen (Lazarides and Lukens, 1971*a*; Harwood *et al.*, 1974*c*). Here it was shown that these polysomes were larger than a 23-ribosomal aggregate but smaller than a 50–60-ribosomal aggregate (Lazarides and Lukens, 1971*a*). On the assumption that the spacing of the ribosomes along the mRNA molecule was the same as found for polysomes synthesizing other proteins, it was calculated that the mRNA molecule was of an appropriate size to code for a single polypeptide chain of about 1000 amino acids. These three lines of evidence tend to exclude the possibility that in the cell systems examined to date collagen is initially synthesized as a single, large polypeptide precursor which is subsequently cleaved to form the three pro-α chains of procollagen.

A considerable body of evidence has shown that procollagen from most tissues contains peptide extensions on the amino-terminal ends of the three pro-α chains, and for some time it was assumed that the only extensions were on the amino-terminal ends. More recent observations, however, indicate that peptide extensions are present on both ends of each pro-α chain. The presence of amino-terminal extensions was first demonstrated by electron microscopy of segment-long-spacing (SLS) aggregates of procollagen (Lenaers *et al.*, 1971; Stark *et al.*, 1971; Dehm *et al.*, 1972; Jimenez *et al.*, 1973*a*). It was initially assumed that the amino-terminal extensions were the only extensions on procollagen for two reasons: (1) Although globular extensions were occasionally seen on the carboxy-terminal ends of SLS segments of procollagen, these appeared to disappear as the procollagen was purified (B. R. Olsen and D. J. Prockop, unpublished observations); and (2) after bacterial collagenase digestion of type I or II procollagen, one appeared to obtain a single fragment which was apparently large enough to account for the difference in size between procollagen and collagen (von der Mark and Bornstein, 1973; Dehm and Prockop, 1973; Dehm *et al.*, 1974; Sherr *et al.*, 1973). It seemed reasonable, therefore, to conclude that this fragment corresponded to the amino-terminal extension seen by electron microscopy and that if carboxy-terminal extensions were also present, they were relatively small in size. More recently, however, the structure of procollagen has been examined with human or tadpole collagenases which specifically cleave collagen at a point about three-fourths of the distance from the amino-terminal end (see Gross, 1974). The results of these studies indicate that peptide extensions are present on both ends of the molecule and that the extensions on the carboxy-terminal ends are larger than those on the amino-terminal ends (Tanzer *et al.*, 1974; Fessler *et al.*, 1975; Davidson and Bornstein, 1975; Olsen *et al.*, 1976). The larger extension on the carboxy-terminal end apparently contains all the interchain disulfide bonds in the molecule and therefore appears to give rise to the disulfide-linked fragment recovered after digestion of procollagen with bacterial collagenase.

C. *Translation of Collagen mRNA in Vitro*

Several laboratories have now developed cell-free systems which synthesize polypeptide chains of collagen *in vitro*. The most successful systems have been those in which polysomes containing mRNA are isolated and then synthesis of collagenous polypeptides is observed when the polysomes are incubated under appropriate conditions with labeled amino acids (Lazarides and Lukens, 1971*a*; Kerwar *et al.*, 1972, 1973; Prichard *et al.*, 1974; Kerwar, 1974). In the first successful experiment with such a system, the polypeptide products obtained were the size of α chains and not as large as pro-α chains (Lazarides and Lukens, 1971*a*), apparently because peptidases were present in the cell-free system. In subsequent experiments with slightly modified systems, polypeptides synthesized were the same size as pro-α chains (Kerwar *et al.*, 1972, 1973; Kerwar, 1974).

It has been far more difficult to carry out experiments in which the mRNA for collagen is isolated and then used for the synthesis of pro-α chains *in vitro.* Although RNA fractions which probably contain collagen mRNA have been identified in a number of systems (see Harwood *et al.*, 1974*c*), successful translation of the isolated message, which is probably the only satisfactory proof of its identity, has been difficult to accomplish. Attempts to translate the collagen mRNA with the reticulocyte system or with Ehrlich's ascites cell system have frequently yielded short peptides which are digestible by collagenase and which therefore are probably of the correct amino acid sequence (see Benveniste *et al.*, 1973). However, the failure to obtain complete polypeptides is disappointing. Recently more promising results have been obtained with the cell-free system derived from wheat germ (Benveniste *et al.*, 1974). At least part of the labeled polypeptides synthesized by this system appear to be the same size as pro-α and α chains.

Obviously, the development of the system for efficient translation of collagen mRNA will be of considerable interest from several points of view. The collagen mRNA is one of the largest mRNAs in eukaryotic cells. Also, because of the unusual amino acid composition of collagen, the mRNA should have an unusual base composition. The first two bases in all four mRNA codons for proline are cytosine, and the first two bases in the four codons for glycine are guanine. Since glycine is about one-third of the collagen polypeptide chain, and since proline together with hydroxyproline account for another one-quarter of the polypeptide chains, the mRNA for collagen should be particularly rich in cytosine and guanine.

D. *Time for the Assembly of the Polypeptides*

Studies on the synthesis of hemoglobin in reticulocytes (Hunt *et al.*, 1969), of β-galactosidase in *E. coli* (Lacroute and Stent, 1968), and of a variety of proteins in liver (Scornik, 1974) suggested that the amino acids are incorporated into polypeptide chains in these three systems at a rate of about 9–15 residues per site per second. If the same rate applied to the synthesis of the pro-α chains of collagen, the time to synthesize one of these chains would be less than 2 min. However, a Dintzis-type experiment in membranous bone by Vuust and Piez (1972) demonstrated that the assembly of amino acids into the polypeptide chains of collagen is unexpectedly slow. The rate they observed was about 3 amino acids per site per second, and the time to assemble the pro-α chains of procollagen was about 5.8 min. A similar experiment carried out subsequently with cartilage instead of membranous bone showed that the assembly to pro-α chains of type II collagen also required about 6 min (Miller *et al.*, 1973). The reasons why the assembly of amino acids into the collagen is slow compared to hemoglobin, β-galactosidase, and liver proteins are not apparent, but this observation may be of considerable importance in considering other aspects of the biosynthetic pathway (see below). One hypothesis which might be examined further is that the relatively long synthesis time of procollagen is related to the necessity for the newly synthesized pro-α chains to pass through the membranes of the endoplasmic reticulum. Measurements of synthesis times have only been carried out with a few proteins, but it may well be that all proteins for export are synthesized more slowly than intracellular proteins such as hemoglobin.

III. *Posttranslational Modifications*

Since the hydroxyproline, hydroxylysine, galactosylhydroxylysine, and glucosylgalactosylhydroxylysine found in collagen are synthesized by reactions which occur after amino acids are incorporated into peptide linkage, it is apparent that these reactions are not directly controlled by mRNA templates and that they represent "post-mRNA" or "posttranslational" modifications of polypeptide chains (Prockop *et al.*, 1972, 1973). Such posttranslational modifications are involved in the synthesis of all glycoproteins and are probably involved in the synthesis of all proteins destined for "export." However, the posttranslational reactions involved in collagen biosynthesis have a number of unique features.

A. *Hydroxylation of Peptidyl Proline*

As has been known for several decades (Stetten, 1949), free hydroxyproline is not incorporated as such into collagen and therefore the hydroxyproline found in the collagen of higher organisms must arise from the hydroxylation of proline which has been incorporated into some intermediate in collagen biosynthesis (for reviews, see Grant and Prockop, 1972; Cardinale and Udenfriend, 1974). Extensive work carried out in the early 1960s demonstrated that this intermediate is not prolyl-tRNA and that hydroxylation of prolyl-tRNA is not a major source of the 4-hydroxyproline found in collagen (see Urivetzky *et al.*, 1966; Grant and Prockop, 1972). Instead, the intermediate is proline already incorporated into peptide linkages, and the hydroxyproline in collagen is synthesized by the hydroxylation of peptidyl proline. Enzyme or enzymes capable of hydroxylating peptidyl proline have now been identified in most tissues of higher organisms, in several lower organisms such as *Ascaris* worms and earthworms, and in several plants with cell walls containing hydroxyproline in the protein known as extensin. The prolyl hydroxylases from several of these sources, such as *Ascaris* worms and earthworms, clearly differ from the major enzymic activity found in higher organisms and will not be discussed here. (For extensive reviews of prolyl hydroxylase in general, see Cardinale and Udenfriend, 1974; for review of prolyl hydroxylase in plants, see Kuttan and Radhakrishnan, 1973).

Prolyl hydroxylases extracted from homogenates of chick embryos and from newborn rat skin have been the most extensively studied forms of the enzyme, and these two forms appear to be similar. For the purposes of the present review, the term prolyl hydroxylase* will be used to refer to the enzymes isolated from chick embryos or newborn rat skin and to the apparently similar but less-well-characterized enzymes found in other vertebrates or tissue cultures of cells from such animals. It might be noted that there is as yet no clear evidence for the presence of more than one kind of prolyl hydroxylase in higher organisms, even though there is suggestive evidence (Berg and Prockop, 1973*a*) that the small amounts of 3-hydroxyproline found in many collagens are synthesized by a different enzyme.

1. *Properties of Prolyl Hydroxylase*

Prolyl hydroxylase was first obtained in relatively pure form by ion-exchange chromatography and gel filtration from chick embryos (Halme *et al.*, 1970*a* Pänkäläinen *et al.*, 1970) and from newborn rat skin (Rhoads and Udenfriend, 1970). Subsequently a rapid affinity-chromatography

* Other names which have been used for the enzyme include proline hydroxylase: collagen proline hydroxylase; protocollagen proline hydroxylase; peptidyl proline hydroxylase; prolyl-glycyl-peptide: 2-oxoglutarate dioxygenase (EC 1.14.11.2).

procedure was developed for obtaining apparently pure enzyme (Berg and Prockop, 1973*a*). In the affinity-chromatography procedure a crude extract of chick embryos is passed through a column containing the unusual collagen from *Ascaris* worms which is deficient in hydroxyproline and which serves as a substrate for prolyl hydroxylase (Fujimoto and Prockop, 1968; Rhoads and Udenfriend, 1968). The enzyme is then eluted with a synthetic peptide substrate (Figure 1). Because of the high affinity of the *Ascaris* collagen for the enzyme (K_d of 10^{-7} M or less), and because the substrate peptide used to elute the enzyme is essentially uncharged, highly purified enzyme is obtained in a single step. More recently a modified procedure has been developed in which an affinity column is prepared with poly-L-proline form II, a competitive inhibitor for the enzyme (Tuderman *et al.*, 1975). The enzyme obtained from chick embryos with the modified procedure appears to be identical to the enzyme obtained with the original affinity column procedure.

The molecular weight of the prolyl hydroxylase initially isolated from chick embryos was 248,000 by sedimentation equilibrium (Pänkäläinen *et al.*, 1970) and a value of 230,000 was subsequently obtained for the enzyme purified from the same source with the affinity column procedure (Berg and Prockop, 1973*a*). The enzyme obtained from chick embryos was shown to consist of subunits of about 60,000 and 64,000 daltons and therefore apparently was a tetramer consisting of two kinds of subunits (Berg and Prockop, 1973*a*). Enzyme purified from newborn rat skin was also shown to consist of subunits of about 65,000 daltons, but the

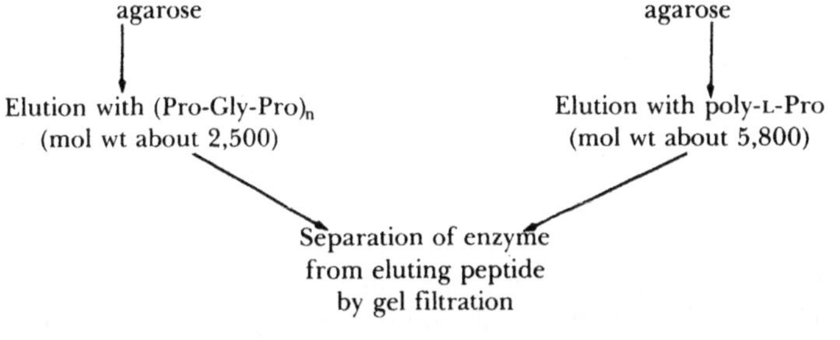

FIGURE 1. Scheme for purification of prolyl hydroxylase by affinity chromatography.

undissociated protein was initially reported to have a molecular weight of between 109,000 and 148,000 (Rhoads and Udenfriend, 1970). Subsequent studies by gel filtration suggested a larger size for the enzyme from newborn rat skin (Cardinale and Udenfriend, 1974) and a value comparable to that of the tetramer from chick embryos (Halme *et al.*, 1970*a*). The initially reported lower value for the undissociated protein from rat skin probably reflects the tendency of the enzyme to dissociate into dimers (see below). Prolyl hydroxylase from mouse skin (Stassen *et al.*, 1974) and from human skin (Kuutti *et al.*, 1975) also appears by gel filtration to have the same size as the enzyme from chick embryos and newborn rat skin. The enzyme from human skin was found to consist of subunits of about 61,000 and 64,000 daltons (Kuutti *et al.*, 1975).

The tetrameric form of prolyl hydroxylase from chick embryos can be dissociated into dimers by dialysis against buffers of low ionic strength or into monomers by reduction with moderate concentrations of reducing agents such as dithiothreitol (Berg and Prockop, 1973*a*). The latter observation suggested that the subunits might be held together by *inter*chain disulfide bonds. More recently, however, it has been shown that alkylation without reduction will convert the tetrameric form of the enzyme to monomers, indicating the subunits are not held together by *inter*chain disulfide bonds (Berg and Prockop, 1976). Dissociation of the enzyme by reducing agents alone is probably, therefore, explained by reduction of *intra*chain disulfide bonds which are essential for the subunits to maintain the native structure necessary for association.

Monomers of the enzyme are inactive. A small amount of activity was observed with dimeric forms of the enzyme from chick embryos but it was not apparent whether this small amount of activity was truly associated with dimers or represented a conversion of dimers to tetramers during the assay of enzymic activity (Berg and Prockop, 1973*a*).

The enzyme purified by affinity chromatography from chick embryos was examined by electron microscopy using a negative staining technique (Olsen *et al.*, 1973 *a*). The monomers were found to be rod-shaped, and the dimers consisted of monomers joined at one end to form a V-shaped structure (Figure 2). The structure of the tetramer was more difficult to visualize, but it appeared to consist of two V-shaped dimers which were interlocked.

Amino acid analyses of the enzyme purified either from chick embryos or rat skin showed the protein to be highly acidic (Halme *et al.*, 1970*a*; Rhoads and Udenfriend, 1970; Berg and Prockop, 1973*a*), and this observation agreed with the measured isoelectric point of 4.4 (Pänkäläinen *et al.*, 1970). Initial attempts to isolate the two different kinds of subunits were frustrated by the tendency of one of the subunits to

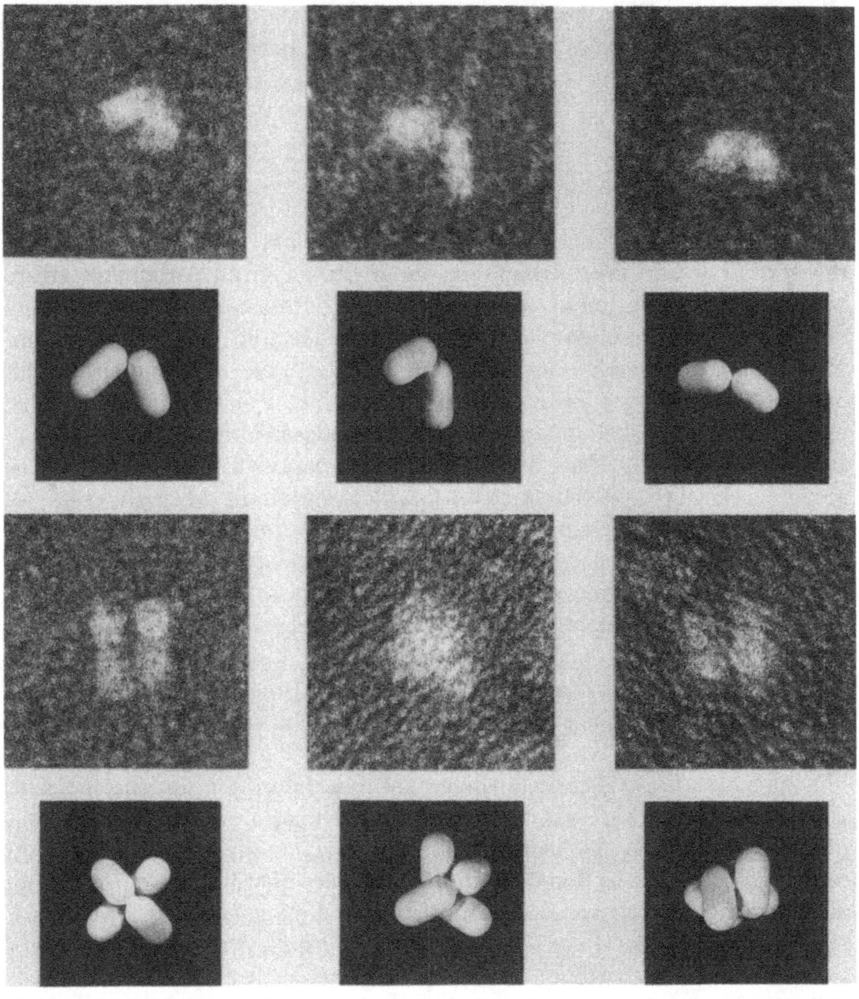

FIGURE 2. Negative-staining electron microscopy of prolyl hydroxylase purified by affinity chromatography from chick embryos. The enzyme was negatively stained with 2% uranyl acetate. Models of the enzyme are presented just below the electron micrographs. First row: Electron micrographs of dimers of the enzyme. Second row: A model for V-shaped dimers of the enzyme presented in different orientations. Third row: Electron micrographs of tetramers of the enzyme. Fourth row: A model for tetramers of the enzyme consisting of two interlocked V-shaped dimers and presented in different orientations. Reproduced with permission from Olsen *et al.* (1973*a*).

aggregate after dissociation, but recently it has been found that both subunits can be obtained in a soluble form if they are reduced and alkylated (Berg and Prockop, 1976). This observation should make it possible to isolate the two kinds of subunits and characterize their chemical composition.

2. *Cofactors and Cosubstrates*

The reaction catalyzed by prolyl hydroxylase is an oxidoreductase reaction in which the oxygen for the hydroxyl group originates from atmospheric O_2 (Fujimoto and Tamiya, 1962; Prockop *et al.*, 1963). The reaction requires nonheme iron (Hurych and Chvapil, 1965; Prockop and Juva, 1965*b*; Kivirikko and Prockop, 1967*a*) and α-ketoglutarate (2-oxoglutarate) (Hutton *et al.*, 1966). The discovery of the highly specific requirement for α-ketoglutarate was of considerable interest (see below), and it was shown that the synthesis of peptidyl hydroxyproline is accompanied by a stoichiometric decarboxylation of α-ketoglutarate to give succinate and carbon dioxide (Rhoads and Udenfriend, 1968; Cardinale *et al.*, 1971). The oxidative decarboxylation of α-ketoglutarate does not involve participation of cofactors, such as flavin, pyridine nucleotide, or thiamine pyrophosphate, or of an enzyme complex such as mitochondrial α-ketoglutarate dehydrogenase (Rhoads and Udenfriend, 1970; Pänkäläinen and Kivirikko, 1971). However, the mechanism by which α-ketoglutarate participates in the reaction has not yet been defined (see below).

Although there is general agreement that the iron in the enzyme is in a nonheme form, there has been some confusion as to how tightly the iron is bound to the enzyme. Dialysis of impure enzyme preparations does not remove all the iron as demonstrated by the fact that 30% or more of the enzyme activity is recovered without addition of exogenous iron (Kivirikko and Prockop, 1967*c*; Takeuchi and Prockop, 1969). Also prolyl hydroxylase purified by the affinity column procedure (Figure 1) retained about 40% of its maximal activity without addition of iron (R. A. Berg, unpublished observation). However, Pänkäläinen and Kivirikko (1971) used atomic absorption spectroscopy to examine purified prolyl hydroxylase and found less than 1 mole of iron per mole of protein. The apparent discrepancy in these results may be explained by the fact that the prolyl hydroxylase used by Pänkäläinen and Kivirikko for their study was purified by ion-exchange chromatography and several other procedures which may remove iron that otherwise remains bound to the enzyme.

One laboratory reported that the enzyme was not completely inhibited by EDTA, α,α'-dipyridyl, and a variety of other chelating agents (Bhatnagar and Liu, 1973; Liu and Bhatnagar, 1973). Other workers, however, have observed complete inhibition with EDTA and α,α'-dipyridyl (Kivirikko and Prockop, 1967*c*; Juva and Prockop, 1969).

Initial observations suggested that the iron required for the reaction was in the ferrous form (Hurych and Chvapil, 1965; Prockop and Juva, 1965*b*; Hutton *et al.*, 1966), but considerations of the reaction mechanism (see below) suggested to Hurych *et al.* (1973) that the iron is converted to ferric form after binding to the enzyme. They obtained some evidence for this possibility by examining the electron absorption spectra of impure preparations of prolyl hydroxylase.

The role of reducing agents in the reaction has been studied extensively, but at the moment the data defy any simple interpretation. Of primary interest has been the question of whether ascorbate has a direct and important role in the enzymic reaction which might help explain the decreased rate of collagen synthesis seen with ascorbic acid deficiency both *in vivo* and *in vitro* (see Robertson and Schwartz, 1953; Robertson and Hewitt, 1961; Stone and Meister, 1962; Peterkofsky, 1972*a,b*; Barnes and Kodicek, 1972; Levene *et al.*, 1974). Early attempts to purify prolyl hydroxylase clearly demonstrated that addition of ascorbate to the reaction system greatly increased the observed enzymic activity (Hutton *et al.*, 1966). Subsequently, however, there has been considerable difficulty in defining the role of ascorbate more precisely.

One problem is that there is no formal requirement for a reducing agent such as ascorbate in the reaction, regardless of which of the two currently proposed mechanisms is assumed to be correct (see below). A second problem is that the requirement for ascorbate during the reaction *in vitro* can partly be replaced by reduced pteridines and several other reducing agents (Peterkofsky and Udenfriend, 1965; Hutton *et al.*, 1967*b*). The specificity for the requirement of ascorbate is therefore in considerable doubt. A third problem is that even when optimal concentrations of ascorbate are used during the enzymic reaction *in vitro*, the observed activity is further increased by addition of several sulfhydryl-containing compounds which may also act as reducing agents in the system and whose effects have not yet clearly been differentiated from the effects of ascorbate. For example, in the presence of optimal concentrations of ascorbate, the observed enzymic activity is frequently enhanced further by addition of appropriate concentrations of dithiothreitol, albumin, and catalase (see below). It should be noted that enhancement of enzymic activity *in vitro* by ascorbate or by these other agents appears to be a separate phenomenon from the conversion of inactive monomers to active tetramers which may occur when ascorbate or lactate are administered to some cultured fibroblasts (see below), since repeated attempts to demonstrate such a conversion with isolated monomers have failed.

The effects of dithiothreitol and similar reagents (Rhoads *et al.*, 1967; Popenoe *et al.*, 1969) suggested that the active center of the enzyme contained free sulfhydryl groups which were essential for the reaction. The suggestion was supported by subsequent demonstrations that the enzyme was inhibited by sulfhydryl reagents such as *N*-ethylmaleimide, *p*-mercuribenzoate, and iodoacetate, and the demonstration that inhibition by *p*-mercuribenzoate was reversed by treating the enzyme with dithiothreitol (Halme *et al.*, 1970*a*). Further experiments demonstrated that inhibition by *N*-ethylmaleimide could be prevented by α-ketoglutarate (Popenoe *et al.*, 1969), and the inhibition by iodoacetate could be prevented by the substrate peptide $(\text{Pro-Gly-Pro})_n$ (Berg and Prockop, 1976), observations which are all consistent with the presence of free sulfhydryls in the active center.

The effects of sulfhydryl-containing compounds such as dithiothreitol on the reaction are probably closely related to the parallel observation by Rhoads *et al.*, (1967) that addition of albumin to the reaction system increases the observed enzymic activity. The effect of albumin was in part explained by the fact that this protein is rich in free sulfhydryl groups, since the albumin lost all its stimulatory effect after it was oxidized with performic acid (Rhoads *et al.*, 1967) and it lost part of its stimulatory effect after treatment with *N*-ethylmaleimide (Popenoe *et al.*, 1969). However, an albumin effect was also observed even in the presence of optimal concentrations of dithiothreitol, and therefore the data suggested that albumin has a "protein effect" which is independent of its effect on sulfhydryl groups (Rhoads *et al.*, 1967).

The effects of dithiothreitol and albumin have not been clearly distinguished from still another set of observations, namely the effects of catalase on the reaction. Because peroxide is readily generated nonenzymatically by solutions of water, O_2, ferrous iron, and ascorbate, enzymic reactions involving these reagents have long been suspected to run the risk of inactivating the enzyme by peroxide generated nonenzymatically. Levin and Kaufman (1961) demonstrated that dopamine hydroxylase activity was preserved by adding catalase to the enzymic reactions, and a similar observation was made in early studies of prolyl hydroxylase (Kivirikko and Prockop, 1967*a*). The effects of catalase, however, were variable in the sense that in many experiments maximal activity was obtained without addition of catalase (Kivirikko and Prockop, 1967*a*). Also, the amounts of catalase required were very large compared to the expected amounts of peroxide generated nonenzymatically (Cardinale and Udenfriend, 1974). It is apparent therefore that in these experiments a clear distinction cannot be made between the effects of catalase as an enzyme on the reaction, its possible "protein effect," and the possible role of sulfhydryl groups supplied by catalase when it is added to the enzymic system.

There are probably several explanations for the continuing uncertainty about the roles of ascorbate, sulfhydryl-containing reagents, albumin, and catalase. Part of the uncertainty is explained by the fact that most of the studies with these reagents were carried out before purified enzyme became available. However, use of purified enzyme may not in itself resolve all of the problems. Because oxygenase reactions involve highly reactive species of oxygen, these reactions are difficult to control. The problem of defining the role of ascorbate is particularly difficult because nonenzymic solutions of ascorbate, ferrous iron, and O_2 are unstable and are known to generate peroxide and free radicals which can participate in hydroxylating reactions (see Bade and Gould, 1969).

A further observation concerning enzymic activity of prolyl hydroxylase is that large amounts of glycine in buffer solutions help to stabilize the enzyme during purification (Halme and Kivirikko, 1968). There is currently no explanation for this observation, and it has not been a consistent finding (Cardinale and Udenfriend, 1974).

3. *Substrate Specificity*

a. Hydroxylation of Prolyl Residues in the Y Position of the Repeating -Gly-X-Y- Sequences. Sequencing the polypeptide chains of various collagens from higher organisms has demonstrated that most of the 4-hydroxyproline is found in the Y position of the repeating -Gly-X-Y- triplets (for review, see Chapter I). This distribution of hydroxyproline is explained by the observation that the prolyl hydroxylases purified from chick embryos and fetal rat skin specifically hydroxylate prolyl residues in the Y position of both protocollagen and synthetic peptides, and they will not hydroxylate prolyl residues in the X position (Kivirikko and Prockop, 1967*a,b*; Hutton *et al.*, 1967*a*; Nordwig and Pfab, 1969).

It is of interest that, although the repeating triplet sequences of collagen are usually written as -Gly-X-Y-, prolyl hydroxylase appears to "read" the peptide substrates by examining the sequences in the order -X-Pro-Gly-. This point was first illustrated by studies on bradykinin which contains a single hydroxylatable sequence of -Pro-Pro-Gly- (Rhoads and Udenfriend, 1969). It was also illustrated by the demonstrations that the simple tripeptides Pro-Gly-Pro and Gly-Pro-Pro (Kivirikko and Prockop, 1967*b*) do not serve as substrates in concentrations of 1–2 mg/ml, whereas the enzyme can hydroxylate the tripeptides Pro-Pro-Gly (Suzuki and

Koyama, 1969; Kikuchi *et al.*, 1969) and Ala-Pro-Gly (Kivirikko *et al.*, 1969).

There are several known exceptions to the statement that the hydroxyproline in collagens is in the Y position. One exception is the cuticle collagen of earthworm in which most of the hydroxyproline is in the X position (Goldstein and Adams, 1968, 1970). Another exception is found near the carboxy-terminal end of the $\alpha 1$ chain of type I collagen in which one of the repeating -Gly-X-Y- triplets is -Gly-Hyp-Hyp- (Fietzek *et al.*, 1972*a*); however, the hydroxyproline in this X position is a 3-hydroxyproline which is probably synthesized by a different enzyme (Berg and Prockop, 1973*a*). A third exception is a sequence of -Gly-Gly-Hyp-Gly-Gly-Hyp- which Fietzek and Rauterberg (1975) have recently identified in two different peptides from type III collagen. Since $(\text{Gly-Pro-Gly})_4$ is a competitive inhibitor for the prolyl hydroxylase from chick embryos (Kivirikko *et al.*, 1969), it would appear that the hydroxyproline in this sequence in type III collagen is not synthesized by the prolyl hydroxylases studied to date. There are however no direct experimental data on this point.

b. Effect of the X-Position Amino Acid and of Peptide Length on the Kinetics of the Reaction. Most of the initial work on prolyl hydroxylase was carried out using as a substrate radioactively labeled protocollagen, the unhydroxylated form of procollagen which can readily be obtained from connective tissues incubated *in vitro* under conditions in which prolyl hydroxylase is inhibited with anaerobiosis or with the iron chelator α,α'-dipyridyl (see below). In comparing various peptide substrates for prolyl hydroxylase, protocollagen can be used as a benchmark, since its turnover number is as high as that of any other substrate studied to date (V_m/E equal to 4–6 sec^{-1}), and its K_m (about 2nM) is clearly the lowest which has been observed with any substrate (Kivirikko and Prockop, 1967*c*; Hutton *et al.*, 1968; Berg and Prockop, 1973*b*; Tuderman *et al.*, 1975; Kishida *et al.*, 1976).

It has been relatively easy to account for the high turnover number of protocollagen since the turnover number obtained with the simple tripeptide Pro-Pro-Gly is of the same order of magnitude if the carboxy-terminal end of the tripeptide is blocked with a group such as *N*-methylamide (Kishida *et al.*, 1976). If the length of tripeptide is extended to a structure such as $(\text{Pro-Pro-Gly})_5$, the turnover number is the same as that of protocollagen. A systematic comparison of peptides analogous to $(\text{Pro-Pro-Gly})_5$ has shown that the turnover number decreases by a factor of about 20 if the prolyl residue in the X position of each tripeptide is replaced by glutamate. The turnover number is reduced to about one third if the X-position amino acid is arginine. These observations suggest the generalization that the most critical determinant for the rate of hydroxylation of any Y-position prolyl residue is the nature of the amino acid in the X position of the same triplet. If the X-position residue is proline, the turnover number is maximal and is the same value as for the most readily hydroxylated prolyl residues in protocollagen.

The extremely low K_m value for protocollagen has been more difficult to account for, but a striking effect of chain length on K_m has been repeatedly observed. Although an appropriately blocked form of the tripeptide Pro-Pro-Gly has a turnover number of the same order of magnitude as protocollagen, its K_m value is at least 10^5 times higher (Table 1). In peptides with more than one hydroxylatable prolyl residue, it is not clear whether K_m values should be calculated on the basis of the molar concentration of peptide or of the molar concentration of hydroxylatable prolyl residues (Kivirikko *et al.*, 1972*a*). However, regardless of which assumption is used for the calculation, the effect of chain length is readily apparent.

At the moment there is no adequate explanation as to why chain length has such a marked effect on the K_m. One reason for the uncertainty here has been that some observations indicate that the length required to achieve the lowest K_m is considerably larger than any single dimension of the enzyme, which has been shown by negative-staining electron microscopy to be about 100 Å on a side (Olsen *et al.*, 1973*a*). For example, the peptide $(\text{Pro-Pro-Gly})_{10}$ has a length of about 86 Å when present in the largely extended form it assumes in a collagen-like triple helix (see below). One might assume, therefore, that it ought to be able to occupy fully any binding site on the enzyme. Its K_m, however, is several orders of magnitude higher than that of protocollagen. Similarly, experiments with polymer fractions of poly(L-Pro)$_n$, which is a competitive inhibitor for the enzyme, suggested that a polymer of more than 150 prolyl residues was required to obtain an optimal K_i value, and therefore optimal binding to the enzyme (Kivirikko *et al.*, 1967*a*; Prockop and Kivirikko, 1969). From the known structure of poly(L-Pro)$_n$ in the crystalline state, one can calculate that a polymer with 150 residues might in solution have an extended length, as a partially flexible helix, of about 450 Å. These observations have raised the possibility that the Michaelis constant for the hydroxylation of large polypeptides by prolyl hydroxylase might be a complex function and that the reaction might involve some unusual mechanism such as a cooperative binding of several enzyme molecules to the substrate or lateral movement of the enzyme along the polypeptide chain as it hydroxylates successive prolyl residues (see below). To date, however, the limited data available tend to speak against such unusual mechanisms (see Prockop and Kivirikko, 1969; Berg *et al.*, 1976).

c. Other Effects of Varying the Amino Acid in the X Position of the Triplet to be Hydroxylated and of Varying Other Parts of the Peptide Chain. The above statements about the effect of the X-position amino acid and the effect of chain length appear to be the major conclusions which can now be reached about peptide substrates for prolyl hydroxylase. However, a number of observations have been made which cannot be fully accommodated within these generalizations and which may be of greater significance than is now apparent.

A series of experiments were carried out by using the homogeneous peptide (Pro-Pro-Gly)$_5$ as a "model" substrate and comparing it to synthetic peptides in which this structure was systematically altered (Kivirikko *et al.*, 1972*a*; Kishida *et al.*, 1976). One of the observations made in these experiments was that insertion of arginine at the amino-terminal end of the peptide or in the X position of the repeating triplets decreased the K_m. In particular, the K_m for Arg-Gly-(Pro-Pro-Gly)$_5$ was about one half the value for (Pro-Pro-Gly)$_5$, and the value for (Arg-Pro-Gly)$_5$ was about one-sixth the value for (Pro-Pro-Gly)$_5$. In contrast, insertion of glutamate or leucine had little effect on K_m in that the peptides Glu-Gly-(Pro-Pro-Gly)$_5$ and (Glu-Pro-Gly)$_5$ had about the same K_m as (Pro-Pro-Gly)$_5$, and the peptide Arg-Glu-(Leu-Pro-Gly)$_5$ had about the same K_m as Arg-Gly(Pro-Pro-Gly)$_5$. The decrease in K_m observed when arginine was introduced into the peptides suggests that positively charged residues might generally enhance binding of substrate peptides to the highly anionic enzyme (see above). However, this suggestion is clearly too simplistic and the failure of glutamate to affect the K_m emphasizes the difficulty of developing a systematic picture with the limited information now available.

An independent series of experiments was carried out using bradykinin as the "model" substrate for the reaction (Rhoads and Udenfriend, 1969; McGee *et al.*, 1971*b*). The K_m for bradykinin was about one-tenth the K_m for the tripeptide Pro-Pro-Gly (Table 1), and the lower K_m for bradykinin may be in part explained by its being a longer peptide of nine amino acids. It may also be in part explained by the fact that arginine is the amino-terminal amino acid immediately preceding the -Pro-Pro-Gly- sequence which is hydroxylated; as indicated above, the presence of arginine near the amino-terminal end of synthetic peptides tends to decrease the K_m values. A variety of derivatives of bradykinin were examined, including forms in which additional amino acids were added to the amino-terminal end (McGee *et al.*, 1971*b*). Most of these derivatives of bradykinin had about the same K_m as the parent compound. The one exception was a bradykinin derivative in which glutamate was added to the amino-terminal position, and in this case the K_m value decreased by one order of magnitude. This result is not consistent with the studies on the

TABLE 1
Effect of Peptide Length on the Michaelis Constant (K_m) for the Hydroxylation of Sequence -X-Pro-Gly-[a]

Peptide substrate	Molecular weight	K_m[b] -X-Pro-Gly- (μM)	K_m[b] Peptide (μM)
Pro-Pro-Gly	269	20,000	20,000
CH_3CO-Pro-Pro-Gly	311	23,000	23,000
Pro-Pro-Gly-$NHCH_3$	282	44,000	44,000
CH_3CO-Pro-Pro-Gly-$NHCH_3$	324	19,000	19,000
Bradykinin	1,060	1,650	1,650
$(\text{Pro-Pro-Gly})_5$	1,274	1,500	300
CH_3CO-$(\text{Pro-Pro-Gly})_5$	1,317	1,400	280
$(\text{Pro-Pro-Gly})_5$-OCH_3	1,289	1,500	300
$(\text{Glu-Pro-Gly})_5$	1,434	1,400	280
$(\text{Arg-Pro-Gly})_5$	1,565	400	80
$(\text{Pro-Pro-Gly})_{10}$	2,531	280	30
$(\text{Pro-Pro-Gly})_{15}$	3,787	50	3
$(\text{Pro-Pro-Gly})_{20}$	5,295	50	2
Ascaris collagen	60,000	2.4	0.04
Protocollagen	120,000	0.2	0.002

[a] Values for the tripeptide Pro-Pro-Gly and derivatives of Pro-Pro-Gly and $(\text{Pro-Pro-Gly})_5$ are taken from the data by Kishida *et al.* (1976). Values for bradykinin are from the data of McGee *et al.* (1971). Values for $(\text{Glu-Pro-Gly})_5$ and $(\text{Arg-Pro-Gly})_5$ are from Kishida *et al.* (1976); values for $(\text{Pro-Pro-Gly})_{10}$, $(\text{Pro-Pro-Gly})_{15}$ and $(\text{Pro-Pro-Gly})_{20}$ are from Kivirikko *et al.* (1972*a*). The value for *Ascaris* collagen is for reduced and carboxymethylated collagen from the cuticle of *Ascaris lumbricoides* (Berg *et al.*, 1976). The value for protocollagen is from Berg and Prockop (1973*b*).

[b] K_m values are presented both in μM concentration of -X-Pro-Gly- triplet units and in μM concentration of the peptide (see text).

homogeneous peptides $(\text{Glu-Pro-Gly})_3$ (Okada *et al.*, 1972), $(\text{Glu-Pro-Gly})_5$ (Kishida *et al.*, 1976), and Glu-Gly-$(\text{Pro-Pro-Gly})_5$ (Kivirikko *et al.*, 1972*a*), and there is at the moment no explanation for this apparent discrepancy.

In a related series of experiments an attempt was made to synthesize derivatives of bradykinin which might be inhibitors of prolyl hydroxylase. The most effective inhibitors were developed by substituting certain proline analogs for the hydroxylatable prolyl residues in the Y position of bradykinin (McGee *et al.*, 1973). Substitution of the analog L-azetidine-2-carboxylic acid did not make the peptide an inhibitor, but inhibitory activity was obtained when the following analogs, in decreasing order of effectiveness, were substituted: L-thiazolidine-4-carboxyl-, *trans*-4-hydroxy-L-prolyl-, 3,4-dehydro-L-prolyl-, and *cis,trans*-4-methyl-DL-prolyl-. The inhibitory activity of several of these analogs was further increased by the addition of L-glutamyl- to the amino-terminal position of bradykinin. Although these inhibitor peptides are of considerable interest, it should be noted that their effects as potential inhibitors for the enzyme are probably not directly related to the effects seen when free proline analogs are incubated with cells synthesizing collagen (see below).

4. *Effect of Substrate Conformation*

One of the most important observations about prolyl hydroxylase has been the demonstration that the enzyme cannot hydroxylate peptide substrates if these substrates are in a triple-helical conformation.

Some early observations with prolyl hydroxylase suggested that the triple-helical conformation might prevent hydroxylation. For example, two laboratories demonstrated that it was difficult to hydroxylate the hydroxyproline-deficient collagen from the cuticle of *Ascaris* worms unless the collagen was first thermally denatured (Fujimoto and Prockop, 1968; Rhoads and Udenfriend, 1968). Also, it was shown that prolyl residues in the Y position of collagens from a variety of tissues are somewhat underhydroxylated (Bornstein, 1967*a,b*; Butler and Ponds, 1971; Balian *et al.*, 1971, 1972; Wendt *et al.*, 1972*a,b*; Fietzek *et al.*, 1972*a,b*), and that these collagens could be further hydroxylated with prolyl hydroxylase only if the collagens were thermally denatured before reaction with the enzyme (Rhoads *et al.*, 1971). However, many investigators did not regard these observations as unequivocal because the helical structure of *Ascaris* cuticle collagen had not been clearly shown to be the same as that of vertebrate collagens. Also, it was difficult to exclude the possibility that native collagen failed to serve as a substrate simply because it had only a few sites available for hydroxylation and because it tended to aggregate into fibers under the conditions used for the enzymic assay.

Definitive proof that the triple-helical conformation prevents hydroxylation was not developed until it was recognized that protocollagen, the substrate used in most of the initial assays for the enzyme, had a melting temperature of about 24° and therefore was nonhelical under the conditions employed for the enzymic reaction (Berg and Prockop, 1973*b–d*; Jimenez *et al.*, 1973*b*; Rosenbloom *et al.*, 1973; Uitto and Prockop, 1973*a,b*, 1974*a*). Once this situation was understood, it was readily demonstrated that if protocollagen is extracted from cells at a temperature below its melting temperature, it is obtained in a triple-helical conformation, and in the triple-helical conformation it does not serve as a substrate for prolyl hydroxylase under conditions in which nonhelical protocollagen is rapidly and completely hydroxylated (Berg and Prockop, 1973*b*,*d*; Murphy and Rosenbloom, 1973).

In retrospect, it is now apparent that early studies indicating that conformation of the substrate had no effect were in error. Experiments with synthetic peptides were complicated by difficulties in thoroughly ensuring that the peptides were completely triple helical while they were being tested in the enzymic system (Kivirikko *et al.*, 1972*a*). Reports concluding that protocollagen in the triple-helical conformation could be hydroxylated (Hutton *et al.*, 1967*b*; Kivirikko *et al.*, 1968; Nordwig and Pfab, 1968) are now known to involve misinterpretations of the data in that it was initially assumed that the melting temperature of helical protocollagen was about the same as collagen.

As discussed below, the failure of prolyl hydroxylase to hydroxylate triple-helical polypeptides is closely related to the fact that hydroxyproline is essential for collagen polypeptides to form a stable triple-helical conformation under physiological conditions.

5. *Mechanism of the Reaction*

Prolyl hydroxylase as well as lysyl hydroxylase (see below) are examples of the new class of mixed function oxygenases involving ferrous iron and α-ketoglutarate (for reviews, see Abbott and Udenfriend, 1973; Cardinale and Udenfriend, 1974). Although it has been amply demonstrated that α-ketoglutarate is stoichiometrically converted to succinate and carbon dioxide during both the prolyl hydroxylase (Rhoads and Udenfriend, 1968; Cardinale *et al.*, 1971) and the lysyl hydroxylase (Kivirikko *et al.*, 1972*b*) reactions, the precise mechanism of such reactions has not been defined. Elucidation of the mechanism could be useful for understanding not only these two enzymic reactions but the reactions of other enzymes in this class which includes γ-butyrobetain hydroxylase (Lindstedt, 1967; Lindstedt *et al.*, 1968; Lindstedt and Lindstedt, 1970), pyrimidine deoxyribonucleoside 2-hydroxylase (Shaffer *et al.*, 1968), thymine 7-hydroxylase (Abbott *et al.*, 1967; Abbott and Udenfriend, 1973), 5-hydroxymethyluracil oxygenase (Abbott *et al.*, 1968), and 5-formyluracil oxygenase (Watanabe *et al.*, 1970). Recent work indicates that the last three of these reactions, in which thymine is converted by three sequential steps to uracil-5-carboxylic acid, are catalyzed by a single protein (Liu *et al.*, 1973). Formally the reaction that synthesizes homogentisic acid from *p*-hydroxyphenpyruvate follows a similar mechanism (Lindblad *et al.*, 1970).

Two distinct mechanisms have been proposed for such reactions.

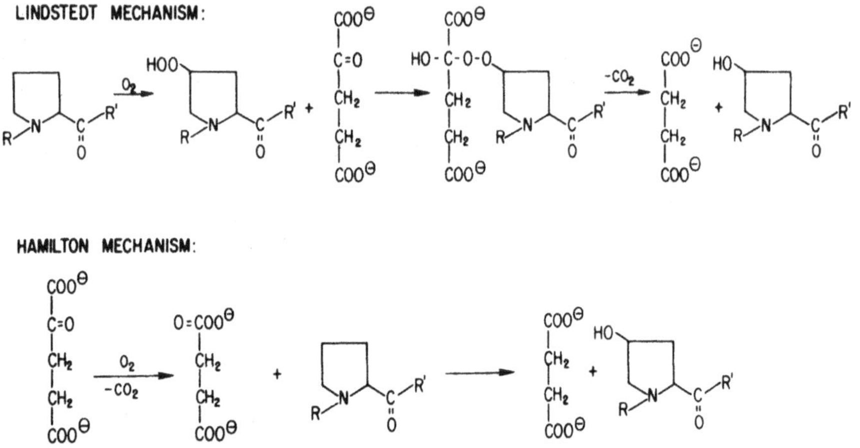

FIGURE 3. Schematic representations for the Lindstedt and Hamilton mechanisms for the hydroxylation of peptidyl proline by prolyl hydroxylase (see text).

Lindstedt and his associates (Holme *et al.*, 1968; Lindblad *et al.*, 1969) proposed that such reactions could proceed by, first, the formation of a complex between the enzyme, O_2, and iron (Figure 3) and this complex could form a hydroperoxide derivative with a substrate such as peptidyl proline. In the second step of the reaction they suggested that the hydroperoxide attacked α-ketoglutarate and then decomposed to the hydroxylated product, succinate and CO_2 (Figure 3).

Hamilton (1971) pointed out that since the substrates of all the oxygenases in this class are simple aliphatic compounds, formation of the hydroperoxide intermediate would be a difficult reaction. On the basis of this and other arguments, he proposed that the reaction might proceed by a mechanism in which persuccinic acid was generated and the persuccinic acid then carried out an oxidative attack on the substrate (Figure 3).

Initial attempts to elucidate the prolyl hydroxylase mechanism by kinetic analysis provided data which were difficult to interpret because of the relatively large number of components in the reaction (Kivirikko *et al.*, 1968). Lindblad *et al.* (1969) obtained evidence in favor of the Lindstedt mechanism by examining the reaction of *t*-butyl hydroperoxide with α-ketoglutarate. They have also supported their mechanism by studies indicating they could not observe uncoupling of α-ketoglutarate decarboxylation from hydroxylation in the γ-butyrobetaine hydroxylase reaction (Lindstedt and Lindstedt, 1970), as would be required by the Hamilton mechanism.

Several observations have, however, provided evidence in favor of the Hamilton mechanism. Cardinale and Udenfriend (1974) showed that decarboxylation of α-ketoglutarate is carried out by partially purified prolyl hydroxylase in the absence of any peptidyl proline. This observation supports the Hamilton mechanism, but Cardinale and Udenfriend (1974) found that 9 moles of carbon dioxide were generated for each mole of enzyme instead of the expected 1 mole. The reason for this discrepancy is not apparent at the moment, but it is possible that in the partial reaction the enzyme generates an unstable form of persuccinic acid. Hurych *et al.* (1973) have also provided support for the Hamilton mechanism on the basis of quantum chemical calculations of the reaction mechanism. It would appear, however, that further experimentation will be necessary to reach a final conclusion as to whether the Lindstedt mechanism or the Hamilton mechanism is the correct one.

A separate aspect of the reaction has been the question of whether superoxide ion (O_2^-) is involved in the reaction. It was observed (Bhatnagar and Liu, 1973; Liu and Bhatnagar, 1973) that prolyl hydroxylase was inhibited with epinephrine and Nitro Blue Tetrazolium, reagents which tend to trap superoxide. However, the reaction was not inhibited by superoxide dismutase (Bhatnagar and Liu, 1973; Liu and Bhatnagar, 1973; Cardinale and Udenfriend, 1974). Since it is difficult in such experiments to rule out nonspecific effects from reagents such as epinephrine and Nitro Blue Tetrazolium, it is not entirely clear at this stage whether superoxide is in fact involved in the reaction.

Still another aspect of the reaction mechanism has been the question of how the reaction proceeds when the enzyme encounters protocollagen or any extended polypeptide with several prolyl residues in the Y position. One approach to this problem has been developed by preparing peptides of the structure $(\text{Pro-Pro-Gly})_5$ (Kivirikko *et al.*, 1971) and $(\text{Pro-Pro-Gly})_{10}$ (Berg *et al.*, 1976) in which specific triplets were labeled with [^{14}C]proline. By measuring the synthesis of [^{14}C]hydroxyproline it was possible to follow the hydroxylation of specific triplets in the peptides. The results demonstrated a remarkable asymmetry in that with $(\text{Pro-Pro-Gly})_5$ the fourth triplet from the amino-terminal end (Kivirikko *et al.*, 1971), and with $(\text{Pro-Pro-Gly})_{10}$ the ninth triplet was hydroxylated much more readily than any other triplet (Berg *et al.*, 1976) (Figure 4). The enzyme therefore appeared to hydroxylate preferentially the penultimate triplet in the peptides. The reasons why the penultimate triplet is most readily hydroxylated have not, however, been determined. One possibility is that the enzyme has a binding site which accommodates the amino-terminal region of peptides such as $(\text{Pro-Pro-Gly})_{10}$ and a reactive site which, after the peptide is

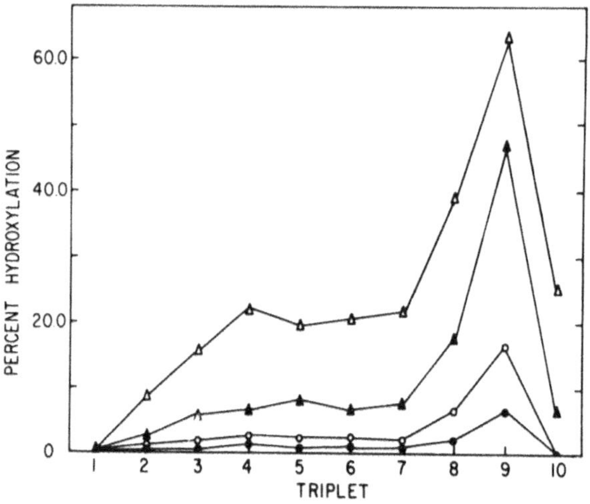

FIGURE 4. Asymmetry in the hydroxylation of (Pro-Pro-Gly)$_{10}$ by prolyl hydroxylase. Nine different ^{14}C-labeled preparations of (Pro-Pro-Gly)$_{10}$ were hydroxylated to varying extents by prolyl hydroxylase, and the degree of hydroxylation in each triplet was estimated by assaying the peptides for [^{14}C]hydroxyproline. Triplet number refers to triplets numbered from the amino-terminal end. The overall degree of hydroxylation of the peptide was 1.3% (●——●), 3.4% (○——○), 10.7% (▲——▲), and 23.4% (△——△). Data by Berg *et al.*, 1976.

bound, preferentially hydroxylates a triplet near the carboxy-terminal end (Figure 5). However, there are a number of other possible explanations for the data.

The results obtained by hydroxylating ^{14}C-labeled (Pro-Pro-Gly)$_{10}$ were also of interest in that they provided a test of the possibility that the enzyme hydroxylates large peptides by means of a zipper-like mechanism in which it moves along the peptides and hydroxylates successive triplets. At low levels of hydroxylation of (Pro-Pro-Gly)$_{10}$, the penultimate triplet was hydroxylated far more than any other triplet. Therefore it was apparent that hydroxylation of this peptide does not involve a zipper-like mechanism in which more than one triplet must be hydroxylated in each productive encounter between enzyme and substrate. However, since the extended length of (Pro-Pro-Gly)$_{10}$ is not much greater than the diameter of the enzyme (Figure 5), the data probably do not provide an adequate test for the possibility that such a zipper-like mechanism is involved in the hydroxylation of longer peptides such as protocollagen. Some evidence against any zipper-like mechanism in the hydroxylation of protocollagen-like peptides was obtained in earlier studies in which enzyme–substrate complexes were isolated by gel filtration (Juva and Prockop, 1969), but more extensive work is probably necessary to exclude this mechanism. The point is worthy of further study, since a zipper-like mechanism might well explain the large effect of chain length on K_m seen with various peptides (Table 1).

6. *Assays for Enzymic Activity and for Enzyme Protein*

Purification and characterization of prolyl hydroxylase has been in large part dependent on the availability of appropriate assays for the enzyme. As summarized in Table 2, it is now possible to assay the enzyme with a variety of substrates and a variety of procedures. Two of the assays depend on the use of radioactively labeled protocollagen. Protocollagen is prepared by incubating connective tissue cells or intact tissues with radioactive proline and either without O_2 or with α,α'-dipyridyl. The protocollagen is then extracted from the cells or tissues and used as a substrate. Several different tissues from chick embryos have been used to prepare the radioactively labeled protocollagen, and these have included using minces of whole chick embryos, isolated tissues such as cartilage, and cells isolated from leg tendons by enzymic digestion under controlled conditions. The results obtained with protocollagen from these various sources are similar, but several precautions are usually necessary: The same preparation of the protocollagen should be used for all samples in a given assay, the substrate should be heat denatured to destroy residual prolyl hydroxylase activity before storage, and repeated freezing and thawing of the protein should be avoided. The primary limitations of assays using radioactively labeled protocollagen are the difficulty in

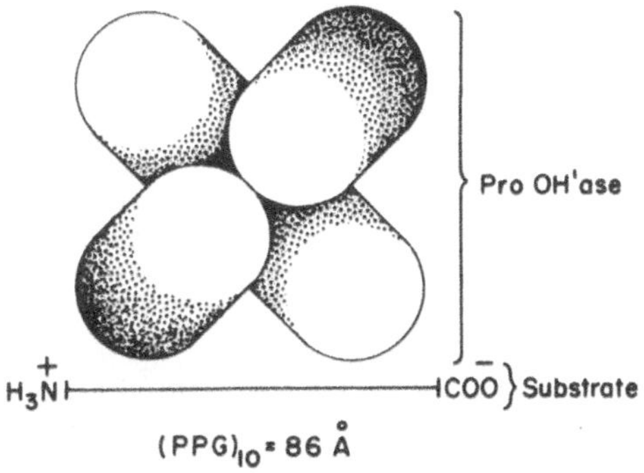

FIGURE 5. Relative dimensions of prolyl hydroxylase and the substrate peptide (Pro-Pro-Gly)$_{10}$. The length of the peptide corresponds to its length in the largely extended conformation the peptide assumes when it forms a collagen-like triple helix.

TABLE 2

Assays of Prolyl Hydroxylase Activity

Substrate	Product assayed	Sensitivity in nanograms of enzyme	Comments	Ref.[d]
[^{14}C]Proline-labeled protocollagen	[^{14}C]Hydroxyproline by chromatography or by radiochemical procedure (e.g., ref. 3)	2[a]	Highly specific; highly sensitive; substrate difficult to prepare; linear over narrow range; relatively tedious	1–6
[^{3}H]Proline-labeled protocollagen	[^{3}H]H_2O released from 4-carbon of proline (e.g., ref. 7)	1–10[b]	Highly sensitive; less specific than [^{14}C]proline-labeled protocollagen assay; substrate difficult to prepare; linear over narrow range; relatively rapid and simple	7,8
(Pro-Pro-Gly)$_n$	Hydroxyproline (e.g., ref. 12)	120–240[c]	Highly specific; sensitivity low; substrate commercially available; linear over broad range; relatively slow and tedious	9–12
(Pro-Pro-Gly)$_n$ or *Ascaris* cuticle collagen	[^{14}C]CO_2 released from [^{14}C]α-ketoglutarate (see ref. 13)	10–60[c]	Nonspecific; can only be used with partially purified enzyme; sensitivity "intermediate;" substrate commercially available; linear over broad range; rapid and simple	11,13,14

[a] Sensitivity is estimated from the observation that the assay can detect the synthesis of 200 dpm of hydroxy[^{14}C]proline. With saturating concentrations of [^{14}C]proline-labeled protocollagen prepared with tendon cells under standard conditions (Berg and Prockop, 1973*b*), 1 ng of prolyl hydroxylase from chick embryos synthesized about 100 dpm of hydroxy[^{14}C]proline in 1 hr at 37°C (Berg and Prockop, unpublished data). The sensitivity can probably be increased slightly by increasing the amount of [^{14}C]proline and decreasing the number of cells used to prepare the [^{14}C]protocollagen (Berg and Prockop, 1973*b*), thereby increasing the specific activity of the substrate.

[b] Sensitivity is estimated from the observation that the assay can detect the release of 100 cpm of [^{3}H]H_2O (See Hutton *et al.*, 1966). About 10 ng of prolyl hydroxylase purified from newborn rat skin will release about 100 cpm of [^{3}H]H_2O from [^{3}H]proline-labeled protocollagen in 60 min at 30°C (Rhoads and Udenfriend, 1970; Stassen *et al.*, 1974). The observed specific activity of the enzyme purified from rat skin (0.8 μg of hydroxyproline synthesized per μg enzyme per hr at 30°C) was about one-tenth the specific activity of the enzyme purified from chick embryos (8 μg of hydroxyproline synthesized per μg enzyme per hr at 37°C). Since the calculation of sensitivity is based on the observed specific activity and since some of the enzyme protein may have been inactivated during purification of the enzyme from newborn rat skin, the sensitivity may be as low as 1 ng. Also, the sensitivity apparently can be improved by increasing the amount [^{3}H]proline used for the preparation of [^{3}H]protocollagen from chick embryo tissues (see Fleckman *et al.*, 1973).

[c] Sensitivity is estimated from the observations that the synthesis of 1 μg (8 nmoles) of hydroxyproline can be detected with the specific chemical assay for this amino acid (Kivirikko *et al.*, 1967*b*; Kivirikko and Prockop, 1967*a*; Halme *et al.*, 1970*a*) and the production of 150 cpm [^{14}C]carbon dioxide (about 2 nmoles) can be detected with the [^{14}C]α-ketoglutarate assay (Berg and Prockop, 1973*a*). When 250 μg per ml (58% of saturating concentration) of a polymer fraction of $(\text{Pro-Gly-Pro})_n$ with an average molecular weight of 2400 is used as a substrate, 240 ng of prolyl hydroxylase synthesizes 1 μg of hydroxyproline per hr at 37°C. When 500 μg per ml (saturating concentration) of the homogeneous peptide $(\text{Pro-Pro-Gly})_{10}$ is used as a substrate, 120 ng of prolyl hydroxylase synthesizes 1 μg of hydroxyproline per hr at 37°C. The sensitivity of the assay using [^{14}C]α-ketoglutarate can be increased to 10 ng by using [^{14}C]α-ketoglutarate of a higher specific activity than that used in published reports (K. I. Kivirikko, unpublished data). Polymers with the structure $(\text{Pro-Gly-Pro})_n$ have been available from Miles, Yeda, Ltd., Rehovot, Israel. Homogeneous peptides $(\text{Pro-Pro-Gly})_5$ and $(\text{Pro-Pro-Gly})_{10}$ can be purchased from the Protein Research Foundation, Minoh, Osaka, Japan or from Beckman Inctruments, Inc., Palo Alto, California, U.S.A.

[d] References: 1. Kivirikko and Prockop (1967*c*); 2. Takeuchi *et al.* (1967); 3. Juva and Prockop (1966*b*); 4. Kivirikko and Prockop (1967*b*); 5. Berg and Prockop (1973*b*); 6. Kishida *et al.* (1976); 7. Hutton *et al.* (1966); 8. Fleckman *et al.* (1973); 9. Halme *et al.* (1970*a*); 10. Kivirikko and Prockop (1967*a*); 11. Berg and Prockop (1973*a*); 12. Kivirikko *et al.* (1967*b*); 13. Rhoads and Udenfriend (1968); 14. Rhoads and Udenfriend (1970).

preparing the substrate and the relatively narrow range over which the assays are linear. The assay in which [^{14}C]hydroxyproline is measured has the advantage of being highly sensitive and highly specific. However, the assay for the [^{14}C]hydroxyproline is relatively tedious. Assay of the reaction is considerably simplified if the protocollagen is labeled with [3,4-^{3}H]proline and the amount of ^{3}H released to water is measured. However, nonspecific release of [^{3}H]H_2O is occasionally encountered, and the amount of [^{3}H]H_2O released appears to vary with different commercial preparations of tritiated proline and with storage of the [^{3}H]protocollagen.

Use of synthetic peptide substrates such as (Pro-Pro-Gly)$_n$ overcomes the inconvenience of preparing protocollagen since these peptides are now commercially available. Assays using synthetic peptides are highly specific if a chemical assay for hydroxyproline is used to follow the reaction, but the chemical assay is somewhat tedious. The assay becomes much easier to carry out if the reaction is followed by the release of [^{14}C]carbon dioxide from [1-^{14}C]α-ketoglutarate, but in this case the reaction becomes less specific and is only reliable if purified enzyme is used. One variation on these assays is to use as substrate the hydroxyproline-deficient collagen from the cuticle of *Ascaris*. This has the advantage of being less expensive if *Ascaris* worms are readily available.

Preparation of antibodies to purified prolyl hydroxylase has made it possible to assay the amount of cross-reacting protein in cells which synthesize collagen (McGee *et al.*, 1971*a*; Berg *et al.*, 1972). This approach has been used most extensively by Udenfriend and co-workers who have shown that many types of cells contain large amounts of cross-reactive protein which is enzymically inactive (see below). The assay developed by McGee and Udenfriend (1972*a*) for measuring cross-reactive protein depended on inhibition of enzymic activity by the antibodies. In the initial procedure the amount of total enzyme protein in the homogenate was assayed by measuring the extent to which extracts of the homogenate displaced active prolyl hydroxylase from enzyme–antibody complexes (McGee and Udenfriend, 1972*a*). In a more recent modification of this procedure, all the prolyl hydroxylase in the samples to be assayed is first heat-inactivated and then the inactivated protein is reacted with antibody (Stassen *et al.*, 1974). The amount of the antibody which is bound by inactive prolyl hydroxylase is then assayed by determining how much unreacted antibody is still available to subsequently inactivate a measured amount of active enzyme. Alternate procedures have recently been developed in which the cross-reacting protein is assayed by hemagglutination inhibition (Olsen *et al.*, 1973*b*, 1975), or the enzyme is first labeled with radioactive iodine and a more standard type of immunoprecipitation test is then carried out (Kao *et al.*, 1975*a,b*).

B. Hydroxylation of Peptidyl Lysine

The lysyl hydroxylase responsible for synthesizing the hydroxylysine in collagen has many similarities to prolyl hydroxylase in terms of cofactors, cosubstrates, and mechanism, but it has been shown to be a separate enzymic protein (Weinstein *et al.*, 1969; Halme *et al.*, 1970*a*; Miller, 1971; Kivirikko and Prockop, 1972; Popenoe and Aronson, 1972).

1. Properties and Cofactors of Lysyl Hydroxylase

Lysyl hydroxylase has not been obtained in pure form, but it has been purified about 500-fold from chick embryo extracts (Kivirikko and Prockop, 1972). The activity of the partially purified enzyme was recovered in two peaks by gel filtration and the elution positions of these two peaks correspond to molecular weights of about 550,000 and 200,000. It was not clear, however, whether these two peaks represent two forms of a single enzyme or two different proteins.* During anion exchange chromatography the lysyl hydroxylase is eluted before prolyl hydroxylase, suggesting that the protein is less acidic than the highly acidic prolyl hydroxylase (Miller, 1971; Kivirikko and Prockop, 1972; Popenoe and Aronson, 1972).

As with prolyl hydroxylase, the lysyl hydroxylase reaction involves molecular oxygen, α-ketoglutarate, iron, and a reducing agent such as ascorbate (Prockop *et al.*, 1966; Kivirikko and Prockop, 1967*c*; Hausmann, 1967; Hurych and Nordwig, 1967). The reaction is also similar to the prolyl hydroxylase reaction in that α-ketoglutarate is stoichiometrically decarboxylated to carbon dioxide and succinate during the reaction (Kivirikko *et al.*, 1972*b*). A further similarity to prolyl hydroxylase is the need for free sulfhydryl groups, since the reaction is inhibited by *p*-mercuribenzoate in a reversible manner (Kivirikko and Prockop, 1972). Also, bovine serum albumin and dithiothreitol, in appropriate concentrations, enhance lysyl hydroxylase activity (Kivirikko and Prockop, 1972; Popenoe and Aronson, 1972).

* There is some circumstantial evidence which suggests that the hydroxylysine found in the nonhelical peptide extension of the α chains of collagen may be synthesized by a separate lysyl hydroxylase (see Barnes *et al.*, 1974).

2. *Substrate Specificity and Effect of Substrate Conformation*

The amount of information about peptide substrates for lysyl hydroxylase is far more limited than comparable information about peptide substrate for prolyl hydroxylase, but a number of similarities have been established. One of the most important observations is that protocollagen is also a highly effective substrate for lysyl hydroxylase (Kivirikko and Prockop, 1967*c*), and the K_m for protocollagen as a substrate for lysyl hydroxylase is essentially the same as the extremely low value seen when the same protein is used as a substrate for prolyl hydroxylase (Ryhänen and Kivirikko, 1974*b*).

Free lysine and the tripeptide Lys-Gly-Pro did not serve as a substrate for the reaction (Kivirikko and Prockop, 1967*b*; Kivirikko *et al.*, 1972*b*). A slow rate of hydroxylysine synthesis was observed with the tripeptide Ile-Lys-Gly, and lysine-vasopressin with the carboxy-terminal sequence of -Pro-Lys-Gly-NH_2 was a substrate for the synthesis of hydroxylysine (Kivirikko *et al.*, 1972*b*). These observations indicated that a single triplet of X-Lys-Gly fulfills the minimum requirement for recognition by the enzyme. Further observations showed that both chain length and the amino acid sequence around the Y-position lysine were critical determinants in the substrate activities of various peptides. For example, the hexapeptide $(\text{Ile-Lys-Gly})_2$ was a much better substrate than the tripeptide Ile-Lys-Gly (Kivirikko *et al.*, 1972*b*). Also, studies on the hydroxylation of peptide fragments from collagen indicated that lysine in the sequence -Ala-Lys-Gly- was a particularly poor substrate (Kivirikko *et al.*, 1973). One intriguing observation was that the collagen from the cuticle of *Ascaris* worms was not hydroxylated even though it contains some 40 lysyl residues per 1000 (Kivirikko *et al.*, 1972*b*). There appear to be two possible explanations for the failure of the *Ascaris* cuticle collagen to serve as a substrate: either all lysyl residues are found in the X position of the repeating -X-Y-Gly- triplets, or the amino acid sequences around Y-position lysyl residues are highly unfavorable for the hydroxylation. Similar arguments might be considered to explain the observation that after maximal hydroxylation of isolated $\alpha 1$ and $\alpha 2$ chains of collagen, less hydroxylysine is found in the $\alpha 1$ chain than in the $\alpha 2$ chain (Kivirikko *et al.*, 1973).

Studies with [^{14}C]lysine-labeled protocollagen directly demonstrated that the triple-helical conformation prevents hydroxylation of this protein (Ryhänen and Kivirikko, 1974*b*). Under the experimental conditions employed, nonhelical protocollagen was readily hydroxylated, whereas the degree of hydroxylation of helical protocollagen was less than one residue per polypeptide chain.

These studies with protocollagen extended and confirmed earlier studies on collagen. Sequencing of the α chains of type I collagen revealed that a relatively large number of lysyl residues in the Y position of the repeating triplets were hydroxylated incompletely or not at all (Butler, 1968, 1972; Bornstein, 1969; Balian *et al.*, 1971, 1972; Wendt *et al.*, 1972*a*,*b*). Subsequently it was shown that the level of hydroxylation of lysyl residues was increased if type I collagen was thermally denatured and then incubated with large amounts of lysyl hydroxylase *in vitro*, but native collagen was not hydroxylated under the same conditions (Kivirikko *et al.*, 1973).

3. *Assays for the Enzyme*

A number of assays have been developed for the lysyl hydroxylase, and the principles used in these assays are very similar to the assays described above for the prolyl hydroxylase.

One of the first assays reported involved the use of [^{14}C]lysine-labeled protocollagen as a substrate and following the reaction by the synthesis of peptidyl [^{14}C]hydroxylysine. Initially the [^{14}C]hydroxylysine was assayed by hydrolysis of the protein and then by ion-exchange chromatography of the free amino acid (Kivirikko and Prockop, 1967*c*). Subsequently a radiochemical assay was developed in which peptidyl [^{14}C]hydroxylysine was oxidized with periodate so that the C-6 carbon was released as [^{14}C]formaldehyde (Blumenkrantz and Prockop, 1969). This particular assay procedure is highly specific and sensitive for measuring enzymic activity (Kivirikko and Prockop, 1972). The shortcomings of the assay are that the substrate is difficult to prepare and to store (see above), and the assay is linear only over a narrow range.

Another assay procedure has been developed using [^{3}H]lysine-labeled protocollagen and measuring [^{3}H]H_2O released during the hydroxylation of lysine (Miller, 1972). The assay is relatively simple and rapid, but is less specific because of the possibility of nonspecific release of [^{3}H]H_2O from the substrate and because of some variability in the distribution of tritium in commercially available [^{3}H]lysine. In addition, the method has many of the same shortcomings as the assay with ^{14}C-labeled protocollagen, and it has only been used in a few studies.

Assays for lysyl hydroxylase have also been developed in which synthetic peptides are used as substrates (Kivirikko *et al.*, 1972*b*). One of these assays has involved measuring the amount of hydroxylysine synthesized by a chemical procedure (Blumenkrantz and Prockop, 1971), but this assay requires relatively large amounts of enzyme and the chemical assay for hydroxylysine is relatively tedious. A second assay using synthetic peptides has been developed using [^{14}C]α-ketoglutarate as one of the cosubstrates and measuring the release of [^{14}C]carbon dioxide (Kivirikko *et al.*, 1972*b*). This procedure is very rapid and simple but, as in the case of the similar assay developed for prolyl hydroxylase, it is relatively nonspecific and can only be used with partially purified enzyme.

C. *Glycosylation of Peptidyl Hydroxylysine*

Since the glucosylgalactose and the galactose found in collagen are linked to the molecule through an *O*-glycosidic bond to hydroxylysine (Butler and Cunningham, 1966; Spiro, 1967, 1969; Cunningham and Ford, 1968), it is clear that the transferase reactions which add the carbohydrates occur after synthesis of hydroxylysine. Less is known about the transferase enzymes than the hydroxylases in part because of the difficulty of developing specific assays for these enzymes.

1. *Properties and Cofactors of the Enzymes*

Enzymes which transfer galactose to peptidyl hydroxylysine and glucose to galactosylhydroxylysine have been shown to be present in a number of connective tissues (Spiro and Spiro, 1971*c*). Both transferases have been purified 15- to 20-fold from rat kidney cortex (Spiro and Spiro, 1971*a,b*) and 50- to 100-fold from chick embryo cartilage (Myllylä *et al.*, 1975*a*). Partial purification of both of these enzymes from guinea pig skin has also been reported (Bosmann and Eylar, 1968*a,b*), but it was subsequently pointed out that the assays used for this work were nonspecific (Spiro and Sprio, 1971*a,b*; Myllylä *et al.*, 1975*a*). Recently, a purification of over 2000-fold was reported for collagen glucosyltransferase from extracts of whole chick embryos (Myllylä *et al.*, 1976). The molecular weight of collagen glucosyltransferase from whole chick embryos and chick embryo cartilage was found to be about 52,000–54,000, when determined by gel filtration (Myllylä *et al.*, 1976).

Galactosyltransferase transfers galactose from UDP-galactose to peptidyl hydroxylysine to form peptidyl *O*-β-D-galactopyranosylhydroxylysine. Glucosyltransferase transfers glucose from UDP-glucose to galactosylhydroxylysine to form 2-*O*-α-glucopyranosyl-*O*-β-D-galactosylpyranosylhydroxylysine (Spiro and Spiro, 1971*a,b*). Both enzymes have been reported to require a divalent metal cofactor, but there has been some disagreement as to the specificity of this requirement. Initial reports indicated that Co^{2+} and Mn^{2+} were found to be the most effective metals, and the requirement for Mn^{2+} was only partly replaceable by several other divalent cations.

2. *Substrate Specificity*

As with prolyl and lysyl hydroxylases, the question of substrate specificity for the two transferases must be considered in terms of both amino acid sequences which favor the action of the enzymes on a peptide substrate and the effect of the conformation of the peptide on the reaction.

Although free hydroxylysine and free galactosylhydroxylysine are not intermediates in collagen biosynthesis, there has been some interest in whether such residues might be glycosylated, since they might serve as convenient substrates for assays of enzymic activity. Also, glycosylation might occur during the terminal stages of collagen degradation (see Kivirikko, 1970).

In initial studies with the galactosyltransferase from guinea pig skin, it was reported that free hydroxylysine could serve as an acceptor (Bosmann and Eylar, 1968*a*). This observation was not, however, confirmed in subsequent studies with the enzyme from rat kidney cortex (Spiro and Spiro, 1971*b*) or chick embryo cartilage (Myllylä *et al.*, 1975*a*). Since the two latter studies were carried out with more specific assay procedures, it seems probable that the enzyme does not act on free hydroxylysine. The glucosyltransferase has been reported to act on galactosylhydroxylysine which is not in peptide linkage (Spiro and Spiro, 1971*a*), but the reaction has not been examined in much detail.

Most of the information about amino acid sequences which favor glycosylation has come from analyses of amino acids around glycosylated hydroxylysyl residues in various collagens. Such studies have demonstrated that most of the glycosylated hydroxylysine in a variety of α chains is found in one of the following kinds of sequences (Cunningham and Ford, 1968; Morgan *et al.*, 1970; Isemura *et al.*, 1972; Aguilar *et al.*, 1973):

-Gly-Met-*Hyl*-Gly-His-Arg-Gly-Phe-*Hyl*-
-Gly-Phe-*Hyl*-Gly-Ile-Arg-
-Gly-Ile-*Hyl*-Gly-His-Arg-
-Gly-Pro-*Hyl*-Gly-Glu-Leu-

At the time when only the first three of those sequences had been determined, it was suggested that the presence of arginine in the next triplet was a critical feature for determining which hydroxylysyl residues in procollagen are glycosylated. The discovery of the fourth sequence (Aguilar *et al.*, 1973) undermines the validity of this hypothesis.

Conflicting data have been reported on the effect of peptide conformation, but recent results indicate that a triple-helical conformation prevents addition of the second sugar residue, glucose. Denatured collagens from a variety of sources served as substrates for both the galactosyltransferase and the glucosyltransferase from rat kidney cortex (Spiro and Spiro, 1971*a,b*) or chick embryo cartilage (Myllylä *et al.*, 1975*a*). Both transferases from rat kidney cortex were reported to glycosylate native collagen as well (Spiro and Spiro, 1971*a,b*), but subsequent studies with collagen glucosyltransferase from chick embryos indicated that triple-helical conformation of collagen prevents glucosylation of galactosylhydroxylysyl residues (Myllylä *et al.*, 1975*b*). The discrepancy between these two observations may well be explained by the fact that the reactions with the transferases from rat kidney cortex were carried out by incubating the system at 37°C for 2 hr (Spiro and Spiro, 1971*a,b*), conditions under which the collagen substrate might have denatured. At the moment there are no data, other than the studies with rat kidney cortex (Spiro and Spiro, 1971*a,b*), as to the effect of triple-helical conformation on galactosyltransferase reaction.

3. *Assays for the Glycosyltransferases*

Development of specific assays for the glycosyltransferases has been a particularly vexing problem. Some of the first studies on the enzymes used a relatively simple assay in which the incorporation of radioactivity from UDP-[^{14}C]galactose or UDP-[^{14}C]glucose into a trichloroacetic acid-precipitable fraction was followed (Bosmann and Eylar, 1968*a,b*; Kirschbaum and Bosmann, 1973). Subsequent studies demonstrated that this procedure was not specific enough and that large amounts of the radioactivity recovered in the precipitate were present in forms other than glycosylated hydroxylysine (Spiro and Spiro, 1971*a,b*; Myllylä *et al.*, 1975*a*). More specific assays of both enzymes were subsequently described by Spiro and Spiro (1971*a,b*) but they involved a paper chromatography step which required six days. More recently a second specific assay procedure has been developed which is simpler, but it still requires at least two days to assay a reasonable number of samples (Myllylä *et al.*, 1975*a*). It is clear that work on the enzymes would be greatly facilitated by more rapid assay procedure.

D. *Synthesis of Disulfide Bonds*

The earliest amino acid analyses of collagens indicated that the most common forms of the protein did not contain cystine or cysteine, and therefore the molecules could not contain either interchain or intrachain disulfide bonds. Accordingly, the discovery of half-cystine residues and disulfide bonds in procollagen was greeted with great interest. Disulfide bonds have now been demonstrated in all the procollagens which have been examined to date. In addition, it has been shown that type III and type IV collagens contain disulfide bonds.

In type I and type II procollagen the three pro-α chains are linked together by interchain disulfide bonds among the three peptide extensions of the molecules (see above). The presence of interchain disulfide bonds in type I and type II procollagen was demonstrated by examining the effects of reduction with agents such as mercaptoethanol on the apparent size of the polypeptide chains (Dehm *et al.*, 1972; Goldberg *et al.*, 1972; Smith *et al.*, 1972; Fessler *et al.*, 1973; Harwood *et al.*, 1973; Monson and Bornstein, 1973; Sherr *et al.*, 1973; Uitto and Prockop, 1973*a*, 1974*b*; Schofield *et al.*, 1974*a,b*). In addition, the presence of the interchain disulfide bonds was demonstrated by experiments in which the collagen portion of the molecule was removed by digestion with bacterial collagenase so that one obtained a trimeric fragment which consisted of three peptide extensions (Sherr *et al.*, 1973; Dehm *et al.*, 1974; Schofield *et al.*, 1974*b*). In the case of type I procollagen synthesized by skin fibroblasts and by embryonic tendon cells, it has also been shown that the pro-α chains contain *intra*chain disulfide bonds in addition to the *inter*chain disulfide bonds, since after extensive reduction of the isolated pro-α chains some of the antigenic determinants were lost (Timpl *et al.*, 1973; Dehm *et al.*, 1974).

In the case of type III procollagen and collagen, the collagen form of the molecule has been shown to contain interchain disulfide bonds among the two half-cystine residues near the carboxy-terminal end of the molecule (Chung and Miller, 1974; Epstein, 1974). Although a type III procollagen has been identified (Byers *et al.*, 1974), it has not been demonstrated whether the procollagen contains interchain disulfide bonds in addition to those present in the collagen form of the molecule. In the case of type IV procollagen and collagen found in tissues synthesizing basement membranes, the collagen form of the protein contains half-cystine residues (Kefalides, 1972, 1973), and there is preliminary evidence that these interchain disulfide bonds are present near the amino-terminal end of the molecule (Alper and Kefalides, 1974). A type IV procollagen has been identified in lens (Grant *et al.*, 1973) and in renal glomeruli (Grant and Harwood, 1974) and shown to contain interchain disulfide bonds, but it is not yet clear whether these are additional disulfide bonds not found in the collagen itself.

From the evidence discussed below, it is apparent that the interchain disulfide bonds of type I, II, and IV procollagens are synthesized after assembly of amino acids into pro-α chains and that the synthesis of these bonds is closely related to the protein assuming a triple-helical conformation (see Grant *et al.*, 1973; Schofield *et al.*, 1974*a*; Uitto and Prockop, 1973*a*, 1974*b*). However, it is not clear from our present information about procollagen, or from our information about interchain disulfide bonds in any protein (see Wetlaufer and Ristow, 1973), whether an enzymic catalyst is necessary to promote the synthesis of the interchain disulfide bonds or whether the bonds form spontaneously after association of the peptide extensions has occurred.

IV. Intracellular Sites for the Biosynthetic Steps

Considerable information is now available as to where the translational and posttranslational steps in collagen biosynthesis occur within cells. However, since there is still some uncertainty and disagreement on several points, it is probably useful to consider the experimental tools and the limitations of the tools which have been employed to develop our current understanding of the problems.

A. Techniques for Studying the Role of Cell Organelles

There are four principal biochemical techniques which have been applied to studying the role of cell organelles in collagen biosynthesis: pulse-label and chase experiments, isolation of subcellular fractions after disruption of the cells or tissues, autoradiographs of cells or tissues incubated with radioactive amino acids, and identification of specific cellular components *in situ* with antibodies which are "labeled" so that they can be visualized by electron microscopy.

The pulse-label and chase technique is one of the oldest procedures for studying cellular functions and one of the most powerful. Although it has been employed *in vivo* and in other complex biological situations, the procedure is most effective for studying collagen synthesis and secretion if several criteria are met: (1) the biological system allows sufficient label to be incorporated with a pulse time which is short relative to the synthesis time for the polypeptide chains; (2) the label can be effectively chased by changing the medium or by adding carrier amino acids in amounts which do not disrupt the biosynthetic machinery; and (3) the cells can be rapidly and effectively separated from any protein secreted into the media. If these criteria are not met, the data can be misleading. For example, in studies with intact connective tissues "lag periods" are frequently encountered in that incorporation of radioactive amino acids does not become linear until some time after the label is added.* In this circumstance, the specific activity of the amino acids incorporated gradually increases during the labeling period, and therefore the data become difficult to interpret. If the criteria listed above are met, the pulse-label and chase technique can be effectively combined with a number of other procedures to provide considerable information about synthesis and secretion of procollagen. One of the special advantages of the technique is that considerable information can be obtained without resort to extensive purification procedures. For example, by extracting all the labeled cellular proteins with a denaturant such as sodium dodecyl sulfate and then examining the size of the peptides by gel filtration or gel electrophoresis, one can follow the assembly of nascent polypeptide chains and the synthesis of interchain disulfide bonds among the chains (see below).

Isolation of subcellular fractions after partial disruption of cells or tissues is generally regarded as one of the most direct approaches to identify the metabolic activities and functions of cellular organelles, and techniques for the isolation of such fractions have now been available for over a decade. This approach has several advantages, including the fact that organelles can usually be isolated without denaturation of proteins, and therefore the isolated organelles or cell fractions can be assayed for specific enzymic activities. The approach, however, has several limitations. One limitation in applying it to collagen biosynthesis is that the presence of a fibrous extracellular matrix frequently makes it difficult to disrupt cells without resorting to vigorous conditions which can badly fragment cellular components.* Also, the extracellular matrix itself can contaminate the isolated fractions. [One recent approach to eliminating the problems presented by the presence of extracellular matrix has been to use matrix-free cells isolated by controlled enzymic digestion of embryonic tissues as a source of cellular fractions (see Harwood *et al.*, 1973).] A second limitation of this approach is that the process by which the cells are disrupted frequently changes the character of the organelles. For example, recent studies with serial sections of bakers' yeast have demonstrated that the

* We might note that in our own early studies on collagen biosynthesis *in vitro*, we employed intact cartilage and encountered considerable difficulty in preparing subcellular fractions or in performing satisfactory pulse-chase experiments (see Rosenbloom *et al.*, 1967). Largely because of these difficulties, we (Dehm and Prockop, 1971, 1972, 1973; Grant *et al.*, 1972*a*) developed improved techniques for preparing connective tissue cells which were free of extracellular matrix and which synthesized procollagen at a rapid rate when incubated in suspension.

mitochondrial functions of this organism are carried out by a single, multibranched mitochondrial structure; the multiple, ovoid mitochondria which have been isolated from the same yeast have been shown to be artifacts of the fractionation procedure (Hoffmann and Avers, 1973). The same situation may also hold for the mitochondrial structures in other cells, including liver (Brandt *et al.*, 1974). A third limitation is that only a few procedures are available for fractionating and separating organelles which are released by disruption of cells. Therefore is it difficult, if not impossible, to obtain clean fractions in good yield. In essence, the only separation procedure generally available is zonal centrifugation and in most circumstances fractions such as the microsomes are either obtained in poor yield compared to the total amount of such structures in cells, or the isolated fractions are clearly contaminated by other components of the cells. For example, even though liver is a relatively easy tissue from which to prepare subcellular fractions and such fractions have been extensively examined for several decades, it is difficult to fractionate liver so as to account for all the ribosomes in the cells and to recover them in clean fractions of free and membrane-bound ribosomes (Adelman *et al.*, 1973). However, several laboratories have recently devoted considerable effort toward obtaining good fractionation of organelles from cells synthesizing procollagen, and the results obtained with these newer procedures appear better than those that were achieved previously (see Guzman and Cutroneo, 1973; Guzman *et al.*, 1974; Diegelmann *et al.*, 1973; Harwood *et al.*, 1974*b*).

Autoradiographs of cells or tissues incubated with radioactive amino acids have been useful in studying collagen synthesis in some situations. The technique has the advantage of being relatively simple and applicable to a number of different circumstances. One serious disadvantage of the technique is that since any amino acid is incorporated into a variety of proteins by most tissues, the data cannot be unambiguously interpreted unless one encounters a fortunate circumstance such as a cell or tissue in which essentially all of the protein synthesized happens to be collagen. Another disadvantage of the autoradiographic technique is that it does not lend itself to quantitation and only dramatic effects can be examined. A third limitation is that although autoradiographs can be used at the level of electron microscopy, the resolution is limited by factors such as the size of the photographic grains, the thickness of the section, and tangential emission of radioactivity (see Salpeter *et al.*, 1969). As a result, labeled proteins cannot be located with a resolution of better than 1000–2000 Å, and the procedure cannot be used to locate proteins in small cellular compartments. However, considerable information has been obtained about the secretion of procollagen by using such autoradiographs in combination with other approaches in cells where secretory organelles are particularly large and morphologically distinct (see Weinstock and Leblond, 1974).

As pointed out by several authors, it has been apparent for some time that specific antibodies might provide an important tool for locating proteins and other materials in cells, if such antibodies were labeled either with enzymes which give rise to electron-dense products or with metalloproteins which can be visualized by electron microscopy (see Singer and Schick, 1961; Nakane and Pierce, 1967; Avrameas, 1972; Kraehenbuhl and Jamieson, 1972). Although this general approach was effectively used in a number of circumstances, there were until recently several serious technical limitations in applying it. For example, when antibody labeled with peroxidase was employed, the elecrtron-dense products of the peroxidase reaction tended to diffuse away from the sites at which the antibody was bound. Attempts to use antibodies labeled with the metalloprotein ferritin were limited by the fact that the chemical procedures for linking ferritin to antibodies were not efficient and gave rise to heterogeneous products. Also, the use of labeled antibodies was in general limited

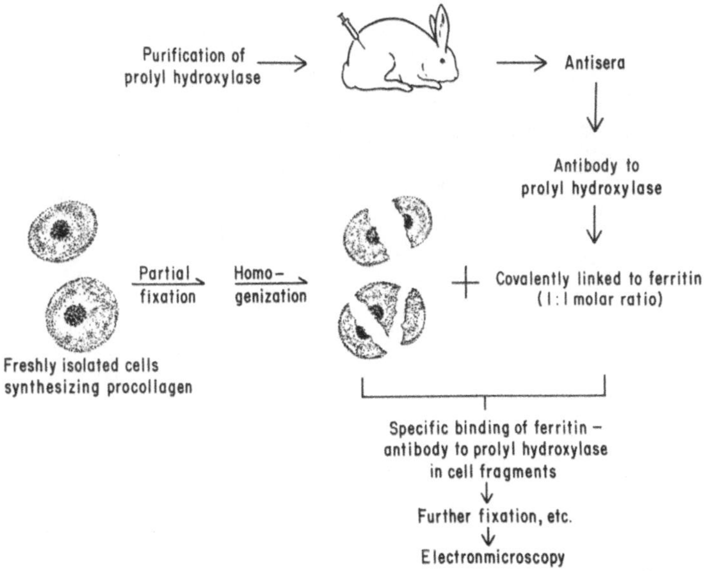

FIGURE 6. Procedures for preparing ferritin-conjugated antibodies and employing the conjugates to locate prolyl hydroxylase in cells synthesizing procollagen. (For detailed description of procedures see Olsen *et al.*, 1973*b*, 1975; Olsen and Prockop, 1974; Kishida *et al.*, 1975.)

by the difficulty of obtaining penetration of the labeled antibody into all the cellular compartments without loss of essential cell morphology or destruction of the antigenicity of the protein to be located. Recently, procedures (Figure 6) for overcoming most of these limitations have been developed (Olsen *et al.*, 1973*b*, 1975; Kishida *et al.*, 1975). One of the critical features of the new procedures is that conditions have been developed for preparing ferritin–antibody conjugates in which the ratio of ferritin to antibody is about 1:1, the ferritin and antibody are monomeric, and most of the immunologic activity of the antibody is retained (Kishida *et al.*, 1975). Another critical feature is that conditions have been developed for partially fixing isolated cells and then fragmenting them by homogenization so that most of the morphological features of the cells are retained but the organelles become permeable to ferritin–antibody conjugates (Figure 7). Thus far, these improved procedures have only been tested with antibodies directed against prolyl hydroxylase (Olsen *et al.*, 1973*b*) and antibodies directed against the peptide extensions on procollagen and protocollagen (Olsen and Prockop, 1974; Olsen *et al.*, 1975; Nist *et al.*, 1975), and therefore they may have limitations which are not yet

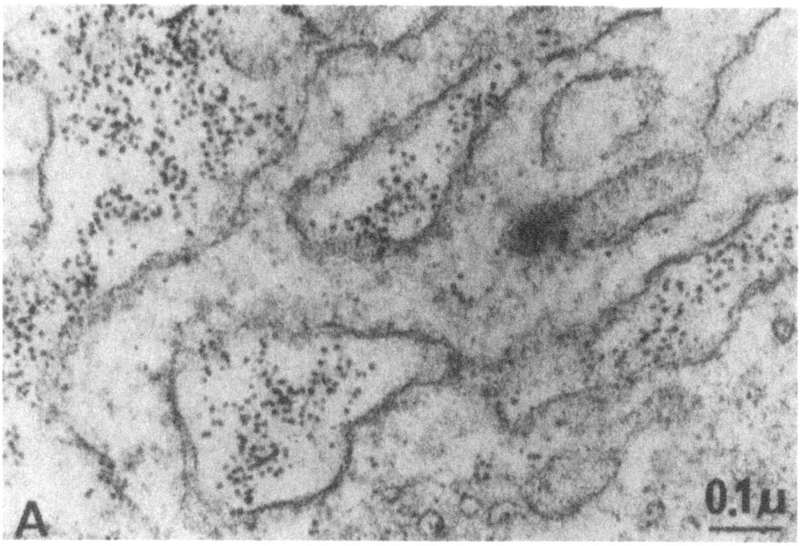

FIGURE 7A. Electron micrographs of fragments of tendon cells stained with ferritin-conjugated antibodies to prolyl hydroxylase. Electron micrograph of a section highly stained with bismuth subnitrate to emphasize the distribution of the ferritin within membrane-bound compartments. Because of the light staining, the ribosomes on the outer surface of the rough endoplasmic reticulum are not sharply defined.

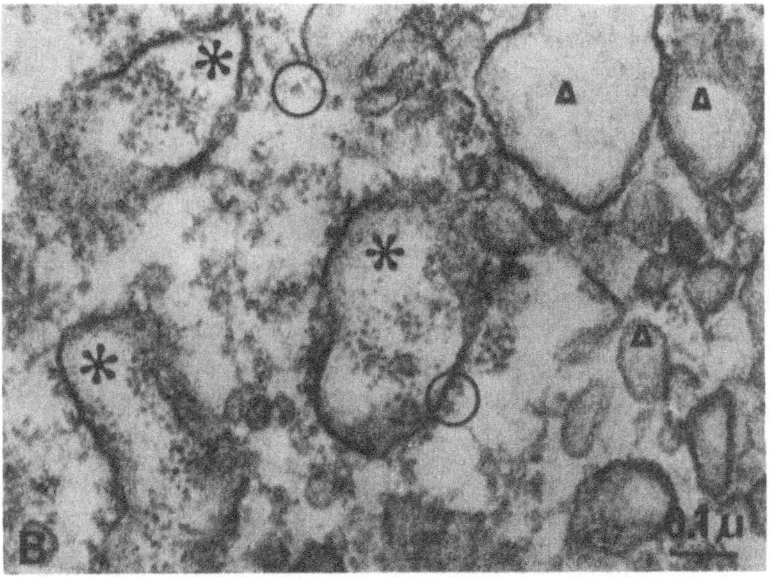

FIGURE 7B. Electron micrograph of a section more heavily stained with bismuth subnitrate to demonstrate that the membrane-bound compartments containing the ferritin consist of rough endoplasmic reticulum. The asterisks indicate rough endoplasmic reticulum containing ferritin, circles indicate ribosomes, and triangles indicate smooth-surfaced elements of the Golgi complex. The Golgi complex was consistently free of ferritin. Reproduced with permission from Olsen *et al.* (1973*b*).

apparent. However, with the two kinds of ferritin–antibody conjugates studied to date several potential problems have been excluded. Most importantly, it has been shown that the isolated cell fragments have about the same amount of cross-reacting protein as whole cell lysates and that most of this cross-reacting protein is accessible to antibody. Therefore the protein located by means of the ferritin–antibody conjugates probably accounts for most of the cross-reacting protein initially present in the cells. Also, since both prolyl hydroxylase and protocollagen were found only in one compartment, the cisternae of the rough endoplasmic reticulum, it was apparent that there was no marked displacement or diffusion of the cross-reacting protein among cellular organelles.

Two minor limitations of the procedures were apparent. One was that local displacement of cross-reacting protein from membranes cannot be excluded (see below). A second limitation is that although the procedures make all cellular organelles accessible to the conjugates, penetration of the Golgi vacuoles is sometimes difficult, probably because the Golgi vacuoles are not a system of interconnected compartments such as the endoplasmic reticulum and each vacuole must be fragmented separately for it to become accessible to the conjugates. Therefore, failure to observe cross-reacting protein in such vacuoles must be interpreted cautiously. In practice this limitation has been overcome by carrying out experiments with cells in which Golgi vacuoles are large and prominent (Nist *et al.*, 1975) or treating cells with colchicine, a procedure which delays the secretion of procollagen (Dehm and Prockop, 1972; Diegelmann and Peterkofsky, 1972; Ehrlich and Bornstein, 1972*b*) and which causes distension of Golgi vacuoles (Olsen and Prockop, 1974).

B. Reactions Occurring within Specific Organelles during Biosynthesis

For a number of years there were conflicting observations (Revel and Hay, 1963; Ross and Benditt, 1965; Cooper and Prockop, 1968; Salpeter, 1968) as to whether the synthesis and secretion of procollagen involved a pathway which substantially differed from the kind of pathway involved in the synthesis and secretion of other proteins destined for "export," such as

TABLE 3
Intracellular Locations of Steps in the Biosynthesis of Procollagen

Biosynthetic step	Apparent role of the biosynthetic step	Intracellular location of the biosynthetic step
Translation	Primary structure	Ribosomes bound to the endoplasmic reticulum
Hydroxylations of peptidyl proline	Essential for stable 3-helix at 37°C	Cisternae of rough endoplasmic reticulum before formation of 3-helix
Hydroxylations of peptidyl lysine	Essential for sugar additions; provides more stable cross-links	Cisternae of rough endoplasmic reticulum before formation of 3-helix
Synthesis of interchain disulfide bonds	Probably essential for helix formation	Cisternae of rough endoplasmic reticulum[a] after chain completion
Additions of sugar to peptidyl hydroxylysine	May affect fiber formation	Cisternae of rough endoplasmic reticulum[a]

[a] As discussed in text, the available data have not rigorously excluded the possibility that some of these processes continue as the protein passes into the smooth endoplasmic reticulum and Golgi vacuoles. Also it is possible that the point at which these processes terminate may vary with experimental conditions or the special nature of the cell.

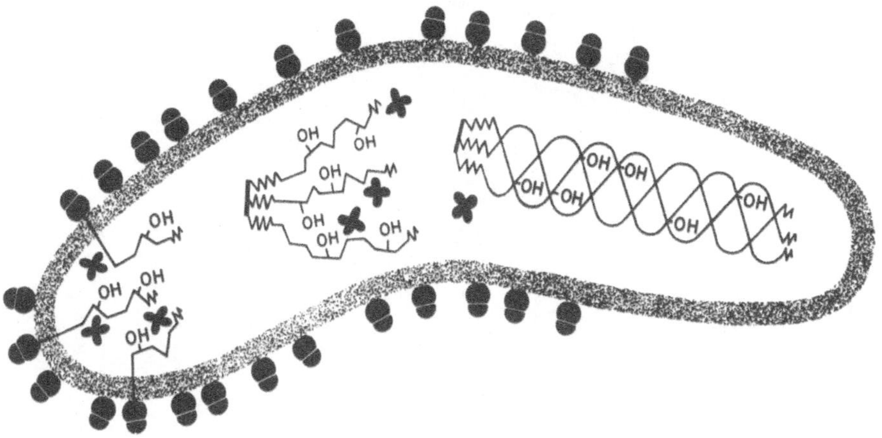

FIGURE 8. Schematic representation of the assembly of procollagen in the rough endoplasmic reticulum. The scheme suggests that the assembly occurs in three stages: (1) Assembly of pro-α chains and partial hydroxylation as the amino-terminal ends of the chains pass into the cisternae of the endoplasmic reticulum. (2) Further hydroxylation of the pro-α chains and the formation of interchain disulfide bonds among the carboxy-terminal extensions. (3) Folding of the collagen portion of the chains into a triple-helical conformation. Dark cloverleaf-like structure is used to represent prolyl hydroxylase (see Figure 2). The scheme is modified slightly from the scheme presented earlier by Schofield and Prockop (1973).

the digestive enzymes secreted by the pancreas (see Palade *et al.*, 1962). It is now apparent that most of the data for an unusual mode of secretion for procollagen, including data developed in our own laboratory (Cooper and Prockop, 1968), were misleading because of the limitations inherent in the techniques employed to develop the data (see below). Within the past several years, data obtained with a variety of techniques have produced a virtually unanimous opinion that procollagen is synthesized and secreted by the same kinds of organelles as are involved in the synthesis and secretion of other proteins for export.

1. *Location of Specific Steps in Synthesis and Secretion*

On the basis of the data now available one can assign specific steps in the biosynthesis of procollagen to the locations indicated in Table 3, and one can consider the overall process in terms of the model scheme shown in Figures 8 and 9. The scheme as shown attempts to present the usual sequence of events when cells or tissues synthesize procollagen under "optimal" conditions and with adequate access to O_2, ferrous iron, and

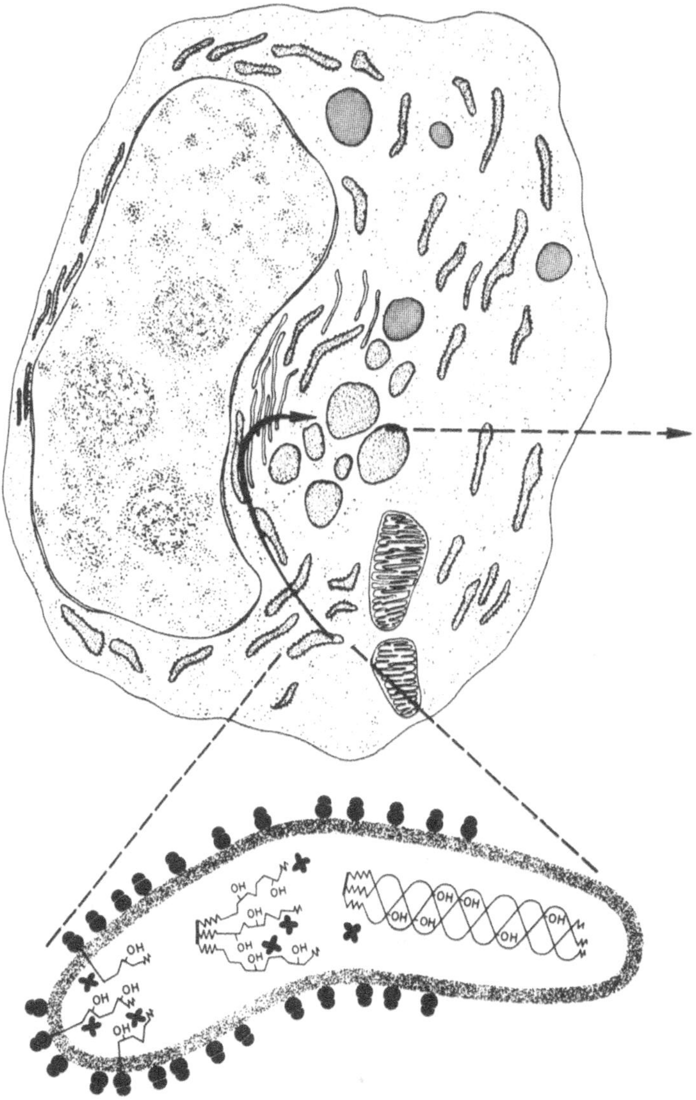

FIGURE 9. Schematic representation of the assembly and secretion of procollagen by a fibroblast such as those isolated by enzymic digestion of tendons from chick embryos (see Dehm and Prockop, 1971, 1972; Olsen *et al.*, 1975). The scheme indicates that after assembly of triple-helical procollagen in the rough endoplasmic reticulum, the protein passes into the Golgi vacuoles and is then secreted.

other required cofactors. As discussed below, the sequence of events can be altered considerably by various manipulations of experimental conditions.

The important features of the scheme are the following: Amino acids are assembled into pro-α chains on polysomes bound to the membranes of the endoplasmic reticulum. As the chains are assembled, the amino-terminal extensions pass through the membranes into the cisternae of the endoplasmic reticulum, and as the polypeptides pass into this compartment, hydroxylations of prolyl and lysyl residues are initiated. However, the hydroxylations are not completed at this stage. They continue for some time after assembly of pro-α chains is completed and probably for some time after the carboxy-terminal ends are released from the polysomes. The interchain disulfide bonds among the peptide extensions of the pro-α chains are not synthesized while the chains are being assembled; they appear only after assembly is completed and probably after release of the pro-α chains into the cisternae. After the hydroxylations of the chain and after synthesis of the interchain bonds, the collagen portions of the polypeptides become triple helical and then the molecule passes from the rough endoplasmic reticulum to the Golgi vacuoles. Galactosyl and glucosyl residues are added to the molecule while it is in the endoplasmic reticulum, but additional glycosylation may occur during or after its passage to the Golgi vacuoles. From the Golgi vacuoles the protein is secreted into the extracellular space.

2. *Evidence for the Location of Specific Steps*

The scheme shown in Figures 8 and 9 has been developed with evidence from a variety of sources and different features of the scheme have been established to varying degrees of certainty.

a. Studies with Subcellular Fractions Indicating the Site of Collagen and Hydroxyproline Synthesis. Since the early 1960s numerous attempts have been made to study procollagen synthesis by preparing subcellular fractions of cells and tissues. (For reviews of most of these experiments see Gould, 1968; Grant and Prockop, 1972.) Many of these experiments were carried out by incubating cells or tissues with radioactive proline and then demonstrating the radioactive peptidyl hydroxyproline or collagenase-sensitive peptides were recovered in microsomal or polysomal fractions from the cells or tissues. (For examples of such experiments, see Prockop *et al.*, 1962; Gould, 1968; Goldberg and Green, 1967; Miller and Udenfriend, 1970; Lazarides *et al.*, 1971). An alternate approach was to show that by appropriate manipulations of the cells or of cell fractions one could demonstrate prolyl hydroxylase activity in microsomal fractions. (For examples, see Peterkofsky and Udenfriend, 1963; Prockop and Juva, 1965*a*.) Most of these studies provided evidence which suggested both that procollagen polypeptides are synthesized on membrane-bound polysomes and that hydroxyproline is synthesized by the hydroxylation of peptidyl proline while procollagen polypeptides are being assembled on such polysomes. However, some investigators felt that the technical difficulties in preparing subcellular fractions in homogeneous form and in high yield made it necessary to interpret these data with caution. Specific criticisms included the following: the isolated microsomal and polysomal fractions were not shown to be free of contamination by newly synthesized collagen from other cellular compartments; relatively long labeling times were necessary to observe the appearance of labeled hydroxyproline in microsomal or polysomal fractions; and half or more of the prolyl hydroxylase in many tissues was recovered as a soluble protein after simple homogenization of the tissues and therefore the enzyme did not appear to be a microsomal one. Even in retrospect it is difficult to conclude with certainty whether these concerns were justified in considering the many published reports which supported essentially the same conclusions, but the continuing concern about these problems is reflected in the fact that a number of different laboratories have continued to publish additional reports in which they have tried to improve both the procedures for preparing subcellular fractions and the techniques for examining such fractions.

One recent experimental approach involved the demonstration that prolyl hydroxylase activity which was initially associated with microsomal fractions was released from the fractions by treatment with detergents (Guzman and Cutroneo, 1973; Diegelmann *et al.*, 1973). Another recent approach was to prepare microsomal fractions and then to examine the effects of puromycin on the release of the nascent, ribosomal polypeptides (Harwood *et al.*, 1974*a*). Treatment with puromycin released the nascent polypeptides from polysomes, but the peptides were not released from the microsomal fractions until the fractions were further treated with detergent. A third recent experimental approach was to show that if microsomal fractions were prepared from cells in which the hydroxylases were inhibited with α,α'-dipyridyl, the microsomal fractions contained protocollagen and the protocollagen was hydroxylated when the fractions were incubated with the required cofactors for prolyl hydroxylase (Guzman *et al.*, 1974).

b. Location of Prolyl Hydroxylase, Procollagen, and Protocollagen with Ferritin–Antibody Conjugates. With ferritin-labeled antibodies to prolyl hydroxylase, the enzyme was located within the cisternae of the endoplasmic reticulum (Figure 7). There was no evidence of any enzyme either in the cytoplasm or in the Golgi vacuoles. The electron micrographs suggested that the enzyme was evenly distributed throughout the cisternae, but because local displacement may have occurred during preparation and fixation of the cell fragments, the results did not rigorously exclude the possibility that the enzyme was bound to the inner surface of the membrane. However, assays with antibodies to prolyl hydroxylase demonstrated that the enzyme is not secreted in any appreciable amount by cells synthesizing and secreting procollagen (Olsen *et al.*, 1975). Therefore it is apparent that as procollagen passes out of the endoplasmic reticulum, prolyl hydroxylase is retained within the compartment by some mechanism which has not yet been clearly defined.

The techniques developed for use of ferritin–antibody conjugates have also been applied to locating procollagen in cells. With conjugates prepared with antibodies directed against the peptide extensions of procollagen (Dehm *et al.*, 1974), procollagen has been located both in the cisternae of the endoplasmic reticulum and in the large Golgi vacuoles (Olsen and Prockop, 1974; Nist *et al.*, 1975).

The antibodies to procollagen (Dehm *et al.*, 1974) prepared for these experiments also reacted with the peptide extensions on protocollagen. Therefore it is possible to use the same ferritin–antibody conjugates to show that in cells in which the hydroxylases were inhibited with α,α'-dipyridyl, protocollagen was found in the cisternae of the endoplasmic reticulum (Olsen *et al.*, 1975). Of special interest was the observation that protocollagen was not found in the Golgi vacuoles under experimental conditions in which procollagen was found in this compartment in control cells (see below).

c. Hydroxylation of Prolyl Residues During Assembly of the Polypeptide Chains. Evidence from several sources has now established that when cells or tissues are incubated under "optimal" conditions, hydroxylation of prolyl residues in the newly synthesized polypeptide chains begins as the nascent polypeptide chains are still being assembled. Initial evidence for this conclusion was obtained by demonstrating that peptidyl hydroxyproline was present in subcellular fractions of microsomes or polysomes, or in the peptides released from such fractions by puromycin (for examples, see Peterkofsky and Udenfriend, 1963; Goldberg and Green, 1967; Miller and Udenfriend, 1970; Lazarides *et al.*, 1971). Initially, results from our own laboratory suggested that the hydroxyproline synthesis did not occur until assembly of the polypeptide chains was essentially complete (Juva and Prockop, 1966*a*; Kivirikko and Prockop, 1967*d*; Bhatnagar *et al.*, 1967*b*; Rosenbloom *et al.*, 1967), but it is now clear that the data were misleading because it was not then known that the time to assemble the polypeptide chains of collagen is unusually slow compared to the time for assembly of the other polypeptide chains (see above). Also, at the time the experiments were carried out, the procedures for separating nascent chains from completed polypeptides by gel filtration were not entirely satisfactory. Subsequent experiments in this laboratory (Uitto and Prockop, 1974c) have provided data which support the suggestion initially made by other investigators in that after cells isolated from embryonic tendon were pulse-labeled with [^{14}C]proline for short periods of time, [^{14}C]hydroxyproline was found in short peptides. With longer labeling periods, these short peptides became pro-α chains and therefore they must represent ribosomal peptides in which some prolyl residues are hydroxylated.

d. Hydroxylation of Prolyl Residues after Assembly of the Polypeptide Chains. The current consensus that hydroxylations of prolyl residues begins while the chains are still being assembled has tended to obscure two further considerations: (1) Even under "optimal" conditions for incubation of the cells, the hydroxylations are not complete during the phase of polypeptide assembly and they continue after the pro-α chains achieve their maximal length. (2) If the prolyl and lysyl hydroxylases are temporarily inhibited, all the required hydroxylations can occur after assembly of the pro-α chains is completed and after the pro-α chains have been released from polysomes.

The conclusion that hydroxylations continue after chain assembly even under "optimal" incubation conditions is based on experiments in which pro-α chains from the total intracellular pool of such polypeptides were isolated. Comparison of such polypeptide chains with pro-α chains of procollagen secreted in the medium by tendon cells indicated that the degree of hydroxylation of the intracellular pro-α chains was on the average about 20% less than that of the secreted procollagen (Uitto and Prockop, 1974c). The results indicated therefore that an increase of 20% in the degree of hydroxylation occurred after the polypeptides had achieved their maximal length but before they were secreted. The data from these experiments did not, however, establish whether the further 20% increase in the degree of hydroxylation occurred before or after the pro-α chains were released from polysomes. It may be that the carboxy-terminal ends of the chains remain attached to polysomes for some time after their assembly is complete and the further hydroxylation may occur during this interval. It seems more likely, however, that the further hydroxylation occurs after the pro-α chains are released.

Hydroxylation of pro-α chains after release of the polypeptides from polysomes was demonstrated in experiments in which the prolyl and lysyl hydroxylases were reversibly inhibited by incubating tissues or cells with α,α'-dipyridyl or under anaerobic conditions. When appropriate concentrations of α,α'-dipyridyl were used, the rate at which pro-α chains of protocollagen were synthesized remained essentially the same as the synthesis of pro-α chains under control conditions even with incubation periods as long as 4 hr (Juva *et al.*, 1966; Bhatnagar *et al.*, 1967*a*; Jimenez *et al.*, 1973*a*; Uitto and Prockop, 1974*a*). However, if after 1 or 2 hr, the α,α'-dipyridyl was washed out of the tissues or cells and replaced with iron, the accumulated pro-α chains of protocollagen were hydroxylated to procollagen. The observation that protein synthesis continued at about the control rate in itself suggested that pro-α chains were released from polysomes at a normal rate in the presence of α,α'-dipyridyl, since otherwise the cellular supply of ribosomes for chain initiation would rapidly be exhausted and protein synthesis would cease. Direct experimental proof for this conclusion was obtained by subcellular fractionation of tissue culture cells incubated with and without α,α'-dipyridyl (Lazarides and Lukens, 1971). Therefore the hydroxylation of previously accumulated protocollagen which occurs after the α,α'-dipyridyl is removed from tissues or cells must represent hydroxylation of pro-α chains already released from polysomes.

Hydroxylation of pro-α chains already released from polysomes is further illustrated by the experiment presented in Figures 10 and 11. When cells from embryonic tendons were incubated under N_2, the cells

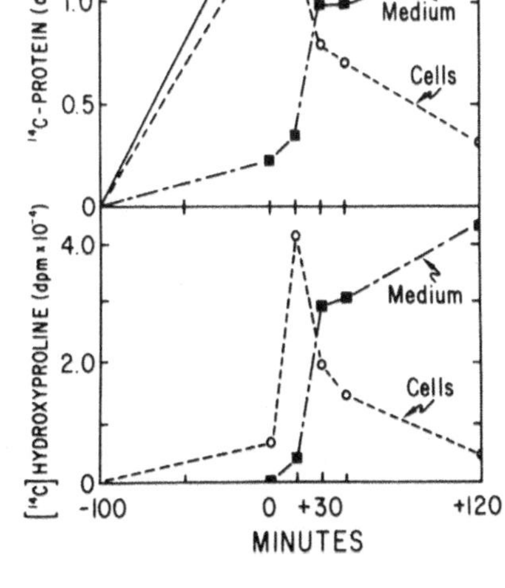

FIGURE 10. Reversible accumulation of [^{14}C]protocollagen under anaerobic conditions. Tendon cells were incubated for 90 min under N_2 with [^{14}C]proline. At −10 min, cycloheximide was injected through the stopper of the vessel, and at zero time the system was exposed to O_2. Aliquots containing 1.5×10^7 cells were removed at the times indicated. The cells and medium were separated by centrifugation and assayed for nondialyzable ^{14}C and for [^{14}C]hydroxyproline. The solid line in the upper frame indicates the total ^{14}C-labeled protein in the cells plus medium (●—●). Reproduced with permission from Uitto and Prockop (1974*a*).

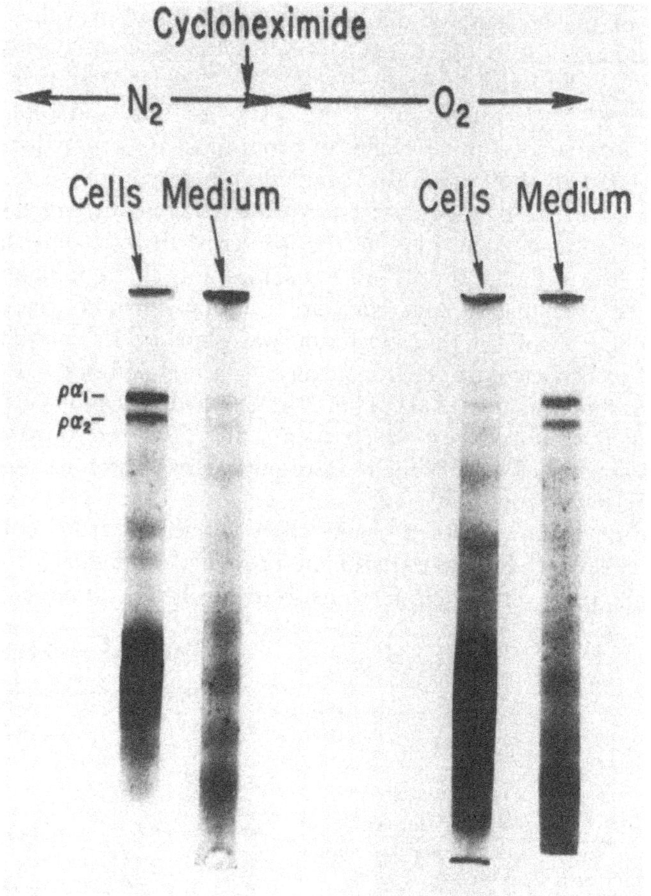

FIGURE 11. Polyacrylamide-gel electrophoresis in sodium dodecyl sulfate of proteins in cells and in medium from the experiment shown in Figure 10. The proteins were reduced with mercaptoethanol prior to electrophoresis, and therefore both protocollagen and procollagen are separated into pro-α1 and pro-α2 chains. In samples removed before the system was exposed to O_2 the cells contained pro-α chains, but none were detected in the medium (left-hand gels). After protein synthesis was stopped with cycloheximide and the system was exposed to O_2, pro-α chains were recovered only in the medium (right-hand gels). As discussed in text, the results indicated that protocollagen synthesized under anaerobic conditions remained intracellular and that the same protein was hydroxylated and secreted after the system was exposed to O_2. Reproduced with permission from Uitto and Prockop (1974*a*).

synthesized protocollagen. When cycloheximide was added to inhibit protein synthesis and then the cells were exposed to O_2, the previously synthesized protocollagen was hydroxylated to procollagen and secreted. Since cycloheximide inhibits protein synthesis in part by preventing the release of polypeptides from polysomes, the polypeptides which were hydroxylated and subsequently secreted in this experiment must have been released from polysomes before the cycloheximide was added and before hydroxylation was initiated by exposing the cells to O_2. (This experiment is further discussed below in considering the effects of conformation on the secretion of protocollagen and procollagen.)

e. Hydroxylation of Lysyl Residues. Less information is available about the synthesis of hydroxylysine than about the synthesis of hydroxyproline, but the current information suggests that the synthesis of hydroxylysine occurs at the same location and at about the same stage in polypeptide synthesis as the synthesis of hydroxyproline. Lysyl hydroxylase is difficult to study in subcellular fractions because the enzyme tends to aggregate (see above), but it was recently demonstrated that the enzyme is found in the same microsomal fractions which contain prolyl hydroxylase (Harwood *et al.*, 1974*b*). Pulse-label experiments with freshly isolated connective tissue cells showed that nascent chains contain hydroxylysine (Uitto and Prockop, 1974*c*; Grant *et al.*, 1975), and therefore the hydroxylation of lysyl residues begins during peptide assembly when cells are incubated under "optimal" conditions.

Evidence for hydroxylation of lysine after chain completion was obtained in the experiments in which the degree of hydroxylation of intracellular and extracellular pro-α chains were compared (Uitto and Prockop, 1974*c*). The degree of hydroxylation of lysine in the intracellular pro-α chains was about 20% less than the same value for pro-α chains in the procollagen secreted by the same cells.

Evidence that synthesis of hydroxylysine can also occur after release of pro-α chains came from experiments in which cartilage was incubated with α,α'-dipyridyl and with either [^{14}C]lysine or [^{14}C]proline (Blumenkrantz *et al.*, 1968). When the α,α'-dipyridyl was removed from the tissue, [^{14}C]lysine in the protocollagen which had accumulated in the tissue was hydroxylated, and the rate of hydroxylation was about the same as for [^{14}C]proline in the protocollagen.

f. Synthesis of Interchain Disulfide Bonds in the Endoplasmic Reticulum. The *inter*chain disulfide bonds in type I, II, or IV procollagens have been shown to be synthesized after translation of the pro-α chains is completed. Also, there is evidence that such bonds in type I and II procollagens are synthesized before the protein leaves the endoplasmic reticulum.

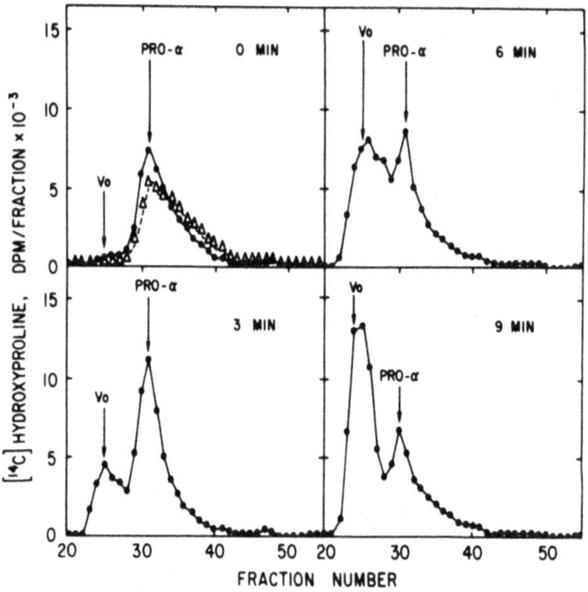

FIGURE 12. Synthesis of interchain disulfide bonds during the biosynthesis of procollagen in cells isolated from chick embryo tendons. The cells were pulse-labeled for 4 min with [^{14}C]proline and then the label was chased by adding [^{12}C]proline. The total [^{14}C]protein from the cells was extracted with sodium dodecyl sulfate and examined by gel filtration in sodium dodecyl sulfate. Procollagen polypeptides were specifically identified in the chromatogram by assaying the fractions for peptide-bound [^{14}C]hydroxyproline. At the end of the 4-min pulse-labeling period (0 min chase), pro-chains containing [^{14}C]hydroxyproline were present in the cells, but the pro-α chains were not linked by interchain disulfide bonds, since the elution pattern was the same whether or not the protein was reduced with mercaptoethanol before gel filtration. After chase periods of 3–9 min, an increasing fraction of the pro-α chains were linked by disulfide bonds and appeared in the void volume of the column (V_0) when the protein was chromatographed without prior reduction with mercaptoethanol. Symbols: Elution pattern after reduction with mercaptoethanol (△——△); elution pattern without reduction (●——●). Reproduced with permission from Schofield *et al.* (1974*a*).

The conclusion that interchain bonds do not form until after translation is completed was first suggested by the observation that cells isolated from embryonic lens contained collagenous polypeptides which appeared to be completed pro-α chains and which were not disulfide-linked (Grant *et al.*, 1973). This observation was developed further by pulse-chase experiments with isolated tendon and cartilage cells (Schofield *et al.*, 1974*a*; Uitto and Prockop, 1974*b*). The results indicated that short, nascent polypeptides did not decrease in size after reduction with mercaptoethanol and therefore did not contain interchain disulfide bonds (Figure 12).

Chasing the label demonstrated that reducible interchain bonds did not appear until sometime after the polypeptides reached the size of pro-α chains (Figure 12). In an independent study, it was shown that after tendon cells were labeled for a short period of time, fractions of the rough endoplasmic reticulum contained pro-α chains which were not linked by disulfide bonds (Harwood *et al.*, 1973). The conclusion that the interchain bonds do not form during translation was further supported by the demonstration that the pro-α chains synthesized by isolated polysomes did not contain interchain disulfide bonds (Kerwar, 1974).

The conclusion that interchain disulfide bonds are synthesized within the cisternae of the endoplasmic reticulum was initially made on the basis of the observation that protocollagen extracted from tendon or cartilage cells contained interchain disulfide bonds (Uitto and Prockop, 1973*a*, 1974*b*; Schofield *et al.*, 1974*a*). Since the protocollagen in such cells was shown to be in the endoplasmic reticulum, it followed that the interchain bonds in the protein must have been synthesized in this compartment. The same conclusion was also developed by the demonstration that when tendon cells were labeled with [^{14}C]proline for 2 hr, some of the pro-α chains found in fractions of the rough endoplasmic reticulum contained interchain bonds and essentially all the pro-α chains in fractions containing smooth endoplasmic reticulum were linked by such bonds (Harwood *et al.*, 1973). Still further support for the same conclusion was provided by the observation that when tendon cells were incubated with one of several proline analogs, the cells were shown to synthesize and accumulate pro-α chains which were nonhelical (see below) but linked by interchain disulfide bonds (Uitto and Prockop, 1974*d*). Since it was subsequently shown by both ferritin-labeled antibodies and subcellular fractionation that these nonhelical pro-α chains were located within the endoplasmic reticulum (Uitto *et al.*, 1975*b*), the results provided further proof of synthesis of interchain bonds in this organelle.

It might be noted that although these observations indicate that the interchain disulfide bonds can be synthesized in the rough endoplasmic reticulum, they do not necessarily demonstrate that all such bonds are synthesized in this compartment. Although it seems unlikely, some of the interchain bonds may be synthesized after the protein moves to other compartments such as the Golgi vacuoles.

g. Formation of the Triple Helix in the Endoplasmic Reticulum. The question of whether the molecule becomes triple-helical in the endoplasmic reticulum or whether it assumes this conformation after reaching the Golgi vacuoles has not been completely resolved, but most of the evidence suggests that helix formation occurs within the endoplasmic reticulum. The principal data which speak to this question come from the experiments with cells which were incubated so that they synthesized either protocollagen or disulfide-linked pro-α chains containing proline analogs. Since these two different kinds of protein are nonhelical in cells kept at 37°, and since they both accumulate within the cisternae of the endoplasmic reticulum, the observations suggest that the procollagen molecule must become triple-helical in order to pass from the rough endoplasmic reticulum to the smooth endoplasmic reticulum and Golgi vacuoles at a normal rate. However, it will be important to develop more direct experimental data for this conclusion before it is accepted without reservation.

h. Addition of Galactose and Glucose to the Molecule. Addition of both galactose and glucose can occur within the cisternae of the endoplasmic reticulum, but there is still some uncertainty as to whether glycosylations can also occur in Golgi vacuoles.

Examination of subcellular fractions of tendon cells has indicated that the rough endoplasmic reticulum contained both galactosyltransferase and glucosyltransferase activity (Harwood *et al.*, 1975). Also, pro-α chains which contained proline analogs and which accumulated in the endoplasmic reticulum (Uitto *et al.*, 1975*b*) were found to contain both galactosylhydroxylysine and glucosylgalactosylhydroxylysine (Rosenbloom and Prockop, 1971; Uitto and Prockop, unpublished data). Most recently, ribosomes isolated from chick embryo fibroblasts were found to contain glycosylated hydroxylysyl residues in nascent peptide chains (Brownell and Veis, 1975). It appears, therefore, that glycosylation of the peptide chains can occur shortly after lysyl residues are hydroxylated in the rough endoplasmic reticulum.

Subcellular fractionation of tendon cells also indicated that Golgi vacuoles contained both galactosyltransferase and glucosyltransferase activities (Harwood *et al.*, 1975). The presence of both enzymic activities suggests that both glycosylations can occur in this organelle, but this conclusion is not consistent with the observation that a triple-helical conformation prevents addition of glucose to collagen (Myllylä *et al.*, 1975*b*), if one assumes that procollagen becomes triple-helical before it passes from the rough endoplasmic reticulum to the Golgi vacuoles. Further work will probably be required to resolve this apparent discrepancy.

i. Passage of Procollagen through the Golgi Vacuoles. Involvement of Golgi vacuoles, or of secretory vacuoles, in the processing and secretion of procollagen has now been established by a variety of techniques, including morphological studies, autoradiography, subcellular fractionation, and labeled antibodies.

Although initial electron micrographic studies on connective tissues suggested the presence of collagen within membranous structures of cells, these observations were difficult to interpret because many of the published photographs probably represent tangential sections of extracellular fibers surrounded by irregular cellular processes. Because of special anatomical conditions, these and other problems were largely avoided in morphological studies on two tissues: corneal epithelium (Trelstad, 1971) and odontoblasts (Frank, 1970; Weinstock and Leblond, 1974). In both of these tissues, protein aggregates comparable to SLS aggregates of procollagen were clearly demonstrated in membranous structures corresponding to Golgi vacuoles. In the case of odontoblasts, the role of the Golgi vacuoles in secretion of procollagen was confirmed by autoradiography (Frank, 1970; Weinstock and Leblond, 1974). The presence of procollagen in Golgi vacuoles was also established by examination of subcellular fractions of tendon cells (Harwood *et al.*, 1973). Finally, the presence of procollagen in Golgi vacuoles was demonstrated by using ferritin-labeled antibodies to procollagen in tendon cells (Olsen and Prockop, 1974) and in corneal fibroblasts (Nist *et al.*, 1975).

In considering the secretion of procollagen through the membranous organelles of cells (Figure 9), it should be noted that several general features of the secretory process are still unexplained. For example, it is generally assumed that regions of the rough endoplasmic reticulum lose their ribosomes and gradually "bud off" to become part of the systems of smooth vacuoles involved in secretion. At the moment, however, there are no indications as to the mechanisms which might control this seemingly complex process. Similarly, there are as yet no detailed suggestions as to mechanisms for control of the process whereby procollagen and other proteins for "export" move from the Golgi vacuoles to the exterior of the cells. "Channels" which pass from Golgi vacuoles near the nucleus to the periphery of fibroblasts synthesizing procollagen have recently been observed (Hay and Dodson, 1973), but there is no information as to whether such channels play any active role in the secretory process.

V. Role of Posttranslational Reactions in the Folding and the Secretion of Procollagen

Since procollagen biosynthesis involves several relatively complex posttranslational reactions, it is of interest to ask: What is the function of these reactions? Recent observations indicate that at least part of the answer to this question is that two of the reactions are essential for the collagen portion of the molecule to become triple-helical and that the triple-helical conformation in turn is essential for the protein to be secreted at a normal rate.

A. The Special Role of Hydroxyproline in Stabilizing the Triple Helix

One of the striking chemical features of collagen is the presence of about 100 residues of hydroxyproline per polypeptide chain. The discovery in the 1940s of this apparently unique imino acid in collagen immediately prompted speculation as to the role it might play in the structure of the protein. In the mid-1950s, Gustavson (1954, 1955) made the specific suggestion that hydroxyproline might stabilize the fiber structure of collagen, perhaps because the hydroxyl group of hydroxyproline allowed the formation of additional hydrogen bonds among adjacent molecules in the fiber. After careful consideration, Gustavson's suggestion was discounted by most investigators, because it was shown that the thermal stabilities of various collagen fibers were more closely proportional to their content of imino acids (hydroxyproline plus proline) than to their hydroxyproline contents (for reviews, see Harrington and von Hippel, 1961; von Hippel, 1967; Traub and Piez, 1971). Also, during the 1960s a consensus gradually developed on a further point which was not explicitly part of Gustavson's suggestion, namely that hydroxyproline did not have any role in stabilizing the triple-helical structure of individual collagen molecules (see Ramachandran, 1967; Traub and Piez, 1971). The reasons for the consensus on this point were the following: (1) The thermal stabilities in solution of various kinds of collagen molecules were not proportional to their hydroxyproline contents and correlated much better with their total contents of imino acids (proline plus hydroxyproline). (2) Techniques for synthesizing polymers resembling collagen were developed, and with the polymers first synthesized there were no apparent differences in thermal stability of the helical structures formed by polymers such as $(\text{Gly-Pro-Hyp})_n$ and $(\text{Gly-Pro-Pro})_n$. (3) Several models for the triple-helical structure of collagen were developed from X-ray diffraction data, but in all of the generally accepted models the γ-hydroxyl of hydroxyproline was on the periphery of the structure and therefore apparently unavailable for chemical bonding. (For further discussion of these observations, see Ramachandran, 1967; von Hippel, 1967; Traub and Piez, 1971).

Experimental data developed in the past few years has superceded these earlier considerations and led to the conclusion that hydroxyproline plays a critical role in stabilizing the triple helix of collagen under physiological conditions. One part of the evidence was provided by collagen-like peptides which Sakakibara and his co-workers (1968) synthesized with a specific modification of the Merrifield technique for the solid-state synthesis of peptides. With the modified procedure, tripeptide units such as -Pro-Pro-Gly- were successively linked together so as to give "polytripeptides" which were of defined molecular weight (Sakakibara *et al.*, 1968). Peptides with the structure $(\text{Pro-Pro-Gly})_n$ were the first ones synthesized with the procedure, and when peptides with $n = 10$, 15, or 20 were examined, they were found to form triple-helical structures similar to the triple-helical structure of collagen (Kobayashi *et al.*, 1970). The same peptides were also found to form microcrystals similar to the fibrous-long-spacing aggregates formed by collagen under specific experimental conditions (Berg *et al.*, 1970; Olsen *et al.*, 1971), and this observation led to the successful preparation of macrocrystals of $(\text{Pro-Pro-Gly})_{10}$ suitable for X-ray diffraction (Sakakibara *et al.*, 1972; Okuyama *et al.*, 1972). Subsequently, the peptides $(\text{Pro-Hyp-Gly})_5$ and $(\text{Pro-Hyp-Gly})_{10}$ were synthesized and were also found to form triple-helical structures in solution (Sakakibara *et al.*, 1973). However, the presence of hydroxyproline greatly increased the thermal stability of the helices, and the midpoint of the thermal transition of helix to coil (T_m) for the peptide $(\text{Pro-Hyp-Gly})_{10}$ was about 35° higher than the T_m of $(\text{Pro-Pro-Gly})_{10}$ under the same conditions.

Another part of the experimental data for a role for hydroxyproline was developed at about the same time by studies on protocollagen. In one series of experiments (Berg and Prockop, 1973*b*,*c*), about 1 mg of protocollagen was extracted from tendon cells incubated with α,α'-dipyridyl, and the protocollagen was purified by limited proteolytic digestion and several extraction and precipitation steps. Amino acid analysis and gel-filtration experiments showed that the protein was essentially the same as collagen except that it contained less than one residue per chain of hydroxyproline and hydroxylysine, and it was correspondingly rich in proline and lysine (Table 4). The protein was triple-helical, as shown by equilibrium sedimentation and optical rotation, but its T_m in dilute acetic acid was about 15° lower than the T_m of a comparable collagen (Figure 13).

At about the same time as these data were obtained, the decreased thermal stability of protocollagen was also demonstrated by using susceptibility to proteolysis as a test of helicity (Uitto and Prockop, 1973*b*, 1974*a*; Jimenez *et al.*, 1973*b*; Berg and Prockop, 1973*d*). Subsequently, the same conclusion has been established by a detailed series of ultracentrifugation studies on protocollagen at different temperatures (Fessler and Fessler, 1974). Further and independent evidence for the conclusion that hydroxy-

TABLE 4
Partial Amino Acid Composition of the Pro-α Chains of Procollagen and Protocollagen Synthesized by Cells Freshly Isolated from Chick Embryo Tendons

	Procollagen[a] (per 1000 residues)	Protocollagen[a] (per 1000 residues)
Glycine residues	326	313
Proline residues	122	208
Hydroxyproline residues	99	0[b]
Lysine residues	18	39
Hydroxylysine residues	17	0[b]
Proline + hydroxyproline	221	208
Lysine + hydroxylysine	35	39

[a] Amino acid compositions were determined on proteins purified after limited proteolytic digestion with chymotrypsin (see Berg and Prockop, 1973*b*).
[b] Each polypeptide chain contained less than one residue of hydroxyproline or hydroxylysine (see (Berg and Prockop, 1973*b*).

proline stabilizes the triple helix was obtained by comparing helical structures formed with α1-CB2 peptides obtained from rat skin and tendon collagen. The triple-helical structure formed by the peptide derived from skin had a higher melting temperature, and the only chemical difference between this peptide and the peptide obtained from tendon collagen was that the peptide from skin had a higher content of hydroxyproline (Ward and Mason, 1973).

In retrospect, it is apparent that the arguments originally advanced for the conclusion that hydroxyproline did not contribute to the stability of the triple helix were not sufficiently rigorous. Collagens which differ in hydroxyproline content also differ in their contents of many other amino acids and differ in their amino acid sequences (see Kulonen and Pikkarainen, 1970). Therefore it is hazardous to use comparisons of such collagens to develop conclusions about the contribution of hydroxyproline to helical stability. Also, the synthetic techniques originally employed to prepare polymers such as $(\text{Gly-Pro-Hyp})_n$ and $(\text{Gly-Pro-Pro})_n$ (see Engel *et al.*, 1966) gave products which were heterogeneous in size and which had some physical properties differing from those of homogeneous peptides with the same sequences but prepared by the modified solid-state procedure (Sakakibara *et al.*, 1968, 1973). Therefore some of the conclusions developed from studies with the polymer preparations are now recognized to be misleading. Finally, although molecular models for collagen (Ramachandran *et al.*, 1973) and the helical structure of the polymer $(\text{Pro-Gly-Pro})_n$ (Traub *et al.*, 1969; Traub, 1974) have now been refined to a considerable degree, the X-ray data have not in themselves provided a unique solution to the structure of collagen. Therefore the molecular models, without supporting kinetic and thermodynamic data, are difficult to use as the sole criteria for determining the contributions of specific amino acid residues to helical stability.

Although there is now general agreement that hydroxyproline stabilizes the triple helix, there is little agreement as to how it produces this effect. As discussed elsewhere in this volume, Ramachandran *et al.* (1973) pointed out that a water molecule they had previously incorporated via two hydrogen bonds into their latest model for collagen was in a favorable position to form a third hydrogen bond to the γ-hydroxyl group of hydroxyproline and thereby help to stabilize the helix further. However, they are of the opinion that the water cannot form the third bond to the γ-hydroxyl unless it takes part in the other two hydrogen bonds. Since this requires an α-amino group in the X position of the adjacent chain, their model cannot account for the increased helical stability of $(\text{Pro-Hyp-Gly})_{10}$ as compared to $(\text{Pro-Pro-Gly})_{10}$ (Sakakibara *et al.*, 1973). Traub (1974) has proposed a variant on his own model for $(\text{Pro-Gly-Pro})_n$ in which a water molecule forms a hydrogen-bonded bridge between the γ-hydroxyl of hydroxyproline and the carbonyl of the preceding glycine in the same

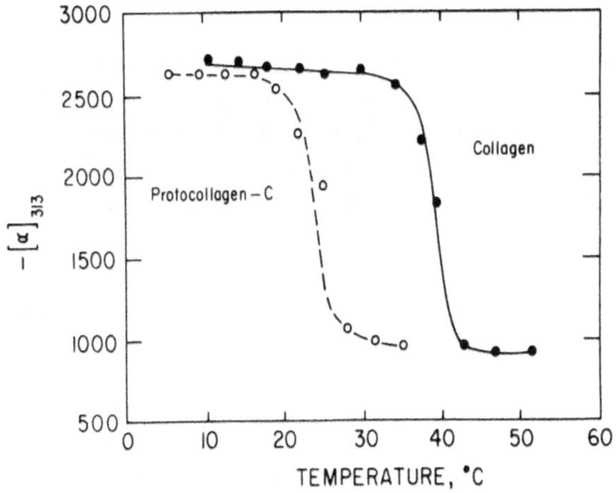

FIGURE 13. Thermal stability of the triple-helical structure of protocollagen and of collagen. Protocollagen was extracted from tendon cells incubated with α,α′-dipyridyl and then purified by limited proteolytic digestion with chymotrypsin. The purified protein (protocollagen-C) corresponded to protocollagen without the nonhelical amino-terminal and carboxy-terminal regions. For comparison, a comparable collagen was prepared by isolating procollagen from the medium of the same kind of cells incubated under control conditions and digesting the procollagen with chymotrypsin. Reproduced with permission from Berg and Prockop (1973*c*).

chain. The two hydrogen bonds to water in the Traub proposal appear to be identical to two of three hydrogen bonds to water proposed by Ramachandran *et al.* (1973), but there seems to be some question as to whether a water molecule with just two such bonds, and apparently exposed to solvent, could make a contribution to the stability of the helix. Unfortunately, the manner in which water binds to any protein has still not been defined (see Kuntz and Kauzmann, 1974), and it may be difficult to devise definitive tests for either proposal. We ourselves have explored the possibility of a direct hydrogen bond between the γ-hydroxyl of hydroxyproline and a carbonyl group in an adjacent chain, and we have proposed a model incorporating such a bond (Berg *et al.*, 1973). One of the merits of the proposal is that it requires a *cis*-peptide bond between glycine and proline in the adjacent chain, and because such a *cis*-peptide bond should be detectable by several techniques, the model can readily be subjected to experimental test. Torchia *et al.* (1975) have recently used [^{13}C]carbon nuclear magnetic resonance to test for the presence of a *cis*-peptide bond in the triple-helical structure formed by α1-CB2 and have not found any evidence for such a bond. It might be noted, however, that the model for a direct hydrogen bond (Berg *et al.*, 1973) is a restrictive one in the sense that it requires that the glycine in the adjacent chain which participates in the hydrogen bond must be followed by proline. The amino acid sequence of α1-CB2 indicates that because of this restriction, a triple helix formed from three α1-CB2 peptides is unlikely to contain the direct hydrogen bond suggested by Berg *et al.* (1973). The model therefore cannot explain the data of Ward and Mason (1973), and it seems unlikely that the direct hydrogen bond it postulates can be formed by all the hydroxyprolyl residues in natural collagens.

B. *The Role of the Peptide Extensions of Procollagen and Interchain Disulfide Bonds in Formation of the Triple Helix*

Although the presence of hydroxyproline is essential for the polypeptides of procollagen or collagen to assume a triple-helical structure which is stable at 37°, the process of helix formation also depends on correct association of the three pro-α chains. Recent observations, in fact, indicate that under most experimental conditions association of the peptide extensions on pro-α chains, and not hydroxylation of prolyl residues, is the rate-controlling step which determines when the triple helix forms and, indirectly, the extent to which prolyl and lysyl residues are hydroxylated.

1. General Considerations as to the Role of the Peptide Extensions in Facilitating Helix Formation

As has been recognized for several decades, α chains of collagen, containing a full complement of hydroxyproline, will renature into triple-helical molecules *in vitro,* but the rate at which helical structures form is slow (see von Hippel, 1967; Beier and Engel, 1966). For example, in one series of experiments to develop optimal conditions for renaturation, only 50% of the maximal degree of helicity was obtained after 24 hr (see Kühn, 1969). In contrast, it has been demonstrated that in tendon fibroblasts newly synthesized pro-α chains become triple-helical within 5–10 min of the time they are synthesized (Uitto and Prockop, 1973*a*; Schofield *et al.,* 1974*a*). On the basis of such observations, as well as on the basis of more general considerations (see Speakman, 1971), it now appears obvious that the peptide extensions on pro-α chains must play a critical role in ensuring that the collagen portion of the chains fold into the correct triple-helical conformation rapidly and efficiently during biosynthesis of procollagen. The most likely sequence of events is that the peptide extensions of the pro-α chains begin to fold into the appropriate three-dimensional structure soon after they are assembled on ribosomes. From our current information about the biosynthesis of globular proteins (see Wetlaufer and Ristow, 1973; Anfinsen, 1973), it seems probable that the folding of the peptide extensions would be governed entirely by their amino acid sequences. Also, by analogy with globular proteins consisting of several subunits, it seems probable that after the peptide extensions have folded, their conformation directs association of the peptide extensions. Finally, it seems highly probable that the correct association of the peptide extensions greatly facilitates the nucleation step in which the first residues in the collagen portion of the chains assume the correct conformation, become linked by interchain hydrogen bonds,* and thereby become the site from which helix formation is propagated throughout the length of the chains (see Beier and Engel, 1966; Harrington and Karr, 1970).

Although most investigators now agree on this general scheme for

* Most, but not all, investigators (see Wetlaufer and Ristow, 1973) believe that a microsomal enzyme discovered by Anfinsen and his associates (Anfinsen, 1972) plays a critical role in the synthesis of proteins containing disulfide bonds. The enzyme catalyzes disulfide exchange and apparently therefore promotes a "shuffling" of disulfide bonds so that inappropriate pairings are rapidly disrupted and the correct pairings are made as the polypeptide chains fold. It seems likely that a similar enzyme is involved in the synthesis of the correct intrachain and interchain disulfide bonds found in the peptide extensions of procollagen.

folding of pro-α chains into the triple helix, we do not yet have conclusive experimental proof for many parts of it. In particular, although small amounts of procollagen and protocollagen have now been available in some laboratories for several years, attempts to renature purified forms of these proteins have been disappointing in that the yields of helical structures have been small and the times for renaturation have been long (Uitto and Prockop, unpublished data; Darnell *et al.*, 1974; Byers *et al.*, 1974). The difficulties with these experiments may indicate that folding of pro-α chains may require some special ancillary mechanism, such as participation of prolyl hydroxylase in the folding process (see Darnell *et al.*, 1974), which is available within cells and which was not present in the experiments with purified protein. However, there may be some relatively trivial explanation for the experimental problems. For example, it may be that under all the conditions tested to date there was disulfide exchange in the peptide extensions of the pro-α chains and that this exchange produced incorrect pairing of disulfide bonds (see Anfinsen, 1972),* which in turn interfered with formation of the triple helix. At the moment, we are inclined to assume that the experimental problems in renaturing protocollagen and procollagen have a relatively trivial explanation.

Unfortunately, our current failure to resolve these experimental problems makes it difficult to resolve an important further question about the role of the peptide extensions in facilitating formation of the triple helix: Is noncovalent association of the extensions sufficient to initiate formation of the triple helix, or are interchain disulfide bonds among the three extensions essential for the process? Most of the recent observations are best explained by assuming that the interchain disulfide bonds are essential, but conclusive data are still not available. The point is worthy of careful consideration, because the conclusion that the interchain bonds are essential indicates a role for such bonds in the folding of procollagen which has not yet been demonstrated in the biosynthesis of any other protein. A considerable body of experimental evidence indicates that in globular proteins containing disulfide bonds, such bonds are not synthesized before the polypeptide chains fold into the correct structure and that disulfide bonds are not of any importance in promoting the folding of proteins during biosynthesis (see Olin and Edelman, 1964; Björk and Tanford, 1971; Anfinsen, 1972; Wetlaufer and Ristow, 1973). Instead, the function of the disulfide bonds in globular proteins appears to be that of simply stabilizing the correct three-dimensional structure of a protein after the correct structure has been formed (see Anfinsen, 1972).

* See footnote on page 223.

2. *Synthesis of Interchain Disulfide Bonds before Formation of the Triple Helix*

Interchain disulfide bonds were present in type I and II protocollagen isolated from cells or tissues incubated at 37° C with α,α'-dipyridyl (Schofield *et al.*, 1973*a*, 1974*a,b*; Fessler and Fessler, 1974; Uitto and Prockop, 1973*a*, 1974*b*). The interchain bonds were present even when extensive precautions were taken to ensure that the collagen portion of the pro-α chains were never allowed to become triple-helical. It is apparent, therefore, that in the case of type I and II procollagen interchain disulfide bonds can be synthesized without complete folding of the polypeptides into their correct three-dimensional structure.

3. *Conditions under Which Helix Formation Is and Is Not Limited by Chain Association and Disulfide Bonding*

The presence of interchain disulfide bonds in protocollagen does not necessarily mean that synthesis of such bonds always precedes formation of the triple helix. However, a number of observations suggest that this is the usual sequence in cells or tissues synthesizing procollagen under "optimal" conditions.

One of the first indications that chain association might limit helix formation during the biosynthesis of procollagen came from studies with cells isolated from embryonic lens which synthesized basement membrane procollagen (Grant *et al.*, 1972*a,b,* 1973). Cells labeled under steady-state conditions were found to contain pro-α chains, but only about 15% of the pro-α chains were in triple-helical conformation as tested by their resistance to pepsin digestion. Also, only a small fraction of the pro-α chains were linked by the interchain disulfide bonds. In contrast, essentially all the pro-α chains secreted by the same cells were triple helical and disulfide linked. The results suggested, therefore, that the synthesis of the interchain bonds and helix formation did not occur until just before the protein was secreted.

In further experiments with pulse-label and chase techniques in tendon cells (Schofield *et al.*, 1974*a*) and in cartilage cells (Uitto and Prockop, 1974*b*), a close correlation was found between the time at which interchain bonds appeared among pro-α chains and the time at which the protein became triple-helical. In the case of cartilage cells, in which the synthesis and secretion of procollagen required a relatively long time, it

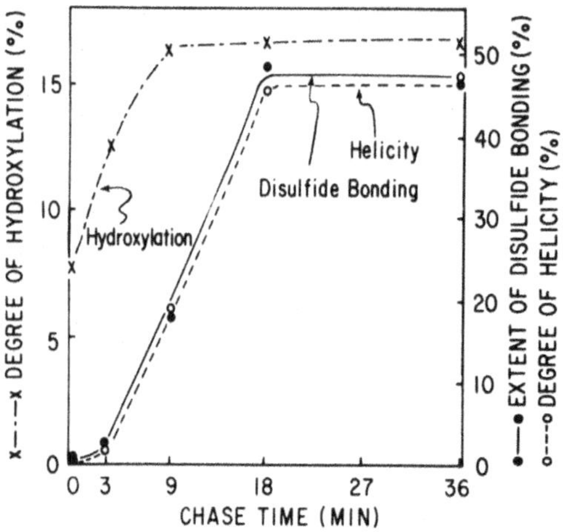

FIGURE 14. Interchain disulfide bonding, formation of the triple helix, and hydroxylation of proline in cartilage cells. Cells isolated by enzymic digestion of chick embryo cartilage were pulse-labeled for 4 min with [^{14}C]proline and then the label was chased by adding [^{12}C]proline. The degree of hydroxylation is expressed as the ratio of [^{14}C]hydroxyproline to total [^{14}C]protein in the system. The extent of disulfide bonding was calculated as a ratio of [^{14}C]hydroxyproline which eluted in the void volume to the total [^{14}C]hydroxyproline in the chromatogram (see Figure 12). The degree of helicity was calculated as the ratio of [^{14}C]hydroxyproline recovered as α chains after limited proteolytic digestion to the total nondialyzable [^{14}C]hydroxyproline in the sample before proteolytic digestion. Reproduced with permission from Uitto and Prockop (1974*b*).

was possible to examine the question of when interchain disulfide bonding occurred in relation to hydroxylation of the pro-α chains and formation of the triple helix (Uitto and Prockop, 1974*b*). The results (Figure 14) demonstrated that hydroxylation of prolyl residues was maximal soon after the chains were assembled, but there was a measurable lag between the hydroxylation and synthesis of the interchain bonds. As indicated, the protein became triple-helical at about the same time as the interchain bonds were synthesized. The results suggested, therefore, that when the cells were incubated under "optimal" conditions, the hydroxylation of prolyl residues did not limit the rate of helix formation. Instead, they suggested that chain association, as reflected by the appearance of the interchain bonds, was the limiting step.

These suggestions were supported by experiments in which cartilage cells were incubated under anaerobic conditions so that they synthesized protocollagen in which the pro-α chains were linked by interchain disulfide bonds. As shown in Figure 15, when the cells were exposed to O_2, the intracellular protocollagen was maximally hydroxylated in less than 10 min, and the protein became triple-helical at about the same rate as the rate of hydroxylation. It appeared, therefore, that once the interchain disulfide bonds were formed, helix formation depended primarily on the hydroxyproline content of the polypeptides and the rate of hydroxylation limited the rate of helix formation.

4. *Evidence Suggesting That Interchain Disulfide Bonds Are Essential for Helix Formation*

Further evidence for a specific requirement for interchain disulfide bonds was provided by additional experiments with intact cells. Although it has been difficult to renature purified procollagen or protocollagen, protocollagen in cells which have synthesized and accumulated this protein becomes triple-helical within 5 min if the cells are cooled to 15° or less (Uitto and Prockop, 1973*a*). [This rapid folding of intracellular protocollagen in large part explained our initial erroneous conclusion that protocol-

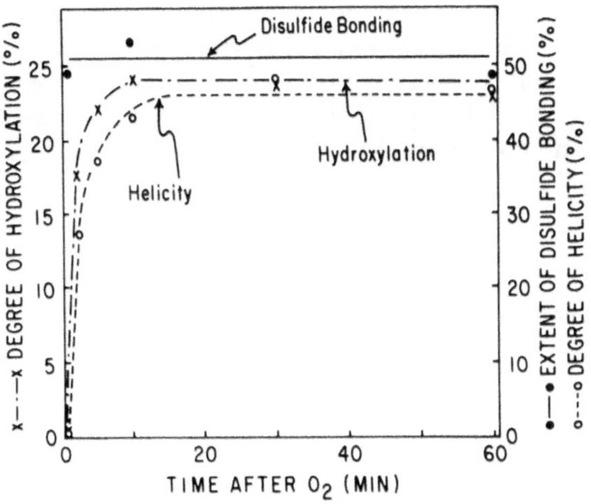

FIGURE 15. Rate of helix formation and hydroxylation of proline in cartilage cells subjected to temporary anoxia and then exposed to O_2. The degree of hydroxylation was estimated as described in the legend of Figure 14. Because the [^{14}C]protocollagen present at zero time did not contain [^{14}C]hydroxyproline, the degree of disulfide bonding was estimated as the fraction of total pro-α chains linked by disulfide bonds. The degree of helicity was estimated as the fraction of pro-α chains recovered as α chains after pepsin digestion under conditions in which triple-helical collagen resists proteolysis. Reproduced with permission from Uitto and Prockop (1974*b*).

lagen was triple-helical in cells synthesizing this protein at 37°C (Jimenez *et al.*, 1973*a*).] Since it was previously shown that treating intact fibroblasts with dithiothreitol probably reduced the disulfide bonds in intracellular proteins (Stassen *et al.*, 1973), it was possible to use cells containing protocollagen to examine the question of whether disulfide bonds were essential for protocollagen to become triple-helical (Uitto and Prockop, 1973*a*). It was found that when cells containing protocollagen were exposed to dithiothreitol, the interchain disulfide bonds in protocollagen were reduced. When such cells were subsequently cooled to 15°, the protocollagen did not become triple-helical. The effect was reversible in that removal of the dithiothreitol by washing the cells in fresh medium was followed by reformation of the interchain disulfide bonds and once these bonds were reformed, a large fraction of the intracellular protocollagen became triple-helical on cooling (Uitto and Prockop, 1973*a*). The observations were therefore consistent with the hypothesis that interchain disulfide bonds must be present among the peptide extensions of pro-α chains for these polypeptides to become triple-helical at a biologically appropriate rate. However, at this juncture most investigators would agree that it is probably necessary to perfect systems for studying the folding of purified procollagen and protocollagen and to carry out the kinds of detailed studies which have been performed with proteins such as immunoglobulins (see Björk and Tanford, 1971) before one can clearly distinguish the role of the interchain bonds from the role of intrachain disulfide bonds and from the role of noncovalent association of the extensions in the formation of the triple helix.

C. The Conformation-Dependent "Barrier" to the Secretion of Nonhelical Procollagen or Protocollagen

The observation that when prolyl and lysyl hydroxylases are inhibited, connective tissue cells synthesize and accumulate protocollagen demonstrates that secretion of procollagen at a normal rate depends on the posttranslational modifications of the protein. For some time it was thought (see Grant and Prockop, 1972) that normal secretion might depend on a single chemical change in the protein, such as the addition of galactose which might provide an essential "tag" required for secretion. This "sugar-tag" hypothesis was originally suggested as a general mechanism whereby cells could distinguish proteins destined for export from proteins synthesized for internal consumption; it was developed from the observation that most extracellular proteins contain small amounts of hexose, whereas no intracellular protein is known to contain hexose (Eylar, 1966). More detailed analysis of the data showed the sugar-tag hypothesis was not generally applicable to exported proteins (see Winterburn and Phelps, 1972), and recent observations suggest that the hexoses normally found on procollagen are not essential for its secretion (see below). Instead, the data indicate that secretion of procollagen depends primarily on the protein assuming a triple-helical conformation.

Most of the direct evidence for this conclusion has been developed by using cells which are isolated by controlled enzymic digestion of tendon or cartilage from chick embryos. When incubated in suspension for up to 6 hr after isolation, the major metabolic activity of the cells is the synthesis and secretion of procollagen (Dehm and Prockop, 1971, 1972, 1973). The important experiments performed with such cells have included the following: (1) examination of the effects of inhibiting the prolyl and lysyl hydroxylases by incubation of the cells under anaerobic conditions or with the iron chelator α,α'-dipyridyl; (2) examination of the effects of several proline analogs which are incorporated into protein; and (3) exploration of the effects of lowered temperature on conformation and secretion of the protein.

1. *Effects of Inhibiting Prolyl Hydroxylase and Lysyl Hydroxylase by Anaerobic Conditions or α,α'-Dipyridyl*

As discussed above, if cells freshly isolated from embryonic tendons are incubated under N_2 or in the presence of about 0.3 mM α,α'-dipyridyl, hydroxylation of peptide prolyl and lysyl residues is prevented and protocollagen is synthesized (see Figures 10 and 11 and Table 4). The protocollagen is not secreted to any appreciable extent in the first 2 hr or so (see Juva *et al.*, 1966; Margolis and Lukens, 1971; Grant and Prockop, 1972), and in the case of tendon cells incubated with α,α'-dipyridyl the amount of intracellular pro-α chains increases over twofold compared to controls (Figure 16).

If cells which have been incubated several hours with α,α'-dipyridyl are then washed extensively with fresh medium containing ferrous iron, the intracellular protocollagen is hydroxylated and secreted (Uitto and Prockop, unpublished data). However, hydroxylation of intracellular protocollagen is simpler to follow with experiments such as the one shown in Figures 10 and 11 in which the hydroxylases are inhibited by incubating cells under N_2 and the inhibition is then reversed by exposing the cells to

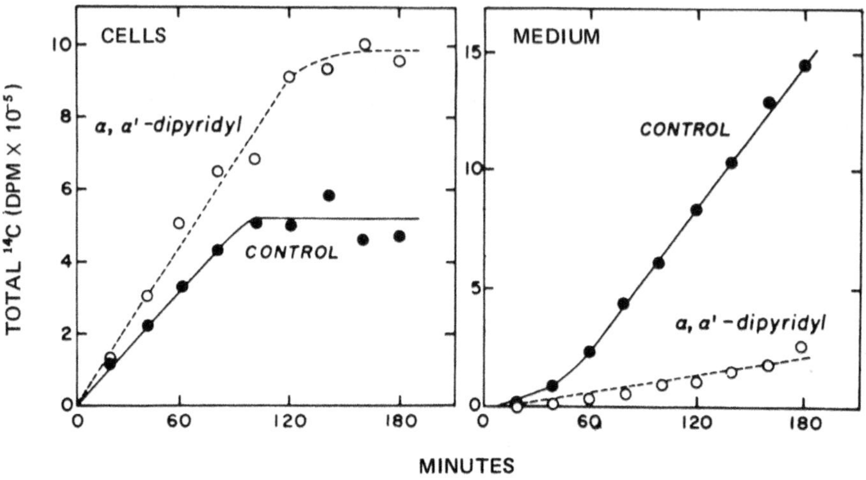

FIGURE 16. Effect of α,α′-dipyridyl on the synthesis and secretion of [^{14}C]protein by cells isolated from chick embryo tendons. The cells were incubated for 30 min with or without 0.3 mM α,α′-dipyridyl and then [^{14}C]proline was added. Most of the [^{14}C]protein which accumulated in the cells incubated with α,α′-dipyridyl was shown to be [^{14}C]protocollagen. Reproduced with permission from Jimenez *et al.* (1973*a*).

O_2. Examination of the conformation of the [^{14}C]protein by limited proteolytic digestion demonstrated that the protocollagen synthesized in the cells under N_2 becomes triple-helical at about the same time it is hydroxylated and secreted into the medium (Uitto and Prockop, 1974*a*). The results suggested, therefore, that secretion depends on the conformation of the protein.

At least two concerns have been raised about such experiments using anaerobic conditions or α,α′-dipyridyl to inhibit prolyl and lysyl hydroxylases in cells: (1) The effects of anaerobiosis and of α,α′-dipyridyl might be nonspecific in that a number of processes other than the hydroxylating reactions might be inhibited. (2) It is patently impossible for cells to continue indefinitely to synthesize and accumulate protocollagen without secreting it.

Incubation under anaerobic conditions (Juva *et al.*, 1966) or with a chelator such as α,α′-dipyridyl (Hurych and Chvapil, 1965) almost certainly alters a number of metabolic functions of cells and tissues. However, many connective tissue cells have only limited access to O_2 *in vivo* (Silver, 1973) and apparently for this reason are relatively resistant to anoxia. Also, it has been discovered that by adjusting the α,α′-dipyridyl concentration appropriately, one can obtain a highly selective effect on prolyl and lysyl hydroxylases. The most critical data here deal with the effects of anaerobic conditions or α,α'-dipyridyl on the rate of protein synthesis. When cells isolated from chick embryo tendons are incubated under anaerobic conditions, the rate of protein synthesis as measured by pulse-labeling with [^{14}C]proline is the same as the control for about 30 min (Uitto and Prockop, 1974*a*). Thereafter the rate of protein synthesis decreases to about half of the control and remains at this level for at least another 3 hr. When the same cells are incubated with about 0.3 mM α,α'-dipyridyl, the rate of [^{14}C]proline incorporation remains the same as the control for 4 hr (Uitto and Prockop, 1974*a*). A similar situation holds for cells isolated from embryonic cartilage, and with these cells it has been shown that 0.3 mM α,α'-dipyridyl does not affect the rate at which [^{35}S]sulfate is incorporated into sulfated macromolecules (probably proteoglycans) or the rate at which sulfated macromolecules are secreted from the cells (Dehm and Prockop, unpublished observations). These observations suggest, therefore, that although anaerobiosis and α,α'-dipyridyl are probably not entirely specific in their effects on connective tissue cells, their major action for the first 2–4 hr on freshly isolated tendon or cartilage cells is to inhibit the hydroxylases.

The dilemma as to how cells can continue to synthesize protocollagen or partially hydroxylated procollagen for many hours has not been completely resolved, and the data suggest that after cells have accumulated a considerable pool of such protein, either it begins to be degraded within the cells or is slowly secreted. Evidence suggesting intracellular degradation was obtained by examining the medium of tendon cells incubated for 2 hr with [^{14}C]proline and α,α'-dipyridyl (Jimenez *et al.,* 1973). The total amount of nondialyzable ^{14}C in the medium was reduced by over 80% and no [^{14}C]polypeptides of the size of pro-α or α chains were present (Figure 16). There were, however, more small ^{14}C-labeled peptides than in the control. Since the small [^{14}C]peptides in the medium could be hydroxylated by prolyl hydroxylase, they were apparently derived from [^{14}C]protocollagen. The observations suggested, therefore, that after the hydroxylases were inhibited for some time and after a critical amount of protocollagen had accumulated in the cells, a new steady state was reached in which the synthesis of protocollagen was balanced by an increase in the rate at which the accumulated protein was degraded to small peptides. Presumably this degradation of protocollagen occurred intracellularly, but since extracellular proteases were present in the cell suspension, the data did not rigorously exclude the possibility that the nonhelical protein was secreted and then rapidly degraded. We have recently observed that if protease inhibitors are added to the medium immediately after it is isolated from cells, most of the nondialyzable ^{14}C is recovered as nonhelical pro-γ chains (Kao, Berg, and Prockop, submitted for publication in

1976). It appears therefore that the degradation previously observed was by extracellular proteases and that after nonhelical pro-γ chains have accumulated for some time, they begin to be secreted from the cells.

Evidence indicating that nonhelical protocollagen can, under some circumstances, be secreted by cells has come from experiments in which isolated tissues or cultured fibroblasts were incubated for 8 hr or more with α,α'-dipyridyl (Müller *et al.*, 1973; Ramaley *et al.*, 1973) or under anaerobic conditions (Ramaley *et al.*, 1973), and from experiments in which cultured fibroblasts have been subjected to ascorbic acid deficiency for several days (see Bates *et al.*, 1972; Ramaley *et al.*, 1973). In these experiments at least part of the [^{14}C]protein recovered from the medium consists of intact pro-α chains (Bates *et al.*, 1972; Müller *et al.*, 1973; Ramaley *et al.*, 1973). In interpreting these data it is difficult to exclude the possibility that after tissues or cell cultures are exposed to α,α'-dipyridyl or anaerobiosis for 8 hr or more, or to ascorbate deficiency for several days, some of the cells may begin to undergo damage to membranes and other structures so that the protein escapes by nonspecific mechanisms. It seems unlikely that such cell damage can account for all the observations, but with isolated tissues or with monolayer cultures of fibroblasts it is difficult to carry out many of the kinds of control experiments with pulse-chase techniques and other manipulations which have been performed with freshly isolated tendon and cartilage cells in suspension (see Dehm and Prockop, 1971, 1972, 1973; Uitto and Prockop, 1974*a*).

Taken as a whole, the observations indicate that if connective tissue cells synthesizing procollagen are subjected to anaerobiosis or appropriate concentrations of α,α'-dipyridyl for 2 hr or so, the major effect is that the cells synthesize protocollagen without either secreting or degrading it to any appreciable extent. If exposure to these conditions is extended well beyond 2 hr, the cells slowly begin to secrete the nonhelical protein. The secretion of nonhelical protein under such circumstances may occur either because the "barrier" or other mechanism for retaining the nonhelical protein is saturated (see below) or because of more extensive changes in the cells.

2. *Effects of Several Proline Analogs Which Are Incorporated into Protein*

Additional evidence on the role of posttranslational reactions in the secretion of procollagen has been developed from studies with analogs of proline which are incorporated into the polypeptide chains of procollagen as well as into other proteins (for a review on proline analogs, see Uitto and Prockop, 1975). Because of the specificity of the amino acid-activating enzymes, most structural analogs of naturally occurring amino acids are not incorporated into protein by cells. Analogs of proline, particularly those with a *cis* substitution on C-4, appear to be an exception to this generalization, and a number of proline analogs have now been shown to be incorporated into procollagen and other proteins. These proline analogs include L-azetidine-2-carboxylic acid, *cis*-4-fluoro-L-proline, 3,4,-dehydroproline, *cis*-4-hydroxy-L-proline, and probably *cis*-4-bromo-L-proline. Incubation of intact connective tissues or of isolated cells with any one of these analogs markedly decreases the rate at which triple-helical procollagen is secreted from cells and deposited in tissues (Figure 17). For some time, the mechanism by which the four or five structurally distinct analogs produced this effect was obscure, but the mystery has now been in large part explained by the observation that incorporation of *cis*-hydroxyproline into newly synthesized pro-α chains prevented the pro-α chains from becoming triple-helical (Uitto *et al.*, 1972*a*). Similar observations were subsequently made with four other proline analogs which affect procollagen synthesis (Uitto and Prockop, 1974*d*; Jimenez and Rosenbloom, 1974). Apparently, because the protein cannot become triple-helical, it is not secreted at a normal rate and accumulates in the cell (Figure 17).

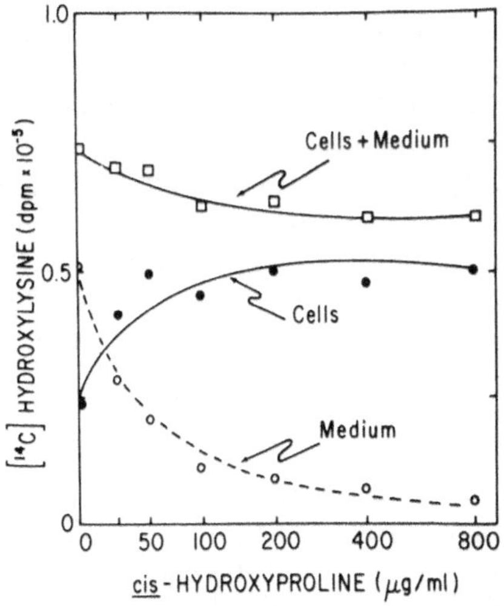

FIGURE 17. Effect of *cis*-hydroxyproline on the synthesis and secretion of procollagen by tendon cells incubated in suspension. The cells were incubated for 2 hr with [^{14}C]lysine and with the concentrations of *cis*-hydroxyproline indicated. The synthesis and secretion of procollagen were followed by assaying the nondialyzable [^{14}C]hydroxylysine in the cells and medium. Reproduced with permission from Uitto and Prockop (1974*d*).

With freshly isolated tendon cells, the concentrations of the four analogs which prevent helix formation, as tested by resistance to limited proteolysis, are about the same as the concentrations required to inhibit procollagen secretion (Uitto and Prockop, 1974*d*). Also, it has been shown that the intracellular protein containing the analogs does not become triple-helical when the cells are cooled below 20°C under conditions in which intracellular protocollagen becomes triple-helical (Uitto *et al.*, 1972*a*; Uitto and Prockop, 1974*d*). Therefore the effect on helix formation is not explained simply on the basis that substitution of a proline analog into the Y position of the pro-α chains decreases the amount of *trans*-hydroxyproline in the pro-α chains.

With tendon cells it has also been possible to carry out a number of critical control experiments with the analogs (Uitto and Prockop, 1974*d*). For example, it has been shown that incubation with any one of the four or five proline analogs does not markedly affect the rate of polypeptide synthesis, and in the case of *cis*-hydroxyproline, the rate of peptide synthesis remains the same as control for up to 4 hr. Also, it has been shown that incubation of cells with the proline analogs does not prevent the formation of interchain disulfide bonds among the three pro-α chains. In addition, it was shown that the analogs in the free form do not affect the rate at which previously accumulated protocollagen is hydroxylated to procollagen or the subsequent secretion of the procollagen from cells.

The effects of proline analogs on the formation of the triple helix was recently extended by studies with synthetic peptides. Sakakibara and his associates (Inouye *et al.*, 1976) synthesized peptides of defined molecular weight and containing *cis*-hydroxyproline (cHyp) in the structures (cHyp-Pro-Gly)$_{10}$ and (Pro-cHyp-Gly)$_{10}$. Examination of these peptides by optical rotation and circular dichroism has demonstrated that neither of them forms a triple-helical structure in water under conditions in which (Pro-Hyp-Gly)$_{10}$ or (Pro-Pro-Gly)$_{10}$ readily becomes helical. The results therefore provided direct evidence that the *trans* configuration of the 4-hydroxyl group on hydroxyproline is critical in allowing hydroxyproline to participate in stabilizing the triple helix. In addition, since both of the peptides had even less tendency to form helical structures than (Pro-Pro-Gly)$_{10}$, the results clearly demonstrated that the presence of the hydroxyl group in the *cis* configuration on C-4 interferes with helix formation.

It appears, therefore, that incubating cells or tissues with one of several proline analogs provides an additional method of manipulating cells so that they synthesize pro-α (and pro-γ) chains which do not become triple helical. The results are similar to those obtained by incubating cells under anaerobic conditions or in the presence of α,α′-dipyridyl in that for 1–2 hr the nonhelical protein continues to accumulate in the cells (Figure 17). When tendon cells are incubated with the analogs for 2 hr or longer, the nonhelical protein is secreted (see Uitto and Prockop, 1974*d*; Uitto and Prockop, submitted for publication in 1976).

3. *Effects of Lowered Temperature on Conformation and Secretion*

The ability to incubate cells so that they accumulate protocollagen has provided a further means of examining the relationship between conformation of the protein and its secretion. One of the critical tests here has been to allow cells to accumulate protocollagen at 37° and then to lower the temperature of the cells.

In one series of experiments it was shown that if the temperature of cells containing protocollagen was lowered to 25°, a fraction of the intracellular protocollagen was secreted (Jimenez *et al.*, 1974). The polypeptide chains isolated from the medium were approximately the same size as pro-α chains, and testing conformation of the protein by its ability to be hydroxylated with prolyl hydroxylase suggested that it was triple-helical. Unfortunately testing the secreted protein by limited proteolysis gave equivocal results as to whether it was triple-helical or not. The data as a whole, however, suggested that because the temperature of the cells was about the same as the T_m of protocollagen, some of the protocollagen became triple-helical and was secreted without hydroxylation either of prolyl or of lysyl residues in the protein.

In another series of experiments, cells were allowed to accumulate protocollagen by incubation under anaerobic conditions at 37°C, the temperature of the cells was dropped to 31°C or 34°C, and then the cells were exposed to O_2 so that the hydroxylating reactions could proceed (Uitto and Prockop, 1974*f*). The critical feature of the experimental design was that because the protocollagen synthesized under anaerobic conditions was already linked by interchain disulfide bonds, the rate of helix formation was limited by the rate of hydroxylation of prolyl residues and not by chain association (see Figures 13 and 14). Therefore, when hydroxylation of the protocollagen was allowed to proceed at 31°C by exposing the cells to O_2 at this temperature, the protein folded into a triple-helical conformation and hydroxylation ceased before the hydroxyproline content reached that found in procollagen synthesized under control conditions (Figure 18). As indicated, the hydroxylysine content of the protein was also markedly reduced because of the "premature" folding of the polypeptides. Of special interest was the observation that the triple-helical, but markedly underhydroxylated, procollagen synthesized under these conditions was secreted from the cells at a normal rate. The results appeared, therefore, to provide convincing proof that helical conforma-

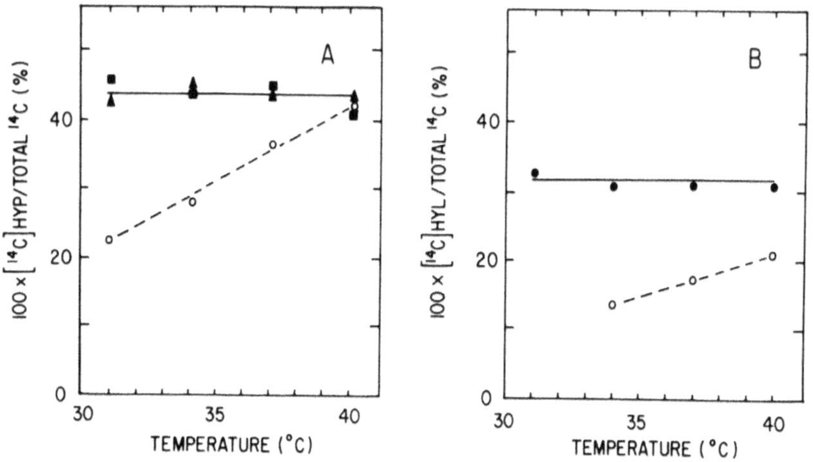

FIGURE 18. Effect of temperature on the extent to which protocollagen is hydroxylated before secretion. Cells isolated by enzymic digestion of embryonic tendons were incubated at 37°C under N_2 so that they synthesized and accumulated protocollagen (see Figures 10 and 11). The temperatures of the incubation system were then adjusted to 31, 34, 37, or 40°C, and the cells were exposed to O_2 for 120 min. The medium was removed from the cells, and the procollagen in the medium was then separated into its pro-α-chain components by polyacrylamide electrophoresis in sodium dodecyl sulfate. (A) Ratio of [^{14}C]hydroxyproline to total [^{14}C]hydroxyproline plus [^{14}C]proline in pro-α chains in the medium of cells initially labeled with [^{14}C]proline. The upper line indicates the value for the ratios in cells incubated under O_2 continuously throughout the labeling period. As discussed in text and elsewhere (Uitto and Prockop, 1974f), lowering the incubation temperature does not in itself affect the degree of hydroxylation apparently because the hydroxylation is complete before the chain association necessary for helix formation occurs. The lower broken line indicates the values obtained from cells which were incubated under N_2 so as to synthesize protocollagen at 37° and the protocollagen was hydroxylated by exposing the cells to O_2 at the temperatures indicated. (B) A similar experiment in which the cells were incubated with [^{14}C]lysine and then the [^{14}C]hydroxylysine in the pro-α chains of the procollagen secreted in the medium was assayed. Reproduced with permission from Uitto and Prockop (1974f).

tion, and not content of hydroxyproline or hydroxylysine, was the critical determinant for allowing secretion of the protein.

In considering the experiments presented in Figure 18, it might be noted that if one simply lowered the temperature of tendon cells from 37°C to 31°C, there was no effect on the degree of hydroxylation of the procollagen synthesized and secreted by the cells. The explanation for this observation is apparently that the hydroxylations, which begin on nascent chains, are largely complete before the chain association necessary for formation of the triple helix can occur. Therefore, helix formation does not occur "prematurely" at the lower temperature, and the hydroxyproline content of the pro-α chains is in excess of the minimal amount required for stable helix formation at that temperature.

In further experiments (Uitto and Prockop, unpublished data), it was shown that manipulation of the cells so that the content of glycosylated hydroxylysine was reduced from about six residues per pro-α chain to 0.8 residues per pro-α did not decrease the rate of secretion. Therefore, the "sugar tag" did not appear to influence secretion. The observations and conclusions were consistent with earlier evidence against the sugar-tag hypothesis which was provided by studies on two patients with hydroxylysine-deficient collagen disease, a variant of the Ehlers–Danlos syndrome (Pinnell *et al.*, 1972). The collagen of one of these patients was shown to contain less than 0.2 residues of hydroxylysine per chain, and therefore it must have contained less than one residue of glycosylated hydroxylysine per molecule. However, the protein was apparently secreted normally from cells.

4. *Speculations as to the Nature of the Conformation-Dependent "Barrier"*

The observation that nonhelical protocollagen or analog-containing procollagen is not secreted at a normal rate indicates that cells synthesizing procollagen contain some mechanism which can distinguish protein on the basis of the triple-helical conformation. At the moment, we have no direct evidence as to the mechanism by which this distinction is made, but at least three possibilities should probably be considered: (1) Cells synthesizing procollagen contain a highly selective structural barrier such as a membrane which allows only the triple-helical protein to pass it at a normal rate. (2) The triple-helical conformation confers on the protein physical properties which are required for passage at a normal rate through the secretory apparatus. (3) The enzymes prolyl and lysyl hydroxylase serve as barriers in the sense that they bind the nonhelical form of the protein and prevent its secretion.

In the absence of further experimental data, none of these three possibilities can be excluded, but at the moment a more convincing argument can be developed for the third than for the first and second. For example, from our current understanding of cellular structure it is difficult to suggest where a membrane barrier might be located so as to delay movement of nonhelical protocollagen from the cisternae of the endoplasmic reticulum to Golgi vacuoles (see above). Similarly, it is possible that procollagen is not secreted at a normal rate until it becomes triple-helical because, for example, the rigidity and larger Stokes radius of this conformation facilitates its flow through some gel-like region of the secretory apparatus. However, it is difficult to develop this concept in much detail. In contrast, reasonable evidence now exists to indicate that prolyl hydroxylase and lysyl hydroxylase could serve as "barriers" to the secretion of nonhelical pro-α and pro-γ chains: Both enzymes have an extremely high affinity for nonhelical protocollagen (Table 1) and both are found within the cisternae of the endoplasmic reticulum, the same compartment in which protocollagen accumulates. Since neither enzyme can hydroxylate triple-helical protocollagen or collagen, one can readily imagine that as soon as the protein becomes triple-helical, it dissociates from the enzymes and therefore becomes available for secretion. Clearly it will be important to try to develop direct tests for this intriguing possibility.

VI. Regulation of Intracellular Steps of Procollagen Biosynthesis

In considering the biosynthesis of most proteins, it is apparent that the rate at which a functional protein is synthesized depends primarily on the rate of polypeptide synthesis, since once amino acids are assembled into a polypeptide with the correct sequence, the polypeptide rapidly folds into the correct three-dimensional conformation (see Wetlaufer and Ristow, 1973, for review). Therefore the question of how synthesis of the protein is regulated is generally answered in terms of what factors control transcription and translation of the appropriate mRNA. However, from the data discussed above, it is apparent that the rate at which cells synthesize triple-helical procollagen can potentially be controlled at three different levels: the synthesis of pro-α chains, the hydroxylation of prolyl residues, and the chain association required for formation of the triple helix. Since the synthesis of pro-α chains is not always the rate-controlling step in procollagen biosynthesis, we are presented with a situation which appears to be unique in protein biosynthesis and which apparently makes the regulation of procollagen biosynthesis more complex than regulation of the synthesis of other proteins.

A. Regulation at the Level of Transcription and Translation

We currently have little information about regulation of procollagen synthesis at the level of either translation or transcription, even though it is clear that such regulation exists and is important. For example, although collagen fibers are among the more stable constituents of higher organisms, there are marked increases in the rate at which collagen accumulates in tissues during specific phases of development and in conditions such as wound healing, pulmonary fibrosis, hepatic cirrhosis, and progressive systemic sclerosis. Such increases in collagen accumulation generally reflect an increase in the rate of procollagen synthesis, but we have little information as to the biochemical events which change the rate of synthesis. (For further discussion of this point, see Grant and Prockop, 1972; Uitto and Prockop, 1974*e*.)

In terms of transcription of procollagen, it is apparent that since at least five different genes are involved in synthesizing the five different α chains found in higher organisms (see above), some mechanism must exist for selecting the genes which are used for collagen biosynthesis. It is frequently assumed that such selection of genes occurs early in the embryonic differentiation of cells, and once the selection is made, the cell is committed to synthesizing a given type of collagen. Recent observations, however, have suggested that the genes used for procollagen can be "switched" after cell differentiation. For example, articular chondrocytes, which synthesize type II collagen *in vivo,* were observed to synthesize type I collagen when they were grown in monolayer cultures (Layman *et al.,* 1972). Also, when cultures of chondrocytes synthesizing type II collagen were treated with bromodeoxyuridine, the cultures apparently began to synthesize type I collagen (Mayne *et al.,* 1973; Levitt and Dorfman, 1974). In addition, it was found that exposure of chondrocytes to lysosomal enzymes for periods as brief as 2 hr changed the type of collagen synthesized by the cells from type II to type I (Desmukh and Nimni, 1973). Some of these observations must be interpreted with caution since the results might be explained by the presence of two cell populations in the cultures, with one of them overgrowing the other as the culture conditions are altered. However, some of the observations have been verified by experiments with clones from single cells (see Mayne *et al.,* 1973). Also, the short time required for the response to lysosomal enzymes (Desmukh and Nimni, 1973) makes it unlikely that this effect can be explained by changes in the cell population. It should also be noted that clones of skin cells (Church and Tanzer, 1973) and of Schwann cells (Church *et al.,* 1973) synthesize two distinct types of collagen. Further study of such cell clones should provide important information as to the factors which control the "switching on" and "switching off" of specific genes for the transcription of mRNA.

Our information about translation is in general as incomplete as our information about transcription, but here two specific suggestions have been made as to mechanisms by which the translation of collagen polypeptides might be regulated. One suggestion is that the free intracellular pool of amino acids such as proline may control the rate of collagen synthesis (Rojkind and DeLeon, 1970). This suggestion was supported by the demonstration that when experimental liver fibrosis was induced in rats, there was an increase in the free pool of proline and a concomitant decrease in the free intracellular pool of glutamate. However, in examining these data, it is difficult to be certain whether changes in the free amino acid pools do in fact control the rate of collagen synthesis or whether they are secondary to other events such as an increase in the number of fibroblasts in fibrotic liver.

A second specific suggestion as to how the translation of collagen might be regulated was developed from studies on the peptide extensions of procollagen. As discussed above, the peptide extensions on procollagen are apparently removed just as the molecule is incorporated into collagen fibers.* Several investigators have, therefore, considered the possibility that after being cleaved from procollagen, the fragment or fragments participate in feedback inhibition of further procollagen synthesis by cells (see Lichtenstein *et al.,* 1973). In support of this possibility it has been shown that fragments which correspond to peptide extensions of procollagen have been found to be loosely associated with abnormal fibers found in the skin of cattle with dermatosparaxis (Kohn *et al.,* 1974). Such fragments have also been found in the sera of chick embryos (Dehm *et al.,* 1974) and in the medium from cultured membranous bone (Pontz *et al.,* 1973; Bornstein *et al.,* 1974) and cultured fibroblasts (Taubman *et al.,* 1974). The fragments therefore did not appear to be immediately degraded and could well fulfill some regulatory function. It should be noted, however, that if the released fragments function in this manner, the mechanism must be relatively complex, since the concentration of released peptide extensions around cells synthesizing procollagen would be proportional only to the rate at which the protein is being assembled into fibers; it would not reflect what would appear to be the more important parameter, namely, the amount of collagen in tissues (see Grant and Prockop, 1972, for further discussion of this point).

* The possibility that the peptide extensions on procollagen may help to direct fiber formation has been suggested by Veis *et al.* (1972, 1973). It is also possible that the peptide extensions on procollagen may play a role in the binding of polysomes synthesizing procollagen to the membranes of the endoplasmic reticulum, much as has been suggested for small amino acid sequences at the amino-terminal ends of light chains of immunoglobulins (Milstein *et al.,* 1972) and of albumin (Judah *et al.,* 1973). For more complete discussions of the functions of procollagen see Schofield and Prockop (1973).

B. Regulation of the Posttranslational Reactions

One of the critical questions about regulation of procollagen synthesis is: How are the posttranslational reactions regulated and controlled?

Regulation of the posttranslational reactions would be relatively simple if all cells synthesizing procollagen contained an excess of the hydroxylating and glycosylating enzymes, and if newly synthesized pro-α chains were exposed to these enzymes under conditions which allowed all the reactions to go to completion before the protein was secreted. Since, however, amino acid sequencing of collagen has clearly demonstrated that the Y position prolyl and lysyl residues are in many instances incompletely hydroxylated and glycosylated (see Bornstein, 1967*b*; Butler, 1968; Fietzek *et al.*, 1972*a,b*), it is apparent that under many conditions *in vivo* one or more of the posttranslational reactions is incomplete at the time the procollagen molecule is secreted. Therefore, regulation at the posttranslational level presents us with a complex mosaic of possible regulatory factors and conditions which might influence both the rate at which procollagen is synthesized and the chemical structures of the molecule in terms of its content of hydroxyproline, hydroxylysine, and glycosylated hydroxylysine.

1. Effects on the Rate of Procollagen Synthesis

Since hydroxylation of prolyl residues and chain association are essential for the synthesis of triple-helical procollagen, complete inhibition of either of these steps will completely inhibit procollagen biosynthesis. It is more difficult, however, to predict the consequences if one of these steps is partially inhibited. As has been apparent for some time, one of the critical questions in such situations is which step in the biosynthetic pathway is rate-controlling. A second critical question is whether the flux of substrate through the rate-controlling step is limited by the availability of substrate, essential cofactors, allosteric modifiers of the enzyme, or enzyme protein. If it is impossible to resolve these two questions, it is generally impossible to interpret much of the data one can readily obtain about regulation of a pathway, since dramatic changes in an enzyme which is not involved in the rate-controlling reaction will have no effect on the overall flux through the pathway. Also, if the flux of substrate through the rate-controlling step is limited by the concentration of substrate or of an essential cofactor, dramatic changes in the activity of the enzyme required for the reaction will have little or no effect on the system.

Unfortunately, it has proven to be surprisingly difficult in most metabolic pathways to determine which reaction is rate-controlling under any given set of experimental conditions. (See Scrutton and Utter, 1968, for a clear discussion of problems in determining rate-controlling steps in glucose metabolism.) In the case of procollagen biosynthesis, the problems are somewhat simplified by the fact that the steps are irreversible. Therefore one can frequently determine which step in the pathway is the slow and rate-controlling one simply by comparing the steady-state levels of the intermediates or by carrying out pulse-chase experiments in which one can estimate the average time required for a given step in the sequence (see above). The problems presented by the procollagen pathway are complicated, however, by the fact that prolyl hydroxylation can occur either before or after chain association and disulfide bonding, and prolyl hydroxylation and chain association are in a sense competing reactions. Therefore, several alternate pathways are open to pro-α chains *en route* to becoming triple-helical procollagen in intermediate states in which prolyl hydroxylation and chain association proceed at about the same rate and neither is clearly the rate-controlling step. A systematic analysis of the problem is beyond the scope of this chapter, but several important aspects can be illustrated by a further consideration of the two extreme situations presented in Figures 14 and 15, i.e., the situation in which chain association is rate-controlling and the situation in which prolyl hydroxylation is rate-controlling.

If chain association is the slow step in procollagen biosynthesis (Figure 14), several predictions can be made about the system. One prediction is that as long as this condition holds, the overall rate of procollagen synthesis will not be influenced by changes in the level of prolyl hydroxylase activity. For example, since the hydroxylation of pro-α chains in cartilage cells incubated under "optimal" conditions is complete well before chain association (Figure 14), one can predict that prolyl hydroxylase activity in such cells can be reduced to well below its normal level without affecting the hydroxyproline content or the rate of procollagen synthesis. Also, one can predict that an increase in the prolyl hydroxylase activity of the cells will have essentially no effect on the rate of procollagen synthesis or on the hydroxyproline content of the procollagen.

A second prediction one can make about a situation in which chain association is the rate-controlling step is that the procollagen synthesized will have more hydroxyproline than the minimum amount required to make the triple helix stable at the temperature at which it is synthesized. It is important to note that this circumstance offers a clear biological advantage. Because the hydroxyproline content is greater than the minimal amount required for helical stability, the procollagen will not unfold if the temperature of the tissue is increased by a few degrees between the time the molecule is assembled and the time at which it is incorporated into the extracellular fiber. (Since the shrinkage temperature of collagen fibers is about 58°C or about 15°C higher than the denaturation temperature for collagen in solution, it is unlikely that moderate variations in the hydroxyproline content of the molecule have any biologically important effect on the thermal stability of extracellular fibers. However, this possibility has not been tested experimentally.)

We might note that the biological advantage of chain association being the limiting step in helix formation raises the possibility that procollagen synthesis involves some mechanism for keeping pro-α chains unassociated and preventing synthesis of interchain disulfide bonds. One mechanism for delaying the formation of interchain disulfide bonds is provided by the discovery that the interchain bonds are apparently confined to the carboxy-terminal extensions (Tanzer *et al.*, 1974; Fessler *et al.*, 1975; Davidson and Bornstein, 1975; Olsen *et al.*, 1976). One can readily imagine that physical separation of the nascent pro-α chains prevents association and interchain bonding among the carboxy-terminal ends until the chains are completed and released from ribosomes. It is also possible, however, that a more elaborate mechanism exists in some cells to delay the synthesis of the interchain bonds until a much later stage in procollagen assembly. This possibility was first suggested by the observation that an unexpectedly long time of about 60 min was required for the synthesis of interchain bonds and helix formation in lens cells (Grant *et al.*, 1972*a*). Also, the suggestion was prompted in part by previous studies on the synthesis of immunoglobulin M which suggested that sulfhydryl groups in newly synthesized subunits of this protein were "masked" so as to prevent premature synthesis of interchain bonds among subunits (Askonas and Parkhouse, 1971). A similar masking of sulfhydryl groups may be required to prevent premature association and disulfide bonding among pro-α chains, but there is as yet no direct evidence for the existence of such a mechanism.

The predictions one can make are considerably different when prolyl hydroxylation is the rate-controlling step. In this situation any experimental condition which increases the rate of prolyl hydroxylation will increase the overall rate of procollagen synthesis. It is important, however, to determine whether the rate of hydroxyproline synthesis is limited by the availability of a cosubstrate such as oxygen, a cofactor such as iron, the unhydroxylated pro-α and pro-γ chains which are the substrates for the reaction, or the active enzyme protein. In the example presented in Figure 15, prolyl hydroxylation is limited by the availability of the cosubstrate oxygen, and therefore exposing the system to air increases the rate of hydroxyproline synthesis and the overall rate of procollagen synthesis and secretion by the cells. Under the conditions of this experiment, and as long as oxygen is the limiting factor, the rate of procollagen synthesis will not be increased by an increase in the rate of pro-α chain synthesis or by an increase in the amount of prolyl hydroxylase in the cells. One can readily imagine, however, a situation in which procollagen synthesis is limited by the amount of active enzyme. It is under this special circumstance that any treatment which increases the prolyl hydroxylase activity of cells or tissues will increase the rate of procollagen synthesis. Also, it is under this special circumstance that assays of the prolyl hydroxylase activity in cell or tissue extracts will provide an accurate measure of the rate of procollagen synthesis (see below).

A further prediction which one can make about the situation in which prolyl hydroxylation is rate-controlling is that the hydroxyproline content of the procollagen synthesized will be only slightly greater than the minimal amount required for stability of the triple helix. This prediction is based on the observations indicating that nonhelical pro-γ chains rapidly become triple-helical when their hydroxyproline content increases to the point where the triple helix is stable at a given temperature, and the observations indicating that hydroxylation cannot continue after the protein becomes triple-helical. The prediction is amply borne out by the experiment illustrated in Figure 18.

2. *Regulation of Prolyl Hydroxylase and of Procollagen Synthesis in Cultured Fibroblasts*

The studies with freshly isolated cells from tendon, cartilage, and lens appear to provide the best demonstration that procollagen synthesis can be regulated at the posttranslational level of either prolyl hydroxylation or chain association. However, a considerable body of additional information suggesting regulation at the posttranslational level has been developed by examining procollagen synthesis in cultured fibroblasts. This approach was made possible by the important discovery of Green and Goldberg (1964) that cultures of some types of fibroblasts synthesize measurable amounts of collagen when grown on plastic surfaces. In their initial studies, Green and Goldberg demonstrated that the rate at which such cultures synthesize collagen, as measured by the appearance of peptide-bound hydroxyproline, was increased by three conditions: an increase in cell density as the cultures grew to a confluent, stationary phase; addition of lactate to the culture medium; and addition of ascorbate to the culture medium. The nature of these effects has been investigated by a number of laboratories over the past decade, and recent data suggest that they are largely explained by the fact that prolyl hydroxylation becomes a rate-controlling step in procollagen synthesis, if the cells are grown at a low cell density and in a medium which is not regularly supplemented with ascorbate.

The effect of cell density has considerable interest because it suggested that surface contact among cells or the transfer of nutrients from one cell to another directly influenced the rate of procollagen synthesis. Initially, there was some confusion about the effect of cell density because a decrease—instead of an increase—in the rate of procollagen synthesis was observed with some fibroblasts as the cells grew from the log phase to the stationary phase (see Peterkofsky, 1972*b*; Levene and Bates, 1973; Kao *et al.*, 1976). This confusion was resolved by the demonstration (Gribble *et al.*, 1969) that the principal effect of cell density is not on the rate of pro-α chain synthesis but on the degree of hydroxylation of prolyl residues. During the log phase, the collagen polypeptides synthesized have a very low hydroxyproline content and the cells contain little or no prolyl hydroxylase activity. As the cells grow to the stationary phase, there is an increase in both the hydroxyproline content of the collagen polypeptides and the prolyl hydroxylase activity. These observations were first made with cultures of L-929 mouse fibroblasts (Gribble *et al.*, 1969), and similar observations have now been made with several other kinds of cultured fibroblasts (see Peterkofsky, 1972*b*; Levene and Bates, 1973). It seems apparent, therefore, that during the log phase of growth in such cell systems, prolyl hydroxylation is rate-controlling for procollagen biosynthesis.

Further studies with L-929 fibroblast cultures raised the possibility that the increase in prolyl hydroxylase activity observed as the cells grow to the stationary phase may in part be explained by the fact that, as the cell density increases, the cultures become more acidic and accumulate lactate. The effect of acidosis was demonstrated by showing that if sonicates of log-phase cells were incubated by pH 6.0 for 60 min, the prolyl hydroxylase activity of the sonicates increased by about 40% (Comstock *et al.*, 1970). More importantly, it was shown that addition of lactate to log-phase cultures of L-929 cells increased the prolyl hydroxylase activity in the cells two- to fivefold (Comstock and Udenfriend, 1970). However, the amount of lactate required (about 40 mM) was larger than the concentration observed in the medium of stationary-phase cultures (about 9 mM). Also, it was subsequently found that if the same L-929 fibroblasts were grown in a medium in which glucose was replaced by galactose, there was a large increase in prolyl hydroxylase activity as the cells grew from the log to the stationary phase, without any increase in the lactate concentration of the medium (Langness and Udenfriend, 1973). However, the increase in prolyl hydroxylase was not as large as that observed in cultures in which the medium contained glucose and in which the lactate concentration was high in the stationary phase. It appears likely, therefore, that cell density increases prolyl hydroxylase activity in such cultures in part by a mechanism which involves lactate and in part by a mechanism which is independent of any lactate effect.

A new perspective concerning the effects of cell density and lactate on collagen synthesis by cultured fibroblasts was introduced by the observation of Peterkofsky (1972*b*) that if ascorbate was added to the medium at frequent intervals, the collagen polypeptides synthesized during the log phase of growth were fully hydroxylated in terms of their hydroxyproline content. This observation indicated, therefore, that the effects of cell density, and presumably of lactate, are observed only if the cultures are grown at low density under conditions in which they become ascorbate-deficient. Ascorbate deficiency readily develops in fibroblast cultures because fibroblasts, even those obtained from animals which can synthesize ascorbate in liver and kidney, cannot synthesize the vitamin, and because ascorbate is rapidly destroyed nonenzymatically when it is added to culture medium and the medium is incubated at 37°C (see Peterkofsky, 1972*b*).

In summary, it appears that many of the attempts to use cultured fibroblasts to study the regulation of procollagen and collagen synthesis have been severely handicapped by the fact that the conditions used to culture the cells inadvertently made the cells ascorbate-deficient during the log phase of growth, and as a result prolyl hydroxylation became the rate-controlling step during the period of most rapid growth. In such cultures, either an increase of cell density or addition of lactate will increase the rate of hydroxylation and thereby the rate of procollagen synthesis. Most of the current evidence suggests, however, that neither cell density nor lactate are of any importance in regulating collagen synthesis by fibroblasts which are not ascorbate-deficient. Unfortunately, we still have little information as to which step becomes rate-controlling and what conditions influence this rate-controlling step when fibroblasts are cultured in medium adequately supplemented with ascorbate.

3. *"Activation" of Prolyl Hydroxylase by Ascorbate and Lactate*

In the course of examining the effects of lactate on procollagen synthesis by L-929 fibroblasts, Comstock *et al.* (1970) made the interesting observation that the increase in prolyl hydroxylase activity observed when lactate was added to the medium did not require protein synthesis, since it occurred even in the presence of the protein inhibitor cycloheximide. The results suggested, therefore, that lactate in some manner "activated" previously synthesized enzyme protein in the cells. This suggestion was supported by further experiments in which it was shown that the increase in enzyme activity was not accompanied by any change in the amount of cellular protein which reacted with specific antibodies to prolyl hydroxylase (McGee *et al.,* 1971*a*; McGee and Udenfriend, 1972*b*). A similar increase in enzymic activity without any increase in cross-reacting protein was also observed when ascorbate was added to log-phase cultures of L-929 fibroblasts (Stassen *et al.,* 1973). Examination of sonicates of the cells by gel filtration demonstrated that log-phase cells contained a considerable amount of cross-reacting protein which was catalytically inactive and of about the same size as inactive subunits of the enzyme (McGee and Udenfriend, 1972*b*). As the cells grew to late log phase, the amount of catalytically inactive enzyme protein decreased and the inactive protein was apparently converted to active enzyme. Assays on a variety of tissues from rat and mouse demonstrated that most of the tissues contained large amounts of inactive enzyme protein, and therefore it was suggested that the inactive protein might serve as a precursor pool of prolyl hydroxylase which could be rapidly activated when more rapid synthesis of procollagen was required (Stassen *et al.,* 1974).

The activation of prolyl hydroxylase by ascorbate and lactate is clearly an important discovery, but several questions remain about the phenomenon. One critical observation is that neither ascorbate nor lactate has any effect when added to purified, inactive monomers of the enzyme (see McGee and Udenfriend, 1972*b;* Berg and Prockop, unpublished observations). A second important observation is that ascorbate apparently does not "activate" prolyl hydroxylase in all cell cultures. For example, it has recently been found that in primary cultures of cells from chick embryo tendons, ascorbate increased the synthesis of procollagen hydroxyproline without increasing the prolyl hydroxylase activity of the cells (Kao *et al.,* 1975*a,b,* 1976). In this system, therefore, the increase in the synthesis of hydroxyproline appeared to be mediated solely by a direct "cofactor" effect of the vitamin. A third important observation is that a wide variety of agents appear to be able to increase prolyl hydroxylase activity in some cell cultures. As part of an extensive series of experiments on the synthesis of procollagen by cultures of 3T6 fibroblasts (see Levene and Bates, 1973; Bates *et al.,* 1972; Levene *et al.,* 1974), it was shown that the prolyl hydroxylase activity in late-log phase cells was increased not only by ascorbate but also by a number of other treatments such as addition of cycloheximide to the medium, omission of glucose from the medium, or by incubation of cells in a low-oxygen atmosphere (Levene *et al.,* 1974). Of special interest was the observation that prolyl hydroxylase activity was also increased by replacing glucose in the medium with galactose, an effect that appears to be the reverse of that seen with L-929 fibroblasts (Langness and Udenfriend, 1973). It is not yet known whether all of these experimental conditions increase prolyl hydroxylase activity in 3T6 fibroblasts by converting the enzyme from an inactive to an active form or whether some of the effects are explained by some other mechanism.

4. *The Hypothesis that Prolyl Hydroxylation Is the Rate-Controlling Step for Procollagen Synthesis in Vivo*

Although the complexities of the problem have been well recognized, a number of investigators have been attracted to the hypothesis that prolyl hydroxylation is the rate-controlling step for procollagen biosynthesis *in vivo.* The hypothesis is attractive in part because, if it is correct, it would greatly simplify any consideration of how collagen synthesis is regulated and controlled. From the considerations presented above, it seems unlikely that prolyl hydroxylation is always the rate-controlling step for procollagen synthesis *in vivo*. However, several arguments can be advanced to support the possibility that it may be the rate-controlling step in some tissues under some conditions.

One argument suggesting that prolyl hydroxylation is rate-controlling in some tissues can be developed from the observation that, although prolyl hydroxylation is much faster than chain association in lens and cartilage cells incubated under "optimal" conditions *in vitro,* the situation is different in isolated tendon cells where chain association and helix formation occur almost as soon as the hydroxyproline content of pro-α chains is maximal (Schofield *et al.,* 1974*a*). Therefore one might easily imagine that because of limitations of oxygen or other circumstances, the flux through the prolyl hydroxylase reaction in tendon cells *in vivo* might be reduced to the point where this step becomes rate-controlling.

A second argument can be developed from the studies on amino acid sequences of collagen indicating that some types of collagen are relatively underhydroxylated in terms of their hydroxyproline content. In particular, the α1-CB2 peptide from rat-tail tendon has been shown to have a much lower hydroxyproline content than the genetically similar peptide from rat skin (Bornstein, 1967*a,b*) or from chick embryo skin (Kang and Gross, 1970). The simplest explanation for these differences in degree of hydroxylation of prolyl residues is that prolyl hydroxylase is rate controlling in rat-tail tendon but not in rat skin or chick embryo skin.

A third argument is that although prolyl hydroxylation may not be rate-controlling under normal conditions, it is very likely to be rate-controlling in scurvy and in pathological conditions which limit the blood supply to tissues. In scurvy, wound healing is clearly impaired, and the impairment of wound healing is almost certainly explained by a decrease in collagen synthesis (see Barnes and Kodicek, 1972). The decreased rate of collagen synthesis is, in turn, probably explained either by the need for ascorbate as a cofactor for prolyl hydroxylation, or by an "activation" of prolyl hydroxylase by ascorbate (see above). Similarly, one can readily expect prolyl hydroxylation to become rate-controlling in any of a variety of disease conditions which are known to reduce the blood supply to tissues to the point where these tissues become essentially anaerobic. For example, the oxygen tension in normal connective tissue of the rabbit ear was found to be about 50 mm Hg, and the oxygen tension dropped to well below 5 mm Hg in the immediate area of a wound (Silver, 1973). Low tissue oxygen tensions of 15–25 mm Hg were also observed in several other kinds of wounds (see Niinikoski and Kivisaari, 1973; Stephens and Hunt, 1971). Since the K_m for oxygen in the prolyl hydroxylase reaction is 10–15 mm Hg (see Grant and Prockop, 1972), it appears that prolyl hydroxylase can readily become rate-controlling during wound healing. This possibility is indirectly supported by the demonstration that the tensile strength of wounds increases more rapidly than normal if animals are placed in hyperbaric oxygen so that the oxygen tension of the wound area is increased (Niinikoski and Kivisaari, 1973; Stephens and Hunt, 1971). It might also be noted that since partial or intermittant occlusion of the blood supply to a tissue, particularly to an extremity, lowers both the oxygen tension and temperature, one might well imagine that conditions similar to those illustrated in Figure 18 would be reproduced and fibroblasts in the tissue would synthesize a procollagen which is relatively underhydroxylated and which will readily unfold if the blood supply to the tissues is restored and the temperature returns to normal. There are no experimental data that support this suggestion, but it would appear worth exploring the possibility that it at least in part explains the breakdown of skin and other connective tissues which is seen in peripheral vascular diseases which restrict blood flow.

A further argument in favor of the hypothesis that prolyl hydroxylation *in vivo* is rate controlling has been developed by assaying prolyl hydroxylase activity in a variety of experimental and pathological states. The results (Table 5) have shown that in general high levels of prolyl hydroxylase activity are seen in conditions which appear to be associated with a rapid rate or collagen synthesis, and low levels of prolyl hydroxylase

TABLE 5

Conditions or Factors Reported to Be Associated with Altered Activities of the Enzymes Catalyzing Intracellular Steps of Collagen Biosynthesis[a]

Enzyme and condition or factor	References
Prolyl hydroxylase	
Age	Mussini *et al.* (1967); Heikkinen and Juva (1968); Uitto *et al.* (1969)
Embryonic development	Halme (1969)
Healing wounds, developing granulomas	Mussini *et al.* (1967); Juva (1968)
Liver injury	Takeuchi *et al.* (1967); Takeuchi and Prockop (1969); Stein *et al.* (1970); Feinman and Lieber (1972)
Pulmonary silicosis	Halme *et al.* (1970*b*)
Uterus during pregnancy	Kao *et al.* (1968); Halme and Jääskeläinen (1970)
Uterus after estradiol treatment	Salvador and Tsai (1973)
Rheumatoid synovial tissue	Uitto *et al.* (1970*b*)
Arteriosclerosis	Langner and Fuller (1969); Fuller and Langner (1970); Lindy *et al.* (972*a*)
Cardiac hypertrophy	Lindy *et al.* (1972*b*)
Scleroderma, active	Uitto *et al.* (1969, 1970*a*); Keiser *et al.* (1971); Uitto (1971); Fleckman *et al.* (1973)
Certain inflammatory skin diseases	Uitto (1971); Fleckman *et al.* (1973)
Skin autografts	Lindy *et al.* (1971)
Keloids and hypertrophic scars	Cohen *et al.* (1971); Fleckman *et al.* (1973)
Mammary gland carcinoma	Cutroneo *et al.* (1972)
Sarcoma	Roberts and Udenfriend (1970)
Hypertension	Ooshima *et al.* (1974)
Lysyl hydroxylase	
Age	Anttinen *et al.* (1973)
Embryonic development	Ryhänen and Kivirikko (1974*a*)
Liver injury	Risteli and Kivirikko (1974)
Hydroxylysine-deficient collagen disease	Krane *et al.* (1972)
Collagen glycosyltransferases	
Age	Spiro and Spiro (1971*c*)
Uterus during pregnancy	Spiro and Spiro (1971*c*)
Kidney in experimental diabetes	Spiro and Spiro (1971*c*)
Liver injury	Risteli and Kivirikko (1974)

[a] For further discussion on diseases involving changes in the structure and metabolism of collagen, see Uitto and Prockop (1974*e*).

activity are seen in conditions which appear to be associated with a slow rate of collagen synthesis. Unfortunately, the techniques for measuring collagen synthesis in most of the conditions are relatively imprecise, and therefore the correlation between prolyl hydroxylase activity and procollagen synthesis can only be made on a qualitative basis. It will clearly be important to establish whether this relationship is a consistent and reliable one, since if it is, assays of prolyl hydroxylase in tissue samples will provide a convenient and accurate measure of the rate of procollagen synthesis under a variety of experimental and pathological conditions. The lack of such a measure of procollagen synthesis has posed a serious obstacle to studies on collagen metabolism *in vivo* (see Kivirikko, 1970). However, it should be noted that this approach makes two assumptions: (1) prolyl hydroxylation is rate controlling and (2) the availability of active enzyme protein is the limiting component for the prolyl hydroxylase reaction. Although both assumptions may be valid in many circumstances, they are unlikely to hold for all situations. For example, the assumptions are clearly inconsistent with the observations by Levene *et al.* (1974) indicating that the prolyl hydroxylase activity increased in 3T6 fibroblasts when the cultures were incubated under a low-oxygen atmosphere or with cycloheximide, two conditions which must certainly have decreased the rate at which the cells synthesized and secreted triple-helical procollagen.

As noted in Table 5, many of the same experimental conditions have also been examined in terms of their effects on the activities of lysyl hydroxylase, glucosyltransferase, and galactosyltransferase in tissues. The results indicate that in general changes in the activities of these enzymes parallel the changes in prolyl hydroxylase activity. For example, in one recent study in which experimental liver fibrosis was induced in rats (Risteli and Kivirikko, 1974), the activities of the four posttranslational enzymes were shown to increase before fibrosis could be demonstrated and they increased further as the fibrosis developed. The four enzymes were not increased to the same degree, and on the basis of this observation it was suggested that there may be specific regulatory changes in the activities of the posttranslational enzymes when fibrosis is induced in liver.

5. *Control Over the Degree of Lysyl Hydroxylation and Glycosylation of Procollagen*

Since the hydroxylation of lysyl residues is apparently not essential for formation of the triple helix or secretion of procollagen, this reaction cannot affect the overall rate of procollagen biosynthesis. In this sense, the lysyl hydroxylase reaction appears to be an "innocent by-stander" in the biosynthetic process. The same appears to hold for the transferase reactions by which galactose and glucose are added to the procollagen molecule.

The extent to which lysyl residues in procollagen are hydroxylated must depend primarily on the concentration of unhydroxylated pro-α chains in the cisternae of the endoplasmic reticulum, the activity of lysyl hydroxylase in the same compartment, and the time which the unhydroxylated pro-α chains remain nonhelical. If chain association is the slow, rate-controlling step (see Figure 14), one can predict that the hydroxylysine content of the procollagen synthesized will generally be high and relatively independent of fluctuations in lysyl hydroxylase activity. If prolyl hydroxylation is rate controlling (Figure 15), the hydroxylysine content of the procollagen will depend critically on the ratio of the lysyl hydroxylase activity to the prolyl hydroxylase activity of cells. It is of interest that marked changes in the hydroxylysine content of collagen are seen during development of chick embryos (see Barnes *et al.*, 1974). Also, increases in the hydroxylysine content of collagen have been observed with vitamin D deficiency (Toole *et al.*, 1972; Barnes *et al.*, 1973). It is also of interest that a survey of various tissues from chick embryos has shown that tissues which contain collagen with a high hydroxylysine content tend to have a high ratio of lysyl hydroxylase to prolyl hydroxylase activity, whereas tissues which contain collagen with a low hydroxylysine content tend to have a low value for this ratio (Ryhänen and Kivirikko, 1974*a*). This correlation suggests that prolyl hydroxylase is rate-controlling and that the level of lysyl hydroxylase activity determines the hydroxylysine content of the collagen synthesized by the tissues. However, it is clearly important to obtain further evidence for these conclusions. At any rate, it seems apparent that changes in the hydroxylysine content of procollagen may have important biological consequences, since evidence from several sources suggests that cross-links of collagen derived from lysyl residues are less stable than those derived from hydroxylysyl residues (see Miller and Robertson, 1973).

We have relatively little information about factors which control the extent to which procollagen is glycosylated before being secreted, but an important clue is apparently provided by the discovery that the transferase which acts the second sugar, glucose, does not act on triple-helical collagen (Myllylä *et al.*, 1975*b*). From this observation it appears that the extent of glycosylation depends on the steady-state concentration of substrate polypeptides, the cofactors and cosubstrates required by the transferases, the activities of the transferases, and the time that the collagen portions of the polypeptide chains remain nonhelical. However, it will clearly be necessary to obtain further information about factors regulating the glycosylating reactions. Also, it may be noted that the biological function of the hexoses found in collagen remains unclear. Several investigators have suggested that the sugar residues help determine the size or other features of collagen fibers (see Spiro, 1969; Grant *et al.*, 1969), but it has been difficult to obtain conclusive data for this suggestion.

6. *Unresolved Questions about Regulation of Procollagen Synthesis*

As our information about procollagen biosynthesis increases, it should be possible to resolve many of the questions discussed above concerning regulation at the translational and posttranslational levels. It should also be possible to resolve the questions posed by a number of interesting phenomena which are still somewhat poorly defined and which have not been discussed here. For example, studies in a number of laboratories have suggested that increased intracellular levels of cAMP may favor collagen synthesis in virus-transformed fibroblasts and other cells lines (see Hsie *et al.*, 1971; Peterkofsky and Prather, 1974). These observations may provide important clues for defining specific regulatory mechanisms. As another example, a large number of cultured cells which appear to be unrelated to fibroblasts have been shown to contain prolyl hydroxylase activity and some of these cells have been shown to synthesize peptidyl hydroxyproline (see Green and Goldberg, 1965; Langness and Udenfriend, 1974). Further study of such cells may reveal a special role for collagen synthesis in the growth and development of many cell types. Alternatively, it may reveal some unexpected function for the enzyme prolyl hydroxylase, such as the possibility that it is an intrinsic protein of the endoplasmic reticulum found in all cells which synthesize proteins for export.

Finally, we might note that further investigation of phenomena such as early embryonic development of connective tissues (see Gross, 1974) may help to answer a number of further questions about regulation, such as the intriguing question of how, in cells which first begin to synthesize procollagen, the translational and posttranslational steps are coordinated so that the cells not only translate the appropriate mRNAs but also acquire the membranous structures and array of posttranslational enzymes within these structures which are essential for the successful assembly and secretion of the molecule.

References

Abbott, M. T., Schandl, E. K., Lee, R. F., Parker, T. S., and Midgett, R. J., 1967, Cofactor requirements of thymine 7-hydroxylase, *Biochim. Biophys. Acta* **132**:525.

Abbott, M. T., Dragila, T. A., and McCroskey, R. P., 1968, The formation of 5-formyluracil by cell-free preparations from *Neurospora crassa*, *Biochim. Biophys. Acta* **169:**1.

Abbott, M. T., and Udenfriend, S., 1973, α-Ketoglutarate-coupled dioxygenases, *in: Molecular Mechanisms of Oxygen Activation* (O. Hayaishi, ed.), pp. 168–214, Academic Press, New York.

Adelman, M. R., Blobel, G., and Sabatini, D. D., 1973, An improved cell fractionation procedure for the preparation of rat liver membrane-bound ribosomes, *J. Cell Biol.* **56:**191.

Aguilar, J. H., Jacobs, H. G., Butler, W. T., and Cunningham, L. W., 1973, The distribution of carbohydrate groups in rat skin collagen, *J. Biol. Chem.* **248:**5106.

Alper, R., and Kefalides, N. A., 1974, Selective cleavage of basement membranes at the aminopeptide linkages of cysteine residues. Isolation of a non-disulfide cross-linkage region in bovine anterior lens capsule, *Biochem. Biophys. Res. Commun.* **61:**1297.

Anfinsen, C. B., 1972, The formation and stabilization of protein structure, *Biochem. J.* **128:**737.

Anfinsen, C. B., 1973, Principles that govern the folding of protein chains, *Science* **181:**223.

Anttinen, H., Orava, S., Ryhänen, L., and Kivirikko, K. I., 1973, Assay of protocollagen lysyl hydroxylase activity in the skin of human subjects and changes in the activity with age, *Clin. Chim. Acta* **47:**289.

Askonas, B. A., and Parkhouse, R. M. F., 1971, Assembly of immunoglobulin M. Blocked thiol groups of intracellular 7S subunits, *Biochem. J.* **123:**629.

Avrameas, S., 1972, Enzyme markers: Their linkage with proteins and use in immunohistochemistry, *Histochem. J.* **4:**321.

Bade, M. L., and Gould, B. S., 1969, Products of proline peroxidation, *FEBS Lett.* **4:**200.

Balian, G., Click, E. M., and Bornstein, P., 1971, Structure of rat skin collagen α1-CB8. Amino acid sequence of the hydroxylamine-produced fragment HA1, *Biochem.* **10:**4470.

Balian, G., Click, E. M., Hermodson, M. A., and Bornstein, P., 1972, Structure of rat skin collagen α1-CB8. Amino acid sequence of the hydroxylamine-produced fragment HA1, *Biochem.* **11:**3798.

Barnes, M. J., and Kodicek, E., 1972, Biological hydroxylations and ascorbic acid with special regard to collagen metabolism, *Vitam. Horm.* **30:**1.

Barnes, M. J., Constable, B. J., Morton, L. F., and Kodicek, E., 1973, Bone collagen metabolism in vitamin D deficiency, *Biochem. J.* **132:**113.

Barnes, M. J., Constable, B. J., Morton, L. F., and Royce, P. M., 1974, Age-related variations in hydroxylation of lysine and proline in collagen, *Biochem. J.* **139:**461.

Bates, C. J., Prynne, A. J., and Levene, C. I., 1972, The synthesis of under-hydroxylated collagen by 3T6 mouse fibroblasts in culture, *Biochim. Biophys. Acta* **263:**397.

Beier, G., and Engel, J., 1966, The renaturation of soluble collagen. Products formed at different temperatures, *Biochem.* **5:**2744.

Bellamy, G., and Bornstein, P., 1971, Evidence for procollagen, a biosynthetic precursor of collagen, *Proc. Natl. Acad. Sci. U.S.A.* **68:**1138.

Benveniste, K., Wilczek, J., and Stern, R., 1973, Translation of collagen mRNA from chick embryo calvaria in a cell-free system derived from Krebs II ascites cells, *Nature* **246:**303.

Benveniste, K., Wilczek, J., and Stern, R., 1974, Translation of collagen messenger RNA in ascites and wheat germ cell-free systems, *Fed. Proc.* **33:**1794.

Berg, R. A., and Prockop, D. J., 1973*a*, Affinity column purification of protocollagen proline hydroxylase from chick embryos and further characterization of the enzyme, *J. Biol. Chem.* **248:**1175.

Berg, R. A., and Prockop, D. J., 1973*b*, Purification of [^{14}C]protocollagen and its hydroxylation by prolyl-hydroxylase, *Biochem.* **12:**3395.

Berg, R. A., and Prockop, D. J., 1973*c*, The thermal transition of a non-hydroxylated form of collagen. Evidence for a role for hydroxyproline in stabilizing the triple helix of collagen, *Biochem. Biophys. Res. Commun.* **52:**115.

Berg, R. A., and Prockop, D. J., 1973*d*, Hydroxylation of native and denatured protocollagen by prolyl hydroxylase, *Abstracts of the Ninth International Congress of Biochemistry,* Stockholm, p. 423.

Berg, R. A., and Prockop, D. J., 1976, Purification of subunits of prolyl hydroxylase (in preparation).

Berg, R. A., Kishida, Y., Kobayashi, Y. K., Inouye, K., Tonelli, A. E., Sakakibara, S., and Prockop, D. J., 1973, A model for the triple-helical structure of $(\text{Pro-Hyp-Gly})_{10}$ involving a *cis* peptide bond and interchain hydrogen-bonding to the hydroxyl group of hydroxyproline, *Biochim. Biophys. Acta* **328:**553.

Berg, R. A., Kishida, Y., Sakakibara, S., and Prockop, D. J., 1976, Preferential hydroxylation of the ninth triplet in $(\text{Pro-Pro-Gly})_{10}$ by prolyl hydroxylase (in preparation).

Berg, R. A., Olsen, B. R., and Prockop, D. J., 1970, Titration and melting curves of the collagen-like triple helices formed from $(\text{Pro-Pro-Gly})_{10}$ in aqueous solution, *J. Biol. Chem.* **245:**5759.

Berg, R. A., Olsen, B. R., and Prockop, D. J., 1972, Labeled antibodies to protocollagen proline hydroxylase from chick embryos for intracellular localization of the enzyme, *Biochim. Biophys. Acta* **285:**167.

Bhatnagar, R. S., and Liu, T. Z., 1973, Mechanism of hydroxylation of proline, *Abstracts of the Ninth International Congress of Biochemistry,* Stockholm, July 1–7, 1973, p. 335.

Bhatnagar, R. S., Prockop, D. J., and Rosenbloom, J., 1967*a*, Intracellular pool of unhydroxylated polypeptide precursors of collagen, *Science* **158:**492.

Bhatnagar, R. S., Rosenbloom, J., Kivirikko, K. I., and Prockop, D. J., 1967*b*, Effect of cycloheximide on collagen biosynthesis as evidence for a post-ribosomal site for the hydroxylation of proline, *Biochim. Biophys. Acta* **149:**273.

Björk, I., and Tanford, C., 1971, Recovery of native conformation of rabbit immunoglobulin G upon recombination of separately renatured heavy and light chains at near-neutral pH, *Biochemistry* **10:**1289.

Blumenkrantz, N., and Prockop, D. J., 1969, A rapid assay for ^{14}C-labeled hydroxylysine in collagen and related materials, *Anal. Biochem.* **30:**377.

Blumenkrantz, N., and Prockop, D. J., 1971, Quantitative assay for hydroxylysine in protein hydrolyzates, *Anal. Biochem.* **39:**59.

Blumenkrantz, N., Rosenbloom, J., and Prockop, D. J., 1968, Sequential steps in the synthesis of hydroxylysine during the biosynthesis of collagen, *Biochim. Biophys. Acta* **192:**81.

Bornstein, P., 1967*a*, The incomplete hydroxylation of individual prolyl residues in collagen, *J. Biol. Chem.* **242:**2572.

Bornstein, P., 1967*b*, Comparative sequence studies of rat skin and tendon collagen. I. Evidence for incomplete hydroxylation of individual prolyl residues in the normal proteins, *Biochemistry* **6:**3082.

Bornstein, P., 1969, Comparative sequence studies of rat skin and tendon collagen. II. The absence of a short sequence at the amino terminus of the skin $\alpha 1$ chain, *Biochemistry* **8:**63.

Bornstein, P., 1970, Structure of α1-CB8, a large cyanogen bromide produced fragment from the α1 chain of rat collagen. The nature of a hydroxyl amine-sensitive bond and composition of tryptic peptides, *Biochemistry* **9:**2408.
Bornstein, P., 1974, The biosynthesis of collagen, *Annu. Rev. Biochem.* **43:**567.
Bornstein, P., Ehrlich, H. P., and Wyke, A. W., 1972, Procollagen: Conversion of the precursor to collagen by a neutral protease, *Science* **175:**544.
Bornstein, P., von der Mark, K., Murphy, W. H., and Garnett, L., 1974, Isolation and characterization of procollagen-derived peptides, *Fed. Proc.* **33:**1595.
Bosmann, H. B., and Eylar, E. H., 1968*a*, Glycoprotein biosynthesis: The biosynthesis of the hydroxylysine–galactose linkage in collagen, *Biochem. Biophys. Res. Commun.* **33:**340.
Bosmann, H. B., and Eylar, E. H., 1968*b*, Attachment of carbohydrate to collagen. Isolation, purification and properties of the glucosyl transferase, *Biochem. Biophys. Res. Commun.* **30:**89.
Brandt, J. T., Martin, A. P., Lucas, F. V., and Vorbeck, M. L., 1974, The structure of normal rat liver mitochondria: A re-evaluation, *Fed. Proc.* **33:**2219.
Brownell, A. G., and Veis, A., 1975, Intracellular location of the glycosylation of hydroxylysine of collagen, *Biochem. Biophys. Res. Commun.* **63:**371.
Butler, W. T., 1968, Partial hydroxylation of certain lysines in collagen, *Science* **161:**796.
Butler, W. T., 1972, The structure of α1-CB3, a cyanogen bromide fragment from the central portion of the α1 chain of rat collagen. The tryptic peptides from skin and dentin collagens, *Biochem. Biophys. Res. Commun.* **48:**1540.
Butler, W. T., and Cunningham, L. W., 1966, Evidence for the linkage of a disaccharide to hydroxylysine in tropocollagen, *J. Biol. Chem.* **241:**3882.
Butler, W. T., and Ponds, S. L., 1971, Chemical studies on the cyanogen bromide peptides of rat skin collagen. Amino acid sequence of α1-CB4, *Biochemistry* **10:**2076.
Butler, W. T., Miller, E. J., Finch, J. E., Jr., and Inagami, T., 1974, Homologous regions of collagen α1(I) and α1(II) chains: apparent clustering of variable and invariant amino acid residues, *Biochem. Biophys. Res. Commun.* **57:**190.
Byers, P. H., McKenney, K. H., Lichtenstein, J. R., and Martin, G. R., 1974, Preparation of type III procollagen and collagen from rat skin, *Biochemistry* **13:**5243.
Cardinale, G. J., and Udenfriend, S., 1974, Prolyl hydroxylase, *Adv. Enzymol.* **41:**245.
Cardinale, G. J., Rhoads, R. E., and Udenfriend, S., 1971, Simultaneous incorporation of ^{18}O into succinate and hydroxyproline catalyzed by collagen proline hydroxylase, *Biochem. Biophys. Res. Commun.* **43:**537.
Chung, E., and Miller, E. J., 1974, Collagen polymorphism: Characterization of molecules with a chain composition $[\alpha 1(III)]_3$ in human tissues, *Science* **183:**1200.
Church, R. L., and Tanzer, M. L., 1973, Identification of two distinct species of procollagen synthesized by a clonal line of calf dermatosparactic cells, *Nature (London) New Biol.* **244:**188.
Church, R. L., Pfeiffer, S. E., and Tanzer, M. L., 1971, Collagen biosynthesis: Synthesis and secretion of a high molecular weight collagen precursor (procollagen), *Proc. Natl. Acad. Sci. U.S.A.* **68:**2638.
Church, R. L., Tanzer, M. L., and Pfeiffer, S. E., 1973, Collagen and procollagen production by a clonal line of Schwann cells, *Proc. Natl. Acad. Sci. U.S.A.* **70:**1943.
Clark, C. C., Tomichek, E. A., Koszalka, T. R., Brent, R. L., and Kefalides, N. A., 1973, Basement membrane biosynthesis in rat embryo parietal yolk sac, *Fed. Proc.* **32:**650.
Cohen, I. K., Keiser, H. R., and Sjoerdsma, A., 1971, Collagen synthesis in human keloid and hypertrophic scar, *Surg. Forum* **22:**488.

Comstock, J. P., and Udenfriend, S., 1970, Effect of lactate on collagen proline hydroxylase activity in cultured L-929 fibroblasts, *Proc. Natl. Acad. Sci. U.S.A.* **66**:552.

Comstock, J. P., Gribble, T. J., and Udenfriend, S., 1970, Further study on the activation of collagen proline hydroxylase in cultures of L-929 fibroblasts, *Arch. Biochem. Biophys.* **137**:115.

Cooper, G. W., and Prockop, D. J., 1968, Intracellular accumulation of protocollagen and extrusion of collagen by embryonic cartilage cells, *J. Cell Biol.* **38**:523.

Cunningham, L. W., and Ford, J. D., 1968, A comparison of glycopeptides derived from soluble and insoluble collagens, *J. Biol. Chem.* **243**:2390.

Cutroneo, K. R., Guzman, N. A., and Liebelt, A. G., 1972, Elevation of peptidyl proline hydroxylase activity and collagen synthesis in spontaneous primary mammary cancers of inbred mice, *Cancer Res.* **32**:2828.

Darnell, J., Jimenez, S., Murphy, L., Harsch, M., and Rosenbloom, J., 1974, Structural changes in procollagen during hydroxylation by prolyl hydroxylase, *Fed. Proc.* **33**:1596.

Davidson, J. M., and Bornstein, P., 1975, Evidence for multiple steps in the limited proteolytic conversion of procollagen to collagen, *Fed. Proc.* **34**:562.

Dehm, P., and Prockop, D. J., 1971, Synthesis and extrusion of collagen by freshly-isolated cells from chick embryo tendon, *Biochim. Biophys. Acta* **240**:358.

Dehm, P., and Prockop, D. J., 1972, Time lag in the secretion of collagen by matrix-free tendon cells and inhibition of the secretory process by colchicine and vinblastine, *Biochim. Biophys. Acta* **264**:375.

Dehm, P., and Prockop, D. J., 1973, Biosynthesis of cartilage procollagen, *Eur. J. Biochem.* **35**:159.

Dehm, P., Jimenez, S. A., Olsen, B. R., and Prockop, D. J., 1972, A transport form of collagen from embryonic tendon: electron microscopic demonstration of an NH_2-terminal extension and evidence suggesting the presence of cystine in the molecule, *Proc. Natl. Acad. Sci. U.S.A.* **69**:60.

Dehm, P., Olsen, B. R., and Prockop, D. J., 1974, Antibodies to chick-tendon procollagen. Affinity purification with isolated disulfide-linked NH_2-terminal extensions and reactivity with a component in embryonic serum, *Eur. J. Biochem.* **46**:107.

Desmukh, K., and Nimni, M. E., 1973, Effects of lysosomal enzymes on the type of collagen synthesized by bovine articular cartilage, *Biochem. Biophys. Res. Commun.* **53**:424.

Diegelmann, R. F., and Peterkofsky, B., 1972, Inhibition of collagen secretion from bone and cultured fibroblasts by microtubular disruptive drugs, *Proc. Natl. Acad. Sci. U.S.A.* **69**:892.

Diegelmann, R. F., Bernstein, L., and Peterkofsky, B., 1973, Cell-free collagen synthesis on membrane-bound polysomes of chick embryo connective tissues and the localization of prolyl hydroxylase on the polysome-membrane complex, *J. Biol. Chem.* **248**:6514.

Ehrlich, H. P., and Bornstein, P., 1972*a*, Further characterization of procollagen: The identification of a pro-α2 chain, *Biochem. Biophys. Res. Commun.* **46**:1750.

Ehrlich, H. P., and Bornstein, P., 1972*b*, Microtubules in transcellular movement of procollagen, *Nature (London) New Biol.* **238**:257.

Engel, J., Kurtz, J., Katchalski, E., and Berger, A., 1966, Polymers of tripeptides as collagen models. II. Conformational changes of poly(L-prolyl-glycyl-L-prolyl) in solution, *J. Mol. Biol.* **17**:255.

Epstein, E. H., Jr., 1974, $[\alpha1(III)]_3$ Human skin collagen, *J. Biol. Chem.* **249**:3225.

Eylar, E. H., 1966, On the biological role of glycoproteins, *J. Theor. Biol.* **10:**89.

Feinman, L., and Lieber, C. S., 1972, Hepatic collagen metabolism. Effect of alcohol consumption in rats and baboons, *Science* **176:**795.

Fessler, L. I., and Fessler, J. H., 1974, Protein assembly of procollagen and effects of hydroxylation, *J. Biol. Chem.* **249:**7637.

Fessler, L. I., Burgeson, R. E., Morris, N. P., and Fessler, J. H., 1973, Collagen synthesis: A disulfide-linked collagen precursor in chick bone, *Proc. Natl. Acad. Sci. U.S.A.* **70:**2993.

Fessler, L. I., Morris, N. P., and Fessler, J. H., 1975, Procollagen processing to collagen via a disulfide-linked triple-stranded intermediate in chick calvaria, *Fed. Proc.* **34:**562.

Fietzek, P. P., and Rauterberg, J., 1975, Cyanogen bromide peptides of type III collagen: First sequence analysis demonstrates homology with type I collagen, *FEBS Lett.* **49:**365.

Fietzek, P. P., Wendt, P., Kell, I., and Kühn, K., 1972*a*, The covalent structure of collagen: Amino acid sequence of α1-CB3 from calf skin collagen, *FEBS Lett.* **26:**74.

Fietzek, P. P., Rexrodt, F. W., Wendt, P., Stark, M., and Kühn, K., 1972*b*, The covalent structure of collagen. Amino-acid sequence of peptide α1-CB6-C2, *Eur. J. Biochem.* **30:**163.

Fleckman, P. H., Jeffrey, J. J., and Eisen, A. Z., 1973, A sensitive microassay for prolyl hydroxylase: Activity in normal and psoriatic skin, *J. Invest. Dermatol.* **60:**46.

Frank, R. M., 1970, Etude autoradiographe de la dentinogènese en microscopie electronique à l'aide de la proline tritiée chez le chat, *Arch. Oral Biol.* **15:**583.

Fujimoto, D., and Prockop, D. J., 1968, Denatured collagen from the cuticle of *Ascaris lumbricoides* as a substrate for protocollagen proline hydroxylase, *J. Biol. Chem.* **243:**4138.

Fujimoto, D., and Tamiya, N., 1962, Incorporation of ^{18}O from air into hydroxyproline by chick embryo, *Biochem. J.* **84:**333.

Fuller, G. C., and Langner, R. O., 1970, Elevation of aortic proline hydroxylase: A biochemical defect in experimental arteriosclerosis, *Science* **168:**987.

Furthmayr, H., Timpl, R., Stark, M., Lapière, C. M., and Kühn, K., 1972, Chemical properties of the peptide extension in the p-α1 chain of dermatosparactic skin collagen, *FEBS Lett.* **28:**247.

Gallop, P. M., Blumenfeld, O. O., and Seifter, S., 1967, Subunits and special structural features of tropo-collagen, *in: Treatise on Collagen* (G. N. Ramachandran, ed.), Vol. 1, p. 339, Academic Press, London and New York.

Goldberg, B., and Green, H., 1967, Collagen synthesis on polyribosomes of cultured mammalian fibroblasts, *J. Mol. Biol.* **27:**1.

Goldberg, B., and Sherr, C. J., 1973, Secretion and extracellular processing of procollagen by cultured human fibroblasts, *Proc. Natl. Acad. Sci. U.S.A.* **70:**361.

Goldberg, B., Epstein, E. H., Jr., and Sherr, C. J., 1972, Precursors of collagen secreted by cultured human fibroblasts, *Proc. Natl. Acad. Sci. U.S.A.* **69:**3655.

Goldstein, A., and Adams, E., 1968, Occurrence of glycylhydroxyprolyl sequences in earthworm cuticle collagen, *J. Biol. Chem.* **243:**3550.

Goldstein, A., and Adams, E., 1970, Glycylhydroxyprolyl sequences in earthworm cuticle collagen: glycylhydroxyprolylserine, *J. Biol. Chem.* **245:**5478.

Gould, B. S., 1968, Collagen biosynthesis, *in: Treatise on Collagen* (B. S. Gould, ed.), pp. 139–189, Academic Press, London and New York.

Grant, M. E., and Harwood, R., 1974, The biosynthesis of glomerular basement membrane collagen, *Biochem. Soc. Trans.* **2:**624.

Grant, M. E., and Prockop, D. J., 1972, The biosynthesis of collagen, *New Engl. J. Med.* **286**:194, 242, 291.

Grant, M. E., Freeman, I. L., Schofield, J. D., and Jackson, D. S., 1969, Variations in the carbohydrate content of human and bovine polymeric collagens from various tissues, *Biochim. Biophys. Acta* **177**:682.

Grant, M. E., Kefalides, N. A., and Prockop, D. J., 1972*a*, The biosynthesis of basement membrane collagen in embryonic chick lens. I. Delay between the synthesis of polypeptide chains and the secretion of collagen by matrix-free cells, *J. Biol. Chem.* **247**:3539.

Grant, M. E., Kefalides, N. A., and Prockop, D. J., 1972*b*, The biosynthesis of basement membrane collagen in embryonic chick lens. II. Synthesis of a precursor form by matrix-free cells and a time-dependent conversion to α chains in intact lens, *J. Biol. Chem.* **247**:3545.

Grant, M. E., Schofield, J. D., Kefalides, N. A., and Prockop, D. J., 1973, The biosynthesis of basement membrane collagen in embryonic chick lens. III. Intracellular formation of the triple helix and the formation of aggregates through disulfide bonds, *J. Biol. Chem.* **248**:7432.

Grant, M. E., Harwood, R., and Schofield, J. D., 1975, Recent studies on the assembly, intracellular processing and secretion of procollagen, *in: Dynamics of Connective Tissue Macromolecules* (A. R. Poole and P. M. C. Burlaigh, eds.), North Holland Publishing Co. (in press).

Green, H., and Goldberg, B., 1964, Collagen and cell protein synthesis by an established mammalian fibroblast line, *Nature* **204**:347.

Green, H., and Goldberg, B., 1965, Synthesis of collagen by mammalian cell lines of fibroblastic and nonfibroblastic origin, *Proc. Natl. Acad. Sci. U.S.A.*, **53**:1360.

Gribble, T. J., Comstock, J. P., and Udenfriend, S., 1969, Collagen chain formation and peptidyl proline hydroxylation in cultures of L-929 fibroblasts, *Arch. Biochem. Biophys.* **129**:308.

Gross, J., 1974, Collagen biology: Structure, degradation and disease, *Harvey Lect.* **68**:351.

Gustavson, K. H., 1954, Hydroxyproline and stability of collagens, *Acta Chem. Scand.* **8**:1298.

Gustavson, K. H., 1955, The function of hydroxyproline in collagens, *Nature* **175**:70.

Guzman, N. A., and Cutroneo, K. R., 1973, Association of prolyl hydroxylase with membranes, *Biochem. Biophys. Res. Commun.* **52**:1263.

Guzman, N. A., Prichard, P. M., Sharaway, M. M., and Cutroneo, K. R., 1974, Isolation of subcellular vesicles containing prolyl hydroxylase and substrate, *Fed. Proc.* **33**:1759.

Halme, J., 1969, Development of protocollagen proline hydroxylase activity in the chick embryo, *Biochim. Biophys. Acta* **192**:90.

Halme, J., and Kivirikko, K. I., 1968, Studies on the stability of protocollagen hydroxylase, *FEBS Lett.* **1**:223.

Halme, J., and Jääskeläinen, M., 1970, Protocollagen proline hydroxylase of the mouse uterus during pregnancy and post-partum involution, *Biochem. J.* **116**:367.

Halme, J., Kivirikko, K. I., and Simons, K., 1970*a*, Isolation and partial characterization of highly purified protocollagen proline hydroxylase, *Biochim. Biophys. Acta* **198**:460.

Halme, J., Uitto, J., and Kahanpää, K., Karhunen, P., and Lindy, S., 1970*b*, Protocollagen proline hydroxylase activity in experimental pulmonary fibrosis of rats, *J. Lab. Clin. Med.* **75**:535.

Hamilton, G., 1971, The proton in biological redox reactions, *in: Progress in Bioorganic Chemistry* (E. T. Kaiser and F. J. Kezdy, eds.), Vol. 1, pp. 83–157, Wiley, New York.

Harrington, W. F., and Karr, G. M., 1970, Collagen structure in solution. I. Analysis of refolding kinetics in terms of nucleation and growth process, *Biochemistry* **9:**3725.

Harrington, W. F., and von Hippel, P. H., 1961, The structure of collagen and gelatin, *Adv. Protein Chem.* **16:**1.

Harwood, R., Grant, M. E., and Jackson, D. S., 1973, The subcellular location of interchain disulfide bond formation during procollagen biosynthesis by embryonic chick tendon cells, *Biochem. Biophys. Res. Commun.* **55:**1188.

Harwood, R., Grant, M. E., and Jackson, D. S., 1974*a*, Secretion of procollagen: Evidence for the transfer of nascent polypeptides across microsomal membranes of tendon cells, *Biochem. Biophys. Res. Commun.* **59:**947.

Harwood, R., Grant, M. E., and Jackson, D. S., 1974*b*, Collagen biosynthesis. Characterization of subcellular fractions from embryonic chick fibroblasts and the intracellular localization of protocollagen prolyl and protocollagen lysyl hydroxylases, *Biochem. J.* **144:**123.

Harwood, R., Connolly, A. D., Grant, M. E., and Jackson, D. S., 1974*c*, Presumptive mRNA for procollagen: Occurrence in membrane-bound ribosomes of embryonic chick tendon fibroblasts, *FEBS Lett.* **41:**85.

Harwood, R., Grant, M. E., and Jackson, D. S., 1975, The association of collagen galactosyl and glucosyl transferases with subcellular fractions of embryonic chick tendon cells, *Biochem. Soc. Trans.* **3:**136.

Hausmann, E., 1967, Cofactor requirements for the enzymatic hydroxylation of lysine in a polypeptide precursor of collagen, *Biochim. Biophys. Acta* **133:**591.

Hay, E. D., and Dodson, J. W., 1973, Secretion of collagen by corneal epithelium. I. Morphology of the collagenous products produced by isolated epithelia grown on frozen-killed lens, *J. Cell Biol.* **57:**190.

Heikkinen, E., and Juva, K., 1968, The activity of protocollagen hydroxylase in rat skin at different ages, *Abstracts of the 5th Meeting of the Federation of European Biochemical Societies*, Prague, 1968, Abstract No. 561.

Hoffmann, H.-P., and Avers, C. J., 1973, Mitochondria of yeast: Ultrastructural evidence for one giant, branched organelle per cell, *Science* **181:**749.

Holme, E., Lindstedt, G., Lindstedt, S., and Tofft, M., 1968, α-Ketoglutarate in biological hydroxylations, *FEBS Lett.* **2:**29.

Hsie, A. W., Jones, C., and Puck, T. T., 1971, Further changes in differentiation state accompanying the conversion of Chinese hamster cells in fibroblastic form by dibutyryl adenosine cyclic 3′:5′-monophosphate and hormones, *Proc. Natl. Acad. Sci. U.S.A.* **68:**1648.

Hunt, T., Hunter, T., and Munro, A., 1969, Control of haemoglobin synthesis: rate of translation of the messenger RNA for the α and β chains, *J. Mol. Biol.* **43:**123.

Hurych, J., and Chvapil, M., 1965, Influence of chelating agents on the biosynthesis of collagen, *Biochem. Biophys. Acta* **97:**361.

Hurych, J., and Nordwig, A., 1967, Inhibition of collagen hydroxylysine formation by chelating agents, *Biochim. Biophys. Acta* **140:**168.

Hurych, J., Hobza, P., Rencová, J., and Zahradnik, R., 1973, An approach to the hydroxylation of collagenous proline, *in: Biology of the Fibroblast* (E. Kulonen and J. Pikkarainen, eds.), pp. 365–372, Academic Press, London.

Hutton, J. J., Jr., Tappel, A. L., and Udenfriend, S., 1966, Requirements for α-ketoglutarate, ferrous iron and ascorbate by collagen proline hydroxylase, *Biochem. Biophys. Res. Commun.* **24:**179.

Hutton, J. J., Jr., Kaplan, A., and Udenfriend, S., 1967*a*, Conversion of the amino acid sequence Gly-Pro-Pro in protein to Gly-Pro-Hyp by collagen proline hydroxylase, *Arch. Biochem. Biophys.* **121**:384.

Hutton, J. J., Jr., Tappel, A. L., and Udenfriend, S., 1967*b*, Cofactor and substrate requirements of collagen proline hydroxylase, *Arch. Biochem. Biophys.* **118**:231.

Hutton, J. J., Jr., Marglin, A., Witkop, B., Kurtz, J., Berger, A., and Udenfriend, S., 1968, Synthetic polypeptides as substrates and inhibitors of collagen proline hydroxylase, *Arch. Biochem. Biophys.* **125**:779.

Inouye, K., Sakakibara, S., and Prockop, D. J., 1976, Effects of the stereo-configuration of the hydroxyl group in 4-hydroxyproline on the triple-helical structures formed by homogeneous peptides resembling collagen *Biochim. Biophys. Acta* **420**:133.

Isemura, M., Ikenaka, T., and Matsushisna, Y., 1973, Comparative study of carbohydrate-protein complexes. I. The structures of glycopeptides derived from cuttlefish skin collagen, *J. Biochem.* **74**:11.

Jacobson, M. F., Asso, J., and Baltimore, D., 1970, Further evidence on the formation of poliovirus proteins, *J. Mol. Biol.* **49**:657.

Jimenez, S. A., Dehm, P., and Prockop, D. J., 1971, Further evidence for a transport form of collagen. Its extrusion and extracellular conversion to tropocollagen in embryonic tendon, *FEBS Lett.* **17**:245.

Jimenez, S. A., Dehm, P., Olsen, B. R., and Prockop, D. J., 1973*a*, Intracellular collagen and protocollagen from embryonic tendon cells, *J. Biol. Chem.* **248**:720.

Jimenez, S., Harsch, M., and Rosenbloom, J., 1973*b*, Hydroxyproline stabilizes the triple helix of chick tendon collagen, *Biochem. Biophys. Res. Commun.* **52**:106.

Jimenez, S., and Rosenbloom, J., 1974, Decreased thermal stability of collagens containing analogs of proline or lysine, *Arch. Biochem. Biophys.* **163**:459.

Jimenez, S. A., Harsch, M., Murphy, L., and Rosenbloom, J., 1974, Effects of temperature on conformation, hydroxylation, and secretion of chick tendon procollagen, *J. Biol. Chem.* **249**:4480.

Judah, J. W., Gamble, M., and Steadman, J. H., 1973, Biosynthesis of serum albumin in rat liver. Evidence for the existence of "proalbumin", *Biochem. J.* **134**:1083.

Juva, K., 1968, Hydroxylation of proline in the biosynthesis of collagen: An experimental study with chick embryo and granulation tissue of rat, *Acta Physiol. Scand. Suppl.* **308**:1.

Juva, K., and Prockop, D. J., 1966*a*, An effect of puromycin on the synthesis of collagen by embryonic cartilage *in vitro*, *J. Biol. Chem.* **241**:4419.

Juva, K., and Prockop, D. J., 1966*b*, Modified procedure for the assay of H^3- or C^{14}-labeled hydroxyproline, *Anal. Biochem.* **15**:77.

Juva, K., and Prockop, D. J., 1969, Formation of enzyme–substrate complexes with protocollagen proline hydroxylase and large polypeptide substrates, *J. Biol. Chem.* **244**:6486.

Juva, K., Prockop, D. J., Cooper, G. W., and Lash, J., 1966, Hydroxylation of proline and the intracellular accumulation of a polypeptide precursor of collagen, *Science* **152**:92.

Kang, A. H., and Gross, J., 1970, Amino acid sequence of cyanogen bromide peptides frmo the amino-terminal region of chick skin collagen, *Biochemistry* **9**:796.

Kao, K. -Y. T., Treadwell, C. R., Previll, J. M., and McGavack, T. H., 1968, Connective tissues XVII. Protocollagen hydroxylase of pig uterus, *Biochim. Biophys. Acta* **151**:568.

Kao, W. W.-T., Berg, R. A., Flaks, J. G., and Prockop, D. J., 1975*a*, Ascorbate can increase hydroxyproline synthesis in cultured fibroblasts without activation or prolyl hydroxylase, *Fed. Proc.* **34**:565.

Kao, W. W.-T., Berg, R. A., and Prockop, D. J., 1975*b*, Ascorbate increases the synthesis of procollagen hydroxyproline by cultured fibroblasts from chick embryo tendons without activation of prolyl hydroxylase, *Biochim. Biophys. Acta* **411**:202.

Kao, W. W.-T., Flaks, J. G., and Prockop, D. J., 1976, Primary and secondary effects of ascorbate on procollagen synthesis and protein synthesis by primary cultures of tendon fibroblasts, *Arch. Biochem. Biophys.*, in press.

Kefalides, N. A., 1972, Isolation and characterization of cyanogen bromide peptides from basement membrane collagen, *Biochem. Biophys. Res. Commun.* **47**:1151.

Kefalides, N. A., 1973, Structure and biosynthesis of basement membranes, *Int. Rev. Connect. Tissue Res.* **6**:63.

Keiser, H. R., Stein, H. D., and Sjoerdsma, A., 1971, Increased protocollagen proline hydroxylase activity in sclerodermatous skin, *Arch. Dermatol.* **104**:57.

Kerwar, S. S., 1974, Studies on the nature of procollagen synthesized by chick embryo polysomes, *Arch. Biochem. Biophys.* **163**:609.

Kerwar, S. S., Kohn, L. D., Lapière, C. M., and Weissbach, H., 1972, *In vitro* synthesis of procollagen on polysomes, *Proc. Natl. Acad. Sci. U.S.A.* **69**:2727.

Kerwar, S. S., Cardinale, G. J., Kohn, L. D., Spears, C. L., and Stassen, F. L. H., 1973, Cell-free synthesis of procollagen: L-929 fibroblasts as a cellular model for dermatosparaxis, *Proc. Natl. Acad. Sci. U.S.A.* **70**:1378.

Kikuchi, Y., Fujimoto, D., and Tamiya, N., 1968, Hydroxylation of poly(L-prolyl-L-prolylglycyl) of defined molecular weights by protocollagen proline hydroxylase, *FEBS Lett.* **2**:221.

Kikuchi, Y., Fujimoto, D., and Tamiya, N., 1969, The enzymic hydroxylation of protocollagen models, *Biochem. J.* **115**:569.

Kirschbaum, B. B., and Bosmann, H. B., 1973, The interaction of folic acid and glycoprotein glycosyl transferase activities of rat kidney, *Biochim. Biophys. Acta* **320**:416.

Kishida, Y., Olsen, B. R., Berg, R. A., and Prockop, D. J., 1975, Two improved methods for preparing ferritin–protein conjugates for electron microscopy, *J. Cell Biol.* **64**:331.

Kishida, Y., Berg, R. A., Sakakibara, S., and Prockop, D. J., 1976, Hydroxylation of -prolyl-prolyl-glycyl- sequences in synthetic peptides by prolyl hydroxylase. Effects of chain length on catalytic constants for the reaction (in preparation).

Kivirikko, K. I., 1970, Urinary excretion of hydroxyproline in health and disease, *Int. Rev. Connect Tissue Res.* **5**:93.

Kivirikko, K. I., and Prockop, D. J., 1967*a*, Hydroxylation of proline in synthetic polypeptides with purified protocollagen hydroxylase, *J. Biol. Chem.* **242**:4007.

Kivirikko, K. I., and Prockop, D. J., 1967*b*, Purification and partial characterization of the enzyme for the hydroxylation of proline in protocollagen, *Arch. Biochem. Biophys.* **118**:611.

Kivirikko, K. I., and Prockop, D. J., 1967*c*, Enzymatic hydroxylation of proline and lysine in protocollagen, *Proc. Natl. Acad. Sci. U.S.A.* **57**:782.

Kivirikko, K. I., and Prockop, D. J., 1967*d*, Partial characterization of protocollagen from embryonic cartilage, *Biochem. J.* **102**:432.

Kivirikko, K. I., and Prockop, D. J., 1972, Partial purification and characterization of protocollagen lysine hydroxylase from chick embryos, *Biochem. Biophys. Acta* **258**:366.

Kivirikko, K. I., Ganser, V., Engel, J., and Prockop, D. J., 1967*a*, Comparison of poly-L-proline I and II as inhibitors or protocollagen hydroxylase, *Hoppe-Seyler's Z. Physiol. Chem.* **348**:1341.

Kivirikko, K. I., Laitinen, O., and Prockop, D. J., 1967*b*, Modifications of a specific assay for hydroxyproline in urine, *Anal. Biochem.* **19**:249.

Kivirikko, K. I., Bright, H. J., and Prockop, D. J., 1968, Kinetic patterns of protocollagen hydroxylase and further studies on the polypeptide substrate, *Biochim. Biophys. Acta* **151**:558.

Kivirikko, K. I., Prockop, D. J., Lorenzi, G. P. and Blout, E. R., 1969, Oligopeptides with the sequences Ala-Pro-Gly and Gly-Pro-Gly- as substrates or inhibitiors for protocollagen proline hydroxylase, *J. Biol. Chem.* **244**:2755.

Kivirikko, K. I., Suga, K., Kishida, Y., Sakakibara, S., and Prockop, D. J., 1971, Asymmetry in the hydroxylation of (Pro-Pro-Gly)$_5$ by protocollagen proline hydroxylase, *Biochem. Biophys. Res. Commun.* **45**:1591.

Kivirikko, K. I., Kishida, Y., Sakakibara, S., and Prockop, D. J., 1972*a*, Hydroxylation of (X-Pro-Gly)$_n$ by protocollagen proline hydroxylase. Effect of chain length, helical conformation and amino acid sequence in the substrate, *Biochim. Biophys. Acta* **271**:347.

Kivirikko, K. I., Shudo, K., Sakakibara, S., and Prockop, D. J., 1972*b*, Studies on protocollagen lysine hydroxylase. Hydroxylation of synthetic peptides and the stoichiometric decarboxylation of α-ketoglutarate, *Biochem.* **11**:122.

Kivirikko, K. I., Ryhänen, L., Anttinen, H., Bornstein, P., and Prockop, D. J., 1973, Further hydroxylation of lysyl residues in collagen by protocollagen lysyl hydroxylase *in vitro, Biochemistry* **12**:4966.

Kobayashi, Y., Sakai, R., Kakiuchi, K., and Isemura, T., 1970, Physicochemical analysis of (Pro-Pro-Gly)$_n$ with defined molecular weight—temperature dependence of molecular weight in aqueous solution, *Biopolymers* **9**:415.

Kohn, L. D., Isersky, C., Zupnik, J., Lenaers, A., Lee, G., and Lalière, C. M., 1974, Calf tendon procollagen peptidase: its purification and endopeptidase mode of action, *Proc. Natl. Acad. Sci. U.S.A.* **71**:40.

Kraehenbuhl, J. P., and Jamieson, J. D., 1972, Solid-phase conjugation of ferritin to Fab-fragments of immunoglobulin G for use in antigen localization on thin sections, *Proc. Natl. Acad. Sci. U.S.A.* **69**:1771.

Krane, S. M., Pinnell, S. R., and Erbe, R. W., 1972, Lysyl-protocollagen hydroxylase deficiency in fibroblasts from siblings with hydrosylysine deficient collagen, *Proc. Natl. Acad. Sci. U.S.A.* **69**:2899.

Kühn, K., 1969, The structure of collagen, *in: Essays in Biochemistry* (P. N. Campbell and G. D. Greville, eds.), pp. 59–87, Academic Press, London and New York.

Kulonen, E., and Pikkarainen, J., 1970, Comparative studies on the chemistry and chain structure of collagen, *in: Chemistry and Molecular Biology of the Intercellular Matrix* (E. A. Balazs, ed.), Vol. 1, p. 81, Academic Press, London and New York.

Kuntz, I. D., and Kauzmann, W., 1974, Hydration of proteins and polypeptides, *Adv. Protein Chem.* **28**:239.

Kuttan, R., and Radhakrishnan, A. N., 1973, Biochemistry of the hydroxyproline, *Adv. Enzymol.* **37**:273.

Kuutti, E.-R., Tuderman, L., and Kivirikko, K. I., 1975, Human prolyl hydroxylase. Purification, partial characterization and preparation of antiserum to the enzyme, *Eur. J. Biochem.* **57**:181.

Lacroute, F., and Stent, G. S., 1968, Peptide chain growth of β-galactosidase in *Esherichia coli, J. Mol. Biol.* **35**:165.

Langner, R. O., and Fuller, R. O., 1969, Elevation of proline hydroxylase in diseased rabbit aorta, *Biochem. Biophys. Res. Commun.* **36**:597.

Langness, U., and Udenfriend, S., 1973, Collagen proline hydroxylase activity and anaerobic metabolism, *in: Biology of Fibroblast* (E. Kulonen and J. Pikkarainen, eds.) pp. 373–377, Academic Press, London and New York.

Langness, U., and Udenfriend, S., 1974, Collagen biosynthesis in non-fibroblastic cell lines, *Proc. Natl. Acad. Sci. U.S.A.* **71**:50.

Layman, D. L., McGoodwin, E. B., and Martin, G. R., 1971, The nature of the collagen synthesized by cultured human fibroblasts, Proc. Natl. Acad. Sci. U.S.A. **68**:454.

Layman, D. L., Sokoloff, L., and Miller, E. J., 1972, Collagen synthesis by articular chondrocytes in monolayer culture, *Exp. Cell Res.* **73**:107.

Lazarides, E., and Lukens, L. N., 1971*a*, Collagen synthesis on polysomes *in vitro*, *Nature (London), New Biol.* **232**:37.

Lazarides, E., and Lukens, L. N., 1971*b*, Collagen polypeptides: Normal release from polysomes in the absence of proline hydroxylation, *Science* **173**:723.

Lazarides, E. L., Lukens, L. N., and Infante, A. A., 1971, Collagen polysomes: Site of hydroxylation of proline residues, *J. Mol. Biol.* **58**:831.

Lenaers, A., Ansay, M., Nusgens, B. V., and Lapière, C. M., 1971, Collagen made of extended α-chains, procollagen in genetically-defective dermatosparactic calves, *Eur. J. Biochem.* **23**:533.

Levene, C. I., and Bates, C. J., 1973, Ascorbic acid and collagen synthesis, *in: Biology of Fibroblast* (E. Kulonen and J. Pikkarainen, eds.), pp. 397–410, Academic Press, London and New York.

Levene, C. I., Aleo, J. J., Prynne, C. J., and Bates, C. J., 1974, The activation of protocollagen proline hydroxylase by ascorbic acid in cultured 3T6 fibroblasts, *Biochim. Biophys. Acta* **338**:29.

Levin, E. Y., and Kaufman, S., 1961, Studies on the enzyme catalyzing the conversion of 3,4-dihyroxyphenylethylamine to norepinephrine, *J. Biol. Chem.* **236**:2043.

Levitt, D., and Dorfman, A., 1973, Control of chondrogenesis in limb-bud cell cultures by bromodeoxyuridine, *Proc. Natl. Acad. Sci. U.S.A.* **70**:2201.

Lichtenstein, J. R., Martin, G. R., Kohn, L. D., Byers, P. H., and McKusick, V. A., 1973, Defect in conversion of procollagen to collagen in a form of Ehlers-Danlos syndrome, *Science* **182**:298.

Lindblad, B., Lindstedt, G. Tofft, M., and Lindstedt, S., 1969, The mechanism of α-ketoglutarate oxidation in coupled enzymatic oxygenations, *J. Am. Chem. Soc.* **91**:4604.

Lindblad, B., Lindstedt, G., and Lindstedt, S., 1970, The mechanism of enzymic formation of homogentisate from *p*-hydroxyphenylpyruvate, *J. Am. Chem. Soc.* **92**:7446.

Lindstedt, G., 1967, Hydroxylation of γ-butyrobetaine to carnitine in rat liver, *Biochem.* **6**:1271.

Lindstedt, G., and Lindstedt, S., 1970, Cofactor requirements of γ-butyrobetaine hydroxylase from rat liver, *J. Biol. Chem.* **245**:4178.

Lindstedt, G., Lindstedt, S., Olander, B., and Tofft, M., 1968, α-Ketoglutarate and hydroxylation of γ-butyrobetaine, *Biochim. Biophys. Acta* **158**:503.

Lindy, S., Brix-Pedersen, F., Turto, H., and Uitto, J., 1971, Lactate, lactate dehydrogenase and protocollagen proline hydroxylase in rat skin autograft, *Hoppe-Seyler's Z. Physiol. Chem.* **352**:1113.

Lindy, S., Turto, H., Uitto, J., Helin, P., and Lorenzen, I., 1972*a*, Injury and repair in arterial tissue in the rabbit. Analysis of DNA, RNA, hydroxyproline, and lactate dehydrogenase in experimental arteriosclerosis, *Circ. Res.* **30**:123.

Lindy, S., Turto, H., and Uitto, J., 1972*b*, Protocollagen proline hydroxylase activity in rat heart during experimental cardiac hypertrophy, *Cir. Res.* **30**:205.

Liu, T. Z., and Bhatnagar, R. S., 1973, Mechanism of hydroxylation of proline, *Fed. Proc.* **32**:613.

Liu, C. K., Hsu, C. -A., and Abbott, M. T., 1973, Catalysis of three sequential dioxygenase reactions by thymine 7-hydroxylase, *Arch. Biochem. Biophys.* **159:**180.

Margolis, R. L., and Lukens, L. N., 1971, The role of hydroxylation in the secretion of collagen by mouse fibroblasts in culture, *Arch. Biochem. Biophys.* **147:**612.

Martin, G. R., Byers, P. H., and Piez, K. A., 1975, Procollagen, *Adv. Enzymol.* **42:**167.

Mayne, R., Schlitz, J. R., and Holtzer, H., 1973, Some overt and covert properties of chondrogenic cells, *in: Biology of Fibroblast* (E. Kulonen and J. Pikkarainen, eds.) pp. 61–68, Academic Press, London.

McGee, J. O'D., and Udenfriend, S., 1972*a*, A general method for the rapid measurement of enzyme precursors by the displacement of enzyme from antibody: The measurement of collagen proline hydroxylase precursor in cultured fibroblasts, *Biochem. Biophys. Res. Commun.* **46:**1646.

McGee, J. O'D., and Udenfriend, S., 1972*b*, Partial purification and characterization of peptidyl proline hydroxylase precursor from mouse fibroblasts, *Arch. Biochem. Biophys.* **152:**216.

McGee, J. O'D., Langness, U., and Udenfriend, S., 1971*a*, Immunological evidence for an inactive precursor of collagen proline hydroxylase in cultured fibroblasts, *Proc. Natl. Acad. Sci. U.S.A.* **68:**1585.

McGee, J. O'D., Rhoads, R. E., and Udenfriend, S., 1971*b*, The substrate recognition site of collagen proline hydroxylase: The hydroxylation of -X-Pro-Gly- sequences in bradykinin analogs and other peptides, *Arch. Biochem. Biophys.* **144:**343.

McGee, J. O'D., Jimenez, M. H., Felix, A. M., Cardinale, G. J., and Udenfriend, S., 1973, Inhibition of prolyl hydroxylase activity by bradykinin analogs containing a prolyl-like residue, *Arch. Biochem. Biophys.* **154:**483.

Miller, E. J. and Matukas, V. J., 1974, Biosynthesis of collagen, *Fed. Proc.* **33:**1197.

Miller, E. J. and Robertson, P. B., 1973, The stability of collagen cross-links when derived from hydroxylysyl residues, *Biochem. Biophys. Res. Commun.* **54:**432.

Miller, E. J. Woodall, D. L., and Vail, M. S., 1973, Biosynthesis of cartilage collagen, Use of pulse labeling to order the cyanogen bromide peptides in the $\alpha 1$(II) chain, *J. Biol. Chem.* **248:**1666.

Miller, R. L., 1971, Chromatographic separation of the enzymes required for hydroxylation of lysine and proline residues in protocollagen, *Arch. Biochem. Biophys.* **147:**339.

Miller, R. L., 1972, Rapid assay for lysyl-protocollagen hydroxylase activity, *Anal. Biochem.* **45:**202.

Miller, R. L., and Udenfriend, S., 1970, Hydroxylation of proline residues in collagen nascent chains, *Arch. Biochem. Biophys.* **139:**104.

Milstein, C., Brownlee, G. G., Harrison, T. M., and Mathews, M. B., 1972, A possible precursor of immunoglobulin light chains, *Nature (London), New Biol.* **239:**117.

Monson, J. M., and Bornstein, P., 1973, Identification of a disulfide-linked procollagen as the biosynthetic precursor of chick-bone collagen, *Proc. Natl. Acad. Sci. U.S.A.* **70:**3521.

Morgan, P. H., Jacobs, H. G., Segrest, J. P., and Cunningham, L. W., 1970, A comparative study of glycopeptides derived from selected vertebrate collagens: A possible role of the carbohydrate in fibril formation, *J. Biol. Chem.* **245:**5042.

Müller, P. K., McGoodwin, E. B., and Martin, G. R., 1973, Intracellular forms of collagen—underhydroxylated collagen, *in: Biology of Fibroblast* (E. Kulonen and J. Pikkarainen, eds.), pp. 349–364, Academic Press, London.

Murphy, L., and Rosenbloom, J., 1973, Evidence that chick tendon procollagen must be denatured to serve as substrate for proline hydroxylase, *Biochem. J.* **135:**249.

Mussini, E., Hutton, J. J., and Udenfriend, S., 1967, Collagen proline hydroxylase in wound healing, granuloma formation, scurvy and growth, *Science* **157**:927.

Myllylä, R., Risteli, L., and Kivirikko, K. I., 1975*a*, Assay of collagen galactosyltransferase and collagen glucosyltransferase activities and preliminary characterization of enzymic reactions with transferases from chick embryo cartilage, *Eur. J. Biochem.* **52**:401.

Myllylä, R., Risteli, L., and Kivirikko, K. I., 1975*b*, Glucosylation of galactosylhydroxylysyl residues in collagen *in vitro* by collagen glucosyltransferase. Inhibition by triple-helical conformation of the substrate, *Eur. J. Biochem.* **58**:517.

Myllylä, R., Risteli, L., and Kivirikko, K. I., 1976, Collagen glucosyltransferase. Partial purification and characterization of the enzyme from whole chick embryos and chick embryo cartilage, *Eur. J. Biochem.* **61**:59.

Nakane, P. H., and Pierce, G. B., Jr., 1967, Enzyme-labeled antibodies for the light and electron microscopic localization of tissue antigens, *J. Cell Biol.* **33**:307.

Niinikoski, J., and Kivisaari, J., 1973, Oxygen and wound healing: A new technique for determining respiratory gas tensions in human wounds, *in: Biology of Fibroblast* (E. Kulonen and J. Pikkarainen, eds.), pp. 591–599, Academic Press, London.

Nist, C., von der Mark, K., Hay, E. D., Olsen, B. R., Bornstein, P., Ross, R., and Dehm, P., 1975, Location of procollagen in chick corneal and tendon fibroblasts by use of ferritin-conjugated antibodies, *J. Cell Biol.* **65**:75.

Nordwig, A., and Pfab, F. K., 1968, Independence of collagen hydroxylation on the conformational state of the precursor substrate, *Biochim. Biophys. Acta* **154**:603.

Nordwig, A., and Pfab, F. K., 1969, Specificities of protocollagen hydroxylase from different sources, *Biochim. Biophys. Acta* **181**:52.

Okada, K., Kikuchi, Y., Kawashiri, Y., and Hiramoto, M., 1972, Synthesis and enzymic hydroxylation of protocollagen model peptides containing glutamyl or leucyl residue, *FEBS Lett.* **28**:226.

Okuyama, K., Tanaka, N., Ashida, T., Kakudo, M., Sakakibara, S., and Kishida, Y., 1972, An x-ray study of the synthetic polypeptide $(\text{Pro-Pro-Gly})_{10}$ *J. Mol. Biol.* **72**:571.

Olin, D. E., and Edelman, G. M., 1964, Reconstitution of 7S molecules from L and H polypeptide chains of antibodies and γ-globulins, *J. Exp. Med.* **119**:789.

Olsen, B. R., and Prockop, D. J., 1974, Ferritin-conjugated antibodies used for labeling of organelles involved in the cellular synthesis and transport of procollagen, *Proc. Natl. Acad. Sci. U.S.A.* **71**:2033.

Olsen, B. R., Berg, R. A., Sakakibara, S., Kishida, Y., and Prockop, D. J., 1971, The synthetic polytripeptides $(\text{Pro-Pro-Gly})_{10}$ form micro-crystalline structures similar to segmental structures formed by collagen, *J. Mol. Biol.* **57**:589.

Olsen, B. R., Berg, R. A., Kivirikko, K. I., and Prockop, D. J., 1973*a*, Structure of protocollagen proline hydroxylase from chick embryos, *Eur. J. Biochem.* **35**:135.

Olsen, B. R., Berg, R. A., Kishida, Y., and Prockop, D. J., 1973*b*, Collagen synthesis: Localization of prolyl hydroxylase in tendon cells detected with ferritin-labeled antibodies, *Science* **182**:825.

Olsen, B. R., Berg, R. A., Kishida, Y., and Prockop, D. J., 1975, Further characterization of embryonic tendon fibroblasts and the use of immuno-ferritin techniques to study collagen biosynthesis, *J. Cell Biol.* **64**:340.

Olsen, B. R., Hoffmann, H.-P., and Prockop, D. R., 1976, Interchain disulfide bonds at the COOH-terminal end of procollagen synthesized by matrix-free cells from chick embryonic tendon and cartilage, *Arch. Biochem. Biophys.*, in press.

Ooshima, A., Fuller, G. C., Cardinale, G. J., Spector, S., and Udenfriend, S., 1974,

Increased collagen synthesis in blood vessels of hypertensive rats and its reversal by antihypertensive agents. *Proc. Natl. Acad. Sci. U.S.A.* **71:**3019.

Palade, G. E., Siekevitz, P., and Caro, L. G., 1962, Structure, chemistry and function of the pancreatic exocrine cell, *in: Ciba Foundation Symposium on the Exocrine Pancreas* (A.V.S. de Reuck and M. P. Cameron, eds.), p. 23 Churchill, Ltd., London.

Pänkäläinen, M., and Kivirikko, K. I., 1971, The absence of iron and certain coenzymes in highly purified protocollagen proline hydroxylase, *Biochim. Biophys. Acta* **229:**504.

Pänkäläinen, M., Aro, H., Simons, K., and Kivirikko, K. I., 1970, Protocollagen proline hydroxylase: Molecular weight, subunits and isoelectric point, *Biochim. Biophys. Acta* **221:**559.

Peterkofsky, B., 1972*a*, Regulation of collagen secretion by ascorbic acid in 3T3 and chick embryo fibroblasts, *Biochem. Biophys. Res. Commun.* **49:**1343.

Peterkofsky, P., 1972*b*, Effect of ascorbic acid on collagen polypeptide synthesis and proline hydroxylation during the growth of cultured fibroblasts, *Arch. Biochem. Biophys.* **152:**318.

Peterkofsky, B., and Prather, W. B., 1974, Increased collagen synthesis in Kirsten sarcoma virus-transformed BALB 3T3 cells grown in the presence of dibutyryl cyclic AMP, *Cell* **3:**291.

Peterkofsky, B., and Udenfriend, S., 1963, Conversion of proline to collagen hydroxyproline in a cell-free system from chick embryo, *J. Biol. Chem.* **238:**3966.

Peterkofsky, B., and Udenfriend, S., 1965 Enzymic hydroxylation of proline in microsomal polypeptide leading to formation of collagen *Proc. Natl. Acad. Sci. U.S.A.* **53:**335.

Pinnell, S. R., Krane, S. M., Kenzora, J. E., and Glimcher, M. J., 1972, A heritable disorder of connective tissue. Hydroxylysine-deficient collagen disease, *N. Engl. J. Med.* **286:**1013.

Pontz, B. F., Müller, P. K., and Meigel, W. N., 1973, A study of the conversion of procollagen. Release and recovery of procollagen peptides in the culture medium, *J. Biol. Chem.* **248:**7558.

Popenoe, E. A., and Aronson, R. B., 1972, Partial purification and properties of collagen lysine hydroxylase from chick embryos, *Biochim. Biophys. Acta* **258:**380.

Popenoe, E. A., Aronson, R. B., and Van Slyke, D. D., 1969, The sulphydryl nature of collagen proline hydroxylase, *Arch. Biochem. Biophys.* **133:**286.

Prichard, P. M., Staton, G. W., and Cutroneo, K. R., 1974, *In vitro* synthesis of collagen peptides on fetal and neonatal rat skin polysomes by rabbit reticulocyte initiation factors, *Arch. Biochem. Biophys.* **163:**178.

Prockop, D. J., and Juva, K., 1965*a*, Hydroxylation of proline in particulate fractions from cartilage, *Biochem. Biophys. Res. Commun.* **18:**54.

Prockop, D. J., and Juva, K., 1965*b*, Synthesis of hydroxyproline *in vitro* by the hydroxylation of proline in a precursor of collagen, *Proc. Natl. Acad. Sci. U.S.A.* **53:**661.

Prockop, D. J., and Kivirikko, K. I., 1969, Effect of polymer size on the inhibition of protocollagen proline hydroxylase by polyproline II, *J. Biol. Chem.* **244:**4838.

Prockop, D. J., Peterkofsky, B., and Udenfriend, S., 1962, Studies on the intracellular localization of collagen synthesis in the intact chick embryo, *J. Biol. Chem.* **237:**1581.

Prockop, D. J., Kaplan, A., and Udenfriend, S., 1963, Oxygen-18 studies on the conversion of proline to collagen hydroxyproline, *Arch. Biochem. Biophys.* **101:**499.

Prockop, D. J., Weinstein, E., and Mulveny, T., 1966, Hydroxylation of lysine in a polypeptide precursor of collagen, *Biochem. Biophys. Res. Commun.* **22:**124.

Prockop, D. J., Dehm, P., Olsen, B. R., Berg, R. A., Jimenez, S. A., and Kivirikko, K. I.,

1972, Collagen synthesis: Relationship between "post mRNA" reactions and secretion of the molecule, *in: Inflammation: Mechanisms and Control* (I. H. Lepow and P. A. Ward, eds.), pp. 43–54, Acadmic Press, New York.

Prockop, D. J., and Dehm, P., Olsen, B. R., Berg, R. A., Grant, M. E., Uitto, J., and Kivirikko, K. I., 1973, Recent studies on the biosynthesis of collagen, *in: Biology of Fibroblast* (E. Kulonen and J. Pikkarainen, eds.), pp. 311–320, Academic Press, London.

Ramachandran, G. N., 1967, Structure of collagen at the molecular level, *in: Treatise on Collagen* (G. N. Ramachandran, ed.), pp. 103–104, Academic Press, London.

Ramachandran, G. N., Bansal, M., and Bhatnagar, R. S., 1973, A hypothesis on the role of hydroxyproline in stabilizing collagen structure, *Biochim. Biophys. Acta* **322:**166.

Ramaley, P. B., Jimenez, S. A., and Rosenbloom, J., 1973, Conformation of underhydroxylated collagen synthesized by 3T6 fibroblasts in culture, *FEBS Lett.* **33:**187.

Revel, J. P., and Hay, E. D., 1963, An autoradiographic and electron microscopic study of collagen synthesis in differentiating cartilage, *Z. Zellforsch. Microsk. Anat.* **61:**110.

Rhoads, R. E., and Udenfriend, S., 1968, Decarboxylation of α-ketoglutarate coupled to collagen proline hydroxylase, *Proc. Natl. Acad. Sci. U.S.A.* **60:**1473.

Rhoads, R. E., and Udenfriend, S., 1969, Substrate specificity of collagen proline hydroxylase: hydroxylation of a specific proline residue in bradykinin, *Arch. Biochem. Biophys.* **113:**108.

Rhoads, R. E., and Udenfriend, S., 1970, Purification and properties of collagen proline hydroxylase from newborn rat skin, *Arch. Biochem. Biophys.* **139:**329.

Rhoads, R. E., Hutton, J. J., and Udenfriend, S., 1967, Factors which stimulate collagen proline hydroxylase, *Arch. Biochem. Biophys.* **122:**805.

Rhoads, R. E., Udenfriend, S., and Bornstein, P., 1971, *In vitro* enzymatic hydroxylation of prolyl residues in the α1-CB2 fragment of rat collagen, J. Biol. Chem. **246:**4138.

Risteli, J., and Kivirikko, K. I., 1974, Activities of prolyl hydroxylase, lysyl hydroxylase, collagen galactosyltransferase and collagen glucosyltransferase in the liver of rats with hepatic injury, *Biochem. J.* **144:**115.

Roberts, N. E., and Udenfriend, S., 1970, Appearance of collagen proline hydroxylase in sera of mice with implanted sarcomas, *J. Natl. Cancer Inst.* **45:**277.

Robertson, W. van B., and Hewitt, J., 1961, Augmentation of collagen synthesis by ascorbic acid *in vitro, Biochim. Biophys. Acta* **49:**404.

Robertson, W. van B., and Schwartz, B., 1953, Ascorbic acid and the formation of collagen, *J. Biol. Chem.* **201:**689.

Rojkind, M., and DeLeon, D. L., 1970, Collagen biosynthesis in cirrhotic rat liver slices. A regulatory mechanism, *Biochim. Biophys. Acta* **217:**512.

Rosenbloom, J., and Prockop, D. J., 1971, Incorporation of *cis*-hydroxyproline into protocollagen and collagen. Collagen containing *cis*-hydroxyproline in place of proline and *trans*-hydroxyproline is not extruded at a normal rate, *J. Biol. Chem.* **246:**1549.

Rosenbloom, J., Bhatnagar, R. S., and Prockop, D. J., 1967, Hydroxylation of proline after the release of proline-rich polypeptides from ribosomal complexes during uninhibited collagen biosynthesis, *Biochim. Biophys. Acta* **149:**259.

Rosenbloom, J., Harsch, M., and Jimenez, S. A., 1973, Hydroxyproline content determines the denaturation temperature of chick tendon collagen, *Arch. Biochem. Biophys.* **158:**478.

Ross, R., and Benditt, E. P., 1965, Wound healing and collagen formation. V. Quantitative electron microscope radioautographic observations of proline-H^3 utilization by fibroblasts, *J. Cell Biol.* **27:**83.

Rubenstein, A. H., and Steiner, D. F., 1970, Proinsulin. The single chain precursor of insulin, *Med. Clin. North Am.* **54:**191.

Ryhänen, L., and Kivirikko, K. I., 1974*a*, Developmental changes in protocollagen lysyl hydroxylase activity in the chick embryo, *Biochim. Biophys. Acta* **343:**121.

Ryhänen, L., and Kivirikko, K. I., 1974*b*, Hydroxylation of lysyl residues in native and denatured protocollagen by protocollagen lysyl hydroxylase *in vitro, Biochim. Biophys. Acta* **343:**129.

Sakakibara, S., Kishida, Y., Kikuchi, Y., Sakai, R., and Kakiuchi, K., 1968, Synthesis of poly-(L-prolyl-L-prolylglycyl) of defined molecular weights, *Bull. Chem. Soc. Jpn.* **41:**1273.

Sakakibara, S., Kishida, Y., Okuyama, K., Tanaka, N., Ashida, T., and Kakudo, M., 1972, Single crystals of $(Pro\text{-}Pro\text{-}Gly)_{10}$, a synthetic polypeptide model of collagen, *J. Mol. Biol.* **65:**371.

Sakakibara, S., Inouye, K., Shudo, K., Kishida, Y., Kobayashi, Y., and Prockop, D. J., 1973, Synthesis of $(Pro\text{-}Hyp\text{-}Gly)_n$ of defined molecular weights. Evidence for the stabilization of collagen triple helix by hydroxyproline, *Biochim. Biophys. Acta* **303:**198.

Salpeter, M. M., 1968, H^3-proline incorporation into cartilage: Electron microscope autoradiographic observations, *J. Morphol.* **124:**387.

Salpeter, M. M., Bachmann, L., and Salpeter, E. E., 1969, Resolution in electron microscope radioautography, *J. Cell Biol.* **41:**1.

Salvador, R. A., and Tsai, I., 1973, Collagen proline hydroxylase activity in the uterus of the rat during rapid collagen synthesis *in vivo, Arch. Biochem. Biophys.* **154:**583.

Schofield, J. D., and Prockop, D. J., 1973, Procollagen-A precursor form of collagen, *Clin. Orthop. Relat. Res.* **97:**175.

Schofield, J. D., Uitto, J., and Prockop, D. J., 1974*a*, Formation of interchain disulfide bonds and helical structure during biosynthesis of procollagen by embryonic tendon cells, *Biochem.* **13:**1801.

Schofield, J. D., Uitto, J., and Prockop, D. J., 1974*b*, Interchain disulphide bonding in procollagen from embryonic chick tendon cells and the formation of the triple-helical structure, *Biochem. Soc. Trans.* **2:**90.

Scornik, O. A., 1974, *In vivo* rate of translation by ribosomes of normal and regenerating liver, *J. Biol. Chem.* **249:**3876.

Scrutton, M. C., and Utter, M. F., 1968, The regulation of glycolysis and gluconeogenesis in animal tissues, *Annu. Rev. Biochem.* **37:**249.

Shaffer, P. M., McCroskey, R. P., Palmatier, R. D., Midgett, R. J., and Abbott, M. T., 1968, The cell-free conversion of a deoxyribonucleoside to a ribonucleoside without detachment of the deoxyribose, *Biochem. Biophys. Res. Commun.* **33:**806.

Sherr, C. J., Taubman, M. B., and Goldberg, B., 1973, Isolation of a disulfide-stabilized, three-chain polypeptide fragment unique to a precursor of human collagen, *J. Biol. Chem.* **248:**7033.

Silver, I. A., 1973, Probes for the measurement of the microenvironment, *in: Biology of the Fibroblast* (E. Kulonen and J. Pikkarainen, eds.), pp. 521–524, Academic Press, London.

Singer, S. J., and Schick, A. F., 1961, The properties of specific stains for electron microscopy prepared by the conjugation of antibody molecules with ferritin, *J. Biophys. Biochem. Cytol.* **9:**519.

Smith, B. D., Byers, P. H., and Martin, G. R., 1972, Production of procollagen by human fibroblasts in culture, *Proc. Natl. Acad. Sci. U.S.A.* **69:**3260.

Speakman, P. T., 1971, Proposed mechanism for the biological assembly of collagen triple helix, *Nature* **229:**241.

Spiro, R. G., 1967, The structure of the disaccharide unit of the renal glomerular basement membrane, *J. Biol. Chem.* **242:**4813.

Spiro, R. G., 1969, Characterization and quantitative determination of the hydroxylysine-linked carbohydrate units of several collagens, *J. Biol. Chem.* **244:**602.

Spiro, R. G., and Spiro, M. J., 1971*a*, Studies on the biosynthesis of the hydroxylysine-linked disaccharide unit of basement membranes and collagens. I. Kidney glucosyltransferase, *J. Biol. Chem.* **246:**4899.

Spiro, M. J., and Spiro, R. G., 1971*b*, Studies on the biosynthesis of the hydroxylysine-linked disaccharide unit of basement membranes and collagens. II. Kidney galactosyltransferase, *J. Biol. Chem.* **246:**4910.

Spiro, R. G., and Spiro, M. J., 1971c, Studies on the biosynthesis of the hydroxylysine-linked disaccharide unit of basement membranes and collagens. III. Tissue and subcellular distribution of glycosyltransferases and the effect of various conditions on the enzyme levels, *J. Biol. Chem.* **246:**4919.

Stark, M., Lenaers, A., Lapière, C. M., and Kühn, K., 1971, Electron optical studies of procollagen from the skin of dermatosparaxic calves, *FEBS Lett.* **18:**225.

Stassen, F. L. H., Cardinale, G. J., and Udenfriend, S., 1973, Activation of prolyl hydroxylase in L-929 fibroblasts by ascorbic acid, *Proc. Natl. Acad. Sci. U.S.A.* **70:**1090.

Stassen, F. L. H., Cardinale, G. J., McGee, J. O'D., and udenfriend, S., 1974, Prolyl hydroxylase and an immunologically related protein in mammalian tissues, *Arch. Biochem. Biophys.* **160:**340.

Stein, H. D., Keiser, H. R., and Sjoerdsma, A., 1970, Proline hydroxylase activity in human blood, *Lancet* **1:**106.

Stephens, F. O., and Hunt, T. K., 1971, Effects of changes in inspired oxygen and carbon dioxide tensions on wound tensile strength: An experimental study, *Ann. Surg.* **173:**515.

Stetten, M. R., 1949, Some aspects of the metabolism of hydroxyproline studied with the aid of isotopic nitrogen, *J. Biol. Chem.* **181:**31.

Stone, N., and Meister, A., 1962, Function of ascorbic acid in the conversion of proline to collagen hydroxyproline, *Nature* **194:**555.

Suzuki, F., and Koyama, E., 1969, Hydroxylation of proline in collagen model peptide, *Biochim. Biophys. Acta* **177:**154.

Takeuchi, T., and Prockop, D. J., 1969, Protocollagen proline hydroxylase in normal liver and in hepatic fibrosis, *Gastroenterology* **56:**744.

Takeuchi, T., Kivirikko, K. I., and Prockop, D. J., 1967, Increased protocollagen hydroxylase activity in the livers of rats with hepatic fibrosis, *Biochem. Biophys. Res. Commun.* **28:**940.

Tanzer, M. L., Church, R. L., Yaeger, J. A., Wampler, D. E., and Park, E. -D., 1974, Procollagen: intermediate forms containing several types of peptide chains and non-collagen peptide extensions at the NH_2 and COOH ends, *Proc. Natl. Acad. Sci. U.S.A.* **71:**3009.

Taubman, M. B., Goldberg, B., and Sherr, C. J., 1974, Radioimmunoassay for human procollagen, *Science* **186:**1115.

Timpl, R., Wick, G., Furthmayr, H., Lapière, C. M., and Kühn, K., 1973, Immunochemical properties of procollagen from dermatosparactic calves, *Eur. J. Biochem.* **32:**584.

Toole, B. P., Kang, A. H., Trelstad, R. L., and Gross, J., 1972, Collagen heterogeneity within different growth regions of long bones of rachitic and non-rachitic chicks, *Biochem. J.* **127:**715.

Torchia, D. A., Lyerla, J. R., Jr., and Quattrone, A. J., 1975, Molecular dynamics and structure of the random-coil and helical states of the collagen peptide, α1-CB2, as determined by ^{13}C-magnetic resonance, *Biochemistry* **14:**887.

Traub, W., 1974, Some stereochemical implications of the molecular conformation of collagen, *Isr. J. Chem.* **12:**435.

Traub, W., and Piez, K. A., 1971, The chemistry and structure of collagen, *Adv. Protein Chem.* **25:**243.

Traub, W., Yonath, A., and Segal, D. M., 1969, On the molecular structure of collagen, *Nature* **221:**914.

Trelstad, R. L., 1971, Vacuoles in the embryonic chick corneal epithelium, an epithelium which produces collagen, *J. Cell Biol.* **48:**689.

Tsai, R. L., and Green, H., 1972, Study of Intracellular collagen precursors using DNA-cellulose chromatography, *Nature (London), New Biol.* **237:**171.

Tuderman, L., Kuutti, E. -R., Kivirikko, K. I., 1975, An affinity-column procedure using poly (L-proline) for the purification of prolyl hydroxylase. Purification of the enzyme from chick embryos, *Eur. J. Biochem.* **52:**9.

Uitto, J., 1971, Collagen biosynthesis in human skin. A review with emphasis on scleroderma, *Ann. Clin. Res.* **3:**250.

Uitto, J., and Prockop, D. J., 1973*a*, Rate of helix formation by intracellular procollagen and protocollagen. Evidence for a role for disulfide bonds, *Biochem. Biophys. Res. Commun.* **55:**904.

Uitto, J., and Prockop, D. J., 1973*b*, Intracellular hydroxylation of accumulated protocollagen and stabilization of the helical structure of collagen, *Abstracts of the Ninth International Congress of Biochemistry*, Stockholm, p. 425.

Uitto, J., and Prockop, D. J., 1974*a*, Intracellular hydroxylation of non-helical protocollagen to form triple-helical procollagen and subsequent secretion of the molecule, *Eur. J. Biochem.* **43:**221.

Uitto, J., and Prockop, D. J., 1974*b*, Biosynthesis of cartilage procollagen. Influence of chain association and hydroxylation of prolyl residues on the folding of the polypeptides into the triple-helical conformation, *Biochem.* **13:**4586.

Uitto, J., and Prockop, D. J., 1974*c*, Hydroxylation of peptide-bound proline and lysine before and after chain completion of the polypeptide chains of procollagen, *Arch. Biochem. Biophys.* **164:**210.

Uitto, J., and Prockop, D. J., 1974*d*, Incorporation of proline analogues into collagen polypeptides. Effects on the production of extracellular procollagen and on the stability of the triple-helical structure of the molecule, *Biochim. Biophys. Acta* **336:**234.

Uitto, J., and Prockop, D. J., 1974*e*, Molecular defects in collagen and the definition of "collagen disease," *in: Molecular Pathology* (R. A. Good, S. B. Day, and J. J. Yunis, eds.), pp. 670–688, Charles C. Thomas, Springfield, Illinois.

Uitto, J., and Prockop, D. J., 1974*f*, Synthesis and secretion of under-hydroxylated procollagen at various temperatures by cells subject to temporary anoxia, *Biochem. Biophys. Res. Commun.* **60:**414.

Uitto, J., and Prockop, D. J., 1975, Inhibition of collagen accumulation by proline analogues: The mechanism of their action, *in: Collagen Metabolism in the Liver* (H. Popper and K. Becker, eds.), p. 139, Stratton Intercontinental Medical Book Corporation, New York.

Uitto, J., Halme, J., Hannuksela, M., Peltokallio, P., and Kivirikko, K. I., 1969,

Protocollagen proline hydroxylase activity in the skin of normal human subjects and of patients with scleroderma, *Scand. J. Clin. Lab. Invest.* **23:** 241.

Uitto, J., Hannuksela, M., and Rasmussen, O., 1970*a*, Protocollagen proline hydroxylase activity in scleroderma and other connective tissue disorders, *Ann. Clin. Res.* **2:**235.

Uitto, J., Lindy, S., Rokkanen, P., and Vainio, K., 1970*b*, Increased protocollagen proline hydroxylase activity in synovial tissue in rheumatoid arthritis, *Clin. Chim. Acta* **30:**741.

Uitto, J., Dehm, P., and Prockop, D. J., 1972*a*, Incorporation of *cis*-hydroxyproline into collagen by tendon cells. Failure of the intracellular collagen to assume a triple-helical conformation, *Biochim. Biophys. Acta* **278:**601.

Uitto, J., Jimenez, S. A., Dehm, P., and Prockop, D. J., 1972*b*, Characterization of the precursor forms of the $\alpha 1$ and $\alpha 2$ chains of collagen from matrix-free tendon cells, *Biochim. Biophys. Acta* **278:**198.

Uitto, J., Timpl, R., and Prockop, D. J., 1975*a*, Characterization of the precursor of type II collagen (in preparation).

Uitto, J., Hoffmann, H.-P., and Prockop, D. J., 1975*b*, Role of triple-helical conformation in the secretion of procollagen: Accumulation of non-helical peptides containing *cis*-hydroxyproline in the cisternae of the rough endoplasmic reticulum, *Science* **190:**1202.

Urivetzky, M., Frei, J. M., and Meilman, E., 1966, Hydroxyprolyl-soluble ribonucleic acid and the biosynthesis of collagen, *Arch. Biochem. Biophys.* **117:**224.

Veis, A., Anesey, J. R., Garvin, J. E., and Dimuzio, M. T., 1972, High molecular weight collagen: A long-lived intermediate in the biogenesis of collagen fibrils, *Biochem. Biophys. Res. Commun.* **48:**1404.

Veis, A., Anesey, J., Yuan, L., and Levy, S. J., 1973, Evidence for an amino-terminal extension in high-molecular-weight collagens from mature bovine skin, *Proc. Natl. Acad. Sci. U.S.A.* **70:**1464.

von der Mark, K., and Bornstein, P., 1973, Characterization of the pro-$\alpha 1$ chain of procollagen. Isolation of a sequence unique to the percursor chain, *J. Biol. Chem.* **248:**2285.

von Hippel, P. H., 1967, Structure and stabilization of the collagen molecule in solution, *in: Treatise on Collagen* (G. N. Ramachandran, ed.), p. 253, Academic Press, London and New York.

Vuust, J., and Piez, K. A., 1970, Biosynthesis of the α chains of collagen studied by pulse-labeling in culture. *J. Biol. Chem.* **245:**6201.

Vuust, J., and Piez, K. A., 1972, A kinetic study of collagen biosynthesis, *J. Biol. Chem.* **247:**856.

Ward, A. R., and Mason, P., 1973, Influence of proline hydroxylation upon the thermal stability of collagen fragment α1CB2, *J. Mol. Biol.* **79:**431.

Watanabe, M. S., McCroskey, R. P., and Abbott, M. T., 1970, The enzymatic conversion of 5-formyluracil to uracil 5-carboxylic acid, *J. Biol. Chem.* **245:**2023.

Weinstein, E., Blumenkrantz, N., and Prockop, D. J., 1969, Hydroxylation of proline and lysine in protocollagen involves two separate enzymatic sites, *Biochim. Biophys. Acta* **191:**747.

Weinstock, M., and Leblond, C. P., 1974, Synthesis, migration, and release of precursor collagen by odontoblasts as visualized by radioautography after [^{3}H]proline administration, *J. Cell Biol.* **60:**92.

Wendt, P., Fietzek, P. P., and Kühn, K., 1972*a*, The covalent structure of collagen: the tryptic, thermolytic and chymotryptic peptides of α1-CB3 from calf skin collagen, *FEBS Lett.* **26:**69.

Wendt, P., von der Mark, K., Rexrodt, F., and Kühn, K., 1972*b*, The covalent structure of collagen, the amino-acid sequence of the 112 residues amino-terminal part of peptide α1-CB6 from calf skin collagen, *Eur. J. Biochem.* **30:**169.

Wetlaufer, D. B., and Ristow, S., 1973, Acquisition of three-dimensional structure of proteins, *Annu. Rev. Biochem.* **42:**135.

Winterburn, P. J., and Phelps, C. F., 1972, The significance of glycosylated proteins, *Nature* **236:**147.

6
Aspects of the Animal Collagenases

JEROME GROSS

I. Introduction

Information accumulated over the past 13 years since the first animal collagenase was detected (Gross and Lapiere, 1962; Lapiere and Gross, 1963) indicates that collagenolysis in biological systems is accomplished by a series of enzymes one of which, operating at physiologic pH and temperature in the extracellular spaces, produces the first and critical cleavage in the helical body of the molecule within the fibril (Gross and Nagai, 1965; Kang *et al.*, 1966; Sakai and Gross, 1967). This step is followed by one or more enzyme activities which reduce the polypeptide fragments to smaller peptides and amino acids. We know considerably more about the enzyme responsible for the initial attack than we do about subsequent dismantling of the fragments. How much of this latter series of events takes place within the cell has yet to be determined. The possibility of a preliminary cleavage of peptide bonds within the terminal nonhelical regions of the molecule resulting in loss of intermolecular cross-linking in insoluble fibrils is a real one, but such an enzyme activity at neutral pH has not yet been unequivocally detected.

Because of the need for close biological control over collagen degradation, one might expect that regulation of collagenolytic activity is accomplished via multiple pathways, for example, hormonal modulation

---

JEROME GROSS · The Developmental Biology Laboratory, Department of Medicine, Massachusetts General Hospital and the Harvard Medical School, Boston, Massachusetts 02114.

of synthesis, control over activation of a zymogen, enzyme inhibition or degradation, and by manipulation of extrafibrillar substances which may protect the substrate from attack. It also would appear that different types of collagen are susceptible to collagenolytic attack in different degrees. Thus far, there is no evidence for separate collagenases for the different collagen types, although the search continues. Whether or not there are different collagenolytic enzymes produced in the various tissues of the same animal is still to be determined. Although it is possible that all cells have the potential for producing collagenase, the evidence thus far suggests that this function is limited to certain cell types in certain organs. This may simply be due to limited sensitivity of current assay techniques or because production of the active enzyme may be transient and closely regulated, much more frequent and extensive for certain cells than for others, and often under close control by "inducers" such as hormones and special cell products.

Although the mechanism of action of all the vertebrate collagenases examined to date appear to be the same in terms of site of cleavage in the initial attack, these enzymes do differ in some characteristics such as susceptibility to inhibitors, immunologic cross-reactivity, molecular weight, and the number of cleavages produced in the substrate.

Because there have been several recent, fairly comprehensive reviews (Perez-Tomayo, 1973; Woessner, 1973; Gross, 1974; Harris and Krane, 1974) of the animal collagenases, this chapter will focus on specific aspects of the subject.

II. Sources of Animal Collagenase

True collagenolytic enzymes (those which attack the helical body of the native collagen molecule under physiologic conditions) have been obtained from numerous animal species, tissues, and recently from cultured cells. In every instance the pH optimum has been in the neutral range, between 7 and 9, with loss of activity below pH 6. Aside from the observation that leukocyte collagenase may be associated with a granule fraction (Robertson *et al.*, 1972*a*; Kruze and Wojtecka, 1972; Oronsky *et al.*, 1973), there is no indication that any of these enzymes are lysosomal in origin. With but few exceptions they have been detected in, and isolated from, serum-free culture media, although they can be extracted in small amounts from some tissues under special conditions. Usually such cultures are short term since serum is excluded from the medium and enzyme production rarely continues beyond a maximum of about 12 days, more commonly disappearing by 5–7 days. Cultures of bovine gingiva have continued producing the enzyme up to 30 days in serum-free medium (Birkedal-Hansen *et al.*, 1974). Recently there have been several reports of collagenase production by maintained and frequently passed strains of fibroblasts from rabbit cornea (Hook *et al.*, 1973) and synovium (Werb and Burleigh, 1974). In these instances the serum was deleted from the medium prior to enzyme collection (to eliminate protease inhibitors) and the collagenase harvested for as long as 8 days, after which serum was restored to continue cell proliferation. Very recently Dayer *et al.* (1976) have measured high enzyme concentrations in cultures of cells from human rheumatoid synovial tissue, grown in the presence of serum, after brief treatment of the medium with trypsin followed by soy bean trypsin inhibitor.

There is good evidence that both epithelial cells and fibroblasts from particular tissues have the capacity for collagenase production. Tadpole tail-fin epithelium (Eisen and Gross, 1965), rabbit and human epidermis, particularly at the edges of healing skin wounds (Grillo and Gross, 1967; Donoff *et al.*, 1971), and gingival epithelium (Fullmer *et al.*, 1969) actively produce collagenolytic enzyme in culture. Using an immunofluorescent technique based on antisera produced against purified human skin collagenase, Reddick *et al.* (1974) have localized the enzyme (or its inactive precursor) to the subepithelial cells of the dermis in human skin. The epithelium does not appear to have the enzyme by this criterion. Table 1 is a current list of tissues from which collagenases have been isolated and characterized. No doubt it will expand with time.

III. *Assay Methods*

The basic techniques for detecting and measuring collagenolytic activity are essentially unchanged from those worked out for the prototype enzyme, tadpole collagenase (Gross and Lapière, 1962; Lapière and Gross, 1963; Nagai *et al.*, 1966). The simplest procedure for detection of activity in an unknown tissue is the visible breakdown of a reconstituted collagen gel substrate under a small living tissue explant in short-term cultures at neutral pH and 37°C. Serum should be omitted from the medium because of the presence of inhibitors, and in most cases the use of antibiotics shown not to interfere with protein synthesis, is advisable. If the explants are uniform and their collagen content and that of the substrate known,

TABLE 1
Sources of Isolated Animal Collagenases

Source	Tissue, cell, or organ	References
Frog tadpole[a,b]	Tail fin, back skin	Nagai *et al.* (1966),[a] Harper *et al.* (1971);[a] Abe and Nagai (1972)[b]
Newt[a]	Regenerating limb	Dresden and Gross (1970)[a]
Human[a,b]	Skin	Eisen *et al.* (1968),[a] (1971)[b]
Rat[a,b]	Skin	Nagai and Hori (1972);[a,b] Tokoro *et al.* (1972)[b]
Guinea pig[a]	Skin	Huang and Abramson (1975)
Chick[a]	Skin	Gross (unpublished)
Guinea pig[a]	Skin wounds	Abramson *et al.* (1975)
Rabbit	Skin wounds	Donoff *et al.* (1971)[a]
Human[a]	Bone	Fullmer and Lazarus (1969)[a]
Mouse[a]	Bone	Shimizu *et al.* (1969), Vaes (1972*b*)
Chick[b]	Bone	Sakamoto *et al.* (1973*b*)
Rat[a,b]	Uterus	Jeffrey and Gross (1970);[a] Ryan and Woessner (1971)[b]
Human	Gingiva	Fullmer *et al.* (1972)[a]
Bovine	Gingiva	Birkedal-Hansen *et al.* (1974)[a]
Human, rheumatoid arthritis[a,b]	Synovia	Evanson *et al.* (1968);[a] Nagai and Hori (1972)[b]
Human, rheumatoid arthritis[b]	Synovial fluid	Harris *et al.* (1969);[b] Abe and Nagai (1972)[b]
Human, rheumatoid arthritis[a]	Nodule, skin	Harris (1972)[a]
Rabbit[a]	Cornea	Hook *et al.* (1971);[a] Berman *et al.* (1973*a,b*)[a]
Human[a]	Cornea	Berman *et al.* (1973*a,b*)[a]
	Free cells	
Human[b]	Leukocytes	Lazarus *et al.* (1968*a,b*);[b] Ohlsson and Olsson (1973)[b]
Rabbit[a]	Macrophages	Wahl *et al.* (1974)[a]
Rat[b]	Liver, Kupfer cells	Fujiwara *et al.* (1973)[b]
	Cultured cells	
Rabbit, human	Corneal fibroblasts	Hook *et al.* (1973);[a] Newsome and Gross (unpublished)[a]
Rabbit	Synovia	Werb and Reynolds (1974)[a]
Rabbit	Skin fibroblasts	Werb and Burleigh (1974)[a]
Human	Skin fibroblasts	Bauer *et al.* (1975)[a]
	Granulomas	
Guinea pig, rabbit	Skin	Perez-Tomayo (1970)
Human	Ear	Abramson and Gross (1971)

TABLE 1—*Continued*

Source	Tissue, cell, or organ	References
Human	Tumors	Yamanishi *et al.* (1972);[b] (1973);[b] Hashimoto *et al.* (1972)[b]; Abramson *et al.* (1975)[a]; Dresden *et al.* (1972)[a]
Rabbit	Tumors	Harris *et al.* (1972)[b]
	Invertebrate	
Crab	Hepatopancreas	Eisen *et al.* (1973)[b]
Flatworm (*Planaria*)	Whole animal	Phillips and Dresden (1973)[b]

[a] Culture medium.
[b] Tissue extract.

this detection system may be used quantitatively, simply by determining the solubilized hydroxyproline after various periods of incubation. This procedure is quite sensitive since it measures the enzyme accumulated over long time periods and probably protects the collagenase from thermal or proteolytic degradation (and perhaps even inhibition) by reason of its immediate and near-continuous association with substrate. For quantitative measurement of collagenolytic activity in culture media or extracts, the most widely used and most effective method is based on degradation at neutral pH of a reconstituted gel of radioactively labeled collagen. This substrate is usually prepared from neutral salt extracts of the skins of isotopically labeled rats or guinea pigs. A simpler, efficient *in vivo* labeling technique has been recently described by Robertson *et al.* (1972*b*), in which chick embryo calvaria are incubated in organ culture with [^{14}C]glycine or proline in the presence of β-aminopropionitrile for 24–48 hr following which the highly radioactive collagen is extracted in dilute acetic acid in the presence of cold carrier collagen obtained from any appropriate animal and purified by the usual methods. There is no problem with the relatively small amount of labeled lathyritic collagen in terms of solubility in the assay system since it will become cross-linked to the large amount of normal cold carrier on incubation at 37° (amino groups in the helical regions of the lathyritic molecules will form the Schiff-base cross-link precursors with the appropriate NH_2-terminal aldehydes of the carrier collagen). Relatively large amounts of heavily labeled collagen, specific activities in excess of 10^6 cpm/mg may be obtained. This collagen may be further diluted to levels of 10^3–10^4 cpm/ml and stored almost indefinitely in dilute acetic acid in the cold. Efforts are being made now in several laboratories to prepare much more heavily labeled tritiated collagen by chemical means.

With labeled collagen prepared by any method, there are, not infrequently, difficulties with high blank readings, i.e., 15–30% above background, particularly if the collagen has been stored cold at neutral pH for more than several weeks. This problem may be handled simply by the following procedure used in our laboratory. All assays are run in 1 ml disposable plastic microcentrifuge tubes. The set of reconstituted gels, prior to adding the enzyme, are sedimented in a Beckman microfuge at room temperature for 3 min, the supernatant fluids are removed and counted, and the pelleted collagen resuspended by agitation in 0.5 ml of fresh warm buffer for a second wash. The gel is then resuspended into 0.1 ml of buffer, and the sample to be assayed is added. With this additional step backgrounds are nearly always reduced to well below 10% of the total and very little radioactivity is released by trypsin. We routinely incubate for 4 hr only to guarantee operating on the linear portion of the time-dependence curve.

Among the various units of activity reported in the literature the most generally useful is the amount of collagen digested per unit time per amount of enzyme at 37°. This is usually mg collagen/hr/mg enzyme, or μg/min/mg enzyme. If the total cpm of substrate in the assay and its specific activity is reported, and the conditions given, then simply cpm/mg enzyme protein per unit of time is adequate for comparative measurements within an experimental system. A less-commonly used assay method involves prevention of fibril formation by preincubation of the reaction mixture of dissolved collagen and enzyme for a fixed time period at 27° followed by inhibition of the enzyme with EDTA or cysteine, then further incubation at 37°C, after which the opacity of the gel is measured (Nagai *et al.*, 1966). Sakamoto *et al.* (1972*c*) have modified this procedure by using radioactively labeled collagen, spinning off the gel formed at 37°, and then measuring the soluble radioactivity remaining in solution. In another recently described assay procedure the enzyme solution is overlaid on microgels of collagen in capillary tubes and incubated at 37°C. Activity is measured by the linear length of gel clearing (Berman and Manabe, 1974). The oldest method for measuring collagenolytic activity, first applied by Gallop *et al.* (1957) for the assay of bacterial collagenase, utilizes the fall in viscosity of a reaction mixture of collagen at neutral pH and temperature between 20 and 37°C. This method is sensitive, quantitative, and can be particularly useful when examination of reaction products is desired (Gross and Nagai, 1965; McCroskery *et al.*, 1975).

Two immunologic assay procedures have been adapted to collagenase, one dependent upon radial diffusion into an agar gel containing antibody (Mancini *et al.*, 1963) and the other a radioimmunoassay (Bauer *et al.*, 1972). Both have a stringent requirement for specific antibody, which means that the immunizing antigen must be pure.

An examination of a new biologic system for collagenolytic activity should always include a demonstration of reaction products indicative of cleavage of peptide bonds within the helical region of the collagen molecule; characteristic polyacrylamide electrophoretic patterns or the electron microscopy of helical products in the SLS form are still appropriate criteria for identifying animal collagenase activity (Gross and Nagai, 1965). Although nearly all the animal collagenases examined to date have revealed the same initial cleavage reaction products, it would not be too surprising if other sites of attack by newly discovered collagenases were observed. The collagenolytic enzyme extracted from the flat worm, *Planaria* (Phillips and Dresden, 1973), is in this category.

IV. Purification of Animal Collagenases

The earliest scheme used for isolation and purification of collagenase from culture medium of tadpole tailfin tissues (Nagai *et al.*, 1966) involved ammonium sulfate precipitation, chromatography on a molecular sieve, followed by starch gel electrophoresis and then chromatography on an anionic exchanger, DEAE cellulose. It was noted that a small amount of caseinolytic activity followed collagenase all the way through the final step and that the loss resulting from this last ion-exchange chromatographic procedure was prohibitive. Harper *et al.* (1972) obtained a fairly high degree of purification and separation from nonspecific proteases by using the fraction precipitated from redissolved, lyophilized culture medium at 30% saturation with ammonium sulfate. The nonspecific proteases remained in solution. The precipitated fraction was then chromatographed on agarose columns, the enzyme fraction often appearing to be homogeneous by SDS polyacrylamide gel electrophoresis and capable of producing just a single cleavage in an isolated denatured α chain. Bauer *et al.* (1971) introduced affinity chromatography for further purification, utilizing native collagen bound to Sepharose beads. Enzyme recoveries and purification using this technique have proved to be quite variable in the hands of different investigators (Werb and Reynolds, 1975; McCroskery *et al.*, 1975), and the recent application of Sepharose-bound cleavage site peptide, α1CB7 (McCroskery *et al.*, 1975) seems to have improved both the yield and specificity. It is quite likely that no one purification procedure will be completely satisfactory for the collagenases from various tissues and modes of original preparation. Table 2 allows comparison of three recently published purification schemes utilized for three different sources of collagenase, one from the medium of tissue explants (Wooley *et al.*, 1975), the second from the media of cell cultures (Werb and Reynolds, 1975), and the third from a rabbit tumor extract (McCroskery *et al.*, 1975). Note that the degree of purification is very different for the different preparations and procedures. The recoveries for the first two appear to be quite similar, and it is interesting to note here that the largest percent loss of enzyme occurs during the last step of gel filtration. Purification of the tumor extract was remarkable in the exceedingly high degree of purification obtained and the increase in yield of total activity. McCroskery *et al.* point out that the first ammonium sulfate precipitation succeeded in removing inhibitory substances, which greatly increased the apparent yield. However, the 100-fold purification factor accomplished by molecular sieve filtration after mild ion-exchange chromatography is not consistent with the experience of the other two laboratories. This is probably explained by the difference in starting material. It is important to note that in this recipe for purification the collagenase was not adsorbed on DEAE Sephadex under the conditions described. It has been the experience of others that attempts to chromatograph the enzyme on this ion exchange resin results in poor recoveries.

The question of enzyme stability during purification remains unsettled. There are various reports of instability of frozen purified enzyme, destruction consequent to lyophilization, and increased stability at ice box temperatures as compared with the frozen state. McCroskery *et al.* (1975) state that the partially purified collagenase is relatively unstable at either 4° or −20°C, whereas the purified enzyme after affinity chromatography on Sepharose-α1CB7 retained nearly all of its activity after storage for one year at −20°C. Clearly, enzyme stability may vary greatly among the different collagenases, and this question of enzyme stability requires much more exploration.

V. The Evidence for Physiologic Function

The known animal collagenases operate to degrade native collagen molecules in solution and in intact native or reconstituted fibrils under physiologic conditions of pH and temperature, although the evidence for their physiologic function remains circumstantial but considerably stronger than in 1970 when this writer last commented on that subject (Gross, 1970).

TABLE 2
Comparison of Purification Schemes for Three Animal Collagenases[a]

	Total protein (mg)	Total activity (units)	Specific activity (units/mg)	Purification (fold)	Recovery (%)
Wooley *et al.* (1975)					
Human rheum. syn. tissue explants					
Tissue culture medium	—	—	0.3	1	—
Concentrate	92	368	4	13	100
Seph. G200	39	312	8	27	85
Seph. QAE, A-50	2.5	175	70	233	47
Seph. G200	0.15	47	312	1040	13
Werb and Reynolds (1975)					
Rabbit synovial fibroblasts					
Cell culture medium	—	1860	—	—	100
Concentrate	60	1550	26	1	83
Seph. G100	10.5	980	93	3.6	52
Seph. QAE	2	762	380	14.7	40
Seph. G-75	0.22	380	1730	66.5	19
McCroskery *et al.* (1975)					
Rabbit tumor extract					
Tissue extract	31,869	438	0.013	1	100
20–50% Am.S. ppt	5100	4370	0.83	65	1000
Seph. DEAE A-50	1062	7550	7.15	533	1720
Biogel A-1.5 m	7.14	5270	738	55,500	1200
Agarose-α1CB7 (Affinity)	0.83	2960	3570	268,780	670

[a] 1 Unit = 1 μg collagen fibrils degraded per min at 35–37°.

The most direct demonstration of physiologic activity would be the identification of the enzyme molecules in contact with collagen fibrils *in vivo* at the time and location of collagenolysis. Gross, Harper, Bloch, and Hayashi (unpublished results) were able to locate collagenase, or its precursors, in epithelial cells and adjacent extracellular regions in tadpole tail-fin and back skin by autoradiography of these tissues after intraperitoneal injection of ^{125}I-labeled immunoglobulin prepared against highly purified enzyme. These preliminary studies need repetition and elaboration with rigorous controls and finer localization, perhaps by electron microscopy. Reddick *et al.* (1974), using fluorescein-labeled antibody to purified human skin collagenase, have observed its binding to fibroblasts and extracellular collagenous tissues in histologic sections of human skin. It was heavily localized in the subepidermal papillary region, shown previously by Eisen (1969) to be the major site of collagenase production in culture. In such studies the evidence for homogeneity of the antigen is critical.

Isolation of the characteristic reaction products from tissues undergoing collagenolysis would add further substantiation. There have been several reports of the presence of the distinguishing substrate fragments, TC^A (Dresden, 1971; Nagai, 1973) and TC^B (Kuboki *et al.*, 1973), in extracts of such tissues undergoing physiologic degradation.

Strong circumstantial evidence for the function of collagenases comes from a wide variety of physiologic experiments wherein there is a close semiquantitative and temporal relationship between the loss of collagen in a living tissue in culture with the appearance of breakdown products and the enzyme in the culture medium. Some examples of such systems are human rheumatoid synovia, bone, postpartum uterus, rabbit skin wounds, alkali burned cornea, and tadpole fin and back skin among others. Sakamoto *et al.* (1975) reported a direct correlation of the degree of bone resorption in mouse calvaria explants in culture (measured by a semiquantitative morphologic scoring technique) with the amount of collagenase isolated from the medium. They also observed a quantitative dose relationship between the amount of parathyroid hormone added to the culture, the extent of bond lysis, and the amount of enzyme released. They had previously reported a direct correlation between the appearance of breakdown products of resorbing bone in the culture medium with the loss of explant collagen (Stern *et al.*, 1963; Kaufman *et al.*, 1965). The rat uterus, 1–3 days postpartum, undergoes extensive degradation of its collagen both *in vivo* and *in vitro*. In the latter situation there is a clear-cut relationship between the appearance of collagen degradation products in the culture medium, along with considerable increase in collagenolytic activity of such media, associated with reciprocally increasing losses of collagen from the tissue explants (Jeffrey *et al.*, 1971*a*). Progesterone levels are known to be high in the serum and tissues during gestation and to fall precipitously in the immediate postpartum period at which time active collagenolysis in the uterus is initiated. Progesterone *in vivo* will block that process, although only in very high concentration and only partially in the rat (Halme and Woessner, 1975) but more extensively in rabbits (Goodall, 1966). Jeffrey and colleagues (1971*b*) reported that progesterone added in physiologic concentrations to tissue cultures of the early rat postpartum uterus almost totally prevented both collagen breakdown and the appearance of collagenolytic enzyme in the medium.

In healing experimental wounds in rabbit skin, collagen removal is an essential feature of the contraction process. Donoff *et al.* (1971) have observed a reciprocal relationship between explant collagen concentration and the increase in collagen breakdown products and collagenase in the culture medium during incubation of tissues from the margins of the contracting wound. The picture is identical for tadpole (*R. catesbiana*) tail-fin and back skin explants in culture where collagen breakdown products and active collagenase appeared in the medium nearly simultaneously with the loss of collagen from the explant. In this situation addition of horse serum containing collagenase inhibitors to the culture greatly diminished the accumulation of breakdown products and collagenase in the medium. The explants healed rapidly and completely and did not disintegrate as invariably occurred in the absence of serum (Gross and Bruschi, 1971). Apparently serum permits the epithelium to completely seal off the cut edges of the tissue before its own collagenase can disrupt the underlying mesenchyme. Perhaps enough serum inhibitor had penetrated the connective tissue before healing to block the collagenolytic process afterward. The action of thyroxine after healing in the presence of serum was not explored. Recently, Davis *et al.* (1975) used healed explants of *R. catesbiana* and *R. pipiens* tail fin in serum-free medium to demonstrate the correlation between thyroxine-induced resorption of collagen and collagenolytic activity secreted into the medium after exposure to thyroxine, again suggestive of the physiologic role of this enzyme.

The observation that tissue culture was required for the production of nearly all the animal collagenases was puzzling to many who wondered whether this characteristic of the system might make it difficult to establish physiologic significance. However, in recent years it has become possible to extract, isolate, and characterize small amounts of characteristic collagenase from a variety of tissues. In none of these circumstances, however, has it been possible to obtain nearly as much enzyme as provided by the culture system. Harris *et al.* (1969) were able to isolate two different collagenases directly from human rheumatoid synovial fluid, both of which had a neutral pH optimum and degraded native collagen to the characteristic two fragments. Prior to this observation collagenase had also been extracted directly from leukocytes (Lazarus *et al.*, 1968*a*; Lazarus 1972) and from the hepatopancreas of the crab (Eisen and Jeffrey, 1969, Eisen *et al.*, 1973). Using an antibody against purified human skin collagenase Eisen *et al.* (1971) detected immunologic cross-reacting protein in simple saline extracts of human skin which yielded active enzyme after molecular sieve filtration. They concluded that they had separated the enzyme from serum inhibitors by this relatively gentle procedure.

The presence of collagen-bound collagenase in the tissue became a possibility when Ryan and Woessner (1971) were able to demonstrate collagenolytic activity by detecting collagen breakdown products in homogenates of rat uterus incubated sterile at 37°C for prolonged periods. Hormonal regulation of the amount of bound enzyme was manifested by the much reduced degradation of collagen in homogenates of explants of cultured uterine tissue previously exposed to progesterone, as compared with untreated controls (Koob and Jeffrey, 1974). Nagai and Hori (1972) attempted to quantitate the relative amounts of collagenase extractable from the connective tissues of three different animals, rat skin, human rheumatoid synovial membrane, and tadpole back skin, for comparison with the amount of enzyme produced in culture. After freeze–thawing, they obtained active collagenase by incubating insoluble tissue pellets at 37° in neutral salt solutions near physiologic ionic strength for periods up to 66 hr. The amount of collagenolytic activity obtained in the supernatant fluid from the synovial tissue, on an equal weight basis, was about 10% of that produced by the same amount of tissue in culture for 3 days (the 25% value reported does not take into account the differences in amounts of tissue used in culture compared to that for extraction). About the same recovery was obtained for the tadpole back skin, although here the amount of tissue extracted for comparison was not provided. It is of interest that additional enzyme was obtained from rat dermis by incubating at elevated temperatures (40–42°C) for up to 3 hr. Under these conditions β components of the collagen substrate were eliminated, suggesting the presence of a neutral protease which removes the telopeptide ends. Nagai and Hori (1972) proposed that relatively considerable amounts of collagenase are present in the tissue at any one time, probably in close association with one or more tissue components, perhaps collagen itself, to be released at 37° as a result of enzyme cleavage of the substrate and escape into the free liquid. They did not believe that the enzyme is bound by α2-macroglobulin in the tissue since this complex is essentially irreversible under physiologic conditions.

These investigators also noted that the amount of collagenase extractable from tadpole back skin was significantly increased when the animals had been previously exposed to thyroxine for 4 days, which correlates well with the extensive remodeling of back skin connective tissue during metamorphosis.

Pardo and Perez-Tomayo (1975) systematically examined the well-known phenomenon of progressive change in properties of purified collagen when stored in solution at neutral pH in the cold for prolonged periods. By gel electrophoresis they examined the polypeptide chain characteristics of purified collagen in neutral solution, and also the same material heat-gelled at 37°C as a function of incubation time, noting progressive increase in the number of low-molecular-weight fragments. They imply that collagen *in vivo* has a small amount of tightly bound collagenase associated with it. There is the alternative possibility that this association only occurs after homogenization of the tissue.

Confidence in the physiologic significance of a tissue enzyme is high if there is an obvious natural tissue substrate for which the enzyme exhibits a high degree of specificity under physiologic conditions. If detection and measurement require a synthetic substrate and if optimal conditions for function are outside the known physiologic range, there is good reason to question its function even if there is temporal or spatial relatedness of enzyme activity with physiologic or pathologic phenomena.

A striking (and unfortunate for the investigator!) feature of these enzymes is the remarkably small amount of purified collagenase available from nearly all tissue systems compared with the large amount of information obtained with regard to its mode of action. The production of 1 mg of highly purified tadpole collagenase requires the culturing of fins and back skins from hundreds of large bullfrog tadpoles. In general this situation prevails for the mammalian collagenases, perhaps with the exception of that obtained by extraction of human granulocytes (Ohlsson and Olsson, 1973) in which case it is reported that 115 mg of highly purified enzyme was obtained from 4 g of protein extracted from 1.1×10^{12} cells. To date there have been no other reports of milligram amounts of purified collagenase extracted from vertebrate tissues. The use of continuous mass cell cultures is promising.

VI. Conditions for Cleavage and Substrate Specificity

Although it is well established that numerous proteases such as pepsin, trypsin, chymotrypsin, papain, and certain ill-defined tissue enzymes can degrade "nonhelical" (telopeptide) amino- and carboxy-terminal ends of the collagen molecule, there is, with few exceptions, little evidence for their cleavage of the helical regions of the molecules in fibrillar form under physiologic conditions. High concentrations of trypsin in 0.5 M $CaCl_2$ appear to cleave the molecule in the region of SLS bands 41–42 (three-fourths the distance from the NH_2 terminus) (Olsen, 1964). Davison and Schmitt (1966) described what appear to be TC^A and TC^B produced under these conditions. The precise site of cleavage is not clear from the electron micrographs. Papain and pronase are known to slowly degrade the molecule in solution from both ends, gradually biting into the helical regions. These proteases also predispose insoluble collagen fibrils in some tissues such as calf skin to solubilization in dilute acid (Nishihara and Miyata, 1962), probably by removing the nonhelical peptides which include cross-linking regions.

The evidence for cleavage of the helical regions of the collagen molecule by cathepsin B_1, or by any other known catheptic enzymes, involves incubation at or near body temperature at low pH, i.e., below 5.5 (Burleigh *et al.*, 1974), conditions known to destabilize the helix and increase susceptibility to cleavage by nonspecific proteolytic attack. The evidence for significant cleavage of the helical regions at neutral pH is unconvincing (Barrett, 1975). For such enzymes to function in the extracellular space, the appropriate pH conditions at the substrate would be required. There is no experimental evidence for extracellular tissue pH below 6.6 and even that of inflammatory exudates is found to be in the neutral range, pH 7.0 or above (Hutchins and Sheldon, 1973). One may always argue that the proper conditions in microregions at the substrate surface might exist. We await this demonstration.

Because collagen fibrils have frequently been seen within the cytoplasmic domains in tissues undergoing resorption, such as in tadpole tail fin (Usuku and Gross, 1965), postpartum uterus, and hair follicles (Parakkal, 1969*a*,*b*, 1972; Brandes and Anton, 1969), it is conceivable that acid pH in the sequestered region might permit lysosomal enzymes such as cathepsin B to attack such fibrils. Simply because collagen fibrils are seen within the cytoplasmic domain is no indication that they are undergoing degradation, even if their structure appears to be disorganized. Electron microscopic studies of tadpole tail fin undergoing resorption indicates the presence of large numbers of abnormal collagen fibrils in the extracellular space, strongly suggesting that they have already been attacked by extracellular collagenolytic enzymes (Gross, 1969). Thus the presence of such disordered collagen within cells may represent a secondary phagocytic process or simply the benign engulfment of fibrils resulting from abnormal extracellular conditions. The likely alternative is that the engulfment of previously attacked fibrils serves the purpose of finishing the process through the activity of intracellular proteases which requires nonhelical polypeptide fragments. This point is discussed in more detail by Perez-Tomayo (1973).

How specific for collagen are the collagenolytic enzymes? Early, but not rigorous, analyses of the susceptibility of hemoglobin and bovine ligamentum nuchae elastin to attack by tadpole collagenase were negative (Sakai and Gross, unpublished results). Of considerable interest is the report by Ohlsson and Olsson (1973) that highly purified leukocyte collagenase will attack fibrinogen and proteoglycans. The collagenase extracted from the crab hepatopancreas and purified (Eisen *et al.*, 1973) also can attack proteins other than collagen, behaving with chymotrypsin-like activity, and can be blocked by protease inhibitors known to be ineffective against all other animal collagenases.

What is the peptide bond specificity of the collagenases? Nagai *et al.* (1966) were unable to remove completely caseinolytic activity from the tadpole enzyme after ammonium sulfate precipitation, molecular sieve filtration, starch-gel electrophoresis, and DEAE-cellulose chromatography. This activity moved with the collagenase and was inhibited by both cysteine and EDTA. These early studies on the specificity of tadpole collagenase attack on denatured collagen (gelatin) indicated that the number of susceptible bonds included Gly-Leu, Gly-Ile, Gly-Val, Gly-Ala, and Gly-Phe, i.e., all glycyl–nonpolar residue peptide bonds (Nagai *et al.*, 1964). However, Gross *et al.* (1974), using the 30% saturated ammonium sulfate precipitate, which did not contain the protease Harper and Gross (1970), for the first stage of purification, succeeded in obtaining a collagenase fraction which cleaved only a single bond in the CB7 region of the $\alpha 1$ chain, and another adjacent bond in $\alpha 2$-CB5. In collaboration with Harris and McCroskery, Gross *et al.* (1974) observed that both the V_2 ascites carcinoma collagenase and that isolated and purified from tadpole tissue culture medium produced a single, and the same, cleavage in separated α chains and in the CNBr peptide, $\alpha 1$-CB7, which contains the cleavage site. No other breaks were made elsewhere in the intact $\alpha 1$ chain or in isolated $\alpha 1$-CB8 which also contains a glycyl–isoleucine bond but with a different amino acid sequence on either side of the sensitive bond from that found in CB7. Gross *et al.* (1974) concluded that the specificity of both the mammalian and amphibian collagenases are identical and, in the $\alpha 1$ chain, is strictly limited to one specific peptide bond, cleaved in both rat skin and chick skin collagen substrates. A similar unique cleavage in the $\alpha 2$ chain at the Gly-Leu bond in $\alpha 2$-CB5 was reported.

Of considerable interest from both biological and structural viewpoints are the differences in susceptibility to collagenolytic attack of the genetically different types of collagen. At first it was thought that cartilage collagen (type II) was refractory to the animal collagenases. However, this proved to be more of a difference in rate of cleavage, cartilage collagen reacting much more slowly but releasing the same reaction products (Harris and Krane, 1975; Nagai, 1973; Davison and Berman, 1973; Harper and Gross, unpublished results; Wooley *et al.*, 1975*a*). Harris *et al.* (1975*a,b*) have compared the relative susceptibilities of types I (skin bone), II (cartilage), and III (fetal skin, blood vessels, leiomyoma) collagens to rheumatoid synovial collagenase and have found their order of decreasing susceptibilities to be III > I > II. Although the sequence data currently available indicates that the cleavage site for α1(I) and α1(II) may be identical (Harris *et al.*, 1975*a,b*), there may be important differences in local stability of the cleavage site regions resulting from sequence differences in adjacent portions of the α chains.

At present we know the sequence on both sides of the residues 772 and 773, -Gly-Ala-, the cleavage site of the α1 chain of Type I collagen (α1(I)). Cleavage occurs in the α2 chain in the presumably adjacent residues, -Gly-Leu-; however the sequences on the COOH terminal side are unknown and only the first three residues, -Leu-Ala-Gly-, are established on the amino terminal side thus far (Gross *et al.*, 1974). Eighteen consecutive residues of the amino terminal sequence of the cleavage sites in type II and type III collagen are established (Miller *et al.*, 1976). In human α1(II) there is only one substitution compared with bovine α1(I), Leu for Val at position 779. Thus the reason for substantial difference in cleavage rate must await sequence analysis of the COOH terminal side. The sequence differences between α1(III) and α1(I) are substantial, eight in the first eighteen residues. The cleaved bond is -Gly-Leu- as in the α2 chain of type I. The next five residues, -Ala-Gly-Leu-Arg-Gly-, are identical with α1(I) and α1(II), but the following two residues, -Ala-Arg-, are different. Although there are five more substitutions in the next ten residues the triplet structure, -Gly-X-Y- is maintained and there are no anomalies suggesting interference or weakening of helical conformation in this region. Of considerable importance, physiologically, is the observation of Miller *et al.* (1976*a*) that trypsin at neutral pH and below the substrate denaturation temperature will extensively and rapidly cleave native type III collagen in solution to reaction products TC^A and TC^B, making one scission in the α chain at the -Ala-Arg- bond, seven peptide bonds beyond the collagenase cleavage site toward the COOH terminal end of the molecule. The nearby -Leu-Arg- bond is not attacked nor are any other -X-Arg- bonds as long as the helix of the two fragments is intact. Probably inclusion of 2 M guanidine prior to denaturing preserves the structure on chromatography. Miller and colleagues note the physiologic implication that degradation of type III collagen might proceed *in vivo* via proteolytic enzymes not collagenolytic for other types of collagen.

Collagenase must function on native fibrils, not just the reconstituted type, since essentially all of the tissue collagen *in vivo* is in fibril form, much of which is insoluble and covalently cross-linked. There is exceedingly little collagen in solution in normal connective tissues (Gross, 1969). Leibovich and Weiss (1971) reported that purified rheumatoid synovial collagenase, which could degrade collagen in solution at neutral pH in characteristic fashion, did not degrade polymeric (insoluble) collagen from human synovial tissues. They proposed that a preliminary attack on the cross-link regions by other proteinases presumed to come from lysosomes was an essential prelude to collagenolytic attack. However, their basic observation has been challenged; Harper and Gross (unpublished results) were able to degrade the same substrate used by Leibovich and Weiss (generously provided by them) to the extent of 50% with semipurified tadpole collagenase under the same environmental conditions employed by Leibovich and Weiss. This could not have been ascribed to preliminary attack on the cross-link region by a contaminating protease since disk electrophoresis of reaction products of acid-extracted collagen (at neutral pH and 25°), reacted with the tadpole enzyme, showed no loss of intramolecular cross-links, i.e., there was no loss of β^A components, indicating that the cross-link regions of the molecule remained intact. Similar observations of effective attack by semipurified synovial collagenase (Harris and Krane, 1973; Wooley *et al.*, 1972) and bone collagenase (Sakamoto *et al.*, 1973*a*) on insoluble cross-linked collagen have been reported. In the case of the synovial enzyme, cartilage collagen in intact bits of tissue, surrounded by the normal complement of proteoglycan, was also degraded at 37° (Harris *et al.*, 1970). Further studies with purified collagenase are needed here.

It is important to note, however, that the kinetics of collagenolytic attack on cross-linked fibrils are very different from that on weakly cross-linked reconstituted gels. Harris and Farrell (1972) have neatly demonstrated that point by measuring the relative rates of degradation of collagens cross-linked with different numbers of methylene bridges introduced with formaldehyde. At 37° collagen gels cross-linked with about three bridges per mole were degraded slowly by synovial collagenase; however, the rate was considerably accelerated at 38.5°. Using native collagen fibrils isolated from human cartilage as a substrate for rheumatoid synovial collagenase, Harris and McCroskery (1974) found a 10-fold increase in degradation rate on raising the temperature from 30 to 36°C. There is a significant rise in temperature in inflamed rheumatoid joints which might be expected to facilitate collagenolysis *in vivo*. Rates of degradation *in vivo* would also depend on local collagenase concentrations, susceptibility of the prevailing collagen type to attack, and the degree of protection of the fibers by "barrier" ground substance or inhibitors.

How important is the helical state of the substrate? The initial publication by Harris (1972) in which highly purified human rheumatoid synovial collagenase was shown to cleave gelatin at 24° into fragments greater than 50,000 mol. wt. did not consider the likelihood that the randomized α chains renatured at 24° even though there was no rise in specific viscosity. At this temperature rapid helix formation can be readily demonstrated by increase in negative optical rotation in the absence of viscosity changes. Perfect triple-helix formation may not be necessary to limit the attack sites of the enzyme. This deficiency was subsequently rectified by McCroskery *et al.* (1973) who observed the same limited cleavage of native collagen and gelatin chains at 27° and 39°C by highly purified rabbit tumor collagenase. Of importance is their observation that the rate of cleavage for native collagen by the tumor enzyme was considerably greater than that of a gelatin substrate, suggesting that either the triple helix is necessary for effective cleavage of the specific bond or, what is more likely, the substrate helical conformation is required for effective enzyme binding. It is possible that a local instability around the cleavage site combined with adjacent firm helical regions are required for ready susceptibility to collagenolytic attack. Analysis of the behavior of synthetic peptides having the same and modified amino acid sequences as that around the cleavage site is bound to shed light on this subject.

How is the further dismantling of large α-chain fragments accomplished? Harris (1972) has isolated a neutral protease from primary cultures of rheumatoid subcutaneous nodules, observing that this enzyme (of higher molecular weight than the collagenase), while having no collagenolytic properties itself, could rapidly degrade the products of collagenolytic attack. It was also stimulated by colchicine and inhibited by chelators, sulfhydryl compounds, and serum. Harris and Krane (1972) have also isolated a peptidase from cultures of human rheumatoid synovia which could not attack native collagen but did degrade gelatin to smaller peptide fragments. Of interest is the observation that this neutral protease (Harris, 1972) degraded the synthetic peptide, Pbz-Pro-Leu-Gly-Pro-DArg (a specific substrate for bacterial collagenase), but could only reduce denatured collagenase digestion products to a mean molecular size of 3000 daltons; only 11% was dialyzable. This peptidase was inhibited by EDTA, cysteine, and serum and had a pH optimum in the neutral range. It could not cleave collagenase-produced fragments in the native state but could do so after denaturation of these fragments at 37°. This enzyme, like collagenase, was inactive below pH 6. The kinetics of its release into the culture medium paralleled that of the collagenase, suggesting that it was synthesized *de novo* or represented activation of a zymogen. The production of both collagenase and neutral protease was stimulated by colchicine and not by sucrose, which is consistent with the likelihood that neither of the enzymes are lysosomally derived. Peptidases capable of cleaving the synthetic peptide described above and operating at neutral pH have been isolated and described for many mammalian tissues (see Aswanikumar and Radhakrishnan, 1973), but the capability for cleavage of large collagen peptides by these enzymes has not been explored in any depth. Thus the enzymatic mechanism for the complete degradation of collagenase reaction products to free amino acids remains to be characterized. It would seem that the complete degradation of such peptides might occur inside the cell, since it is less likely that the entire constellation of required enzymes would be found in the extracellular space.

VII. Regulation of Collagenase Activity

How is collagenolytic activity synchronized in time and space with collagen synthesis and deposition in the building and remodeling of tissue structure? Which biologic signals control its synthesis and activation? How are serum-derived and fixed tissue components involved in modulating its activity in the extracellular space? From a therapeutic viewpoint how may we, pharmacologically, manipulate collagenase function? It is likely that at least some fibrotic and ulcerative diseases involve either diminished or excessive collagen degradation. For most of these questions the answers are not in sight. Experimental approaches to some have begun.

A. Inhibition of Collagenase Activity

With the first detection and isolation of the animal collagenases, the inhibitory activity of EDTA, cysteine, and serum were described (Lapière and Gross, 1963; Nagai *et al.*, 1966). The tadpole collagenase was reversibly inhibited by EDTA and irreversibly blocked by cysteine. We assumed that inhibition by EDTA implied dependence on calcium as a cofactor and that cysteine inhibited the enzyme by reducing disulfide bonds essential for function. The inhibitor effect of EDTA on tadpole collagenase may be reversed simply by dialysis, and the evidence for the calcium cofactor has never been adequately documented for this enzyme.

On the other hand the rat uterine collagenase was irreversibly inhibited by 10^{-3} M EDTA and clearly required calcium; however, the addition of an excess of calcium to the inhibited enzyme did not restore its activity (Jeffrey and Gross, 1970). Again, in contrast with the tadpole enzyme, that from the postpartum rat uterus was not affected by 10^{-3} M cysteine. The collagenase isolated from the medium of cultured rheumatoid synovial tissue was irreversibly inhibited by EDTA; the addition of excess calcium did not restore activity (Evanson *et al.,* 1968).

An interesting comparison can be made between the collagenases obtained from rat skin and uterus with respect to their behavior toward inhibitors and their requirement for calcium. The rat skin enzyme, in contrast with that of uterus, did not lose its activity on dialysis but did so after removal of calcium by ion exchange, following which, however, the addition of calcium completely restored activity (Tokoro *et al.,* 1972). Also, the addition of calcium to the EDTA-inhibited skin enzyme partially restored activity, again in contrast with the behavior of the uterine enzyme. Zinc, copper, manganese, and magnesium were not substitutes for calcium. Cysteine inhibited both skin and uterus enzymes at concentrations of 10^{-1} M and not at all at 10^{-3} M. In this regard it is of interest that two collagenases isolated from epithelium and granulation tissue of healing rabbit skin wounds (Donoff *et al.,* 1971) were different not only in molecular weight but also in their behavior toward cysteine. The epithelial enzyme was stimulated by this amino acid, whereas that from granulation tissue from the same wound was inhibited; both behaved similarly toward Na-EDTA.

Rabbit corneal collagenase was inhibited by Na-EDTA, Ca-EDTA, and cysteine (Berman *et al.,* 1973*a,b*), but this inhibitory effect could be reversed for all three agents by addition of excess calcium. Berman and Manabe (1973) reported that simple addition of calcium to a Ca-EDTA-inhibited corneal enzyme did not restore activity, whereas addition of calcium by dialysis did do so; apparently removal of Ca-EDTA was necessary. These investigators explored the effect of other metals on corneal collagenase, finding that zinc completely restored the activity of the enzyme inhibited by 1,10-orthophenanthraline and concluding that Zn may be a part of the enzyme structure. Inhibitors with low affinity for calcium but with a high binding constant for transition metals such as zinc were potent inhibitors of corneal collagenase. Dithizone and 8-hydroxyquinoline-5-sulfonic acid, which are known inhibitors of zinc metalloenzymes, were also potent inhibitors of corneal collagenase (Berman and Manabe, 1973). Incubation of ulcerating cornea explants in the presence of ^{65}Zn produced zinc-labeled collagenases in the culture medium, and the radioactivity followed enzyme activity through repeated cycles of gel filtration. The authors point out, however, that the enzyme was not pure and that some of the metal might possibly be associated with impurities. These data remain strongly suggestive and further experimentation with pure enzymes is essential. Of interest is the observation of Berman and Manabe (1973) that calcium EDTA in the presence of equimolar amounts of calcium ion still completely inhibits corneal collagenase and even in the presence of a 50-fold excess of calcium is still partially inhibitory. This observation further strengthens the idea that EDTA acts by stripping a metal from the enzyme rather than by simply depriving it of calcium (acetyl cysteine and penicillamine were also inhibitors of corneal collagenase). The collagenolytic enzyme isolated from cultures of human rheumatoid arthritis skin nodules by Harris (1972) was inhibited by metal chelators such as α,α'-dipyridyl, penicillamine, 1,10-orthophenanthroline (10^{-4} M) and L-histidine (10^{-2} M), as well as EDTA, dithiothreitol (DTT), and cysteine. Berman and Dohlman (1975) discussed in detail the action of inhibitors on corneal collagenase, proposing that thiols such as cysteine and DTT may function, at least in part, by metal capture and, perhaps less so, by making disulfide bonds. However, oxygenation of DTT-inhibited enzyme restored activity, and this reversal was prevented by prior alkylation (Hook *et al.*, 1972).

The actual discovery of the animal collagenases depended importantly on the fact that in the initial experiment serum was omitted from the culture medium, since it was soon observed that serum is a potent inhibitor of animal collagenase (Nagai *et al.*, 1966). Subsequently it was noted that enzymes obtained from cultures of human skin (Eisen *et al.*, 1968) and rheumatoid synovial tissue (Evanson *et al.*, 1968) were also inhibited by serum, the active principals of which were found to be the α_2-macroglobulin and α_1-antitryptic factor (Eisen *et al.*, 1971). It would appear now that the latter is less certainly an inhibitor of collagenase than the former (Nagai 1973; Sakamoto *et al.*, 1972*a,b*; Berman *et al.*, 1973*a*). Serum α_2-macroglobulin (α_2M), a potent inhibitor for almost all the known proteases acting at neutral pH including those whose activities are based on active-site serine, metal, or thiols, blocks nearly all the known animal collagenases by binding these enzymes almost irreversibly under physiologic conditions (Werb *et al.*, 1974).

Recent studies on the mechanism of α_2M inhibition indicates cleavage by bound protease of susceptible bonds in this large molecule (720,000 daltons), which is said to trigger a conformational change, sterically trapping the enzyme. This idea originally derived from work of Laskowski (1972) and Laskowski and Sealock (1971) on non-serum protease inhibitors. Apparently the activity is not just limited to neutral protease since cathepsins D and B_1 are both inhibited by α_2M; however, pepsin, exopeptidases, nonproteolytic hydrolases, and zymogens of proteases are not blocked (Harpel, 1973; Barrett and Starkey, 1973). Small peptide substrates were still cleaved by α_2M–protease complexes but macromolecular substrates were not, thus indicating that the active site of the bound enzyme was still operative but sequestered away from approach by large substrate molecules (Harpel, 1973). Werb *et al.* (1974) reported that rabbit synovial cell collagenase is bound by α_2M in a manner seemingly identical with that of trypsin and thermolysin and also makes what appears to be a single peptide-bond cleavage near the center of the 185,000-mol.-wt. subunit. This is of considerable interest since a high degree of cleavage-site specificity exists for the collagenase (Gross *et al.*, 1974), i.e., a specific Gly-Ile peptide bond, which is quite different from the sequence specificities for the other two proteases. How then is this very broad susceptibility of α_2M to a variety of specific proteases explained? It is clear that sequence analysis of the α_2M fragments released by the several different proteases (after reduction) is desirable. The fact that proenzymes are not bound by α_2M provides a useful tool for separating zymogens from active enzymes, and indeed, it has been used recently by Birkedal-Hansen (personal communication) to separate α_2M-bound collagenase and a true zymogen from cultures of bovine gingiva. Sakamoto *et al.* (1972*c*) observed that trypsin can block the inhibition of collagenase by serum when added prior to mixing, probably by forming a more stable bond with α_2M than with collagenase. It is likely that the binding constants are different for different proteases. The reported inability of serum to inhibit leukocyte collagenase (Lazarus *et al.*, 1968*a,b*) and the high-molecular-weight collagenase fraction obtained from human rheumatoid synovial fluid (enzyme A) (Lazarus *et al.*, 1968*a,b*; Harris *et al.*, 1969) has been explained by Werb and associates (1974) in terms of a slower reaction rate of these enzymes with α_2M rather than by a failure to bind.

In trying to ascertain the nature of an experimental complex between α_2M and tadpole collagenase, Abe and Nagai (1972) adsorbed the enzyme to an anti-α_2M Sepharose column but could not elute the enzyme with the usual high-salt or low-pH eluents. Basing their next step on the earlier reports of the effectiveness of 3 M thiocyanate in separating antigen–antibody complexes (deSaussure and Dandliker, 1969), they stripped the column with this denaturing agent, liberating both active collagenase and denatured α_2M; however, only 5–10% of the enzyme activity was recovered, whereas treatment of enzyme alone with thiocyanate reduced activity by only 50%. Thiocyanate, and also repeated freeze–thawing of α_2M completely eliminated the inhibitory activity of α_2M and its ability to bind collagenase without interfering with its antigenicity.

Using dialysis against 3 M thiocyanate to release the collagenolytic activity of rheumatoid synovial fluid, Abe and Nagai (1973) reported very little activity before thiocyanate with considerable increase 16 hr later following dialysis and removal of thiocyanate. Relatively little effect was seen in nonrheumatoid joint fluids. Although they found correlation between the amounts of α_1-at, α_2M, and released collagenase activity in nonrheumatoid fluids, there appeared to be no correlation in rheumatoid arthritis. This suggested to them that there may be "an unidentified principle" other than α_1-at and α_2M, the inhibitory activity of which is also abolished by thiocyanate. Since then, Wooley *et al.* (1975*c*) have detected an additional collagenase inhibitor in serum unrelated to α_2M (molecular weight difference) or to α_1 antitryptic factor (immunologically distinct) with molecular weight 40,000. It inhibited a variety of human collagenases but not trypsin or papain, just the reverse of the inhibitory activity of α_1-at. Bauer *et al.* (1975) reported an inhibitor (molecular weight somewhat less than 40,000) in serum-free medium from skin fibroblast cultures, which they concluded is derived from the cells. McCroskery *et al.* (1975) also reported an inhibitor of molecular weight 40,000–50,000 daltons in extracts of rabbit V_2 carcinoma which is separable by ammonium sulfate fractionation from collagenase.

It is important to note in the report by Abe and Nagai (1973), in which rheumatoid synovial fluid was chromatographed on an anti-α_2M affinity column, that a very small proportion of the total measurable enzyme passing through the column was associated with α_2M (demonstrated after treatment of this fraction with 3 M NaSCN); the bulk of it passed through the column early with the wash fluid (their Figure 3). This is in keeping with the earlier studies of Harris *et al.* (1969) who reported large amounts of free enzyme activity in rheumatoid synovial fluids. Recently Sakamoto and colleagues (1975) found that active collagenase could be separated from the medium of mouse bone cultured in the presence of serum by passing it through a Sepharose–heparin column. Apparently the enzyme was selectively and strongly adsorbed from the medium, presumably dissociated from the tight complex with α_2M. Nagai *et al.* (1975) report that collagenase has a far higher affinity for α_2M than for collagen fibers, an observation of no small physiological interest.

Recently another class of protein inhibitors of collagenase have been described, including lysozyme and other basic proteins such as histones and protamines (Sakamoto *et al.*, 1974). Mixtures of egg-white lysozyme or calf thymus histone with collagenase in roughly equal proportions by weight resulted in about 50% inhibition in the radioactive fibril assay. Protamine as the free base was about 1/10 as active, whereas its salts were totally inactive. Difficult to interpret was the observation that the inhibitory properties of lysozyme were displayed in the fibril assay systems but not in collagen in solution, except at very high lysozyme concentrations. Lysozyme did not significantly inhibit *Cl. histolyticum* collagenase whereas histone and protamine did interfere. Sakamoto *et al.* (1974) point to the fact that lysozyme is found in relatively high concentrations in cartilage, particularly in the endochondrial ossification zone in growing bone. They wonder whether lysozyme in the tissue may play some significant role in regulating collagen degradation. Eisenstein *et al.* (1973) had already noted that cartilage explanted either to the rabbit cornea or chick embryo chorioallantoic membrane was resistant to vascularization as compared with other tissues.

In addition to the now well-recognized inhibitory components of the serum, there is a reasonable probability of finding regulating factors fixed in the connective tissues, perhaps in close association with collagen fibers. Hook *et al.*, (1971) were able to inhibit corneal collagenase activity by adding "physiological concentrations" of bovine cartilage and cornea proteoglycan extracted in guanidine–HCl to the assay collagen preceding fibril formation. The same concentrations of purified corneal keratan sulfate had no blocking action; other glycosaminoglycans free of protein were not examined. The observations of Lapière and Gross (1963), indicating a selective sparing of newly synthesized collagen during thyroxine-initiated resorption of tadpole tail fin, could be explained by the presence of relatively high concentrations of inhibitory proteoglycan around newly synthesized collagen. Studies on the effect of hyaluronidase treatment of corneal stroma on susceptibility to collagenolytic degradation (Gnädinger *et al.*, 1969) suggested that "mucopolysaccharides may provide a protective function to collagen fibers." It is conceivable that proteases and hyaluronidases may function in removing protective extrafibrillar substances from the fibers preparatory to the enzymatic dismantling of collagen.

An intriguing aspect of the possible physiologic significance of indigenous tissue inhibitors of collagenase is raised by the angiogenesis-stimulating properties of invasive neoplasms. New blood vessels actively penetrate the extracellular matrices during normal processes of organ vascularization and also appear to play a critical role in the growth of malignant tumors (Folkman, 1974). This group of investigators (Brem and Folkman, 1975) subsequently reported that living cartilage, or cartilage extract, strongly inhibited the growth and proliferation of ingrowing capillaries in rabbit cornea serving as host to a rabbit V_2 carcinoma explant or in the chick embryo chorioallantoic membrane carrying a Walker carcinosarcoma fragment. It is quite likely that capillaries penetrate dense connective tissues by enzymatic dissolution of extracellular matrix, probably including collagen (although this has not yet been demonstrated). If so, these experiments reporting the inhibition of tumor vascularization may be indicating the production by cartilage of inhibitors of collagenase and other lytic enzymes functioning in the extracellular space.

B. Stimulation of Collagenase Activity

Of equal interest are the substances which, when added to cultured tissues, enhance collagenase activity. It was noted by Harris and Krane (1971) that colchicine in concentrations of 0.1 μg/ml in the culture medium increased the amount of measurable collagenolytic activity 10-fold in cultures of rheumatoid synovium. The agent had no effect on the medium itself but required living cells. There was no evidence of mitotic arrest, but there was an increase in labeled leucine incorporation in several proteins, including collagenase. There appeared to be an increase in fragmentation of lymphocytes and plasma cells in colchicine-treated synovial tissues, and Harris and Krane (1971) suggested the possibility of a release of "activator materials" from lymphocytes which might stimulate collagenase synthesis, a perceptive speculation which foreshadowed more recent developments with regard to macrophage activity. The stimulatory effect of colchicine in cultures of human skin has also been observed by Coffey and Salvador (personal communication). Harper and Gross (unpublished data) were unable to detect any effect of colchicine on collagenase production in cultures of tadpole tissue. The experience of Raisz *et al.* (1973) to the effect that bone resorption stimulated by parathyroid hormone or vitamin D metabolites was effectively blocked by low concentrations of colchicine indicates a complexity in the action of this agent that is yet to be resolved. The mechanism whereby total bone resorption is coupled to demineralization mediated by parathyroid hormone is important to understand; there is good reason to believe that in bone, collagen degradation does not occur in the absence of mineral resorption.

The possible biologic effect of heparin on bone resorption was suggested by Griffith *et al.* (1965) and Jaffe and Willis (1965). Following up these clinical observations with *in vitro* experiments, Sakamoto *et al.* (1973*a*) have reported that heparin, dextran sulfate, and other sulfated polysaccharides enhanced collagenolytic activity threefold, which could be correlated with approximately a 50% increase in the amount of collagen breakdown products in the medium released from mouse bone explants in culture. Heparin added to either semicrude or highly purified mouse collagenase also increased enzyme activity as much as 150%. However, as in the case of the inhibitory behavior of lysozyme, the enhancing effect of heparin depended on the state of aggregation of the assay substrate; the effect was noted only with reconstituted fibrils and not with dissolved collagen. They also observed a significant increase in collagenase digestion of native insoluble collagen obtained from decalcified mouse calvaria.

Perhaps related to the pharmacologic effect of heparin is the appearance of large numbers of mast cells during remodeling and replacement of bone trabeculi in healing experimental fractures (Severson, 1969). The same excessive accumulation of mast cells adjacent to the gingival supporting tissues of teeth was found when jaw bone resorption occurs in periodontal disease (Wislocki and Sognnaes, 1950). However, in areas of cartilage resorption there were no mast cells evident. Perhaps they are not necessary here because of the large amount of sulfated polysaccharides.

The possibility that phagocytosis of particulate material might stimulate collagenase production was raised by experiments in which cultured synovial cells which had ingested either latex particles or particulate fungicide material (Werb and Reynolds, 1974) increased their output of both collagenase and a neutral protease into the culture medium by as much as 12-fold. The increase in rate of secretion of the two enzymes was correlated with the amount of latex ingested. The type of particle apparently was not specific since mycostatin and dextran sulfate could accomplish the same effect. As Werb and Reynolds point out, these phenomena may be related to the stimulatory effect of heparin on collagenase production in mouse bone explants (Sakamoto *et al.,* 1974). The secretory process differs from that of the release of lysosomal enzymes in that the latter occurs within minutes or hours after ingestion of stimulating material, whereas increased collagenase secretion is delayed for at least a day, continuing for days or weeks without further phagocytosis. This phenomenon is not unique to collagenase secretion; for example, macrophages which have ingested agar *in vivo* secrete plasminogen-activating proteases in culture (Unkless *et al.,* 1974).

The likelihood of macrophages being involved in the dissolution of collagen in chronic inflammatory lesions has long been considered. Collagenolytic activity has been ascribed to macrophages in diseased gingiva (Senior *et al.,* 1972; Robertson *et al.,* 1974). Wahl and colleagues (1974) have examined macrophages in an experimentally induced peritoneal exudate in rabbits which had been stimulated by exposure to bacterial endotoxin. Control (unstimulated) cultures released no collagenolytic activity, whereas those exposed to *E. coli* or *Salmonella* lipopolysaccharide actively secreted collagenase into the culture medium over a period of 5 days. The active fraction contained lipid, whereas a lipid-free polysaccharide moiety was not stimulatory. Enhancement could be blocked by the addition of cycloheximide to the cells anytime during a 24-hr period after exposure. Disk electrophoresis of the collagenolytic reaction products demonstrated the characteristic fragments, described, however, as 62 and 67% the size of α chains rather than the usual 75%, certainly a difficult estimation to make without electron micrographs of SLS. The enzyme was inhibited by serum, EDTA, and cysteine. Since enzyme was not obtained from extracts of the stimulated macrophages and its secretion or synthesis could be stopped by cycloheximide, it would seem to be produced *de novo* or activated from a zymogen. In an accompanying study Wahl *et al.* (1975) observed that previously activated guinea pig lymphocytes when stimulated by the antigen, or concanavalin A, produced substances (lymphokines) which in turn could stimulate peritoneal macrophages to secrete collagenase. The stimulated spleen lymphocytes did not themselves produce the enzyme, and unstimulated but activated spleen cells were not capable of inducing collagenase production by macrophages. As with endotoxin-stimulated macrophages, cycloheximide blocked the appearance of the enzyme. As these investigators pointed out, it is not unlikely that stimulation of macrophages by bacterial endotoxins is indirect, having to pass first through lymphocytes which may contaminate the macrophage population. These in turn produce lymphokines which then stimulate macrophages. In chronic inflammatory diseases such as rheumatoid arthritis, macrophages might be involved, although the evidence from the work of Harris and Krane suggests that a fixed tissue cell in the synovium is responsible.

There are several observations in the literature which suggest that wounding or other injury is a stimulus to collagenase production. Grillo and Gross (1967) observed considerably greater collagenolytic activity in the epithelium and granulation tissues at the edge of clean healing skin wounds in guinea pigs as compared with skin at increasing distances from the wounded region. Indeed, there was some indication of an epithelial-mesenchymal interaction stimulating the production of a collagenase. In an analogous situation the annular alkali burn produced at the periphery of the rabbit cornea was associated with an increased collagenolytic activity by the central uninjured corneal tissue (Pfister *et al.*, 1971). Injury evokes remodeling of connective tissues which must involve collagen degradation. The movement of epithelia across new surfaces may require removal of basement membrane. Huang and Abramson (1975) made use of this information to increase the output of collagenase from guinea pig skin by mild excoriation of the skin prior to explantation.

C. *Hormonal Regulation of Collagenase Activity*

If animal collagenases are directly involved in physiologic degradation of collagen *in vivo* there should be clear-cut evidence for hormonal control in those biologic systems in which hormone activity is known to have a clearly demonstrable effect on collagen metabolism. The influence of parathyroid hormone on bone resorption, the role of progesterone in maintaining the robust structure of the uterus particularly during the last stages of pregnancy, the reported adverse effects of cortisone and its derivatives on the course of ulceration in the injured cornea, and the influence of thyroxine on tail resorption in the metamorphosing tadpole are examples.

The evidence for bone resorption under the influence of parathyroid hormone need not be reviewed. Direct evidence for collagen breakdown in culture under the influence of this hormone was provided by Stern *et al.* (1963) who reported a marked increase in the release of labeled hydroxyproline-containing peptides in the culture medium of mouse calvaria under the influence of parathyroid extract and high oxygen tension; the latter itself was insufficient. The direct demonstration of an increase in collagenolytic activity in bone cultures of rats previously treated *in vivo* with parathyroid hormone was furnished by Walker *et al.* (1964). The stimulatory action of the combination of parathyroid hormone and heparin on collagenolytic activity when added directly to bone cultures was described by Shimizu *et al.* (1969), and from the same laboratory Sakamoto *et al.* (1973*b*) subsequently purified and characterized bone collagenase. Fullmer and Lazarus (1969), however, have obtained active collagenase from the medium of cultures of human bone in the absence of added stimulating factors, substantiated both by quantitative assay and demonstrated by the characteristic collagen reaction products. Little is known of the mechanism of parathormone stimulation of collagenolytic activity in bone cultures. Since the function of this hormone is closely linked with demineralization the latter process may in some manner trigger collagenolysis.

Perhaps one of the most dramatic examples of physiologic collagen resorption in the mammal is observed in the postpartum uterus where, for example in the rat, during the first 3 days after delivery of the fetus uterine collagen is rapidly resorbed with a half-life of about 1½ days. Collagen increases about 10-fold during the course of pregnancy and is totally resorbed in 72 hr (Harkness and Moralee, 1956; Woessner 1962). Jeffrey and Gross (1970) obtained an active collagenase from cultures of 12-hr postpartum rat uterus, partially characterizing it in terms of specificity of attack on native collagen substrates, its neutral pH optimum, and characterization of reaction products. Jeffrey and colleagues (1971*a*) subsequently wrote that the levels of this enzyme in culture medium correlated well with the loss of collagen in the explant and also described its ebb and flow during the 3-day postpartum period. The control of collagen resorption in this system appears to be under close hormonal control. Estradiol and progesterone levels in uterine tissue fall rapidly after delivery. In a corollary experiment Goodall (1966) found that progesterone administered to rabbits immediately following delivery markedly retarded uterine involution, in contrast to the ineffectiveness of estradiol. There does appear to be an important species difference in sensitivity of the uterine collagenolytic system to progesterone (Halme and Woessner, 1975). The addition of this hormone or its potent analog 6α-methyl-17α-acetoxyprogesterone (Provera) in physiologic concentrations to postpartum rat uterine cultures strongly inhibited the appearance of collagenase in the medium of tissues explanted within 20 hr after delivery. Beyond this time neither progesterone nor Provera were inhibitory (Jeffrey *et al.,* 1971*b*). These hormones have no direct effect on the enzyme. In comparison with a near-total block of collagenolytic activity when the hormone is added to the cultures, *in vivo* administration during this period resulted only in an average 40% reduction in the loss of collagen from the involuting uterus. On more detailed examination of this phenomenon Jeffrey and Koob (1973) observed a spotty pattern of collagen resorption in this organ; in some areas there was no evidence of loss, whereas in others removal of collagen appeared to be complete, the hormone having no inhibitory effect. The anatomical or histologic distribution of this variation was not described. Jeffrey and Koob also observed an inhibitory effect of progesterone on collagen resorption from crush-injured uterine tissue in which considerable proliferation of collagen had occurred. In the absence of the hormone rapid and massive resorption of collagen followed the preceding accumulative phase. Koob and Jeffrey (1974) in a further study on the mechanism of progesterone inhibition examined the effect of cyclic AMP and its dibutryl analog, finding only partial inhibition of collagenolysis and collagenase production *in vitro* even at high concentrations. However, the combination of subminimal inhibitory levels of progesterone plus low levels of cAMP or theophylline could completely abolish collagenolytic function with a significantly greater than additive effect. Also of interest was their observation that both the hormone and cAMP added to the culture could partially inhibit collagen degradation in an incubated pellet of homogenate of the uterine explants, a variant of the method of Ryan and Woessner (1971). The authors do not conclude that cAMP and progesterone are necessarily parts of the same pathway in regulating collagenolysis. They do propose the possibility that progesterone acts in a manner analogous to actinomycin D in blocking the synthesis of a transcribed factor needed for biosynthesis of collagenase.

Of additional interest are the observations of Jeffrey *et al.* (1975) that hydrocortisone (10^{-7} M) and dexamethasone (10^{-8} M) completely suppress collagenase activity in cultures of human skin without blocking the enzyme directly. Dibutyryl cAMP (1mM) had almost the same inhibitory effect.

Based on Jeffrey's observation that progesterone interferes with the synthesis of uterine collagenase, Gross and Newsome speculated that the hormone might be useful therapeutically to prevent pathologic tissue ulceration and chose to explore its effect on alkali-induced perforations of the rabbit cornea. Earlier studies had shown that collagenase produced by corneal tissues in response to a variety of injuries, both in rabbits and humans, was closely associated with ulceration and perforation of the cornea. Ophthalmologists had found that the near-continuous topical application of the collagenase inhibitors, cysteine or EDTA, over a period of several weeks could block the progression of corneal damage to perforation (Dohlman and Pfister, 1972; Brown and Hook, 1971). It seemed to us that interference with collagenase production would be a more efficient method of accomplishing the same end result. Without considering the likelihood that there may not be progesterone receptors in corneal tissue (an afterthought), we embarked on a series of experiments designed to test the possible therapeutic usefulness of locally applied and systematically administered Provera in alkali burns of the rabbit cornea (Newsome and Gross, 1975). A highly significant number of Provera treated eyes in all replicated experiments failed to perforate or to develop ulcers to the descemetocele stage as compared with controls. The capacity of Provera to prevent collagenase production was demonstrated qualitatively and quantitatively by culturing standard tissue samples of treated and control cornea on collagen gels. Cyclohexamide and freeze–thawing prevented collagenase production by control tissues, indicating that the released enzyme was neither stored in leukocytes nor on collagen fibers. We also noted that dexamethasone added to the culture substrate almost totally blocked the appearance of collagenolytic activity in burned but untreated corneas, consistent with the observations of Koob *et al.* (1974) on the inhibitory effect of hydrocortisone and dexamethasone on the production of collagenase in normal human skin, rheumatoid synovium, and rat uterus. The possibility that progestational hormones may be of therapeutic value in a variety of human diseases and secondary complications involving collagenolysis has not escaped us.

Many investigators have wondered whether the tissue collagenases may participate significantly in the invasive behavior of malignant tumors. Riley and Peacock (1967) detected collagenolytic activity in a wide variety of human tumors by culturing them on reconstituted collagen gels. Since that time there have been a number of other reports of collagenases isolated from culture medium and extracts of human and animal tumors which had essentially the same characteristics as those obtained from normal tissues. This literature has been recently reviewed by Harris and Krane (1974) and Gross (1974). Since it has not been possible to clearly separate tumor cells from host cells, the question as to whether or not malignant cells make collagenase is unsettled. Harris *et al.* (1972) made the issue clear when they noted that tissue explants consisting primarily of tumor produced rapidly decreasing amounts of enzyme in the culture medium beginning at time of explantation as compared with the surrounding host stroma which, in contrast, released increasing amounts of enzyme into the culture medium with time of incubation. In addition, actinomycin D had no influence on the release of collagenase by the former whereas it effectively blocked enzyme release by the latter. The possibility exists that the tumor cells stimulate active production of collagenolytic enzyme by adjacent host cells, the enzyme then being stored either in the tumor cells or by adsorption to the indigenous collagen fibrils. Numerous intriguing questions may be raised here. Do neoplastic cells release substances which stimulate host cells to produce collagenase, or substances which activate procollagenases produced by the host cells? The recent purification of highly active rabbit carcinoma collagenase by McCroskery *et al.* (1975) and the production of monospecific antisera to purified animal collagenases (Werb and Reynolds, 1975*a*) plus our growing knowledge of collagenase precursors, their mode of activation and inhibitors of this enzyme, should provide the necessary tools for exploring these questions in depth.

D. Procollagenases and Their Activation

Since the earliest observation of collagenolytic activity in cultures of tadpole tissue and its abolition by freeze–thawing the explant or by the addition of inhibitors of protein synthesis, the question of *de novo* synthesis or the activation of a zymogen was moot (Lapière and Gross, 1963). The question was complicated by the nearly simultaneous observation of the inhibitory activity of serum and the subsequent demonstration of an inactive collagenase–α_2-macroglobulin complex partially reversible by denaturing agents such as 3 M thiocyanate. Thus, the inactive form of collagenase might be explained by either the existence of an activator of a zymogen or the physiologic separation of an enzyme inhibitor complex, or both. The evidence is stronger for the existence of both mechanisms in several different systems. In serum-free medium of cultures of bovine gingival tissue (Birkedal-Hansen *et al.,* 1975), in human rheumatoid synovial tissue (Nagai, 1973), in rheumatoid synovial fluid (Harris *et al.*, 1969), and in human skin (Eisen *et al.*, 1971) there appear to be very high-molecular-weight materials which show immunologic identity with collagenase but which are enzymatically inactive. In those studies in which enzyme activity appears after exposure to trypsin, it is not altogether clear whether this represents displacement of collagenase from an α_2M inhibitor complex or activation of a procollagenase.

If there are tissue inhibitors of relatively low molecular weight, i.e., 10–50,000 daltons, the demonstration of proteolytic activation of a precursor will require evidence for cleavage of a peptide bond or at least significant conformational change within the single molecule. The problem would have been simpler if α_2-macroglobulin were the sole, or at least major, inhibitor, since this molecule has a molecular weight of the order of 700,000 daltons and can be identified immunologically. As noted earlier, we now know that there are lower-molecular-weight inhibitors in serum and tissues. If a presumptive precursor molecule can be highly purified, is immunologically homogeneous, has a molecular weight somewhat greater than or equal to that of the active enzyme, and can be activated by proteolytic cleavage, the evidence is strong for its identity as a procollagenase. One would like to see the presumptive zymogen migrate as a single band on SDS electrophoresis, i.e., under strong dissociating conditions, and after exposure to its activator to move again in the same system as a single component with different mobility. If activation is a result of conformational change within a molecule it may be difficult to demonstrate alterations in molecular weight; more subtle measurements of primary or secondary structural alterations will be necessary.

In the case of the tadpole collagenolytic system, Harper *et al.* (1971) detected, by means of antisera prepared to highly purified tadpole collagenase from culture media, an immunologically cross-reactive but enzymatically inactive component in simple saline extracts of tail fin and back skin tissues. Using the same scheme for purification of the enzyme from culture media, they isolated and purified to electrophoretic (SDS-gel electrophoresis) and immunologic (immunodiffusion) homogeneity a protein of about 110,000 daltons which would not bind to fibrous collagen in solution. In contrast, the purified enzyme obtained from tissue culture medium had a molecular weight of about 100,000 and bound strongly to collagen. Exposure to trypsin, chymotrypsin, or plasmin at 37° did not activate the presumptive zymogen, but mixing with third-day culture medium (in contrast to first-day) after removal of all active collagenase did result in the production of active enzyme which increased as a function of time of incubation and with temperature and which could bind to collagen. Although enzyme activity was not usually found in first-day culture medium, a large amount of immunologically cross-reactive material could be measured (Harper and Gross, 1972). There was no evidence for the presence of inhibitory substances in the medium. In addition, a rough quantitative correlation could be made between the decrease in procollagenase with the increase in both active enzyme and activator material as a function of days of incubation in culture. These data were considered (Harper *et al.*, 1971, Harper and Gross, 1972) to be strong evidence for the existence of tadpole procollagenase. A simple demonstration of increase in active enzyme by exposure to protease such as trypsin is not adequate evidence for the presence of a zymogen since, as shown by Sakamoto *et al.* (1972*b*) in the media of mouse bone cultures and later by Birkedal-Hansen and colleagues (1975) in gingival culture medium, trypsin may block the inhibitory activity of serum α_2M competitively. It should be noted here that Nagai *et al.* (1975) have failed to find a high-molecular-weight collagenase in tadpole tissue culture medium, but only one peak of activity at about 40,000 daltons. However, they report recovering only 10% of the total enzyme activity of the medium, in the total eluate of a G200 column. They believe a high-molecular-weight enzyme may reflect adsorption to other tissue components. Birkedal-Hansen and colleagues (1976) report the presence of three peaks of collagenase activity of different molecular weight in alveolar macrophage culture medium which can be reduced to one low-molecular-weight peak by 1 M NaCl, providing evidence for reversible polymerization. This point remains to be conclusively settled.

While the demonstration of significant increases in collagenolytic activity by exposure of crude or semipurified leukocyte collagenase to rheumatoid synovial fluids is suggestive of activation of a proenzyme (Kruze and Wojtecka, 1972; Oronsky *et al.*, 1973), because of the complexity of crude preparations and the absence of any data characterizing either the presumptive zymogen or activator material there still exists the possibility of modulation of an enzyme–inhibitor complex. Similarly, in the case of mouse bone collagenase (Vaes, 1972*b*) there is a need to isolate the presumptive zymogen for characterization before and after activation. The situation is somewhat clearer in the case of the bovine gingival collagenolytic system (Birkedal-Hansen *et al.*, 1975). Second-day culture medium contained no active collagenase, but upon molecular sieve chromatography (Sephadex G-150) two well-separated peaks of activatable collagenase appeared. That in the void volume was associated with α_2-macroglobulin (immunologic identification) and could be activated by exposure to thiocyanate or trypsin. The well-defined fraction eluting slightly ahead of *bona fide* gingival collagenase (mol. wt. 63,000) could be activated only by exposure to trypsin. Of some importance is the fact that tryptic activation required very brief exposures and low concentrations at room temperature, i.e., 4 μg/ml of trypsin for 1 min. This component, however, could not be activated with thiocyanate. Unfortunately the activated material was not rechromatographed to determine whether there was a shift in molecular weight, nor was it purified sufficiently to examine by SDS-gel electrophoresis. Thus, in the gingival collagenolytic system both modes of regulation appeared to exist. Regarding the enzyme–inhibitor complex, one might ask whether this exists *in vivo* or whether the active enzyme and inhibitor are well separated, the complex only forming artifically on mixing in the culture medium. It is likely that an additional collagenolytic moiety is present in the tissue bound to collagen and released on prolonged incubation at 37°C, as has been found in postpartum uterus (Ryan and Woessner, 1971; Jeffrey and Koob, 1973), or isolated by extraction in high salt concentrations, i.e., 1 M NaCl from tadpole tissue, human skin, rheumatoid synovia (Nagai, 1973), and chick and mouse bone (Sakamoto *et al.,* 1973*b*, 1975).

From the technical side, it is worth noting that Birkedal-Hansen could recover 30% of the initial collagenase activity of serum-inhibited enzyme with thiocyanate and 60% by exposure to trypsin. In contrast with synovial collagenase (Abe and Nagai, 1973) where dialysis against thiocyanate inactivated 50% of the collagenase, only 15% of gingival enzyme was destroyed. Another technical point, Woessner (personal communication) was able to release significant amounts of uterine collagenase by a 4-min exposure of rat postpartum uterine homogenate to temperatures of 60° in the presence of 0.1 M CaCl . Nagai (1973) accomplished a similar "heat-shock" release of tadpole collagenase by brief exposures of tissue homogenates to temperatures of 40–42°C.

Of particular interest to this writer is the potential for regulation via the activation of an inactive collagenase precursor. Such a regulatory mechanism is consistent with that known for a variety of enzymatic systems, both the relatively simple digestive enzymatic mechanisms and more complex regulatory processes such as those involved in the activation of complement, in the clotting mechanism, and in the plasminogen–plasmin system. Activation of a tissue enzyme precursor by the digestive enzymes such as trypsin can only be a model system. It does suggest strongly, however, that proteolytic activity is involved. It was therefore with some amusement that the author noted a discussion in the paper by Birkedal-Hansen *et al.* (1975) of the apparent inhibition of gingival collagenase by a mixture of trypsin and excess soybean trypsin inhibitor, from which they conclude that "it is therefore likely that the soybean trypsin inhibitor prevents *contaminating proteases* in the preparation from partly activating the latent collagenase during the radioassay."

It is quite likely that the control of collagen degradation operates through a number of pathways. Rates of collagenolysis are dependent upon collagen type (McCroskery *et al.*, 1975); type I from skin, type II from cartilage, and type III, a fetal collagen from skin, are degraded by highly purified rabbit tumor collagenase at markedly different rates. Tissue temperature is of importance, particularly in a region such as the knee joint which is maintained normally at about 34°C and rises to 38°C in inflamed states (Harris and McCroskery, 1974). Harris and colleagues (1972, 1974) had previously shown a marked correlation in susceptibility of substrate to collagenolytic attack dependent upon both the degree of cross-linking and the temperature. Changes in susceptibility of collagen to collagenase degradation might also be modified as a function of time of development, anatomic location, physiologic states, and pathology because of significant differences in the types and amounts of ground-substance components, such as proteoglycans, in close association with collagen fibers. The tissue complexity is very likely manipulated by enzymatic systems having nothing to do with collagen itself, such as hyaluronidases, noncollagenolytic proteases, and factors regulating rates and types of proteoglycan synthesis.

The role of collagenase inhibitors, both from the serum and possibly indigenous to the tissues, must be explored in greater depth. For example, the tissue distribution of α_2-macroglobulin in remodeling systems may be of considerable significance. We do not as yet know how much, if any, of this serum component is normally or abnormally present in the extracellular space. In the presence of serum, how do cultured bone explants resorb rapidly under the influence of parathyroid hormone or heparin? Since active resorption occurs *in vivo* there must be segregation of serum inhibitors from the enzymes in the tissues. Under what conditions are the collagenases accessible to serum inhibitors *in vivo*?

By what mechanisms do hormones modulate collagen resorption? One might guess that control of synthesis of zymogen or activators is a likely route, either via direct hormonal interaction with activator components or with their precursors at a postsynthetic step. How is biosynthesis of collagen synchronized with degradation in the orderly progression of morphogenesis?

The wide range of biological systems in which collagen metabolism plays an important role and which are also susceptible to modern experimental approaches augurs well for interesting developments to come.

Acknowledgment

Many thanks to Barbara Dewey and Deborah Scharf for their invaluable assistance with this manuscript. This is publication No. 674 of the Robert W. Lovett Memorial Group for the Study of Diseases Causing Deformities. Portions of the work from this laboratory described herein have been supported by National Institutes of Health research grant AM 03564, training grant AM 05067, and generous assistance from Hoffmann La Roche Co., Nutley, New Jersey.

References

Abe, S., and Nagai, Y., 1972, Interaction between tadpole collagenase and human α_2-macroglobulin, *Biochim. Biophys. Acta* **278:**125.

Abe, S., and Nagai, Y., 1973, Evidence for the presence of a complex of collagenase with α_2-macroglobulin in human rheumatoid synovial fluid: A possible regulatory mechanism of collagenase activity in vivo, *J. Biochem.* **73:**897.

Abramson, M., and Gross, J., 1971, Further studies on a collagenase in middle ear cholesteatoma, *Ann. Otol. Rhinol. Laryngol.* **80:**177.

Abramson, M., Huang, C.-C., Schilling, R. W., and Salome, R. G., 1975, Collagenase activity in epidermoid carcinoma of the oral cavity and larynx, *Ann. Otol. Rhino. Laryngol.* **84:**158.

Aswanakumar, S., and Radhakrishnan, A. N., 1973, Studies on a peptidase acting on a synthetic collagenase substrate. Development pattern in rat granuloma tissue and distribution of enzyme in the tissues of various animals, *Biochim. Biophys. Acta* **292:**210.

Barrett, A. J., 1975, The enzymatic degradation of cartilage matrix, *in: Dynamics of Connective Tissue Macromolecules* (P. M. C. Burleigh and A. R. Poole, eds.), p. 188, Amer. Elsevier, New York.

Barrett, A. J., and Starkey, P. M., 1973, The interaction of α_2 macroglobulin with proteinases, *Biochem. J.* **133:**709.

Bauer, E. A., Jeffrey, J. J., and Eisen, A. Z., 1971, Preparation of three vertebrate collagenases in pure form, *Biochem. Biophys. Res. Commun.* **44:**813.

Bauer, E. A., Eisen, A. Z., and Jeffrey, J. J., 1972, Radioimmunoassay of human collagenase, *J. Biol. Chem.* **247:**6679.

Bauer, E. A., Stricklin, G. P., Jeffrey, J. J., and Eisen, A. Z., 1975, Collagenase production by human skin fibroblasts, *Biochem. Biophys. Res. Commun.* **64:**232.

Berman, N., and Dohlman, C., 1975, Collagenase inhibitors, *Arch. Ophthalmol. (Paris)* **35:**95.

Berman, M. B., and Manabe, R., 1973, Corneal collagenases: Evidence for zinc metalloenzymes, *Ann. Ophthalmol.* **5:**1193.

Berman, M. B., Barber, J. C., Talamo, R. C., and Langley, C. E., 1973*a*, Corneal ulceration and the serum antiproteases I. α_1 Antitrypsin, *Invest. Ophthalmol.* **10:**759.

Berman, M. B., Kerza-Kwiatecki, A. P., and Davison, P. F., 1973*b*, Characterization of human corneal collagenase, *Exp. Eye Res.* **15:**367.

Birkedal-Hansen, H., Cobb, C. M., Taylor, R. E., and Fullmer, H. M., 1974, Bovine gingival collagenase. Demonstration and initial characterization, *J. Oral Pathol.* **3:**232.

Birkedal-Hansen, H., Cobb, C. M., Taylor, R. E., and Fullmer, H. M., 1975, Activation of latent bovine gingival collagenase, *Arch. Oral. Biol.* **20:**681.

Birkedal-Hansen, H., Taylor, R. E., and Fullmer, H. M., 1976, Rabbit alveolar macrophage collagenase: Evidence of polymeric forms, *Biochim. Biophys. Acta* **420:**428.

Brandes, D., and Anton, E., 1969, An electron microscopic cytochemical study of macrophages during uterine involution, *J. Gerontol.* **24:**55.

Brem, H., and Folkman. J., 1975, Inhibition of tumor angiogenesis mediated by cartilage, *J. Exp. Med.* **141:**427.

Brown, S. I., and Hook, C. W., 1971, Treatment of corneal destruction with collagenase inhibitors, *Trans. Am. Acad. Ophthalmol. Otol.* **75:**1199.

Burleigh, M. C., Barrett, A. J., and Lazarus, G. S., 1974, *Biochem. J.* **137:**387.

Davis, B. F., Jeffrey, J. J., Eisen, A. Z., and Derby, A., 1975, The induction of collagenase by thyroxine in resorbing tadpole tailfin in vitro, *Dev. Biol.* **44:**217.

Davison, P. F., and Berman, M., 1973, Corneal collagenase: Specific cleavage of types $(\alpha 1)_1\alpha 2$ and $(\alpha 1)_3$ collagens, *Connect. Tissue Res.* **2:**57.

Davison, P. F., and Schmitt, F. O., 1968, The enzymatic digestion of tropocollagen, Hoppe-Seylers Z. *Physiol. Chem.* **349:**119.

Dayer, J. M., Krane, S. M., Russell, G. G., and Robinson, D. R., 1976, Production of collagenase and prostoglandins by isolated adherent rheumatoid synovial cells, *Proc. Natl. Acad. Sci (USA)* **73:**945.

deSaussure, V. H., and Dandliker, W. B., 1969, Ultracentrifuge studies on the effects of thiocyanate ion on antigen-antibody systems, *Immunochemistry* **6:**77.

Dohlman, C. H., and Pfister, R. R., 1972, Management of chemical burns of the eye, *in: Symposium on the Cornea* p. 105, C. V. Mosby, Co. St. Louis.

Donoff, R. B., McLennan, J. E., and Grillo, H. C., 1971, Preparation and properties of collagenases from epithelium and mesenchyme of healing mammalian wounds, *Biochim. Biophys. Acta* **227:**639.

Dresden, M., 1971, Evidence for the role of collagenase in collagen metabolism, *Nature (London), New Biol.* **231:**55.

Dresden, M. H., and Gross, J., 1970, The collagenolytic enzyme of the regenerating limb of the newt, *Triturus viridescens, Develop. Biol.* **22:**129.

Dresden, M. H., Heilman, S. A., and Schmidt, J., 1972, Collagenolytic enzymes in human neoplasms, *Cancer Res.* **32:**993–996.

Eisen, A. Z., 1969, Human skin collagenase: Localization and distribution in normal human skin, *J. Invest. Dermatol.* **52:**442.

Eisen, A. Z., and Gross, J., 1965, Role of epithelium and mesenchyme in the production of a collagenolytic enzyme and a hyaluronidase in the anuran tadpole. *Dev. Biol.* **12:**408.

Eisen, A. Z., and Jeffrey, J. J., 1969, An extractable collagenase from crustacean hepatopancreas, *Biochim. Biophys. Acta* **191:**517.

Eisen, A. Z., Jeffrey, J. J., and Gross, J., 1968, Human skin collagenase. Isolation and mechanism of attack on the collagen molecule, *Biochim. Biophys. Acta* **161:**637.

Eisen, A. Z., Bauer, E. A., and Jeffrey, J. J., 1971, Human skin collagenase. The role of serum α-globulins in the control of activity in vivo and in vitro, *Proc. Natl. Acad. Sci. U.S.A.* **68:**248.

Eisen, A. Z., Henderson, K. D., and Jeffrey, J. J., 1973, A collagenolytic protease from the Hepatopancreas of crab Uca pugilator: purification and properties, *Biochemistry* **12:**1814.

Eisenstein, R., Sorgente, N., Sable, C. W., Miller, A., and Kuttner, K. E., 1973, The resistance of certain tissues to invasion; penetrability of explanted tissues by vascularized mesenchyme, *Am. J. Pathol.* **73:**765.

Evanson, J. M., Jeffrey, J. J., and Krane, S. M., 1968, Studies on collagenase from rheumatoid synovium in tissue culture, *J. Clin. Invest.* **47**:2639.

Folkman, J., 1974, Tumor angiogenesis, in: *Advances in Cancer Research* (G. Klein and S. Weinhauser, eds.), Acad. Press, New York, **19**:331.

Fullmer, H. M., and Lazarus, G. S., 1969, Collagenase in bone of man, *J. Histochem. Cytochem.* **17**:793.

Fullmer, H. M., Gibson, W. A., Lazarus, G. S., Bladen, H. A., and Whedon, K. A., 1969, The origin of collagenase in periodontal tissues of man, *J. Dent. Res.* **48**:646.

Fullmer, H. M., Taylor, R. E., and Guthrie, R. W., 1972, Human gingival collagenase. Purification, molecular weight and inhibitor studies, *J. Dent. Res.* **51**:349.

Fujiwara, K., Sakai, T., Oda, T., and Igarashi, S., 1973, The presence of collagenase in Kupffer cells of the rat liver, *Biochem. Biophys. Res. Commun.* **54**:531.

Gallop, P. M., Seifter, S., and Meilman, E., 1957, Studies on collagen I. The partial purification assay, and mode of activation of bacterial collagenase, *J. Biol. Chem.* **227**:891.

Gnädinger, M. C., Itoi, M., Slansky, H., and Dohlman, C. H., 1969, The role of collagenase in the alkali-burned cornea, *Am. J. Ophthalmol.* **68**:478.

Goodall, F. R., 1966, Progesterone retards post partum involution of the rabbit myometrium, *Science* **152**:356.

Griffith, G. C., Nichols, G., Jr., Asher, J. D., and Flanagan, B., 1965, Heparin osteoporosis, *JAMA* **193**:91.

Grillo, H. C., and Gross, J., 1967, Collagenlytic activity during mammalian wound repair, *Dev. Biol.* **15**:300.

Gross, J., 1969, Some aspects of collagen metabolism, *in: Thule International Symposia. Aging of Connective and Skeletal Tissue,* p. 33, Nordiska Bokhandelns Forlag, Stockholm.

Gross, J., 1970, The animal collagenases, *in: Chemistry and Molecular Biology of the Intercellular Matrix* (E. A. Balazs, ed.), Vol. 3, p. 1623, Academic Press, New York.

Gross, J., 1974, Collagen biology: Structure, degradation and disease, *Harvey Lect.* **68**:351.

Gross, J., and Bruschi, A. B., 1971, The pattern of collagen degradation in cultured tadpole tissues, *Dev. Biol.* **26**:36.

Gross, J., and Lapière, C. M., 1962, Collagenolytic activity in amphibian tissues: A tissue culture assay, *Proc. Natl. Acad. Sci. U.S.A.* **48**:1014.

Gross, J., and Nagai, Y., 1965, Specific degradation of the collagen molecule by tadpole collagenolytic enzyme. *Proc. Natl. Acad. Sci. U.S.A.* **54**:1197.

Gross, J., Harper, E., Harris, E. D., Jr., McCroskery, P., Highberger, J. H., Corbett, C., and Kang, A. H., 1974, Animal collagenases: Specificity of action, and structure of the substrate cleavage site, *Biochem. Biophys. Res. Commun.* **61**:605.

Halme, J., and Woessner, F., 1975, Effect of progesterone on collagen breakdown and tissue collagenolytic activity in the involuting rat uterus, *J. Endocrinol.* **66**:357–362.

Harkness, R. D., and Moralee, B. E., 1956, The time course and route of loss of collagen from the rats uterus during post partum in solution, *J. Physiol. (London)* **132**:502.

Harpel, P. D., 1973, Studies on human plasma $\alpha 2$ macroglobulin–enzyme interactions, *J. Exp. Med.* **138**:508.

Harper, E., and Gross, J., 1970, Separation of collagenase and peptidase activity of tadpole tissues in culture, *Biochim. Biophys. Acta* **198**:286.

Harper, E., and Gross, J., 1972, Collagenase, procollagenase and activator relationships in tadpole tissue cultures, *Biochem. Biophys. Res. Commun.* **48**:1147.

Harper, E., and Gross, J., 1973, The activator of tadpole procollagenase, *Fed. Proc.* **32**:614.

Harper, E., Bloch, K., and Gross, J., 1971, The zymogen of tadpole collagenase, *Biochemistry* **10:**3035.

Harris, E. D., 1972, A collagenolytic system produced by primary cultures of rheumatoid nodule tissue, *J. Clin. Invest.* **51:**2973.

Harris, E. D., and Farrell, M. E., 1972, Resistance to collagenase: A characteristic of collagen fibrils crosslinked by formaldehyde, *Biochim. Biophys. Acta* **278:**133.

Harris, E. D., and Krane, S. M., 1971, Effects of colchicine on collagenase in culture of rheumatoid synovium, *Arthritis Rheum.* **14:**669.

Harris, E. D., and Krane, S. M., 1972, An endopeptidase from rheumatoid synovial tissue culture, *Biochim. Biophys. Acta* **258:**566.

Harris, E. D., and Krane, S. M., 1973, Cartilage collagen: Substrate in soluble and fibril form for mammalian collagenase, *Trans. Assoc. Am. Physicians* **86:**82.

Harris, E. D., and Krane, S. M., 1974, Collagenases, *N. Engl. J. Med.* **291:**557, 605, 652.

Harris, E. D., and McCroskery, P. A., 1974, The influence of temperature and fibril stability on degradation of cartilage collagen by rheumatoid synovial collagenase, *N. Engl. J. Med.* **290:**1.

Harris, E. D., DiBona, D. R., and Krane, S., 1969, Collagenases in human synovial fluid, *J. Clin. Invest.* **48:**2104.

Harris, E. D., Evanson, J. M., DiBona, D. R., and Krane, S. M., 1970, Collagenase and rheumatoid arthritis, *Arthritis Rheum.* **13:**83.

Harris, E. D., Faulkner, C. S., and Wood, S. Jr., 1972, Collagenase in carcinoma cells, *Biochem. Biophys. Res. Commun.* **48:**1247.

Harris, E. D., McCroskery, P. A., Miller, E. J., and Butler, W. T., 1975*a*, Primary structure at the cleavage site of 1(II) by a specific mammalian collagenase, *Proc. III Int. Cong. Rheum.* (in press).

Harris, E. D., Faulkner, C. S., II, and Brown, F. E., 1975*b*, Collagenolytic systems in rheumatoid arthritis, *Clin. Orthop. Relat. Res.* **110:**303.

Harris, E. D., Reynolds, J. J., and Werb, Z., 1975*c*, Cytochalasin B increases collagenase production by cells in vitro, *Nature* **257:**243.

Hashimoto, K., Yamanishi, Y., Maeyesn, E., Dabbous, M. K., and Kanzaki, T., 1973, Collagenolytic activities of squamous cell carcinoma of the skin. *Cancer Res.* **33:**2790.

Hook, C. W., Brown, S. I., Iwanij, V., and Nakanishi, I., 1971, Characterization and inhibition of corneal collagenase, *Invest. Ophthalmol.* **10:**496.

Hook, C. W., Bull, F. G., Iwanij, V., and Brown, S. I., 1972, Purification of corneal collagenases, *Invest. Ophthalmol.* **11:**728.

Hook, R. M., Hook, C. W., and Brown, S. I., 1973, Fibroblast collagenase partial purification and characterization, *Invest. Ophthalmol.* **12:**771.

Huang, C. C., and Abramson, M., 1975, Purification and characterization of collagenase from guinea pig skin, *Biochim. Biophys. Acta* **384:**484.

Hutchins, G. M., and Sheldon, W. H., 1973, pH of inflammatory exudates in granulocytopenic rabbits, *Proc. Soc. Exp. Med.* **143:**1014.

Jaffe, M. D., and Willis, P. W., 1965, Multiple fractures, associated with long term sodium heparin therapy, *JAMA* **193:**158.

Jeffrey, J. J., and Gross, J., 1970, A collagenase from rat uterus: Isolation and partial characterization, *Biochemistry* **9:**268.

Jeffrey, J. J., and Koob, T. J., 1973, Hormonal regulation of collagen catabolism in the uterus, *Endocrinology: Excerpta Med. Int. Cong. Ser.* **273:**1115.

Jeffrey, J. J., Coffey, R. J., and Eisen, A. Z., 1971*a*, Studies on uterine collagenase in tissue culture. I. Relationship of enzyme production to collagen metabolism, *Biochim. Biophys. Acta* **252:**136.

Jeffrey, J. J., Coffey, R. J., and Eisen, A. Z., 1971*b*, Studies on uterine collagenase in tissue culture. II. Effect of steriod hormones on enzyme production, *Biochim. Biophys. Acta* **252:**143.

Jeffrey, J. J., Koob, T. J., and Eisen, A. Z., 1975, Hormonal regulation of mammalian collagenases, *in: Dynamics of Connective Tissue Macromolecules* (P. M. C. Burleigh and A. R. Poole, eds.), p. 147, Amer. Elsevier, New York.

Kang, A. H., Nagai, Y., Piez, K. A., and Gross, J., 1966, Studies on the structure of collagen utilizing a collagenolytic enzyme from tadpole, *Biochem.* **5:**509.

Kaufman, E. J., Glimcher, M. J., Mechanic, G. L., and Goldhaber, P., 1965, Collagenolytic activity during active bone resorption in tissue culture, *Proc. Soc. Exp. Biol. Med.* **120:**632.

Koob, T. J., and Jeffrey, J. J., 1974, Hormonal regulation of collagen degradation in the uterus: Inhibition of collagenase expression by progesterone and cyclic AMP, *Biochim. Biophys. Acta* **354:**61.

Koob, T. J., Jeffrey, J. J., and Eisen, A. Z., 1974, Regulation of human skin collagenase activity by hydrocortisone and dexamethasone, *Biochem. Biophys. Res. Commun.* **61:**3.

Kruze, D., and Wojtecka, E., 1972, Activation of leucocyte collagenase proenzyme by rheumatoid synovial fluid, *Biochim. Biophys. Acta* **285:**436.

Kuboki, Y., Shimokawa, H., Ono, T., and Sasaki, S., 1973, Detection of collagen degradation products in bone, *Calcif. Tissue Res.* **12:**303.

Lapière, C. M., and Gross, J., 1963, Animal collagenase and collagen metabolism, *in: Mechanisms of Hard Tissue Destruction* (R. F. Sognnaes, ed.), No. 75, p. 663, Am. Assoc. Adv. Science, Washington, D.C.

Laskowski, M., Jr., 1972, Interaction of proteinase with proteinase inhibitors, *in: Pulmonary Emphysema and Proteolysis* (C. Mittman, ed.), p. 311, Academic Press, New York.

Laskowski, M., and Sealock, P. W., 1971, Protein–proteinase inhibitors—molecular aspects, *in: The Enzymes*, 3rd ed. (P. Boyer, ed.), Vol. 3, p. 375, Academic Press, New York.

Lazarus, G. S., 1972, Role of granulocyte collagenase in collagen degradation, *Am. J. Pathol.* **68:**565.

Lazarus, G. S., Daniels, J. R., Brown, R. S., Bladen, H. A., and Fullmer, H. M., 1968*a*, Degradation of collagen by a human granulocyte collagenolytic system, *J. Clin. Invest.* **47:**2622.

Lazarus, G. S., Decker, J. L., Oliver, H., Daniels, J. R., Multz, C. V., and Fullmer, H. M., 1968*b*, Collagenlytic activity of synovium in rheumatoid arthritis, *N. Engl. J. Med.* **279:**914.

Leibovich, S. J., and Weiss, J. B., 1971, Failure of human rheumatoid synovial collagenase to degrade either normal or rheumatoid arthritic polymeric collagen, *Biochim. Biophys. Acta* **251:**109.

McCroskery, P. A., Wood, S., and Harris, E. D., 1973, Gelatin: A poor substrate for a mammalian collagenase, *Science* **182:**70.

McCroskery, P. A., Richards, J. F., and Harris, E. D., 1975, Purification and characterization of a collagenase extracted from rabbit tumors, *Biochem J.* **152:**131.

Mancini, G., Vaerman, J-P., Carbonara, A. O., and Heremans, J. F., 1963, A single-radial diffusion method for the immunological quantitation of proteins, *in: Protides of Biological Fluids* (H. Peeters, ed.), p. 370, Proc. 11th Colloquim Bruges, Elsevier, Amsterdam.

Miller, E. J., Finch, J. E., Chung, E., and Butler, W. T., 1976*a*, Specific cleavage of the native type III collagen with trypsin. Similarity of the cleavage products to collagenase cleavage site, *Arch. Biochem. Biophys.* **173:**631.

Miller, E. J., Harris, E. D., Chung, E., Finch, J. E., Jr., McCroskery, P. A., and Butler, W. T., 1976*b*, Cleavage of type II and III collagens with mammalian collagenase: Site of cleavage and primary structure at NH_2 terminal positions of the smaller fragments released from both collagens, *Biochemistry* **5:**787.

Nagai, Y., 1973, Vertebrate collagenase: Further characterization and the significance of its latent form in vivo, *Mol. Cell Biochem.* **1:**137.

Nagai, Y., and Hori, H., 1972, Vertebrate collagenase: Direct extraction from animal skin and human synovial membrane, *J. Biochem.* **72:**1147.

Nagai, Y., Lapière, C. M., and Gross, J., 1964, Purification and characterization of amphibian collagenolytic enzyme, *Sixth Internat. Cong. Biochem.* **II:**135.

Nagai, Y., Lapière, C., and Gross, J., 1966, Tadpole collagenase: Preparation and purification, *Biochemistry* **5:**3123.

Nagai, Y., Hori, H., Kawamoto, T., and Romiya, M., 1975, A regulation mechanism of collagenase activity *in vitro* and *in vivo*, *in: Dynamics of Connective Tissue Macromolecules* (P. M. C. Burleigh and A. R. Poole, eds.), p. 171, Amer. Elsevier, New York.

Newsome, D., and Gross, J., 1975, Treatment of alkali-burned rabbit corneas with medroxyprogesterone, Assoc. for Research in Vision and Ophthal., Inc., Sarasota, Florida (Abstract), p. 23.

Nishihara, T., and Miyata, T., 1962. Effect of proteases on the soluble and insoluble collagen and the structure of the insoluble collagen fiber, *Transactions of the Collagen Research Society of Japan* **3:**66.

Ohlsson, K., and Olsson, I., 1973, The neutral proteases of human granulocytes. Isolation and partial characterization of two granulocyte collagenases, *Eur. J. Biochem.* **36:**473.

Olsen, B. R., 1964, Electron microscopic studies on collagen. III. Tryptic digestion of tropocollagen macromolecules, *Z. Zellforsch.* **61:**913.

Oronsky, A. L., Perper, R. J., and Schroder, H. C., 1973, Phagocytic release and activation of human leucocyte procollagenase, *Nature* **246:**417.

Parakkal, P. F., 1969*a*, Involvement of macrophages in collagen resorption, *J. Cell Biol.* **41:**345.

Parakkal, P. F., 1969*b*, Role of macrophages in collagen resorption during hair cycle growth, *J. Ultrastruct. Res.* **29:**210.

Parakkal, P. F., 1972, Macrophages: Time course and sequence of their distribution in the post-partum uterus, *J. Ultrastruct. Res.* **40:**284.

Pardo, A., and Perez-Tomayo, R., 1974. The collagenase of carrageenan granuloma, *Conn. Tiss. Res.* **2:**243.

Pardo, A., and Perez-Tomayo, R., 1975, The presence of collagenase in collagen preparations, *Biochim. Biophys. Acta* **389:**121.

Perez-Tomayo, R., 1970, Collagen resorption in carrageenan granulomas, *Lab. Invest.* **22:**137.

Perez-Tomayo, R., 1973, Collagen degradation and resorption: Physiology and pathology, *in: Molecular Pathology of Connective Tissues* (R. Perez-Tomayo and M. Rojkind, eds.), p. 323, Marcel Dekker, New York.

Pfister, R. R., McCulley, J. P., Friend, J., and Dohlman, C. H., 1971, Collagenase activity of intact corneal epithelium peripheral alkali burns, *Arch. Ophthalmol.* **86:**308.

Phillips, J., and Dresden, M. H., 1973, A collagenase in extracts of the invertebrate *Bipalium kewense*, *Biochem. J.* **133:**329.

Raisz, L. G., Holtrop, M. E., and Simmons, H. A., 1973, Inhibition of bone resorption by colchicine in organ culture, *Endocrinology* **92:**556.

Reddick, M. I., Bauer, E. A., and Eisen, A. Z., 1974, Immunocytochemical localization of collagenase in human skin and fibroblasts in monolayer culture, *J. Invest. Dermatol.* **62:**361.

Robertson, P. B., Ryel, R. B., Taylor, R. E., Shyu, K. W., and Fullmer, H. M., 1972*a*, Collagenase: Localization in polymorphonuclear leukocyte granules in the rabbit, *Science* **177:**64.

Robertson, P. B., Taylor, R. E., and Fullmer, H. M., 1972*b*, A reproducible quantitative collagenase radiofibril assay, *Clin. Chim. Acta* **42:**43.

Robertson, P. B., Cobb, C. M., Taylor, R. E., and Fullmer, H. M., 1974, Activation of latent collagenase by microbial plaque, *J. Periodont. Res.* **9:**81.

Ryan, J. N., and Woessner, J. F., Jr., 1971, Mammalian collagenase: Direct demonstration in homogenates of involuting rat uterus, *Biochem. Biophys. Res. Commun.* **44:**144.

Sakai, T., and Gross, J., 1967, Some properties of the products of reaction of tadpole collagenase with collagen, *Biochemistry* **6:**518.

Sakamoto, S., Goldhaber, P., and Glimcher, M. J., 1972*a*, Further studies on the nature of the components in serum which inhibit mouse bone collagenase, *Calcif. Tissue Res.* **10:**280.

Sakamoto, S., Goldhaber, P., and Glimcher, M. J., 1972*b*, Maintenance of mouse collagenase activity in the presence of serum protein by addition of trypsin, *Proc. Soc. Exp. Biol. Med.* **139:**1038.

Sakamoto, S., Goldhaber, P., and Glimcher, M. J., 1972*c*, A new method for the assay of tissue collagenase, *Proc. Soc. Exp. Biol. Med.* **139:**1057.

Sakamoto, S., Goldhaber, P., and Glimcher, M. J., 1973*a*, Mouse bone collagenase. The effect of heparin on the amount of enzyme released in tissue culture and on the activity of the enzyme, *Calcif. Tissue Res.* **12:**247.

Sakamoto, S., Sakamoto, M., Goldhaber, P., and Glimcher, M. J., 1973*b*, Isolation of tissue collagenase from homogenates of embryonic chick bones, *Biochem. Biophys. Res. Commun.* **53:**1102.

Sakamoto, S., Sakamoto, M., Goldhaber, P., and Glimcher, M. J., 1974, The inhibition of mouse bone collagenase by lysozyme, *Calcif. Tissue Res.* **14:**291.

Sakamoto, S., Sakamoto, M., Goldhaber, P., and Glimcher, M. J., 1975, Collagenase and bone resorption: Isolation of collagenase from culture medium containing serum after stimulation of bone resorption by addition of parathyroid hormone extract, *Biochem. Biophys. Res. Commun.* **63:**172.

Senior, R. M., Bielefeld, D. R., and Jeffrey, J. J., 1972, Collagenolytic activity in alveolar macrophages, *Clin. Res.* **20:**88.

Severson, A. R., 1969, Mast cells in areas of experimental bone resorption and remodelling, *Br. J. Exp. Pathol.* **50:**17.

Shimizu, M., Glimcher, M. J., Travis, D., and Goldhaber, P., 1969, Mouse bone collagenase: Isolation, partial purification, and mechanism of action, *Proc. Soc. Exp. Biol. Med.* **130:**1175.

Stern, B. D., Mechanic, G. L., Glimcher, M. J., and Goldhaber, P., 1963, The resorption of bone collagen in tissue culture, *Biochem. Biophys. Res. Commun.* **13:**137.

Tokoro, Y., Eisen, A. Z., and Jeffrey, J. J., 1972, Characterization of a collagenase from rat skin, *Biochim. Biophys. Acta* **258:**289.

Unkless, J. C., Gordon, S., and Riech, E., 1974, Secretion of plasminogen activator by stimulated macrophages, *J. Exp. Med.* **139:**834.

Usuku, G., and Gross, J., 1965, Morphologic studies of connective tissue resorption in the tailfin of metamorphosing bullfrog tadpole, *Dev. Biol.* **11:**352.

Vaes, G., 1972*a*, Multiple steps in the activation of the inactive precursor of bone collagenase by trypsin, *FEBS Lett.* **28:**198.

Vaes, G., 1972*b*, The release of collagenase as an inactive proenzyme by bone explants in culture, *Biochem. J.* **126:**275.

Wahl, L., Wahl, S., Mergenhagen, S., and Martin, G. E., 1974, Collagenase production by endotoxin activated macrophages, *Proc. Natl. Acad. Sci. U.S.A.* **71:**3598.

Wahl, L. M., Wahl, S. M., Mergenhagen, S., and Martin, G. E., 1975, Collagenase production by lymphokine-activated macrophages, *Science* **187:**261.

Walker, D. G., Lapiere, C. M., and Gross, J., 1964, A collagenolytic factor in rat bone promoted by parathyroid extract, *Biochem. Biophys. Res. Commun.* **15:**397.

Werb, Z., and Burleigh, M. C., 1974, A specific collagenase from rabbit fibroblasts in monolayer culture, *Biochem. J.* **137:**373.

Werb, Z., and Reynolds, J. J., 1974, Stimulation by endocytosis of the secretion of collagenase and neutral proteinase from rabbit synovial fibroblasts, *J. Exp. Med.* **140:**1482.

Werb, Z., and Reynolds, J. J., 1975*a*, Purification and properties of a specific collagenase from rabbit synovial fibroblasts, *Biochem. J.* **151:**645.

Werb, Z., and Reynolds, J. J., 1975*b*, Immunochemical studies with a specific antiserum to rabbit fibroblast collagenase, *Biochem. J.* **151:**655.

Werb, Z., and Reynolds, J. J., 1975*c*, Rabbit collagenase—immunological identity of the enzymes released from cells and tissues in normal and pathological conditions, *Biochem. J.* **151:**665.

Werb, Z., Burleigh, M. C., Barrett, A. J., and Starkey, P. M., 1974, The interaction of α2-macroglobulin with proteinases binding and inhibition of mammalian collagenases and other metal proteinases, *Biochem. J.* **139:**359.

Wislocki, A. B., and Sognnaes, R. F., 1950, Histochemical reactions of normal teeth, *Am. J. Anat.* **87:**239.

Woessner, J. F., 1962, Catabolism of collagen and non-collagen protein in the rat uterus during post-partum involution, *Biochem. J.* **83:**304.

Woessner, J. F., 1973, Mammalian collagenases, *Clin. Orthop.* **96:**310.

Woolley, D. E., Glanville, R. W., and Crossley, M. J., 1972, Purification and some properties of rheumatoid synovial collagenase, *Scand. J. Lab. Invest.* (Suppl. 29) **123:**37.

Woolley, D. E., Lindberg, K. A., Glanville, R. W., and Evanson, J. M., 1975*a*, Action of rheumatoid synovial collagenase on cartilage collagen different susceptibilities of cartilage and tendon collagen to collagenase attack, *Eur. J. Biochem.* **50:**437.

Woolley, D. E., Glanville, R. W., Crossley, M. J., and Evanson, J. M., 1975*b*, Purification of rheumatoid synovial collagenase and its action on soluble and insoluble collagen, *Eur. J. Biochem.* **54:**611.

Woolley, D. E., Roberts, D. R., and Evanson, J. M., 1975*c*, Inhibition of human collagenase activity by a small molecular weight serum protein, *Biochem. Biophys. Res. Commun.* **66:**747.

Yamanishi, Y., Dabbous, M. K., and Hashimoto, K., 1972, Effect of collagenolytic activity in basal cell epithelioma of the skin on reconstituted collagen and physical properties and kinetics of the crude enzyme *Cancer Res.* **32:**2551.

Yamanishi, Y., Maeyens, E., Dabbous, M. K., Ohyama, H., and Hashimoto, K., 1973, Collagenolytic activity in malignant melanoma physiocochemical studies, *Cancer Res.* **33:**2507.

7
Immunological Studies on Collagen

RUPERT TIMPL

I. Introduction

The first unequivocal evidence that collagen is an antigenic protein was provided by the work of Watson *et al.* (1954). Subsequent to that study there was a steadily increasing interest in this topic (for previous review articles, see Kirrane and Glynn, 1968; O'Dell, 1968). Later on, the rapid progress made on structural aspects of collagen provided the background for more sophisticated immunological approaches. Thus, to date collagen can be considered as a well-characterized protein antigen (Furthmayr and Timpl, 1976)

Many of the previous efforts were initiated in order to elucidate the possible autoantigenic nature of collagen and its relation to chronic inflammatory processes. So far this approach has not been particularly successful. More recently, immunological studies have been mainly based on a demand for specific antibodies which could be used as tools to characterize structural and metabolic properties of collagen. Due to its unique structural features, collagen may also be considered as a model antigen for basic immunological studies (Timpl *et al.*, 1973*b*). These topics will be discussed in the present chapter; the review of literature is, for the most part, restricted to articles which have been published since 1967.

RUPERT TIMPL · Max-Planck-Institut für Biochemie, Martinsried b. München, Germany.

II. Diversity and Localization of Antigenic Determinants on the Collagen Molecule

A. Attempts to Classify Distinct Groups of Antigenic Determinants

For large protein antigens like collagen, it is to be expected that the size and complexity in primary structure is reflected by a comparable diversity of antigenic determinants. Restriction of this heterogeneity in terms of considering certain structural regions as operationally distinct and independent antigenic domains is a convenient approach preceding the more sophisticated studies of single antigenic sites. It was in this context that Sela *et al.* (1967) introduced the terms "sequential antigenic determinants," to characterize sites solely dependent on amino acid sequence, and "conformational antigenic determinants," to describe antigenic regions which also require the exact juxtaposition of amino acid side chains in the three-dimensional space. Applying this scheme to collagen, it is obvious that conformational antigenic determinants should reside in the triple-helical structure, whereas sequential antigenic determinants might be detected by the interaction of antibodies with uncoiled α chains. Another structural feature which lends itself to a subdivision of the sequential antigenic sites in collagen is furnished by the observation that three different sequence regions are found in the α chains: (1) A large central portion of 1011 amino acids which contains glycine in each third position is especially enriched in imino acids and provides the basis for the triple-helical assembly of the constituent chains. (2) Two short regions composed of 9–25 amino acid residues which do not exhibit these sequence regularities are found at both extremities of the α chains. These "appendages" are considered to be nonhelical because they can be removed from the α chains by proteolytic digestion or by specific chemical means without affecting the triple-helical structure of the major part of the molecule.

Based on these structural features, a distinction between three different groups of antigenic determinants of the collagen molecule was made in a recent study (Beil *et al.*, 1973) and is shown schematically in Figure 1. Helical antigenic determinants are those restricted to the triple-helical areas and cannot be demonstrated in uncoiled α chains, as for example, obtained by heat denaturation. Terminal antigenic determinants are considered to be located in the nonhelical sequence regions, and one way to identify them is by their lability to proteolytic digestion (Schmitt *et al.*, 1964). Another group of sequential antigenic determinants is apparently located in the sequence region which forms the triple helix, and

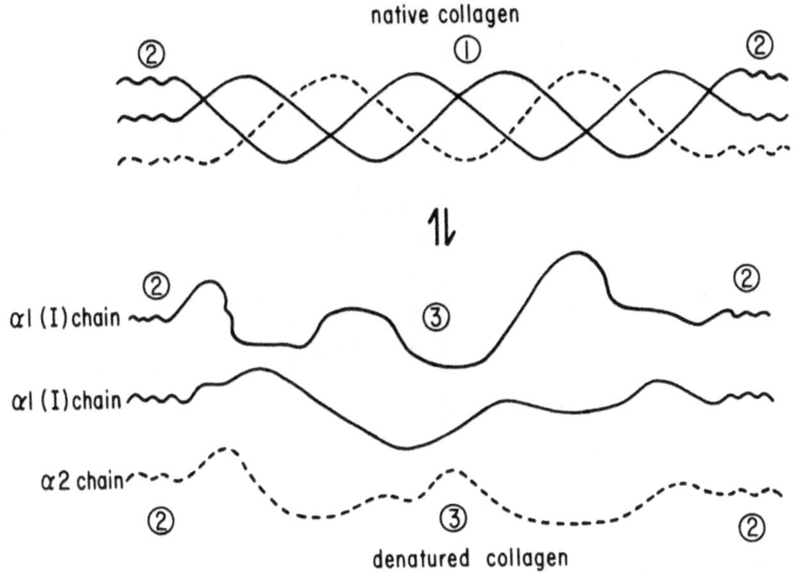

FIGURE 1. Schematic outline of the localization of three different groups of antigenic determinants on the collagen molecule (according to Beil *et al.*, 1973): (1) helical antigenic determinants; (2) terminal antigenic determinants; (3) central antigenic determinants.

these sites are therefore stable against proteases. Since these sites are essentially masked in the triple-helical assembly but become exposed upon denaturation, the term "central antigenic determinants" has been suggested for them.

In earlier studies, a classification of antigenic determinants of the collagen molecule according to interspecies cross-reaction patterns and to reactivity of protease-treated collagen was proposed by Steffen *et al.* (1968, 1971*a*). Antigenic determinants referred to as P-specific obviously correspond to sites in nonhelical sequences, whereas antigenic determinants considered as A- or S-specific are probably located in uncoiled helical sequences (Steffen *et al.*, 1970; Pontz *et al.*, 1970). The presence of antigenic sites uniquely related to the three-dimensional structure of collagen was not considered in this scheme. Since deductions from cross-reaction studies might mainly reflect evolutionary relationships rather than indicate substantial structural differences between groups of antigenic determinants, this experimental approach seems to have some limitations.

A third approach to distinguish between different sequential antigenic determinants was furnished by characterizing the antigenic activity

TABLE 1

Reactivity of Antibodies to Helical Collagen of the Type $[\alpha 1(I)]_2\alpha 2$ *against Chromatographically Purified* α *Chains*

Source of collagen	Antibodies produced in	Reactivity for α chains		Reference
		Strong	Weak or absent	
Rat skin	Rabbit	$\alpha 2$	$\alpha 1(I)$	Michaeli *et al.*, 1969
Rat skin	Rabbit	$\alpha 1(I)$ and $\alpha 2$		Timpl *et al.*, 1970
Rat skin	Rabbit	$\alpha 1(I)$ and $\alpha 2$		Lindsley *et al.*, 1971
Rat skin	Chicken	$\alpha 1(I)$ and $\alpha 2$		Furthmayr *et al.*, 1972
Human skin	Rabbit	$\alpha 2$	$\alpha 1(I)$	Michaeli and Epstein, 1971
Human dura mater	Rabbit	$\alpha 1(I)$	$\alpha 2$	Furthmayr *et al.*, 1971
Calf skin	Rabbit		$\alpha 1(I)$ and $\alpha 2$	LeRoy, 1968
Calf skin	Rabbit	$\alpha 1(I)$ and $\alpha 2$		Pontz *et al.*, 1970
Guinea pig skin	Rabbit	$\alpha 2$	$\alpha 1(I)$	Michaeli *et al.*, 1969
Carp swim bladder	Rabbit	$\alpha 1$ and $\alpha 2$		Meigel *et al.*, 1971

of individual α chains derived from molecules of the chain composition $[\alpha 1(I)]_2\alpha 2$. Studies on the antibody response to guinea pig, rat, and human skin collagen (Michaeli *et al.*, 1969; Michaeli and Epstein, 1971) have indicated the outstanding importance of $\alpha 2$ chains for antigenicity. Other investigators using the same or different collagens demonstrated a considerable antigenic activity of the $\alpha 1(I)$ chain as well. A summary of the present findings is given in Table 1. Although α chains can be obtained in highly purified form, thus providing a reasonable basis for such studies, some arguments against the general value of this approach should be indicated: The finding of a considerable sequence homology between $\alpha 1(I)$ and $\alpha 2$ chains may cause strong interchain cross-reactions at least between sites located in helical sequences (Timpl *et al.*, 1971; Furthmayr *et al.*, 1972*a*). Also the limitation of this approach is obvious for collagens composed of a single type of α chain (Miller, 1973; Kefalides, 1973; Chung and Miller, 1974).

Even though precise information on antigenic sites of collagen is already available, further investigations at this gross molecular level should still be of value for biological studies and for the initial structural characterization of collagen antigens. These experimental approaches have also been of considerable aid in the development of serological techniques and procedures to provide anticollagen antisera with restricted specificity. Immunoadsorption with various collagens was found to be the method of choice for this purpose (Timpl *et al.*, 1967, 1968*a*; Beil *et al.*, 1973). The use of antigen columns containing individual α chains, cyanogen bromide peptides, or synthetic peptides has been suggested as an improvement for the separation of antibody reagents (Lindsley *et al.*, 1971; Timpl *et al.*, 1972; Maoz *et al.*, 1973*b*).

B. *Triple-Helical Structure and Antigenic Specificity*

Although the triple-helical structure of collagen has been known for a long time, it was only recently that unequivocal evidence was obtained for its specific recognition by antibodies (Beil *et al.*, 1973). Rat antisera obtained by immunization with helical calf collagen of the chain composition $[\alpha 1(I)]_2\alpha 2$ showed no reactivity against denatured collagen molecules in an agglutination and inhibition assay, but renaturation of the α chains to the triple-helical structure was accompanied by a large recovery of the original antigenic activity. Treatment of this helical collagen by protease in order to remove nonhelical sequences had no effect on the activity of these antigenic determinants.

In continuation of this study, Hahn and Timpl (1973) demonstrated lack of cross-reaction or only very weak cross-reactions for artificial triple helices of the composition $[\alpha 1(I)]_3$ or $[\alpha 2]_3$ which were prepared by renaturation of the respective purified α chains. Treatment of $[\alpha 1(I)]_2\alpha 2$ molecules with bacterial collagenase (Stark and Kühn, 1968) produced a set of overlapping triple-helical fragments from the central region which still possessed a considerable portion of the original antigenic determinants. Cross-reaction studies with skin collagen from various species indicated a considerable species specificity of the helical antigenic determinants, although some weak reactions, especially with human collagen, also suggested the occurrence of common antigenic sites. No cross-reaction was observed with rat skin collagen. The data suggested the involvement of amino acid side chains from at least two different chains in providing some of the discrete helical antigenic determinants and the expression of multiple specificities along the triple-helical axis of the molecule. The apparently limited interspecies cross-reaction of the helical determinants, as opposed to the broad cross-reaction patterns found for central antigenic determinants in uncoiled α chains, indicates that new species-specific features are acquired by the α chains upon triple-helical folding.

A specific antibody response can also be elicited in the rat against artificial molecules of the composition $[\alpha 1(I)]_3$ and against $[\alpha 1(II)]_3$ molecules which are found in cartilaginous tissues (Hahn *et al.*, 1974). Lack of cross-reaction between both of these antigens, as well as for $[\alpha 1(I)]_2\alpha 2$ collagen, again suggested that triple-helical structure by itself is insufficient to confer antigenic specificity to collagen, but it also requires an exact juxtaposition of discrete amino acid side chains.

Although the apparent restriction of the rat antibody response to helical antigenic determinants endows this model with certain advantages for immunochemical studies, antibodies of comparable specificity could also be detected in chickens immunized with rat collagen (Beil *et al.*, 1972, 1973). These antibodies showed a weak but definite cross-reaction with denatured collagen. Because of the simultaneous occurrence of antibodies to central antigenic determinants, it is not clear as yet whether the results reflect incomplete separation of these antibodies by immunoadsorption or a less precise recognition of helical antigenic determinants by chicken antibody. The easy and reversible denaturation of collagen certainly imposes on such studies technical difficulties which might be overcome to some extent by working at temperatures at which one or the other conformations of collagen are stabilized (Beil *et al.*, 1973). Recent studies (Nowack *et al.*, 1975*b*) showed that the antibody response of mice is also mainly directed to helical antigenic determinants. Adelmann *et al.* (1973), who have used a radioimmune assay, provided similar evidence for the guinea pig.

Studies by Davison *et al.* (1967) on the specificity of rabbit antibodies to calf collagen also demonstrated a dependence of serological activity on triple-helical structure. However, treatment of the antigen by pepsin or pronase in this case destroyed the antigenic activity, suggesting the localization of the antigenic sites at the very terminal sites.

C. *Antigenic Determinants of the Terminal, Nonhelical Regions*

The observation of Schmitt *et al.* (1964) that a prior pepsin treatment of calf collagen largely abolished its reactivity with rabbit antibodies was the first indication of the presence of important antigenic sites in the terminal regions of collagen. This result was confirmed in many subsequent studies (Davison *et al.*, 1967; Steffen *et al.*, 1968; Pontz *et al.*, 1970; Timpl *et al.*, 1970*a*; Furthmayr *et al.*, 1971). Since SLS segments of pepsin-treated collagen did not show an appreciable shortening of the molecule, Pontz *et al.* (1970) concluded that the C-terminal antigenic region of calf collagen is located within the last 15 amino acid residues. This conclusion was later confirmed by precise sequence studies (Becker *et al.*, 1972). More recent investigations using monospecific antibody reagents demonstrated that pepsin, although readily acting on nonhelical C-terminal sequences, did not destroy the N-terminal antigenic region in rat collagen $\alpha 2$ chain (Beil *et al.*, 1973). Proteases of broader specificity, like pronase, are therefore recommended for such studies (Davison *et al.*, 1967; Beil *et al.*, 1973), although they might also act at sites on the helical body (Kühn and Eggl, 1966).

Studies on terminal antigenic determinants have mainly been restricted to $[\alpha 1(I)]_2\alpha 2$ collagens and their corresponding rabbit antibodies. Since collagen of this type contains four different nonhelical regions, characterization of the protease lability of the antigenic sites did not provide sufficient information on their exact location. Most, if not all, of these terminal antigenic determinants are of sequential nature and could therefore be demonstrated on individual α chains (Michaeli *et al.*, 1969; Pontz *et al.*, 1970; Timpl *et al.*, 1970*a*; Lindsley *et al.*, 1971; Furthmayr *et al.*, 1971; Michaeli and Epstein, 1971; Timpl *et al.*, 1972; Rauterberg *et al.*, 1972). The characterization of the serologic activity of CNBr peptides derived from the α chains was first applied by Michaeli *et al.* (1969) and is now considered as the most convenient way to identify and localize

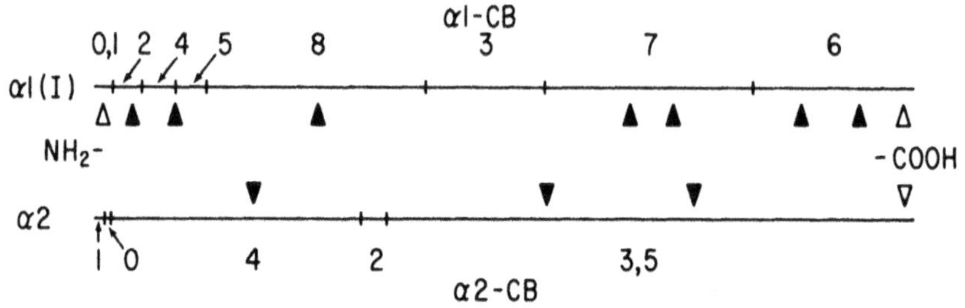

FIGURE 2. Order and relative size of the cyanogen bromide peptides in calf-skin collagen α1(I) and α2 chain and approximative localization of terminal (△) and central (▲) antigenic determinants. The immunological data are taken from Pontz *et al.* (1970), Timpl *et al.* (1971*a*), and Rauterberg *et al.* (1972).

terminal antigenic determinants. In Figure 2, the order of the CNBr peptides in calf collagen α chains, as well as the localization of their terminal antigenic determinants, is illustrated. Table 2 summarizes the present findings on collagens obtained from different species. These results suggest that each nonhelical sequence region is equivalent in being recognized as an antigenic determinant in the antibody response. Whether or not some apparent differences in dependence on the species source reflect insufficient search for respective antibodies or is due to a high

TABLE 2
Localization of Terminal Antigenic Determinants on [α1(I)]$_2$α2 Collagen

CNBr peptide	Location	Species source of collagen	References
α1-CB(0,1)	N terminal α1(I) chain	Calf	Rauterberg *et al.*, 1972
		Human	Furthmayr *et al.*, 1971; Michaeli and Epstein, 1971
α2-CB1	N terminal α2 chain	Guinea pig	Michaeli *et al.*, 1969
		Rat	Lindsley *et al.*, 1971; Timpl *et al.*, 1972
α1-CB6	C terminal α1(I) chain	Calf	Pontz *et al.*, 1970
		Rat	Timpl *et al.*, 1970; Lindsley *et al.*, 1971
		Human	Furthmayr *et al.*, 1971
α2-CB5	C terminal α2 chain	Calf	Pontz *et al.*, 1970; Becker *et al.*, 1972
(α2-CB(3,5))		Rat	Timpl *et al.*, 1970; Lindsley *et al.*, 1971

similarity of sequence to rabbit collagen (Bornstein and Nesse, 1970) is not yet clear as will be discussed to some extent later on in this chapter.

Initial studies with CNBr peptides used these peptides as antigens to inhibit agglutination of α-chain-coated red cells (Michaeli *et al.*, 1969; Pontz *et al.*, 1970; Michaeli and Epstein, 1971). This approach probably detects major antigenic sites only. Subsequent studies employed either monospecific antibody solutions (Timpl *et al.*, 1972), red cells coated with individual CNBr peptides (Timpl *et al.*, 1970*a*; Furthmayr *et al.*, 1971), or labeled peptides in radioimmunoassays (Lindsley *et al.*, 1971). The more elaborate techniques used in these experiments demonstrated the possibility of a simultaneous antibody response against different terminal antigenic determinants in an individual animal. Appropriate cross-reaction studies revealed the uniqueness of each terminal antigenic region within the same collagen molecule (Michaeli *et al.*, 1969, 1971; Furthmayr *et al.*, 1971; Lindsley *et al.*, 1971; Timpl *et al.*, 1972; Rauterberg *et al.*, 1972), a finding which is in accordance with the available data on the amino acid sequences of the nonhelical regions (Traub and Piez, 1971). Some doubts still exist about a possible cross-reaction between the C-terminal sites in the α1(I) and α2 chain (Pontz *et al.*, 1970; Timpl *et al.*, 1970*a*; Lindsley *et al.*, 1971). Since the corresponding amino acid sequence of the α2 chain has not yet been elucidated, it might be that differences in the immunological results reflect degradation of this site.

The demonstration of serological activity for the short N-terminal peptides α1-CB(0,1) (or α1-CB1) and α2-CB1 is usually considered as good evidence for a localization of the antigenic determinant in the nonhelical region because only three amino acids in those peptides originate from the helical sequence. The nonhelical region of the C-terminal peptides α1-CB6 and α2-CB(3,5) (or α2-CB5) accounts for just 10% or less of the entire sequence. Additional proof either by demonstrating protease lability (Pontz *et al.*, 1970) or by further fragmentation to appropriate smaller peptides (Becker *et al.*, 1972) is therefore required to obtain definite evidence for location of the antigenic sites in the nonhelical region. The presence of antigenic sites involving parts of the nonhelical region, as well as adjacent helical areas (Timpl *et al.*, 1971; Stoltz *et al.*, 1973), may sometimes complicate a clear distinction between terminal and central or helical antigenic determinants.

Several studies using either protease lability (Davison *et al.*, 1967; Timpl *et al.*, 1968*b*) or the localization of the terminal antigenic determinants on respective CNBr peptides (Michaeli *et al.*, 1969; Timpl *et al.*, 1972; Rauterberg *et al.*, 1972) as criteria agreed on a limited interspecies cross-reaction between terminal antigenic sites. So far no cross-reaction with rabbit collagen could be demonstrated for rabbit antibodies directed to terminal sites on foreign collagen. These observations seem to be understandable with respect to the known high interspecies variability in amino acid sequence of the nonhelical regions (Traub and Piez, 1971). Recent observations on a possible degradation of nonhelical sequences during extraction indicate, however, that some of the reported findings may reflect loss of potentially cross-reacting antigenic determinants rather than lack of a cross-reaction. Precise structural studies on a C-terminal antigenic determinant (see next paragraph) have already led to a revision of previous observations (Timpl *et al.*, 1968*b*).

Becker *et al.* (1976) recently identified a terminal antigenic determinant in bovine type III collagen. This determinant is located in a nonhelical segment of about 14 amino acids at the aminoterminal end of $\alpha 1$(III) chain and showed only little cross-reaction with $\alpha 1$(I) and $\alpha 2$ chain. Preliminary evidence also exists for terminal determinants in pepsin-treated type II collagen (Hahn *et al.*, 1975*a*).

Although the localization of certain terminal antigenic determinants on fragments as small as pentapeptides (Furthmayr and Timpl, 1972; Rauterberg *et al.*, 1972) left no doubt as to their sequential nature, there is still no clear evidence whether the triple-helical structure can influence their binding with antibodies. The experimental information ranges from showing an equal accessibility on helical and uncoiled collagen (Beil *et al.*, 1973) and better accessibility on uncoiled collagen (Lindsley *et al.*, 1971) to a strict dependence on intact conformation, at least in the case of calf collagen (Davison *et al.*, 1967). The latter finding is in contradiction with the results of Becker *et al.* (1972) who located the major antigenic site of calf collagen entirely in a nonhelical sequence region. In the absence of better structural information, one is tempted to conclude that Davison *et al.* (1967) have investigated a different antigenic site which, as in rat collagen (Stoltz *et al.*, 1973), might involve part of the adjacent helical region.

D. *Antigenic Sites Exposed by Unfolding Collagen*

The demonstration of residual serological activity on α chains obtained from protease-treated collagen (Davison *et al.*, 1967; Steffen *et al.*, 1968; Timpl *et al.*, 1968*a*) was the first evidence for presence of antigenic sites in the central sequence region of uncoiled α chains. The strong reaction of respective antibodies with denatured collagen, but not with helical collagen (Steffen *et al.*, 1967*b*; Beil *et al.*, 1973), suggested that these antigenic determinants became essentially unmasked upon unfolding the triple-helical structure. For that reason, the term "central antigenic determinants" has been introduced (Beil *et al.*, 1973) to distinguish them from the helix-dependent antigenic sites.

Strong support for the existence of such antigenic determinants came from demonstrating the serologic activity of CNBr peptides of the central portion of α chains in calf and rat collagens (Timpl *et al.*, 1971; Furthmayr *et al.*, 1972*a*), human collagen (Michaeli and Epstein, 1971; Furthmayr *et al.*, 1971), and guinea pig collagen (Michaeli *et al.*, 1971). The results so far indicate that almost every CNBr peptide carries central antigenic determinants, and the respective findings for calf collagen are shown in Figure 2. Since antibodies against these sites are found in very low concentration in rabbit antisera (Timpl *et al.*, 1971), it might explain the failure of Lindsley *et al.* (1971) to obtain evidence for these antigenic sites by radioimmune assay. Chicken antisera obviously provide a better source for such antibodies (Beil *et al.*, 1972).

Structural approaches to these antigenic sites are still limited (Timpl *et al.*, 1971; Furthmayr *et al.*, 1972*a*) and point to the following conclusions: Each CNBr peptide from the central region contains at least one unique antigenic determinant, but the number of different antigenic determinants per peptide is certainly greater than one. Probably, there is a large variety of central antigenic determinants which are yet insufficiently recognized. A few studies (Pontz *et al.*, 1970; Timpl *et al.*, 1971; Michaeli and Epstein, 1971; Furthmayr *et al.*, 1972*a*) agreed on a distinctly or moderately better antigenic activity of $\alpha2$ chains than that of $\alpha1$(I) chains, but opposite findings were reported as well (Furthmayr *et al.*, 1971). There is also some indication on a possible cross-reaction between $\alpha1$(I) and $\alpha2$ chains which might be expected on the basis of large stretches of identical sequences present in them (Fietzek and Kühn, 1976). In spite of a weak reaction with antibodies to central antigenic determinants by passive cutaneous anaphylaxis, inhibition tests did not reveal measurable activity for polymeric tripeptides of the sequence Gly-Pro-Pro or Gly-Pro-Ala (Timpl *et al.*, 1971). This is at variance with conclusions based on studies employing multiple proteolytic digestion of collagen (Steffen and Timpl, 1970) which suggested that such typical collagen sequences are important in central antigenic determinants.

Despite the relative unimportance of central antigenic determinants on skin (type I) collagen (Timpl *et al.*, 1971), a strong response against such sites was observed in the rabbit after immunization with cartilage (type II) collagen (Hahn *et al.*, 1975*a*). One of the major antigenic sites could be detected on the CNBr peptide $\alpha1$(II)-CB11 which is homologous to $\alpha1$(I)-CB8 (see Figure 2).

Experimental work to date has shown a high interspecies cross-reactivity for antibodies to central antigenic determinants (Timpl *et al.*, 1968; Steffen *et al.*, 1971; Furthmayr *et al.*, 1972; Hahn *et al.*, 1975*a*), and this is in agreement with the high evolutionary stability of helical collagen sequences (Traub and Piez, 1971; Fietzek and Kühn, 1976). A considerable reactivity was also found for denatured collagen or α chains from the animal species used to produce these antibodies. This observation was instrumental in introducing the term "antigenic determinants of general collagen specificity" (Steffen *et al.*, 1968, 1971*a*). Studies at the level of α-chain fragments, however, demonstrated that cross-reaction is far from being complete for each antigenic determinant (Pontz *et al.*, 1970; Timpl *et al.*, 1971).

A recent study (Beil *et al.*, 1973) also indicated that chicken antibodies which showed strong reaction for CNBr peptides from the central region (Furthmayr *et al.*, 1972*a*) could react with helical collagen to a limited extent. This demonstrates that an absolute distinction between central and helical antigenic determinants cannot be made for every model. The high imino acid content of the helical sequences certainly imposes, even in the uncoiled state, some constraints on the flexibility of the α chains which may account for these findings. Immunization of chicken with denatured instead of helical collagen (Beil *et al.*, 1973) which did not reveal measurable reaction with helical collagen might provide a better source for such antibodies.

E. *Immunological Specificity of Fish and Invertebrate Collagen*

Several immunological studies with collagen from the carp swim bladder (Rothbard and Watson, 1965; Davison *et al.*, 1967; Wolff *et al.*, 1970; Meigel *et al.*, 1971) agreed on an exclusive or prevalent specificity of the antibodies for fish collagen but a poor discrimination between collagens from various fish species. Evidence could also be obtained for some minor antibody populations still cross-reacting with mammalian collagen (Wolff *et al.*, 1970; Meigel *et al.*, 1971). The lack of dependence of the antigenic sites on triple-helical structure led to the suspicion that impurities were involved in the serological reaction (Davison *et al.*, 1967). The finding of an equal reactivity for chromatographically purified $\alpha 1(I)$ and $\alpha 2$ chains in another study (Meigel *et al.*, 1971) did not support this opinion.

Meigel *et al.* (1971) also investigated collagen from two invertebrate species, the liver fluke and sea anemone. The triple helix of the collagen from the latter source is obviously assembled from only a single type of α

chain (Nordwig *et al.*, 1973). For both antigens, a highly specific antibody response could be detected with uncoiled collagen. The cross-reactions between both invertebrate collagens as well as with mammalian and fish collagen were quite weak. An antigenic determinant of sea anemone collagen could be located on a CNBr peptide from the helical region of the α chain (Nowack *et al.*, 1974). The isolation of a serologically active tryptic peptide allowed the conclusion that two structural pecularities of sea anemone collagen, i.e., its high content of 3-hydroxyproline and of carbohydrate attached to hydroxylysine, cannot play a role in this antigenic determinant.

A few studies have been concerned with the immune response to *Ascaris* cuticle collagen which is quite different from all the collagens as yet discussed in terms of conformation and subunit assembly. McBride and Harrington (1967*a,b*) suggested a reverse folding of polypeptide chains of 62,000 mol. wt. to a triple-helical arrangement; linking of this subunit by disulfide bridges provides the final, native collagen molecule with a molecular weight of 900,000 daltons. Fuchs and Harrington (1970) demonstrated complete cross-reaction between the native collagen and the reduced subunits by the aid of antibodies to native *Ascaris* collagen. Maoz *et al.* (1971) found no cross-reaction of antibodies to the reduced subunits with the disulfide-linked, native molecule. Two different subunits could be separated by chromatography but gave identical serologic results. Both kinds of antisera exhibited no cross-reaction with mammalian collagen (Michaeli *et al.*, 1972). In a recent study of Maoz *et al.* (1973*b*), strong cross-reactions with polymeric Pro-Gly-Pro indicated the involvement of such sequences in the antigenic determinants of *Ascaris* cuticle collagen.

F. Procollagen

Procollagen, the precursor form of collagen, contains additional peptide segments connected to the C- and N-terminal ends of the α-chains, the function of which is still not elucidated (reviewed in Chapter 5, and by Schofield and Prockop, 1973; Martin *et al.*, 1975). Hence, the constituent polypeptide chains are designated as pα chains or pro-α chains to distinguish them from the collagen α chains. Chemical studies on isolated peptide extensions (Furthmayr *et al.*, 1972*b*; von der Mark *et al.*, 1973; Sherr *et al.*, 1973; Kohn *et al.*, 1974) demonstrated large differences when compared to the helical sequences of the α chains and a moderately high content of cystine. These findings stimulated immunological studies on procollagen, not only from the viewpoint of searching for procollagen-specific antibodies, but also in an effort to answer the question of how an attached globular protein unit might modify the immune response to collagen.

One of the first immunological studies described the antibody response to procollagen which accumulates in the skin of dermatosparactic calves (Timpl *et al.*, 1973*a*). The pro-α chains of this material are unlike procollagen used in subsequent studies. They are not cross-linked by disulfide bonds which probably reflects partial degradation in the extracellular space. An immediate and prevalent response to this kind of collagen and only weak reactions with $[\alpha 1(I)]_2\alpha 2$ collagen was found, and antibodies specific for procollagen could be obtained upon immunoadsorption. These antibodies reacted equally strong with pro-α1 and pro-α2 chains, although some indications were presented on their serologic dissimilarity. The characteristic antigenic determinants of the pro-α1 chain could be recovered on a collagenase-derived peptide extension which, after complete reduction and alkylation, lost its antigenic activity entirely. Results of gel precipitation were interpreted as indicating the presence of only a few antigenic determinants on this extension.

Von der Mark *et al.* (1973) used purified pro-α1 chains from embryonic chick cranial bones as antigen and observed at later stages in the immunization course a specific response to this antigen without cross-reaction with α chains. In inhibition and gel precipitation tests the antigenic determinants were localized on peptide extensions obtained from pro-α1 chains by cyanogen bromide and collagenase cleavage. The frequent observation of a better reactivity of reduced peptide extensions obviously reflects the use of reducing agents during the extraction of pro-α1 chains. Thus, these antibodies might recognize some antigenic determinants on uncoiled sequences but, since they reacted with disulfide-bonded procollagen, apparently also exhibit specificity for the native antigen.

In their studies on human procollagen, Sherr and Goldberg (1973) used serum-free medium from fibroblast cultures for immunization. Such antisera precipitated 45% of the labeled procollagen from the medium, and the study of dissolved immunoprecipitates indicated the disulfide-linked nature (pro-γ chains) of the antigen. Since no reaction with degraded pro-α chains and with α chains was observed, the authors concluded that the antigenic determinants are localized near the N-terminal end of the procollagen extension. In a subsequent study (Sherr *et al.*, 1973), three antigens could be demonstrated by the antibodies in the culture medium, but only one antigen was labeled. The isolated disulfide-linked peptide extension showed only a single band in gel diffusion.

Native chicken tendon procollagen which contained a high proportion of disulfide-linked pro-α chains was the antigen used by Dehm *et al.*

(1974) to produce antibodies in the rabbit. These antibodies were purified on an immunoadsorbent prepared from the disulfide-linked peptide extension and revealed a high degree of procollagen specificity. About 60% of the procollagen found in culture medium of tendon cells could combine with this antibody, which corresponds to the proportion of disulfide-bonded procollagen in such preparations. Partial reduction of the peptide extension to monomeric moieties did not effect the antigenic activity, but a complete cleavage of the disulfide bonds abolished their reaction with the antibodies. Staining by peroxidase-labeled antibodies of electrophoretically separated, partially reduced pro-α chains indicated a preferential localization of the antigenic determinants on the pro-α1 chain

Recently, convincing evidence was obtained (Byers *et al.*, 1975) that the disulfide-bonded procollagen extension is connected to the C-terminal portion of chains instead of to the N-terminal end. Thus, in the studies of Sherr and Goldberg (1973) and Dehm *et al.* (1974) apparently antibodies were detected which are directed against the C-terminal procollagen peptide, whereas antibodies to dermatosparactic procollagen or to chick procollagen pro-α1 chain (Timpl *et al.*, 1973; von der Mark *et al.*, 1973) react against determinants in the N-terminal extension which is not disulfide linked. Previous observations on the lack of cross-reaction between disulfide bonded extension and pro-α1 chain (Sherr and Goldberg, 1973; Nist *et al.*, 1975) may now be interpreted as indicating different sets of antigenic determinants on both the N-terminal and C-terminal extension. Antigenic determinants on both ends of the molecule were indicated in a recent study on procollagen isolated from dermatosparactic cell cultures (Park *et al.*, 1975). Taubman *et al.* (1974) could also demonstrate by radioimmune assay a free C-terminal extension in human serum.

More comprehensive data on the structure and antigenicity of the N-terminal segments became available during the last year. In pro-α1(I) chain the precursor-specific segment consists of a globular and a collagenous domain (Becker, Timpl, Helle, and Prockop, in preparation). Antibodies could be produced by immunization with the globular peptide and related antigens were detected in biological fluids by radioimmune assay (Rohde *et al.,* 1976). Characterization of the procollagen peptide from amniotic fluid showed that it was also composed of a globular and collagenous domain (Nowack *et al.*, 1976*b*). Only little antigenic activity was found in this collagenous region. Beside conformational determinants which are lost upon reduction, the globular part possesses also weak sequential determinants. The precursor-specific segment of pro-α2 chain consists mainly of a collagen-like sequence and lacked immunogenic activity (Rohde, Becker and Timpl, unpublished). Antibodies could also be prepared against type III procollagen and showed no or only weak reactions with type I procollagen (Lee *et al.*, 1975; Nowack *et al.*, 1976*c*). Like procollagen from dermatosparactic animals type III procollagen from fetal bovine skin has an intact N-terminal segment but lacks the C-terminal region (Timpl *et al.*, 1975). The determinants unique to type III procollagen are located in its N-terminal segment, which consists of three identical chains connected to each other by disulfide bridges. Complete reduction of disulfide bonds destroyed antigenicity, which indicates that the antigenic determinants are located in the globular but not in the collagenous domain of the precursor-specific sequences (Nowack and Timpl, unpublished).

G. Chemically Modified Collagen

Attachment of tyrosyl residues onto collagen (Kirrane and Robertson, 1968) and gelatin (Sela and Arnon, 1960) was found to improve the immunogenicity of these weak antigens. Depending on the degree of substitution, the antibody response against tyrosylated gelatin became mainly directed towards these new groupings (Arnon and Sela, 1960; Givol and Sela, 1964). The easy change in antigenic specificity of collagen by limited structural modifications was also found by Jasin and Glynn (1965), who reported on some cross-reaction between antibodies to acetylated collagen and synthetic polymers containing acetylated hydroxyproline. Steffen *et al.* (1967*a*) showed that rabbit antisera to hydroxylamine-treated collagen are apparently specific for the cleavage sites since the antibodies could be effectively inhibited by aspartic acid hydroxamate. The high inhibitor concentrations required suggest the involvement of further adjacent amino acids in these antigenic sites.

Collagen seems to be a particularly good model for the study of the effect of chemical treatments on antigenic specificity since, unlike the globular proteins, even extensive modification of acidic and basic amino acid side chains does not seriously affect the conformation of collagen (Rauterberg and Kühn, 1968). Studies on the immune response against heavily succinylated or methylated helical collagen revealed a prevalence of antibodies towards haptenic groups which could be inhibited by succinylated ϵ-amino caproic acid or by γ-methyl glutamic acid (Timpl *et al.*, 1970*b*). The reduced response to the genuine antigenic determinants of collagen, even though they were not blocked by the chemical treatment, suggested antigenic competition in favor of the new haptens. In another study, Paz and Seifter (1972) reacted collagen with the drug hydralazine (1-hydrazinophthalazine) which apparently binds to lysine-derived aldehydes. In spite of a low substitution (0.5–0.8 groups per α chain), part of the antibodies obtained against this conjugate seemed specific for the hapten and cross-reacted also with hydralazine coupled onto different protein carriers. As concluded for the succinyl and methyl haptens, peptide sequences around the hydralazine determinant obviously determine the specificity of these antigenic sites to some extent.

H. *Role of Carbohydrate Moieties*

Although collagen, depending on tissue origin, contains variable amounts of glucosylgalactosyl or galactosyl residues exclusively attached to hydroxylysine, no detailed study has as yet been undertaken to define their function as antigenic determinants in experimental models. At least in rabbit and chicken antisera to $[\alpha1(I)]_2\alpha2$ collagen only a weak or moderate reaction with the CNBr peptides α1-CB5 and α2-CB4 was found (Timpl *et al.*, 1971; Furthmayr *et al.*, 1972*a*). Since those peptides include the main sites of carbohydrate attachment (Aguilar *et al.*, 1973), these results do not support a significant role of glycosyl residues in the antibody response to this type of collagen, but neither do they exclude their involvement. More likely candidates to detect such antigenic determinants are antibodies to cartilage (Hahn *et al.*, 1975*a*) or basement membrane collagen (Denduchis and Kefalides, 1970), antigens known to contain much more of these carbohydrate units.

Evidence for a potential antigenic role of the collagen type of carbohydrate came from recent work of Mahieu *et al.* (1973) on antibodies from patients with Goodpasture's syndrome. These antibodies could be specifically inhibited by a glycopeptide containing glucosylgalactosyldisaccharide attached to hydroxylysine in the sequence Hyl-Gly-Glu-Asp-Gly. Galactosylhydroxylysine was also found to be an equally effective inhibitor indicating a quite small size for the antigenic determinant. Weak cross-reactions with the collagen $\alpha1(I)$ chain were explained by the low content of this carbohydrate and/or the differences in the adjacent amino acid sequence.

III. *Amino Acid Sequence of Antigenic Determinants*

The correlation between amino acid sequence and immunological activity is considered as one of the ultimate goals in investigating protein antigens. The feasibility of this approach for collagen was indicated in studies which demonstrated persisting activity for binding of antibody on small peptides obtained by degradation of collagen with collagenase, trypsin, or other proteases (Steffen and Timpl, 1965; Kettman *et al.*, 1967; Kirrane and Robertson, 1968; Steffen *et al.*, 1970). It was, however, the restriction of antigenic determinants to distinct CNBr peptides which paved the way for a more effective analysis. These studies suggested that the terminal antigenic determinants are the most promising target because of their probable restriction in heterogeneity and because of a relatively strong antibody response to these sites in the rabbit.

At present, the more precise localization of the following terminal antigenic determinants is known: in the N-terminal region of rat collagen $\alpha2$ chain (Furthmayr and Timpl, 1972), in the N-terminal region of the calf and human collagen $\alpha1$(I) chain (Rauterberg *et al.*, 1972), and in the C-terminal region of the calf collagen $\alpha1$(I) chain (Becker *et al.*, 1972). The basic findings are illustrated in Figure 3. The methods applied involved degradation of the respective CNBr peptides by proteases and comparison of the fragments in inhibition assays. Thus, a possible minimal size of the sequence required for efficient binding with the antibodies could be defined. Lack of appropriate proteases in some cases raises ambiguity as to the involvement of one or more adjacent amino acids in particular antigenic sites. A possible maximal size was found to involve five or six amino acids for some of the sites studied which is in accordance with findings described for other protein antigens (Benjamini *et al.*, 1972). This suggests that the regions indicated in Figure 3 indeed represent single antigenic determinants. In connection with other studies (Timpl *et al.*, 1972; Beil *et al.*, 1973), no measurable decrease in serologic activity was observed when the antibodies were reacted either against helical collagen, α chains, CNBr peptides, or small proteolytic fragments.

A further interesting finding was the variability in the recognition of distinct amino acid sequences in the antibody response of individual animals. With eight different antisera, three overlapping antigenic regions could be defined at the N-terminal end of the rat collagen $\alpha2$ chain (Furthmayr and Timpl, 1972). Recent results on the cross-reaction between C-terminal antigenic determinants of the $\alpha1$(I) chain (Timpl *et al.*, 1973*b*) may be interpreted in the same manner. This variation in the immunological recognition of the nonhelical areas may reflect genetic dependence. Anticipating that the response to single antigenic determinants is still of polyclonal nature, individual recognition may well reflect variations in the proportions of the products of different immune cell clones. This would then probably relate to complex events in the regulation of the antibody response rather than to genetic differences.

α2-CB1	rat	PCA-Tyr-Ser-Asp-Lys-Gly-Val-Ala-
	rabbit	PCA-Phe-Asx-Gly-Lys-Gly,Gly,Gly-
α1-CB(0,1)	calf	PCA-Leu-Ser-Tyr-Gly-Tyr-Asp-Glu-Lys-Ser-Thr-Gly-Ile-
	rabbit	PCA-Met-Ser-Tyr-Gly-Tyr-Asp-Glu-Lys-Ser-Ala-Gly-Val-
α1-CB6	calf	-Tyr-Asp-Leu-Ser-Phe-Leu-
	rabbit	-Phe-Asp-Phe - - - Ile-Met-

FIGURE 3. Comparison of nonhelical sequences in rat collagen α2-chain and calf collagen α1(I) chain containing terminal antigenic determinants with homologous regions of rabbit collagen. The lines above the rat and calf sequences indicate the minimal size of distinct antigenic determinants as found in studies of Furthmayr and Timpl (1972), Rauterberg *et al.* (1972), and Becker *et al.* (1972). The data on the rabbit sequence are from Becker *et al.* (1975*b*).

The definition of the size of antigenic determinants does not necessarily imply that each amino acid contained in this region is involved in the binding to the antibody combining site. Additional information could be obtained by chemical and serological comparison of homologous regions on collagen from other species. Because of the lack of cross-reaction with rabbit collagen, these amino acid sequences have been of particular attraction (Bornstein and Nesse, 1970; Becker and Timpl, 1972; Becker *et al.*, 1975*b*), and the available data are included in Figure 3. For each antigenic region at least one substitution is found on rabbit collagen, but the difference may be either as small as a single replacement of leucine by methionine or may essentially involve the whole region.

From 13 individual antisera which recognize the C-terminal antigenic determinant on calf collagen, only 11 and 8 sera still reacted with the corresponding nonhelical region on rat and human collagen, respectively (Timpl *et al.*, 1973*b*). An explanation for these observations was provided by chemical studies which indicated the substitution of one leucine residue in the antigenic region of calf collagen (Figure 3) in the rat collagen sequence (Stoltz *et al.*, 1972) and besides the change of one leucine residue, the replacement of tyrosine by phenylalanine in the human sequence (Furthmayr and Timpl, unpublished results). The insertion of an additional glycine residue between glycine and isoleucine in the central part of the N-terminal sequence in calf α1(I) chain (see Figure 3) is the only difference observed between these nonhelical regions in calf and human collagen (Click and Bornstein, 1970; Rauterberg *et al.*, 1972), and it caused a decrease in binding with specific antibodies.

A second approach is related to antigenic determinants which involve lysine and is based on comparison with peptides containing lysine-derived aldehyde instead of lysine. Up to a 60-fold reduction in serologic activity could be demonstrated for the oxidized peptide derivative of the N-terminal antigenic site in rat collagen α2 chain (Furthmayr and Timpl, 1972). However, other antisera containing antibodies against the same antigenic site did not reveal any difference between both peptides. A single antiserum against a N-terminal site in the calf collagen α1(I) chain (Rauterberg *et al.*, 1972) also did not discriminate between both forms of the peptide.

In the first immunochemical study on collagen, Schmitt *et al.* (1964) suggested the considerable importance of tyrosine for the antigenic activity of this protein. This seems to be verified by the more recent findings since, with the exception of a single antigenic region, all the others contained tyrosine (Figure 3). The replacement of tyrosine by phenylalanine in the rabbit collagen has already been discussed by Bornstein and Nesse (1970) as a possible reason for the particular features of the antibody response of rabbits towards collagen. In view of the data on protein determinants which do not contain tyrosine (Benjamini *et al.*, 1972), it remains to be determined how crucial the role of tyrosine might be and if it can be replaced by other, structurally related, amino acids. In any event, the cross-reaction studies with human collagen did not support the idea that a change from tyrosine to phenylalanine is accompanied by a complete loss of antigenic activity.

During their study on the C-terminal antigenic region of calf collagen, Becker *et al.* (1972) observed a new antigenic specificity on a second, apparently smaller, CNBr peptide α1-CB6[b]. Structural investigations demonstrated the loss of 19 amino acid residues in α1-CB6[b] from the very C-terminal end of the α1(I) chain. This occurred by cleavage in the region of the original antigenic determinant; -Tyr-Asp-Leu↓Ser-Phe-. Since extraction of collagen α chains under denaturing conditions did not reveal the presence of α1-CB6[b], accidental shortening of the collagen molecule during extraction with neutral salts or acidic buffers seems the most likely explanation. A second example of an "artificial" antigenic determinant was found by Stoltz *et al.* (1973) at the C-terminal end of α1(I) chains from neutral salt-extracted rat-skin collagen. The shortening occurred in a similar position, probably in the sequence -Tyr-Asp↓X-Ser-Phe-. Both antibody reagents did not react or only weakly cross-reacted with peptides containing the entire nonhelical sequence. The particular importance of the new carboxyl group either from leucine in calf collagen or from aspartic acid in rat collagen is therefore indicated for these "artificial" antigenic determinants. Lack of cross-reaction between both shortened peptides emphasized the crucial role of a single amino acid residue. The data also indicated that tyrosine is involved in both antigenic sites. The dependence of serologic activity on the peptide size (Stoltz *et al.*, 1973) was interpreted as to the involvement of some amino acid residues from the helical sequence. This seems not unlikely because the tyrosine involved is just the fourth residue beyond the end of the helical sequences (Fietzek and Kühn, 1976).

Recent immunization studies with rat collagen $\alpha 1$(I) chains possessing the whole nonhelical sequence provided evidence for the recognition of a different C-terminal antigenic determinant (Timpl *et al.*, 1973*b*). This site appeared to be located in the same region as the antigenic determinant of calf collagen (see Figure 3), e.g., involves at least one additional amino acid beyond the aspartic acid which is terminal in the shortened sequence. These findings demonstrate that extraction conditions can dramatically influence the antigenic specificity of collagen α chains.

As yet, no precise data is available for the structure of antigenic determinants in the helical sequence regions. Circumstantial evidence from cross-reactions of anti-*Ascaris* collagen with polymeric (Pro-Gly-Pro)$_n$ (Maoz *et al.*, 1973*b*) probably reflect a special kind of conformation in this peptide. Kettman *et al.* (1967) described precipitation of anti-guinea pig collagen antibodies with the synthetic peptide Gly-Pro-Gly-Pro-Pro-Gly-Ala-Lys. Since it is difficult to understand how this small peptide can act as a precipitating antigen, the information from this model is still to be considered as preliminary. In this context, it should be emphasized that the helical collagen sequence is far from being represented by simple repetition of a few tripeptide units (Fietzek and Kühn, 1976). Approaches on the basis of synthesized peptides are therefore certainly limited and can by no means replace the more direct way of identification of antigenic determinants on appropriate fragments derived from the collagen molecule itself.

IV. The Specificity of Cell-Mediated Immune Reactions

The characterization of delayed-type hypersensitivity against collagen as a means of investigating the specificity of thymus-derived T cells receives its attraction from the outstanding role of this cell type in transplantation rejection, in autoimmune processes, and because of its regulatory function on antibody-producing cells. Except for a few earlier studies (Steffen, 1965; Kirrane and Robertson, 1968; Adelmann *et al.*, 1968) which demonstrated that collagen can indeed activate T cells without inducing a concomitant antibody production, most of the pertinent data is of a recent origin. Guinea pig has been generally used as the standard model in these investigations except in one case which utilized arthritic rats (Steffen and Wick, 1971). A single injection of collagen in complete Freunds' adjuvant usually produces a long lasting hypersensitivity state of the delayed type. Antibody production is either weak or absent but may be increased by multiple antigen injections (Steffen and Wick, 1971; Adelmann *et al.*, 1972; Senyk and Michaeli, 1973). Measurement of the cell-mediated immune reaction was accomplished in most studies by skin tests. A concise description of several parameters of this reaction and suggestions for statistical evaluation of the results may be found in the report of Adelmann *et al.* (1972). Cell-mediated immunity was also measured *in vitro* by inhibition of macrophage migration (Michaeli *et al.*, 1972; Adelmann *et al.*, 1972), and the results obtained with this assay were in agreement with those of the skin tests. Stimulation of thymidine incorporation of lymphocytes by antigen *in vitro* appeared to be a less sensitive test and less specific for T cells (Senyk and Michaeli, 1973). Selective binding to collagen fibers has been reported for immune cells of mice immunized against collagen or $(\text{Pro-Gly-Pro})_n$ (Maoz and Fuchs, 1974), but it was not established whether these antigen-reactive cells are of T- or B-cell origin.

In guinea pigs immunized with native, triple-helical collagen of the chain composition $[\alpha 1(\mathrm{I})]_2\alpha 2$, both native and denatured collagen elicited cell-mediated reactions of equal strength (Senyk and Michaeli, 1973). A decreased reactivity for denatured collagen was reported by Adelmann (1973) and $\alpha 1(\mathrm{I})$ chains appeared to be less reactive than $\alpha 2$ chains (Adelmann, 1972). Sensitization with denatured collagen or α chains led to positive skin reactions with triple-helical collagen, indistinguishable from those observed with the sensitizing antigens. A clear distinction between $\alpha 1(\mathrm{I})$ and $\alpha 2$ chains was again observed in this model (Adelmann, 1972, 1973).

Studies with pepsin-treated collagen aimed at elucidating the role of nonhelical sequences did not reveal any difference with untreated collagen in the ability to sensitize the animals and to elicit skin reactions (Adelmann and Kirrane, 1973). Differences in these properties, however, were noticed between α chains either obtained from neutral salt-extracted or from urea-extracted collagen (Adelmann, 1972). Since both kinds of α

chains are considered to differ only in the length of their nonhelical sequences (Stoltz *et al.*, 1972) it is difficult to correlate these findings.

Distinction between collagens from different mammalian species, e.g., rat and calf, was achieved in delayed-type hypersensitivity regardless of whether the antigens were compared in their native or uncoiled conformation (Adelmann, 1973). Skin tests with mammalian collagen after sensitization with fish collagen were negative (Adelmann and Kirrane, 1973). Michaeli *et al.* (1972) found, however, strong cross-reactions between *Ascaris* cuticle collagen and human collagen but not with mouse collagen.

Attempts to characterize antigenic determinants involved in the cell-mediated immune response at the level of fragments smaller than α chains are still in a preliminary stage. The feasibility of this approach is indicated by positive skin reactions found for CNBr-digested collagen (Adelmann *et al.*, 1972). The higher reactivity in skin tests of the peptides when compared with triple-helical collagen was related to their higher diffusibility. Kirrane and Robertson (1968) observed no reaction of collagenase-digested collagen in delayed-type skin tests, but the peptides were still inhibitory for antibodies. Whether this observation really indicates that a larger size of the antigenic determinants is required in interaction with T cells remains an open question. The assumption of a different nature and distribution of the sites reacting either with cells or with antibodies seems to be equally acceptable. Cross-reactions with polymeric $(\text{Pro-Gly-Pro})_n$ of guinea pigs sensitized against various triple-helical collagens were found to be quite weak (Adelmann and Kirrane, 1973).

Comparison of these results with present information on antibody specificity is still difficult. This reflects the dilemma encountered in other studies on the specificity of T cells and results from experimental limitations which exist in measuring antigen recognition by cells in a more quantitative way. However, one gets the impression that any cell-mediated response to either native or denatured collagen is always accompanied by a strong cross-reaction with the other conformational form of the antigen. Whether this reflects less precise recognition of conformational features by the T-cell receptors or the presence of at least two subsets of T-cell populations, each different in specificity, cannot be decided. Anticipating that denatured collagen cannot renature to triple-helical structures after injection into guinea pigs, the first possibility appears more likely. This is reminiscent of the observations of Thompson *et al.* (1972) who found a strong cross-reaction between native and reduced lysozyme for the cell-mediated immune response but not with antibodies. In this context, characterization of the antibody specificity in the guinea pig should be a major concern but is as yet in a preliminary stage of investigation (Adelmann *et al.*, 1973). The data available indicate a response to species-specific, helical antigenic determinants and only weak reactions with denatured collagen.

V. Immunology of Collagen-like, Synthetic Polypeptides

Synthetic polypeptides which exhibit a collagen-like sequence and/or conformation have been increasingly used as models to evaluate special immunological features of the collagen molecule. Polyproline and random copolymers of proline, glycine, and/or hydroxyproline were found to be antigenic in the guinea pig and other animals (Jasin and Glynn, 1965*a,b*; Brown and Glynn, 1968; Gurari *et al.*, 1973). Antibodies produced against this polymer were in a few instances highly specific and, for example, distinguished between the right- and left-handed helix of poly-L-proline (Jasin and Glynn, 1965*a*). To date no cross-reactions have been observed with mammalian collagens.

More sophisticated approaches were based on the use of polymerized tripeptide moieties which, as in collagen, contained glycine in every third position of the sequence. Borek *et al.* (1969) were the first to show a definitive although weak immunogenicity of $(\text{Pro-Gly-Pro})_n$ in the guinea pig and rabbit. The antibody response could be considerably increased when this peptide was used as conjugate onto various carrier proteins. This improvement allowed a concise immunochemical investigation of $(\text{Pro-Gly-Pro})_n$ (Maoz *et al.*, 1973*a*). The dependence of immunological activity on the molecular weight of the polymer and the decrease in activity after collagenase treatment were interpreted as indicating the importance of helical conformation for the antigenic sites. Support for this idea was obtained by the finding of weak or absent cross-reactions with the tripeptide Pro-Gly-Pro and the random copolymer $\text{Pro}^{66}\text{Gly}^{34}$. It is known that neither of them have a collagen-like conformation. The antibodies to $(\text{Pro-Gly-Pro})_n$ showed a decreasing affinity to a series of polymeric hexapeptides of the general formula $(\text{Gly-X-Y-Gly-X-Y})_n$ in which proline residues in the X or Y position were replaced by alanine. The largest reduction in activity (by a factor of about 10,000) was observed for the peptide $(\text{Gly-Ala-Pro-Gly-Pro-Ala})_n$.

Rabbit antibodies against $(\text{Pro-Gly-Pro})_n$ showed considerable cross-reaction with *Ascaris* cuticle collagen in precipitation and phage-neutralization tests (Maoz *et al.*, 1973*b*). No reaction was observed with rat or guinea pig collagen. This finding was corroborated by the demonstration that 20–38% of the antibodies in anti-*Ascaris* collagen antisera cross-reacted with polymeric (Pro-Gly-Pro)$_n$. Investigations on the specificity of these cross-reacting antibodies using the random polymer $Pro^{66}Gly^{34}$ and the polymerized hexapeptides mentioned above revealed the same order of decreasing activity, although the differences appeared to be less marked. Guinea pig antibodies against (Pro-Gly-Pro)$_n$ which cross-reacted with mammalian collagen and *vice versa* could be detected by skin testing (Maoz *et al.*, 1973*b*) or by passive cutaneous anaphylaxis (Borek *et al.*, 1969). Despite the weak reactions observed, it is noteworthy that antibodies to $Pro^{66}Gly^{34}$ did not show any cross-reaction with collagen.

In spite of the very elegant immunochemical studies, it is not clear which particular conformation is recognized by the antibodies to polymeric (Pro-Gly-Pro)$_n$. Physical studies by Engel *et al.* (1966) did not decide the question of whether the conformational structure of this polymer is mainly due to a backfolding of a single chain or reflects the parallel alignment of three different chains in a triple-helical assembly. Since a reverse folding was suggested for *Ascaris* cuticle collagen (McBride and Harrington, 1967*b*), the immunological findings seem to support the first possibility. Synthetic models for collagen have recently been prepared by successive condensation of the tripeptide unit Pro-Pro-Gly (Kobayashi *et al.*, 1970). These peptides have been shown to form triple-helical structures involving three individual polypeptide chains, and such peptides should therefore be more appropriate models to study cross-reactions with mammalian types of collagen.

Polymers of the sequence (Gly-Pro-Ala)$_n$, (Pro-Ala-Gly)$_n$, or (Ala-Pro-Gly)$_n$ were shown to elicit antibodies in the guinea pig, but no evidence could be obtained for cross-reactions with rat collagen by skin tests (Brown and Glynn, 1973). In this and other studies (Jasin and Glynn, 1965*a*; Brown and Glynn, 1968; Borek *et al.*, 1969; Maoz *et al.*, 1973*a*) the induction of an immune response of the cell-mediated type was reported for such collagen-like peptides. As measured by skin reactions this response was manifested earlier than the antibody response but appeared restricted to specific time periods. The cell-mediated response was dependent on the molecular weight of polyproline (Brown and Glynn, 1968) or of (Pro-Gly-Pro)$_n$ (Maoz *et al.*, 1973*a*). No dependence on size was found for (Ala-Gly-Pro)$_n$ (Brown and Glynn, 1973). Cross-reactions with collagen were observed only after sensitization with (Pro-Gly-Pro)$_n$, and the intensity of the skin reaction increased with increasing molecular weight of the polymer (Maoz *et al.*, 1973*b*). Another study (Adelmann and Kirrane, 1973) did not find such cross-reactions.

Synthetic peptides which, because of their limited size, probably do not exhibit a collagen-like helical structure were first used by Kettman *et al.* (1967). In a continuation of this study Benjamini *et al.* (1973) demonstrated a specific antibody response in the rabbit against Gly-Pro-Gly-Pro-Pro-Gly-Ala-Lys by conjugating this peptide onto a macromolecular carrier. Inhibition studies with smaller peptides indicated the restriction of the antigenic site to the C-terminal pentapeptide portion. Inhibition tests also revealed a significant cross-reaction with polymeric (Pro-Gly-Pro)$_n$ and with various mammalian collagens either in a helical or uncoiled state. Only weak cross-reactions could be observed by passive hemagglutination. Antisera against mouse or rabbit collagen also showed quite low agglutination titers for this octapeptide in spite of a strong precipitating cross-reaction reported earlier (Kettman *et al.*, 1967). In this context it is of interest that the sequence Gly-Pro-Pro-Gly-Ala-Lys does not occur in the calf collagen α1(I) chain (Fietzek and Kühn, 1976) even though cross-reactions might be explained by similar sequences like Gly-Gln-Hyp-Gly-Ala-Lys.

Specific antibodies to the tripeptide Pro-Gly-Pro were described by Maoz *et al.* (1973*a*) and were also obtained by immunization with an appropriate hapten–carrier conjugate. The antibodies showed weak cross-reactions with the ordered polytripeptide (Pro-Gly-Pro)$_n$, and this result was explained by the antibodies reacting with terminal amino acids in the polymer. No information was given as to the cross-reaction of these antibodies with collagen.

VI. Cellular and Structural Basis for the Induction of an Immune Response to Collagen

Current concepts of the cellular systems involved in immunological functions hold that three different types of cells are required for effective antibody production: B cells which can be triggered to antibody synthesis and secretion, T cells which exhibit a regulatory function, and macrophages which might be involved as a third interacting partner. Although the mechanism of cellular cooperation is yet not known, recognition of distinct antigenic determinants by receptors on B and T cells and perhaps on macrophages are considered as the primary events which are controlled in a very complex manner. For a comprehensive discussion of these topics the reader is referred to recent articles by Good (1972), Katz and Benacerraf (1972), and Gershon (1974).

The role of macrophages in the immune response to collagen is not yet known, but they might be crucial for processing the antigen. Active participation of macrophages in resorption of tissue collagen (Schwarz and Güldner, 1967; Parakkal, 1969) strongly suggests the involvement of these cells in phagocytosing soluble collagen antigens. More recent information on the possible degradation of collagen by a macrophage collagenase is based on studies by Wahl *et al.* (1975). Interestingly, this collagenase is induced by lymphocyte-derived factors which are released *in vitro* upon interaction with lectin. The lectin can be replaced by antigens like DNP-ovalbumin if the lymphocyte donors are presensitized against this antigen. It is tempting to speculate that this type of collagenase induction also occurs *in vivo* after immunization with collagen. Binding of native or denatured collagen onto macrophages provided the first hints of a possible role in cell interaction (Hopper *et al.,* 1976). Addition of specific antisera caused an increase in binding, indicating the presence of cytophilic antibodies in the antisera.

Antibody responses to most protein antigens depend on an efficient T cell help (Gershon, 1974). It is thought that helper T cells interact with carrier determinants and thus trigger B cells for antibody production against haptenic determinants located in the same antigenic molecule. Controversial data were reported for collagen antigens. Fuchs *et al.* (1974*b*) found T cell independence for the antibody response in inbred mice to polymeric (Pro-Gly-Pro)$_n$, to native rat skin and Ascaris cuticle collagen, but not to denatured rat collagen or polymeric Pro^{66}Gly^{34}. This conclusion was based on the observation of full restoration of immune capacity upon transfer of bone marrow (B) cells into irradiated recipients. However, native calf (type I) collagen apparently belongs to the T cell dependent antigens as shown by similar cell transfer experiments and by studies in thymusless, "nude" mice (Nowack *et al.,* 1976*a*). Polymeric forms of collagen as obtained by limited pepsin digestion of tendons were also nonimmunogenic in "nude" mice (Nowack, Zimmermann, and Timpl, unpublished).

Strong support for the concept that T cells regulate immune responses to collagen came from immunogenetic studies in mice. Antibody levels to bovine type I collagen are under control of an immune response (Ir) gene located in the IA subregion of the major histocompatibility (H-2) locus. Other Ir genes may also be involved (Hahn *et al.,* 1975*b*; Nowack *et al.,* 1975*a*). Procollagen can correct low responsiveness to collagen in some but not all mouse strains. The data were interpreted as indicating recognition of new carrier determinants in the procollagen-specific region which now allows an efficient cooperation between B and T cells (cf. McDevitt, 1972). The antibody response to the procollagen-specific region itself is also T cell dependent and governed by an Ir gene located in the IA,B subregion of H-2. This Ir gene is different from those controlling the response to collagen (Nowack, Rohde, Götze, and Timpl, in preparation). Yet another Ir gene was found to determine immune responsiveness against type II collagen from cartilaginous tissue (Nowack *et al.,* 1975*b*). Differences exist between various inbred strains of mice in their response potential to rat collagen or collagen-like polypeptides (Fuchs *et al.,* 1974*a*). The present data therefore indicate that collagen as well as the procollagen-specific regions possess carrier and haptenic determinants. Observations on delayed hypersensitivity reactions in guinea pigs and on carrier properties of chemically modified collagens discussed in preceding paragraphs are compatible with this interpretation.

Because of the awareness of the weak immunogenicity, almost all immunization experiments have been carried out with collagens which were incorporated into complete Freunds' adjuvant. Recent evidence on T-cell activation by adjuvants (reviewed by Allison, 1974) raises the question of whether this experience reflects the need for T-cell cooperation in the antibody response to collagen. Comparative studies in the rabbit demonstrated decreased antibody levels against certain but not all antigenic determinants when the collagen was injected without adjuvant (Stoltz *et al.*, 1973). Adelmann *et al.* (1972) found that the type of adjuvant used directed the immune response in the guinea pig either towards antibody production (incomplete Freunds' adjuvant) or cellular immunity (complete Freunds' adjuvant). Inhibition of the delayed-type reaction was observed when antibody production was induced prior to the challenge aimed at eliciting a state of cell-mediated immunity (Gentner and Adelmann, personal communication). This type of interference was dependent on the time interval between both injections and obviously resembles previous immune deviation models.

Any discussion on the role of T and B cells in the immune response to collagen would be incomplete without considering the potential regulatory effect of tissue collagen in the immunized animals. This might occur by functional elimination of those immune cells which are able to recognize self-antigens (self-tolerance). Chiller *et al.* (1971) and Benjamini *et al.* (1972) discussed the hypothesis that because of an absolute requirement of B cells for T-cell cooperation, self-tolerance at the level of T cells is sufficient to prevent any autoimmune response against the body's own proteins. Inaccessibility of tissue collagen to the immune system is still another possibility (Allison, 1974).

Steffen (1965) reported on delayed skin reactions in guinea pigs immunized with guinea pig collagen, but screening for antibodies in these animals, as well as in rabbits immunized with rabbit collagen, gave virtually negative results. Cell-mediated immunity to homologous guinea pig collagen after immunization with guinea pig or human collagen was also observed by Senyk and Michaeli (1973). Unresponsiveness against homologous collagen could be induced applying the cyclophosphamide technique, and this tolerant state persisted after cross-immunization with human collagen. The evidence for antigen-reactive cells which are able to recognize a self-antigen was interpreted as indicating a state of nonimmunity for tissue collagen rather than self-tolerance. It should be noted that Adelmann *et al.* (1972) could not sensitize guinea pigs against their homologous collagen. This unresponsiveness did not change after denaturation or succinylation of the antigen (Adelmann, 1973). Furthermore, only a weak cross-reaction with guinea pig collagen which was of transient appearance was found after immunization with calf collagen. Evaluation of these results is difficult because (1) negative observations may reflect the reaction of antibodies and/or immune cells with the tissue's collagen and (2) extracted collagen is usually degraded to a limited extent and thus may acquire new antigenic properties. The second possibility gained recent support by the detection of a moderate agglutinating antibody response in the chick against acid-soluble chick bone collagen. These antibodies did not, however, react in immunofluorescence tests with chick tissue collagen (Wick, Furthmayr, Timpl, and Miller, unpublished observations).

Until 1972 essentially all of the information on antibody response to collagen was based on investigations carried out in rabbits. Many immunochemical studies documented the preferential recognition of nonhelical, terminal sequences in the immune response of this species to foreign collagen (see Table 2). Because of the considerable interspecies variability in amino acid sequence known for the nonhelical regions of collagen (Traub and Piez, 1971), this observation was accepted as highly satisfactory (cf. Figure 3). Identity or a high sequence homology between rabbit and rat collagen α1-CB1 (Bornstein and Nesse, 1970) as well as between rabbit and calf collagen α2-CB1 (Becker *et al.*, 1975*b*; Fietzek *et al.*, 1974) furnished a sufficient explanation for the absence of antibody response to these particular sites (Lindsley *et al.*, 1971; Rauterberg *et al.*, 1972). However, the failure to detect such antibodies in the serum may also be caused by absorption of the antibodies on rabbit tissue collagen.

A change of this generally accepted picture originated in recent studies on the antibody response of the chicken, rat, guinea pig and mouse against rat or calf collagen (Beil *et al.*, 1972, 1973; Hahn and Timpl, 1973; Adelmann *et al.*, 1973; Nowack *et al.*, 1975*b*; Hahn *et al.*, 1974). Serologic evaluation of these antisera demonstrated an outstanding response to helical structures of collagen and failed to detect measurable quantities of antibodies against terminal antigenic determinants. Apparently, the nonhelical sequences in the collagens of the immunized species and the immunogens are different enough so that lack of "foreignness" can be precluded in accounting for this observation. On the other hand, antibodies to helical antigenic determinants could discriminate between collagens of different species (Hahn and Timpl, 1973; Adelmann *et al.*, 1973). Thus, the change in the immune response pattern does not necessarily involve a change to less species-specific recognition. Evidence for antibodies to helical antigenic determinants in rabbit antisera, even though of relatively low concentration, was provided recently (Beil *et al.*, 1973; Hahn *et al.*, 1975*a*).

Although studies on the specificity of the cell-mediated immune response (Adelmann, 1973) add further weight to the immunological importance of helical collagen structure, one should avoid formulating generalizations. To date the investigations are too small in number, and the use of different collagen antigens as well as variations in the experimental conditions might result in a pattern quite different than hitherto described. Furthermore, the removal of nonhelical sequences by endogeneous proteases prior to the antigen recognition step cannot be excluded and may provide a nonimmunological explanation for the exclusive immune response to the helical antigenic determinants of collagen.

The expression of particular structural properties of collagen in its immunogenic activity has been the subject of numerous studies. One of the problems considered was whether insoluble collagen is immunologically inert or not. The clinical importance of this question is documented by an enormous literature on the medical use of collagen transplants (reviewed by Chvapil *et al.*, 1973; Stenzel *et al.*, 1974), but essentially no information is available on possible immunological complications. Previous attempts to produce antibodies to insoluble collagen have been reviewed by Kirrane and Glynn (1968) and by O'Dell (1968). One should emphasize in this context that positive serologic reactions with insoluble collagen have to be viewed with caution, especially since the involvement of antibodies to noncollagen contaminants cannot rigorously be ruled out. In a few experiments with extracted collagen which was made less soluble by cross-linking or tanning, it was found that these treatments considerably decreased the immunogenic potency of collagen (Miyata *et al.*, 1971; Brunner *et al.*, 1971; Chvapil *et al.*, 1973).

Occasional reports have appeared on immunological differences between collagens obtained by different extraction procedures, e.g., by neutral salt or acidic solutions (Lustig *et al.*, 1969; Needleman and Stefanovic, 1969). Sufficient evidence for the collagen specificity of the antisera was not included in this work. Based on recent structural studies one may predict differences which have their origin in extraction artifacts.

For example, $\alpha 1(I)$ chains when obtained from neutral salt-extracted rat skin collagen lack four amino acids of their N-terminal and 20 amino acids of their C-terminal sequence (Stoltz *et al.*, 1972, 1973). These $\alpha 1(I)$ chains are very weak immunogens in the rabbit (Lindsley *et al.*, 1971; Stoltz *et al.*, 1973). A marked increase in immunogenic activity was observed if the shortening of the $\alpha 1(I)$ chains was prevented by extraction under denaturing conditions (Timpl *et al.*, 1973*b*). A further possible effect related to extraction conditions may involve intramolecular cross-linking of α chains to β components. Although $\beta 12$ components were reported to be better immunogens than $\alpha 1(I)$ chains (Lindsley *et al.*, 1971), there is yet no evidence for a distinct immunological specificity of β components (Michaeli *et al.*, 1969; Michaeli and Epstein, 1971; Meigel *et al.*, 1971).

Destruction of the triple-helical conformation of collagen by denaturation to uncoiled α-chains is accompanied by a change in immunogenicity and/or antigenic specificity, even though the effects appear less pronounced than that observed after denaturation of other protein antigens. The cell-mediated immune response has apparently the same strength against both forms of the collagen antigen (Adelmann, 1972, 1973). A comparable immunogenicity was also found in studies on the antibody response (Michaeli *et al.*, 1971; Beil *et al.*, 1973). However, a concise analysis of the antibody dynamics revealed a retarded onset of the response to the denatured antigen (Timpl *et al.*, 1973*b*). Antisera against denatured collagen may differ in specificity from those against native collagen because of recognition of new antigenic sites (Michaeli *et al.*, 1971) or the lack of antibodies to distinct terminal antigenic determinants (Lindsley *et al.*, 1971; Stoltz *et al.*, 1973; Timpl *et al.*, 1973*b*). One should also expect that denatured collagen is incapable of eliciting antibodies to helical antigenic determinants (cf. Beil *et al.*, 1973). Several inbred strains of mice failed to respond against denatured collagen (Nowack *et al.*, 1975*b*).

As in the case of most other protein antigens (Benacerraf *et al.*, 1974), the antibody response to collagen is certainly controlled by additional, quite complex regulatory mechanisms. Phenomena ascribed to this regulation involve the sequential appearance of antibodies having distinct specificities, competition between different antigenic determinants, and a distinct unresponsiveness in certain individual animals (Timpl *et al.*, 1970, 1972, 1973*b*; Rauterberg *et al.*, 1972). Even though these phenomena are far from being understood, they at least indicate that results based on studies with a single antiserum are hardly representative of the complex pattern in the anticollagen response. Antibodies to collagen or collagen-like peptides belong mainly to the IgG class, but significant amounts of IgM antibodies have also been found even in the secondary or hyperimmune response (Timpl *et al.*, 1967, 1971; Beil *et al.*, 1972; Maoz *et al.*, 1973*a*). Yet, there was no correlation of a distinct antibody specificity with a particular antibody class.

The weak immunogenicity of collagen has been known since the earliest attempts to provoke an immune response against that antigen. Recently, quantitative data were obtained by precipitation and immunoadsorption studies and indicated antibody levels between 100 and 500 μg/ml in rabbit and chick antisera to rat, calf, or chick collagen (Timpl *et al.*, 1967; Beil *et al.*, 1972; Pawlowski *et al.*, 1974; Hahn *et al.*, 1974*c*). Concentrations up to 3000 μg/ml were reported for rabbit antisera against native *Ascaris* collagen (Fuchs and Harrington, 1970). Antibody levels comparable to those in antisera against mammalian collagen are, however, observed after immunization with reduced *Ascaris* collagen (Maoz *et al.*, 1971). Even lower antibody concentrations of 7–10 μg/ml are found in antisera to $(\text{Pro-Gly-Pro})_n$ but could be augmented by immunization with peptide–carrier conjugates (Maoz *et al.*, 1973*a*). Compared with other proteins like serum albumin, the antibody concentration (Beil *et al.*, 1972) as well as the strength of the cell-mediated response (Adelmann *et al.*, 1972) is significantly lower if collagen is used as antigen.

Hyperimmunization has been frequently used in previous studies for the purpose of producing high antibody titers to collagen. More recently it could be demonstrated that a single antigen injection is sufficient to obtain a measurable antibody response which did not significantly change after a booster injection or after hyperimmunization (Kirrane and Robertson, 1968; Beil *et al.*, 1972; Stoltz *et al.*, 1973).

In spite of the available comprehensive structural information on the immunological properties of collagen, its low immunogenicity still remains a puzzle. Previous ideas on the requirement of "rigid" structures like tyrosine were considered obsolete even in earlier reviews (O'Dell, 1968) and have not found support in more recent results. The meaning of the term rigidity is not quite clear especially since collagen itself is apparently a fairly rigid molecule. A high sequence homology between the collagens of different animal species may be considered as another reason for low immunogenicity. The crossing of phyletic barriers, e.g., by immunizing rabbits with chick instead of with mammalian collagen, was interpreted as increasing the antibody response (Kirrane and Glynn, 1968), but this assumption is not supported by a recent immunoadsorption study (Pawlowski *et al.*, 1974). Previous data on the rabbit response to invertebrate collagen (Meigel *et al.*, 1971; Maoz *et al.*, 1971), as well as on the antibody levels in chicken immunized with rat collagen (Beil *et al.*, 1972), are also not in favor of this concept.

Experimental data from a few recent reports, even though not explicitly described by the authors to explain the weak immunogenic potency of collagen, may nonetheless stimulate further approaches towards an understanding of this phenomenon. Adelmann *et al.* (1973) claimed a low avidity of guinea pig antibodies to helical collagen. Thus, one might assume that the repertoire of collagen-specific immune cell receptors is of extremely low affinity, perhaps due to the regulatory influence of the body collagen. However, Benjamini *et al.* (1973) observed, for antibodies to a collagen-like peptide sequence, affinity constants of the order of 10^7 liter·mol^{-1} which are as high as those known for haptens or other protein antigens. Slow metabolism of the collagen antigen was recently introduced by Fuchs *et al.* (1974*a*) as a possible explanation for the observed T-cell independence of the antibody response. Although the matter of T-cell involvement is still controversial, peculiar features of the collagen catabolism might be the key to understanding its low immunogenicity. Findings of an inducible collagenase in macrophages (Wahl *et al.*, 1975) are relevant in this context and are certainly worth a detailed experimental investigation.

VII. The Possible Role of Collagen as an Autoantigen

Many investigations on the immunological properties of collagen were guided by the consideration of a putative autoantigenic role of this antigen. An initial indication of this possibility originated in studies by Maurer (1954) who discovered components in normal human and animal sera capable of reacting with gelatin. However, even more recent studies have failed to clarify whether antibodies or other serum constituents are responsible for this phenomenon. Wolff *et al.* (1967) reported on positive reactions of human sera with gelatin (denatured acid-soluble collagen) in passive hemagglutination and provided preliminary evidence for the involvement of two different serum components. One of these components might be IgM. This agglutinating antigelatin factor may resemble the factor previously described by Maurer due to its broad reactivity towards different kinds of gelatin (Wolff and Timpl, 1968). Using the same agglutination technique, Bray *et al.* (1969) observed antigelatin factors sensitive to 2-mercaptoethanol in normal human individuals. This factor was absent in sera of patients with severe burns. A high-molecular-weight component reacting with denatured collagen was also demonstrated by a radioimmune assay in nonimmune guinea pig and mouse serum (Adelmann *et al.*, 1973; Nowack *et al.*, 1975*b*). The significance of these findings is not yet clear, especially with respect to the role of denatured tissue collagen in triggering such antigelatin factors.

A possible immunological etiology of rheumatoid arthritis and related diseases has been an intriguing stimulus for numerous experimental and clinical investigations. The histological evaluation of the rheumatoid synovium is highly suggestive of a local immune response because of a dense infiltration by plasma cells and lymphocytes. Long-lasting inflammatory lesions could be produced in animals by intra-articular injection of antigens not related to connective tissue components (Glynn, 1969; Cooke *et al.*, 1972). These models stress the possibility of the "unspecific" nature of the immune processes required to induce and maintain chronic diseases. Injuries to collageneous fibers of the cornea by unrelated immune complexes have been also demonstrated in many instances (cf. Mohos and Wagner, 1969).

Chronic appearance of the diseases may, however, also require the participation of endogenous antigens residing in the target tissue (Glynn, 1969). Searches for a specific autoantigen have been centered on collagen since the earliest immunological studies (Waksman and Mason, 1949). Steffen and Timpl (1963) were first to demonstrate collagen-specific antibodies in sera from rheumatoid patients by using the antiglobulin consumption test. Steffen and co-workers (1971*b*, 1972, 1973), as well as other investigators (Kriegel *et al.*, 1970), attempted in subsequent studies to obtain more convincing evidence for the autoantibody nature of these antibodies and their pathogenetic significance. However, the role of the antibodies in maintaining the chronic process is still obscure, and their specificity for collagen seems to be questionable. The antigen used in most of these studies was insoluble collagen which contains about 15% noncollagen protein (Timpl, 1966). Nevertheless, a hypothesis on the central role of collagen as autoantigen in rheumatoid arthritis has been put forward by Steffen (1970).

An attempt to demonstrate antibodies to purified denatured collagen revealed in rheumatoid sera agglutination titers which were not higher than those found in normal sera (Wolff *et al.*, 1967). A recent study by Michaeli and Fudenberg (1974), however, claimed significant titer differences against denatured and native collagen between such sera. The antibody titers observed in rheumatoid sera were quite low (average titer about 1:8). Characterization of the specificity of the antibodies with CNBr peptides revealed the main antigenic determinant in the N-terminal region of the $\alpha 1(\mathrm{I})$ chain.

In addition to delayed-type reactions found for native and denatured collagen, such reactions were also found for connective tissue glycoproteins in adjuvant-induced arthritis in rats (Steffen and Wick, 1971).

Inclusion of collagen during injection of the adjuvant increased the incidence of delayed-type reactions but did not change the severity of the disease. Antibodies to collagen could not be demonstrated in the sera of arthritic rats.

The high incidence of antibodies to basement membrane components in some human renal diseases is a well-established fact, but identification of collagen as one of the antigens involved is still a controversial matter. Mahieu *et al.* (1973) demonstrated antibodies in patients with Goodpasture's syndrome that were apparently specific for the carbohydrate portion of collagen-like peptides. This glycopeptide could inhibit the migration of leukocytes obtained from such patients. A collagen-like glycoprotein was also identified by McIntosh and Griswold (1971) as the antigen in Goodpasture's syndrome by histological means. Marquardt *et al.* (1973*a,b*) reported results which disagree with these claims, but their evidence that all of the nephritogenic antigens are completely unrelated to moieties of basement membrane collagen was not conclusive. Experimental nephritis can be induced by injecting renal basement membranes into guinea pigs. Autoantibodies which apparently reacted with collagenous components could be identified in these sera, but reaction with closely associated antigens could not be excluded (Lehman *et al.*, 1974).

A considerable frequency of antibodies to native collagen with agglutination titers between 1:2 and 1:64 was also reported for patients with IgA deficiency (Wells *et al.*, 1973). Dietary collagen was considered as a possible stimulus for these antibodies. The relation of these collagen antibodies to reticulin antibodies observed in various gastrointestinal diseases (Wright and Alp, 1971; Essen *et al.*, 1972) is not yet clear.

In view of the continuous efforts to characterize collagen as an autoantigen, the situation has not changed much since the last reviews in 1968. Methods developed in recent studies on experimentally induced antibodies, e.g., radioimmune assays and the use of purified collagen α chains or CNBr peptides, have been applied so far only in a few studies. Such methods may provide improved tools for forthcoming progress in this field. Studies on a possible autoimmune response at the level of T cells is a further interesting possibility for which data are yet essentially lacking.

VIII. Methods Used to Detect and Evaluate the Specificity of Anticollagen Antibodies

The appropriate choice of serological techniques is considered of some importance in immunochemical investigations on protein antigens, and collagen is in that sense certainly not an exception. Although independent studies on the major antigenic sites of rat collagen either by hemagglutination inhibition or radioimmune assays (Timpl *et al.*, 1970, 1972; Lindsley *et al.*, 1971) showed considerable agreement, other parallel studies like that on calf collagen which were carried out by complement fixation or hemagglutination (Davison *et al.*, 1967; Pontz *et al.*, 1970; Becker *et al.*, 1972) arrived at quite different conclusions. Since differences in the results may depend on the methods applied, the use of several techniques seems recommended in equivocal situations and makes it well worthwhile to discuss present findings fitting into this context.

Semiquantitative complement fixation varying either the antiserum and/or the antigen concentration (Watson *et al.*, 1954; Lustig *et al.*, 1969; Rothbard and Watson, 1972) revealed titers between 1:64 to 1:1600 after hyperimmunization. This method was also used to determine cross-reactions (Rothbard and Watson, 1965; Lustig *et al.*, 1969). A quantitative variant was applied for more structural investigations (Davison *et al.*, 1967; Kirrane and Robertson, 1968; LeRoy, 1968). Maximum complement binding was observed with 0.8–100 μg antigen; the higher values are thought to indicate reactions with impurities (Davison *et al.*, 1967). A possible anticomplementary activity of the antigens, like that described for denatured or digested collagen (Kirrane and Robertson, 1968), and the necessity of having a doublet of IgG antibodies in close proximity for effective complement fixation (Cooper, 1973) are certainly limitations of this technique. Loss of antigenic activity upon denaturation of collagen therefore may not necessarily reflect lack of binding with antibodies (Davison *et al.*, 1967) but rather physical separation of antigenic determinants.

Passive hemagglutination is also widely used to investigate anticollagen antisera (Steffen *et al.*, 1967*b*; Michaeli *et al.*, 1971, 1972; Kirrane and Robertson, 1968; Lustig *et al.*, 1969) and titers between 1:100 and 1:16,000 have been reported. The sensitivity of this technique depends on appropriate coating of the red cells, and optimal concentrations, which are quite low (Steffen *et al.*, 1967*b*), have to be determined empirically. This limits the method for direct comparison of different antigens, especially if they are quite different in size and/or conformation. Dependence of titers on the conjugation procedure has been reported (Beil *et al.*, 1972, 1973). Since, as for many other antigens, IgM antibodies have a higher agglutinating potency than IgG antibodies (Timpl *et al.*, 1971), evaluation of agglutination titers should also include a check for a possible preponderance of IgM antibodies (Beil *et al.*, 1972; Stoltz *et al.*, 1973).

Inhibition of antibodies by antigens in a fluid-phase reaction followed by addition of antigen-coated red cells to measure the titer reduction has been applied in many of the structural studies. In this inhibition assay serial antibody dilutions were incubated with a constant amount of inhibitor, and the titer decrease is recorded in comparison to a buffer control (Steffen *et al.*, 1967*b*; Kirrane and Robertson, 1968; Michaeli *et al.*, 1971; Timpl *et al.*, 1970*a*, 1972). In a second variant, constant amounts of antibodies are preincubated with different antigen concentrations and the minimal amount of antigen just preventing agglutination is determined (Furthmayr and Timpl, 1972; Rauterberg *et al.*, 1972; Stoltz *et al.*, 1973). Although such inhibition assays have been successfully used to approach the structure of antigenic determinants, the semiquantitative nature of this method with a supposed error of about ±50% should be kept in consideration.

Radioimmune assays in which complexes between labeled antigen and antibodies are precipitated by antibodies against immunoglobulins were first used by Lindsley *et al.* (1971) to study the antigenic nature of ^{125}I-labeled α chains and CNBr peptides. Since this quantitative test also allows inhibition assays with unlabeled antigens, it seems to be the method of choice for further immunochemical studies. A possible modification of the antigenic structure by the label (e.g., tyrosine residues in the terminal antigenic determinants) was considered unlikely in this study (Lindsley *et al.*, 1971) but has to be considered for each new system. Labeling with ^{125}I was also applied in immunological studies with helical collagen (Adelmann *et al.*, 1973). The authors reported a high sensitivity of this assay and could detect antibodies in guinea pig antisera even in a dilution of $1{:}10^6$.

Collagen labeled *in vivo* by injecting [^{14}C]glycine obviously does not gain a sufficiently high specific activity and requires antibody solutions concentrated by immunoadsorption for its serologic use (Lindsley *et al.*, 1971). However, procollagen from cell cultures which were labeled with [^{14}C]glycine, [^{14}C]proline, or [^{3}H]proline served as an excellent antigen in radioimmune tests with antiprocollagen antibodies (von der Mark *et al.*, 1973; Dehm *et al.*, 1974; Sherr and Goldberg, 1973). One may predict that such antigens would be useful in studies on antibodies to collagen and thus overcome difficulties possibly raised by external labeling.

Quantitative precipitation in a fluid medium was successfully applied in several studies with rabbit antibodies and helical collagen, denatured collagen, or α chains (Michaeli *et al.*, 1968, 1969; Fuchs and Harrington, 1970; Maoz *et al.*, 1971). With rabbit antibodies, low ionic strength and low temperatures favor the reaction against helical collagen (Michaeli *et al.*, 1968). An increase in salt concentrations gave better results with chicken antibodies (Beil *et al.*, 1972). The consideration of hydroxyproline in the precipitates as indication of the collagen specificity of the reaction seems questionable in view of the results of LeRoy (1968). Collagenase digestion of such precipitates is, however, one possible way to obtain purified antibodies (Givol and Sela, 1964; Fuchs and Harrington, 1970; Beil *et al.*, 1972).

The value of gel-precipitation studies is still not clear. Agreement exists on positive and apparently specific reactions with α chains, CNBr peptides, or the procollagen extension (Lindsley *et al.*, 1969; Furthmayr *et al.*, 1972; Timpl *et al.*, 1973*a*; Sherr *et al.*, 1973; Dehm *et al.*, 1974; Kohn *et al.*, 1974). Positive results can be obtained in the chicken after only a single injection with collagen (Beil *et al.*, 1972). In the rabbit, the response is less regular and probably requires hyperimmunization (Lindsley *et al.*, 1969, 1971). A gel-precipitation reaction was also observed for collagen denatured with guanidine or urea (Rothbard and Watson, 1972, 1973; Engel and Catchpole, 1972).

Results with helical collagen are considerably more equivocal. In chicken antisera which exhibited comparable levels of antibodies precipitating with helical or uncoiled collagen, only the uncoiled product gave positive reactions in gel diffusion (Beil *et al.*, 1972). A rational understanding of this observation was provided by Dehm *et al.* (1974) who showed that ^{14}C-labeled helical procollagen is incapable of penetrating agar gels except after removal of the rigid helical portion by collagenase. Both studies suggest that the rodlike collagen molecules cannot be studied by gel precipitation. This opinion is at variance with a few reports on a positive precipitation reaction of native collagen in agar or agarose gels (Lustig *et al.*, 1969; Needleman and Stefanovic, 1969; Denduchis and Kefalides, 1970; Heinrich and Struck, 1967; Oh *et al.*, 1968). In the absence of a rigorous proof of collagen specificity in some of these studies, it is difficult to evaluate the results. One should also consider the possibility that tightly bound impurities might dissociate under conditions of the test and thus lead to positive reactions (LeRoy, 1968). The use of an unusually high antigen concentration of 20 mg/ml (Lustig *et al.*, 1969) certainly favors the involvement in this reaction of impurities and/or small amounts of denatured collagen.

One of the most sensitive serological assays, the neutralization of antigen-coated phages by antibody, has been used in a single study on the cross-reaction with synthetic peptides (Maoz *et al.*, 1973*b*). For this purpose, passive cutaneous anaphylaxis has been also applied (Borek *et al.*, 1969). The value of this technique for collagen is denied in another study because of the strong skin irritations provoked by this antigen (Kirrane and Robertson, 1968) but has been used with *Ascaris* collagen (Maoz *et al.*, 1971). Kirrane and Robertson (1968), however, claim the possibility of measuring skin reactions of the immediate type of sensitized guinea pigs.

Immunofluorescence has been applied occasionally to assess the specificity of the antisera (Steffen *et al.*, 1971; Rothbard and Watson, 1965).

The weak immunogenicity of collagen has been considered in previous discussions (Kirrane and Glynn, 1968; O'Dell, 1968) as mandatory for a rigorous and sophisticated control of the specificity of the antisera. Predictions on interference by more immunogenic contaminants were supported in a few studies. Steffen *et al.* (1967*c*) found considerable antibody titers for serum albumin and IgG in antisera against calf collagen of an apparently high purity. These titers increased during hyperimmunization. Attempts to detect albumin in this collagen preparation indicated a concentration of probably less than 0.1%. Subsequent studies confirmed these observations (Wick *et al.*, 1968; LeRoy, 1968; Steffen *et al.*, 1971*a*). Nonserum contaminants were characterized by LeRoy (1968, 1969) as the main antigens recognized in antisera to soluble calf collagen. These acidic proteins appeared after CM-cellulose chromatography in the front peak and could thus be separated from α chains and β components. The front-peak material still contained some hydroxyproline, and its serologic activity could be destroyed by collagenase. In spite of these findings, there was neither a structural nor a serologic relationship to α chains. Later studies which probably used better purified collagens did not find serologic activity in the front-peak material but in the α chains (Michaeli *et al.*, 1969; Pontz *et al.*, 1970).

Fuchs *et al.* (1964*b*) reported that simultaneous injection of collagen circumvented the need for T-cell help in the antibody response to ovalbumin. This observation might provide an explanation of the ease of eliciting antibodies to minute amounts of noncollagen contaminants.

Cross-reactions of anticollagen sera with serum proteins have been occasionally interpreted as indicating small amounts of collagen antigen in serum (Frey *et al.*, 1965; Oh *et al.*, 1968). Approaches along this line certainly require anticollagen antibodies of well-defined specificity. The interest in this question was recently revived by evidence for the occurrence of collagen-like sequences in the complement component Clq (Reid *et al.*, 1972).

As indicated by these studies, it is probably impossible to purify triple-helical collagen to a degree which will prevent any antibody response against noncollagen impurities. The sensitivity of serologic activity of the antigens towards a collagenase treatment was usually considered good evidence for the collagen specificity of the antibodies. This appears now questionable since even purified, commercially available enzyme preparations were found to contain other proteases (Peterkofsky and Diegelmann, 1971). In addition it has to be considered that even an extremely pure collagenase may cleave peptide bonds in proteins other than collagen or change the conformation of associated contaminant proteins upon destruction of the collagen molecule.

Fortunately, the recent developments in the experimental methods and the increasing understanding of the architecture of the collagen molecule provided a new set of applicable criteria. Thus, any study on the serologic specificity of anticollagen antisera should include either purified α chains or smaller fragments like CNBr peptides. If the studies are concerned with conformational antigenic determinants, it is recommended that renaturation products prepared from these purified constituents should be used. A different situation is faced if the antibodies are designed for studying cells or tissues. Since purification of the antigens is not intended, one has to purify the antibodies, e.g., by immunoadsorption. The use of highly purified collagen for this purpose is probably sufficient since a preferential binding of anticollagen of antibodies to impurities should be expected on such columns. Further improvement in this method can certainly be achieved by the application of purified collagen subunits as the adsorbing antigens.

IX. *Antibodies as Tools to Study Structure and Metabolism of Collagen*

Structural and immunological studies on collagen in the few last years exhibited a relationship of mutual benefit. The prediction of nonhelical sequences in the collagen molecule was mainly based on immunological findings (Schmitt *et al.*, 1964; Pontz *et al.*, 1970) and could be subsequently verified by demonstrating such sequences at the N-terminal and C-terminal sites of the α-chains (reviewed by Traub and Piez, 1971). Intensive work on the sequence of these terminal regions provided, on the other hand, the background for elucidating the structure of antigenic determinants.

Since the nonhelical regions are important for cross-linking (see Chapter 4), antibodies may be potential tools for characterizing cross-linked peptides derived from insoluble collagen. Thus, the involvement in cross-linking of the nonhelical sequence in the C-terminal CNBr peptide α1-CB6 was mainly deduced from immunological evidence (Zimmermann *et al.,* 1973). In another study Becker *et al.* (1975*a*) demonstrated terminal antigenic determinants on tryptic peptides cross-linked to sites which originate from the helical portion of the molecule. Apparently, the antibodies to terminal antigenic determinants so far studied do not distinguish between nonhelical regions either in their free or cross-linked form. Changes in serologic specificity observed upon oxidation of lysine residues to aldehydes (Furthmayr and Timpl, 1972) are the first indications that such discriminating antibody reagents might become available. Evidence is also lacking for the presence of antigenic sites unique to cross-linked peptides, e.g., involving two different polypeptide chains held together by the cross-link. Antibodies against such sites should improve the arsenal of powerful tools designed for screening pathologic changes occurring in collagen at the molecular level.

The discovery of unique antigenic determinants on the procollagen extension stimulated the application of antibodies against these sites in studies on the metabolism of collagen. Kohn *et al.* (1974) identified free extensions which were cleaved off from dermatosparactic procollagen by procollagen peptidase. This process must be accompanied by the uncovering of a new antigenic determinant, e.g., that at the very N-terminal site of the $\alpha 1$-chain (see Fig. 3) which is essentially masked in the pro-$\alpha 1$-chain (Timpl *et al.*, 1973*a*).

With the aid of antibodies to chick procollagen Dehm *et al.* (1974) detected a fetal antigen in chick serum showing complete cross-reaction with the procollagen extension. Whether this antigen resembles a free extension which accumulates during certain stages of embryonic development is not yet clear. Purified antibodies to chick collagen were also applied to identify nascent polypeptide chains on collagen-synthesizing polysomes (Pawlowski *et al.*, 1975).

Localization of procollagen by purified antibodies to dermatosparactic calf procollagen in normal animal and human tissues was attempted by applying the immunofluorescence technique (Timpl *et al.*, 1973*a*). In spite of strong staining of interstitial connective tissue in muscle and kidney, the glomeruli were apparently not involved. In the skin, as demonstrated in Figure 4, only a small subepithelial zone beneath the basement membrane reacted with antiprocollagen. Antibodies to a C-terminal antigenic determinant of the $\alpha 1(\mathrm{I})$ chain stained the entire dermis. Changes in localization of the procollagen antigen were found in the skin of patients suffering from psoriasis or lichen ruber and in some keloids (Wick *et al.*, 1974, 1976*a*). Olsen and Prockop (1974) succeeded in determining the intracellular location of chick procollagen by electron microscopy. This approach was based on a new technique for coupling ferritin to antibodies and a method allowing improved penetration of the reagent into cells (Olsen *et al.*, 1973). As shown in Figure 5, heavy ferritin staining was found in the cisternae of the rough endoplasmic reticulum, but other organelles and the cytoplasm remained essentially unstained. Since under

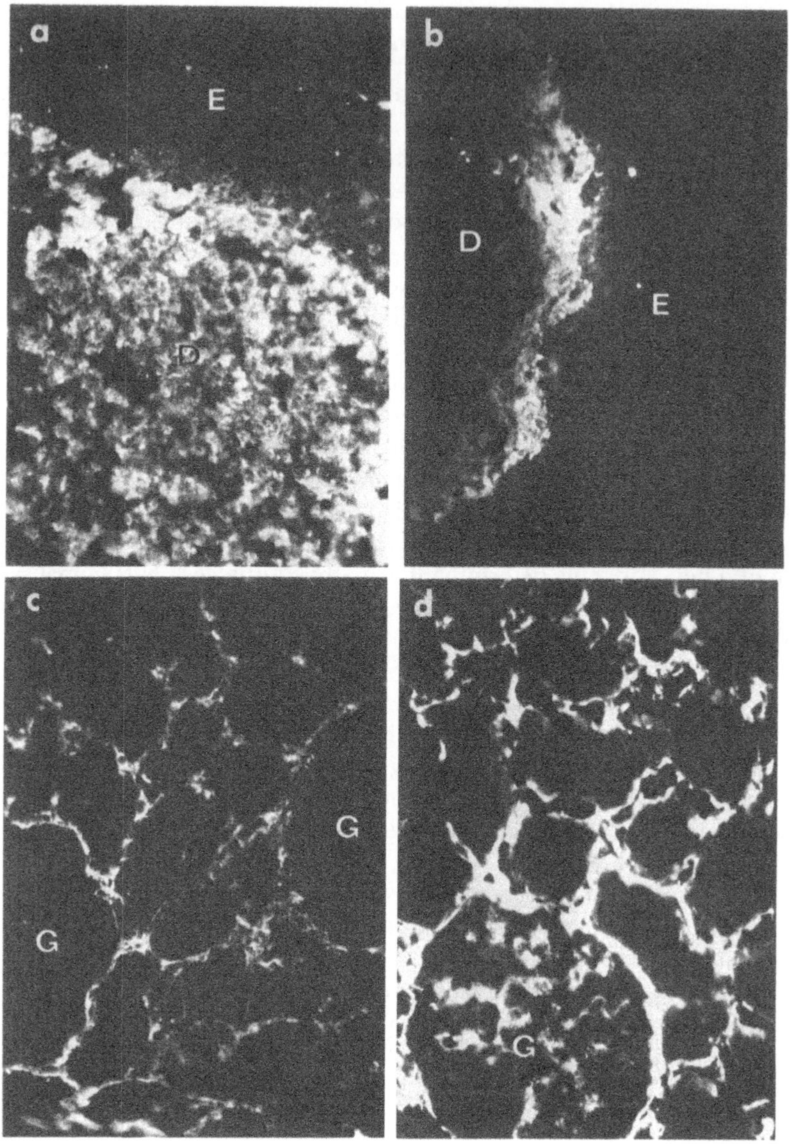

FIGURE 4. Indirect immunofluorescence patterns on frozen tissue sections of purified antibodies to collagen and procollagen. Top: human dermis stained with rabbit antibodies to a C-terminal antigenic determinant in calf collagen (a) or with antibodies to the N-terminal extension in dermatosparactic calf procollagen (b). Bottom: rat kidney stained with chick antibodies to helical rat skin collagen which were purified on a rat collagen adsorbent (c) or with residual antibodies of unknown specificity in the same antiserum from which anticollagen antibodies have been removed by repeated adsorption on helical and uncoiled rat collagen (d). Note in the latter case the heavy staining of the glomeruli. D = dermis, E= epidermis, G = glomerulus. The photographs were obtained by the courtesy of Dr. Georg Wick.

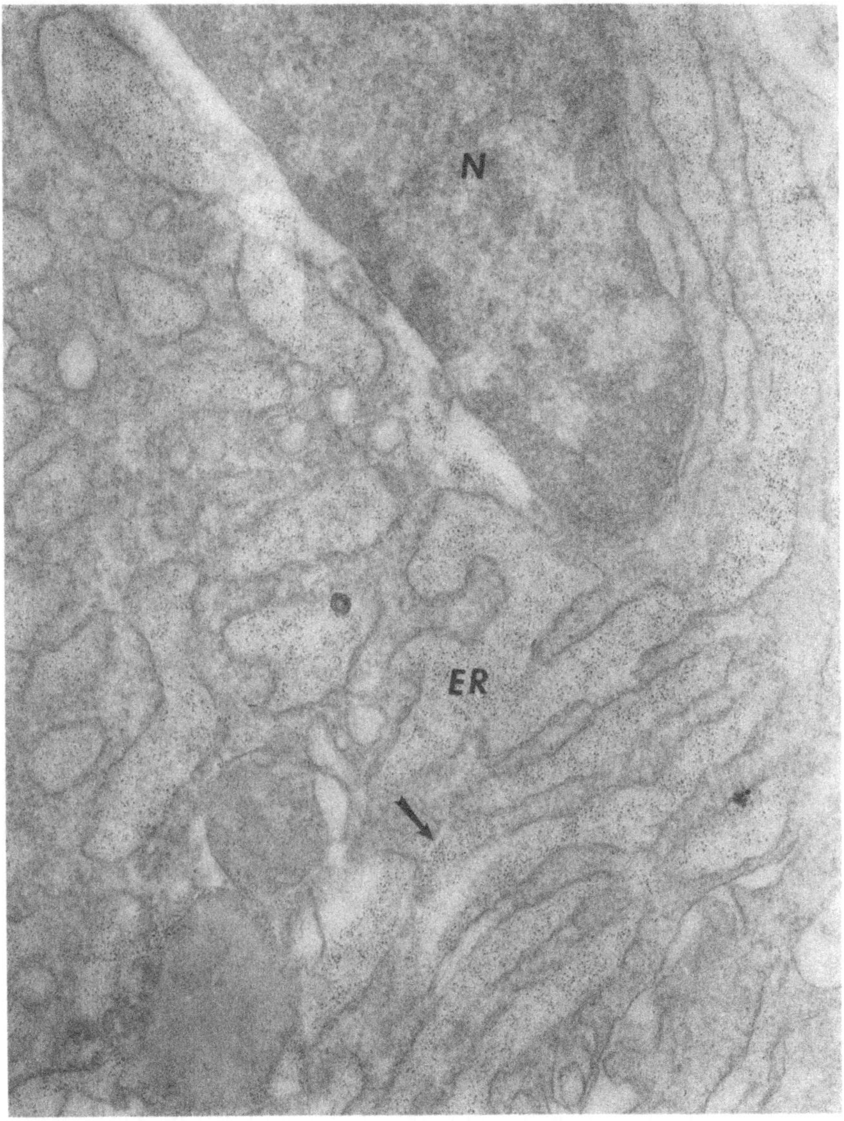

FIGURE 5. Electron micrograph showing part of the tendon fibroblast incubated with ferritin-labeled antiprocollagen antibodies. The fibroblasts were obtained from 17-day-old chick embryonic tendons and partially fragmented by homogenization (for further technical details see Olsen and Prockop, 1974). Note the heavy labeling of the rough endoplasmic reticulum (ER), the lack of labeling of the cell nucleus (N), and the absence of labeling in the cytoplasm except in places where the membranes of the endoplasmic reticulum are ruptured (arrow). (Magnification: × 50,000.) The photograph was obtained by the courtesy of Dr. Bjørn R. Olsen

certain conditions some stain was also observed in the Golgi apparatus, the value of these antibodies in studying the intracellular transport of procollagen is obvious (Nist *et al.*, 1975).

A considerable literature exists on the use of antisera to skin, tendon, or dura mater collagen in immunofluorescence studies. The results agree on a staining of extracellular spaces in kidney, muscle, spleen, skin, and various other organs (Mancini *et al.*, 1965; Rothbard and Watson, 1965, 1967, 1969, 1972; Steffen *et al.*, 1971*a*; Engel and Catchpole, 1972). Staining of fibroblasts was also reported (Lustig *et al.*, 1969). Except in a single study (Mancini *et al.*, 1965), the antisera also showed strong reaction with glomerular basement membranes. More recently, Wick *et al.* (1975) provided evidence that staining of the basement membrane may be an artifact due to antibodies against noncollagen impurities. As shown in Figure 4 purified anticollagen antibodies react in a restricted way with interstitial tissue of kidneys only. On the other hand, anticollagen antisera deprived of anticollagen antibodies by repeated immunoadsorption still reacted with the same region in the kidney sections but also quite distinctly with glomeruli. This observation apparently makes the use of purified antibodies imperative in immunohistological studies on collagen and procollagen.

An electron microscopical investigation with peroxidase-conjugated anti-chick collagen antibodies (Lustig, 1971) revealed intensive cellular and extracellular staining in skin and tendon. The demonstration of collagen in the cytoplasm is in apparent disagreement with results on the localization of procollagen (Olsen and Prockop, 1974). This finding, as well as the lack of uniformity in extracellular staining, might mirror the technical objections against the reliability of the peroxidase technique.

Several immunological studies indicated that it is possible to distinguish between type I, II, and III collagens and procollagens using immunological methods (Hahn *et al.*, 1974; 1975*a*; Becker *et al.*, 1976; Nowack *et al.*, 1976*c*). Some evidence exists also for a distinct specificity of antibodies to basement membrane (type IV) collagen (Denduchis and Kefalides, 1970). By using purified antibodies it could be shown that type III collagen is a component of reticulin (Gay *et al.*, 1975*b*; Becker *et al.*, 1976). Type III procollagen could also be demonstrated in reticulin and might have a structural function since it persists extracellularly (Nowack *et al.*, 1976*c*). The distribution of type I procollagen is much more restricted, e.g., in skin to the stratum papillare (Wick *et al.*, 1975) and in liver to a small zone in the capsula (Nowack *et al.*, 1976*c*). Antibodies to type II collagen react with the matrix of hyaline cartilage (von der Mark *et al.*, 1976; Wick *et al.*, 1976*b*) but do not react with type I collagen in the perichondrial tissue. Antibodies specific for distinct types of collagen have also been used to study connective tissue diseases such as liver cirrhosis and arteriosclerosis (Gay *et al.*, 1975*a,b*). A localized synthesis of type I collagen was found in the lesions in the articular cartilage of osteoarthrotic joints (S. Gay, personal communication). The synthesis of a wrong type of collagen might as suggested by Deshmukh and Nimni (1973) be correlated with the development of osteoarthrosis.

Collagen synthesis by cells in culture was also studied by these antibody reagents. Depending on the culture conditions chondrocytes may switch in synthesis from type II to type I collagen (Müller *et al.*, 1975*a*). Recent studies (Gay, Martin, Müller, Timpl, and Kühn, in preparation) showed simultaneous synthesis of type I and III collagen in single fibroblasts. These observations are illustrated by a double-staining experiment shown in Fig. 6. No staining by antitype III collagen antibodies was observed in cells from a patient with Ehlers–Danlos syndrome type IV, which corresponds to previous data showing that these cells do not synthesize type III collagen (Pope *et al.*, 1975). Staining by antibodies to

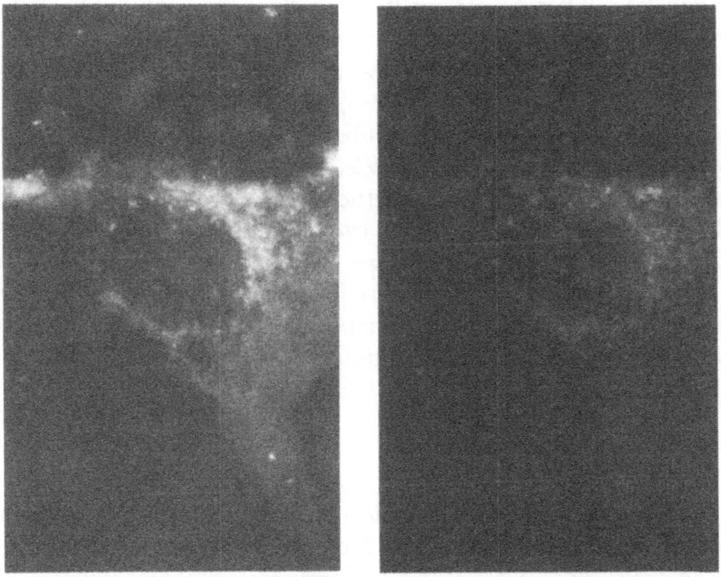

FIGURE 6. Simultaneous staining of a fibroblast from fetal human skin by rabbit antibodies to type III collagen (left) and mouse antibodies to type I collagen (right). The antibodies were visualized by fluorescein isothiocyanate–conjugated goat antibodies to rabbit immunoglobulin (left) or by tetramethylrhodamine isothiocyanate-conjugated goat antibodies to mouse immunoglobulin (right). Magnification ×480. The photographs were obtained by the courtesy of Dr. Steffen Gay.

type III collagen was also used to characterize fibroblasts from a patient with osteogenesis imperfecta (Müller *et al.*, 1975*b*). Von der Mark *et al.* (1976) used antibodies to type I and II collagen to study developmental aspects of collagen biosynthesis in the chick embryo.

In several other recent immunofluorescence studies with fibroblasts, antisera were used instead of purified antibodies. Faulk *et al.* (1975) reported on a capping of the anticollagen stain similar to those cappings observed for immunoglobulin receptors and histocompatibility antigens and they interpreted the results as indicating membrane-bound collagen. It remains, however, to be explained why the reactive antigen is destroyed by pronase but not by collagenase. Duksin *et al.* (1975) could stain both osteoblasts and skin fibroblasts by an antiserum to $(\text{Pro-Gly-Pro})_n$ while only the fibroblasts could be lysed after addition of complement.

Studies on cytotoxic effects of anticollagen antisera on fibroblasts were initiated some 20 years ago (Robbins *et al.*, 1955). Basic findings of this study, like the dependence on complement, were confirmed in a recent, more extended report (Lustig, 1970). Besides some doubt on the exclusive collagen specificity of the antisera, there is also no certainty as to whether the cell damage occurs by interaction of the antibodies either with membrane-bound collagen or with secreted, extracellular collagen. Observations of Maoz *et al.* (1973*c*) of a cytotoxic effect on fibroblasts of antibodies to polymeric $(\text{Pro-Gly-Pro})_n$ strengthened the probability of the cytotoxic action of anticollagen antibodies. The antibodies to the polymer also killed cells from striated muscle, but it cannot be excluded that this effect occurred via reaction with the gelatin layer used to cultivate such cells. The possibility of achieving cytotoxic action and/or interference of fibril formation by antibodies *in vivo* was indicated by the study of Wick *et al.* (1968) who inhibited granuloma formation in rats by local application of anticollagen antisera. Unfortunately in this, as well as in many other studies on biological effects, antisera were used rather than purified antibodies.

ACKNOWLEDGMENTS

This article was written while the author was visitor at the Department of Biochemistry, Rutgers Medical School. I want to thank Dr. Darwin J. Prockop and his staff for providing all the facilities required to complete this work. I wish to thank also Drs. Gad Avigad, Edward J. Miller, and Heinz Furthmayr for a critical and careful reading of the manuscript. I further appreciated the help of many colleagues who provided results and manuscripts prior to publication.

References

Adelmann, B. C., 1972, The structural basis of cell-mediated immunological reactions of collagen. Reactivity of separated α-chains of calf and rat collagen in cutaneous delayed hypersensitivity reactions, *Immunology* **23:**739.

Adelmann, B. C., 1973, The structural basis of cell-mediated immunological reactions of collagen. Recognition by the cutaneous delayed hypersensitivity reaction in guinea pigs of conformational alterations of rat and calf skin collagen, *Immunology* **24:**871.

Adelmann, B. C., and Kirrane, J., 1973, The structural basis of cell-mediated immunological reactions of collagen. The species specificity of the cutaneous delayed hypersensitivity reaction, *Immunology* **25:**123.

Adelmann, B. C., Glynn, L. E., and Kirrane, J., 1968, Delayed hypersensitivity reactions of native and denatured rat and calf skin collagen in sensitized guinea pigs, *Fed. Proc.* **27:**263.

Adelmann, B. C., Kirrane, J. A., and Glynn, L. E., 1972, The structural basis of cell-mediated immunological reactions of collagen. Characteristics of cutaneous delayed hypersensitivity reactions in specifically sensitized guinea pigs, *Immunology* **23:**723.

Adelmann, B. C., Gentner, G. J., and Hopper, K., 1973, A sensitive radioimmunoassay for collagen, *J. Immunol. Methods* **3:**319.

Aguilar, J. H., Jacobs, H. G., Butler, W. T., and Cunningham, L. W., 1973, The distribution of carbohydrate groups in rat skin collagen, *J. Biol. Chem.* **248:**5106.

Allison, A. C., 1974, The roles of T and B lymphocytes in self-tolerance and autoimmunity, *Contemp. Top. Immunbiol,* **3:**227.

Arnon, R., and Sela, M., 1960, Studies on the chemical basis of the antigenicity of proteins. 2. Antigenic specificity of polytyrosyl gelatins, *Biochem. J.* **75:**103.

Becker, U., and Timpl, R., 1972, Cyanogen bromide peptides of the rabbit collagen α1-chain, *FEBS Lett.* **27:**85.

Becker, U., Timpl, R., and Kühn, K., 1972, Carboxyterminal antigenic determinants of collagen from calf skin. Localization within discrete regions of the non-helical sequence, *Eur. J. Biochem.* **28:**221.

Becker, U., Furthmayr, H., and Timpl, R., 1975*a*, Tryptic peptides from the cross-linking regions of insoluble calf skin collagen, *Hoppe Seyler's Z. Physiol. Chem.* **356:**21.

Becker, U., Fietzek, P. P., Furthmayr, H., and Timpl, R., 1975*b*, Non-helical sequences of rabbit collagen. Correlation with antigenic determinants detected by rabbit antibodies in homologous regions on rat and calf collagen, *Eur. J. Biochem.* **54:**359.

Becker, U., Nowack, H., Gay, S., and Timpl, R., 1976, Production and specificity of antibodies against the amino-terminal region in type III collagen, *Immunology,* in press.

Beil, W., Furthmayr, H., and Timpl, R., 1972, Chicken antibodies to soluble rat collagen. I. Characterization of the immune response by precipitation and agglutination methods, *Immunochemistry* **9:**779.

Beil, W., Timpl, R., and Furthmayr, H., 1973, Conformation dependence of antigenic determinants on the collagen molecule, *Immunology* **24:**13.

Benacerraf, B., Katz, D. H., Kapp, J. A., and Pierce, C. W., 1974, Regulatory mechanisms in the immune response, *in: Cellular Selection and Regulation in the Immune Response* (G. M. Edelman, ed.), pp. 265–281, Raven Press, New York.

Benjamini, E., Scibiensky, R. J., and Thompson, K., 1972, The relationship between antigenic structure and immune specificity, *Contemp. Top. Immunochem.* **1:**1.

Benjamini, E., Michaeli, D., Leung, C. Y., Wong, K., and Scheuenstuhl, H., 1973, Immunochemical studies with synthetic peptides related to collagen, *Immunochemistry* **10:**629.

Borek, F., Kurtz, J., and Sela, M., 1969, Immunological properties of a collagen-like synthetic polypeptide, *Biochim. Biophys. Acta* **188:**314.

Bornstein, P., and Nesse, R., 1970, The comparative biochemistry of collagen: The structure of rabbit skin collagen and its relevance to immunochemical studies of collagen, *Arch. Biochem. Biophys.* **138:**443.

Bray, J. P., Estess, F., and Bass, J. A., 1969, Anticollagen antibodies following thermal trauma, *Proc. Soc. Exp. Biol. Med.* **130:**394.

Brown, P. C., and Glynn, L. E., 1968, Antigenicity and tolerance of some synthetic polypeptides related to collagen, *Immunology* **15:**589.

Brown, P. C., and Glynn, L. E., 1973, The antigenicity of sequential polypeptides. II. The antigenicity of some sequential polymers including several related to collagen, *Immunology* **25:**251.

Brunner, H., Dichtl, M., and Steffen, C., 1971, Immunogenität und Spezifität von chromstabilisiertem Kalbskollagen, *Z. Immunforsch.* **141:**460.

Byers, P. H., Click, E. M., Harper, E., and Bornstein, P., 1975, Interchain disulfide bonds in procollagen are located in a large COOH-terminal domain, *Proc. Natl. Acad. Sci. U.S.A.* **72:**3009.

Chiller, J. M., Habicht, G. S., and Weigle, W. O., 1971, Kinetic differences in unresponsiveness of thymus and bone marrow cells, *Science* **171:**813.

Chung, E., and Miller, E. J., 1974, Collagen polymorphism: Characterization of molecules with the chain composition $(\alpha 1(III))_3$ in human tissues, *Science* **183:**1200.

Chvapil, M., Kronenthal, R. L., and van Winkle, W., 1973, Medical and surgical applications of collagen, *Int. Rev. Connect. Tissue Res.* **6:**1.

Click, E. M., and Bornstein, P., 1970, Isolation and characterization of the cyanogen bromide peptides from the α1 and α2 chains of human skin collagen, *Biochemistry* **9:**4699.

Cooke, T. D., Hurd, E. R., Ziff, M., and Jasin, H., 1972, The pathogenesis of chronic inflammation in experimental antigen-induced arthritis. II. Preferential localization of antigen-antibody complex to collageneous tissues, *J. Exp. Med.* **135:**323.

Cooper, N. R., 1973, Activation of the complement system, *Contemp. Top. Mol. Immunol.* **2:**155.

Davison, P. F., Levine, L., Drake, M. P., Rubin, A., and Bump, S., 1967, The serologic specificity of tropocollagen telopeptides, *J. Exp. Med.* **126:**331.

Dehm, P., Olsen, B. R., and Prockop, D. J., 1974, Antibodies to chick tendon procollagen. Affinity purification with the isolated disulfide-linked NH_2-terminal extensions and reactivity with a component in embryonic serum, *Eur. J. Biochem.* **46:**107.

Denduchis, B., and Kefalides, N. A., 1970, Immunochemistry of sheep anterior lens capsula, *Biochim. Biophys. Acta* **221:**357.

Deshmukh, K., and Nimni, M. E., 1973, Effects of lysosomal enzymes on the type of collagen synthesized by bovine articular cartilage, *Biochem. Biophys. Res. Commun.* **53:**424.

Duksin, D., Maoz, A., and Fuchs, S., 1975, Differential cytotoxic activity of anticollagen serum on rat osteoblasts and fibroblasts in tissue culture, *Cell* **5:**83.

Engel, J., Kurtz, J., Katchalski, E., and Berger, A., 1966, Polymers of tripeptides as collagen models. II. Conformational changes of poly (L-prolyl-glycyl-L-prolyl) in solution, *J. Mol. Biol.* **17:**255.

Engel, M. B., and Catchpole, H. R., 1972, Collagen distribution in the rat demonstrated by a specific anti-collagen antiserum, *Am. J. Anat.* **134:**23.

Essen, R. von, Savilahti, E., and Pelkonen, P., 1972, Reticulin antibody in children with malabsorption, *Lancet* **1:**1157.

Faulk, W. P., Conochie, L. B., Temple, A., and Papamichail, M., 1975, Immunobiology of membrane-bound collagen on mouse fibroblasts, *Nature* **256:**123.

Fietzek, P., and Kühn, K., 1976, The primary structure of collagen, *Int. Rev. Connect. Tissue Res.* **7:**1.

Fietzek, P. P., Breitkreutz, D., and Kühn, K., 1974, Amino acid sequence of the aminoterminal region of calf skin collagen, *Biochim. Biophys. Acta* **365:**305.

Frey, J., Hoa, N., Gras, J., Henry, J. C., and Gudefin, Y., 1965, Immunoelectrophoresis reveals collagen solubility in human serum, *Science* **150:**751.

Fuchs, S., and Harrington, W. F., 1970, Immunological properties of *Ascaris* cuticle collagen, *Biochim. Biophys. Acta* **221:**119.

Fuchs, S., Mozes, E., Maoz, A., and Sela, M., 1974*a*, Thymus independence of a collagen-like synthetic polypeptide and of collagen, and the need for thymus and bone marrow-cell cooperation in the immune response to gelatin, *J. Exp. Med.* **139:**148.

Fuchs, S., Maoz, A., and Mozes, E., 1974*b*, The effect of T-independent antigens on the requirement for cell cooperation in the immune response to T-dependent antigens, *Isr. J. Med. Sci.* **10:**1116.

Furthmayr, H., and Timpl, R., 1972, Structural requirements of antigenic determinants in the aminoterminal region of the rat collagen α2 chain, *Biochem. Biophys. Res. Commun.* **47:**944.

Furthmayr, H., and Timpl, R. 1976, Immunochemistry of collagens and procollagens, *Int. Rev. Connect. Tissue Res.* **7:**61.

Furthmayr, H., Beil, W., and Timpl, R., 1971, Different antigenic determinants in the polypeptide chains of human collagen, *FEBS Lett.* **12:**341.

Furthmayr, H., Stoltz, M., Becker, U., Beil, W., and Timpl, R., 1972*a*, Chicken antibodies to soluble rat collagen. II. Specificity of the reactions with individual polypeptide chains and cyanogen bromide peptides of collagen, *Immunochemistry* **9:**789.

Furthmayr, H., Timpl, R., Stark, M., Lapière, C. M., and Kühn, K., 1972*b*, Chemical properties of the peptide extension of the pα1 chain of dermatosparactic skin procollagen, *FEBS Lett.* **28:**247.

Gay, S., Fietzek, P. P., Remberger, K., Eder, M., and Kühn, K., 1975*a*, Liver cirrhosis: immunofluorescence and biochemical studies demonstrate two types of collagen, *Klin. Wschr.* **53:**205.

Gay, S., Balleisen, L., Remberger, K., Fietzek, P. P., Adelmann, B. C., and Kühn, K., 1975*b*, Immunohistochemical evidence for the presence of collagen type III in human arterial walls, arterial thrombi and in leucocytes incubated with collagen *in vitro*, *Klin. Wschr.* **53:**899.

Gershon, R. K., 1974, T cell control of antibody production, *Contemp. Top. Immunbiol.* **3:**1.

Givol, D., and Sela, M., 1964, Isolation and fragmentation of antibodies to polytyrosyl gelatin, *Biochemistry* **3:**444.

Glynn, L. E., 1969, Aetiology of rheumatoid arthritis with regard to its chronicity, *Ann. Rheum. Dis. (Suppl.)* **28:**3.

Good, R. A., 1972, Structure-function relations in the lymphoid system, *in: Clinical Immunbiology* (F. H. Bach and R. A. Good, eds.), Vol. 1, pp. 1–28, Academic Press, New York and London.

Gurari, D., Ungar-Waron, H., and Sela, M., 1973, Antigenicity of dinitrophenyl polyproline, *Eur. J. Immunol.* **3:**196.

Hahn, E., and Timpl, R., 1973, Involvement of more than a single polypeptide chain in the helical antigenic determinants of collagen, *Eur. J. Immunol.* **3**:442.

Hahn, E., Timpl, R., and Miller, E. J., 1974, The production of specific antibodies to native collagens with the chain compositions, $(\alpha 1(I))_3$, $(\alpha 1(II))_3$ and $(\alpha 1(I))_2\alpha 2$, *J. Immunol.* **113**:421.

Hahn, E., Timpl, R., and Miller, E. J., 1975*a*, Demonstration of a unique antigenic specificity for the collagen $\alpha 1$(II) chain from cartilaginous tissue *Immunology* **28**:561.

Hahn, E., Nowack, H., Götze, D., and Timpl, R., 1975*b*, H-2-linked genetic control of antibody response to soluble calf skin collagen in mice, *Eur. J. Immunol.* **5**:288.

Heinrich, S., and Struck, H., 1967, Serologische Untersuchung zur Antigenität und Spezifität löslicher Kollagen-fraktionen. III. Serologisches Verhalten von neutralsalzlöslichem und saurelöslichem Kollagen aus Kalbhaut in der Agar-Gel-Präzipitation nach Ouchterlony, *Z. Immunitätforschg.* **133**:479.

Hopper, K. E., Adelmann, B. C., Gentner, G., and Gay, S., 1976, Recognition by guinea-pig peritoneal exudate cells of conformationally different states of the collagen molecule, *Immunology* **30**:249.

Jasin, H. E., and Glynn, L. E., 1965*a*, The antigenicity of some synthetic poly-imino acids. I. Antigenic properties of poly-L-proline, *Immunology* **8**:95.

Jasin, H. E., and Glynn, L. E., 1965*b*, The antigenic properties of some synthetic poly-imino acids. II. The antigenicity of polypeptides related to collagen; peptides containing hydroxyproline and acetyl-hydroxyproline, *Immunology* **8**:260.

Katz, D. H., and Benacerraf, B., 1972, The regulatory influence of activated T cells on B cell responses to antigen, *Adv. Immunol.* **15**:1.

Kefalides, N. A., 1973, Structure and biosynthesis of basement membranes, *Int. Rev. Connect. Tissue Res.* **6**:63.

Kettman, J. R., Benjamini, E., Michaeli, D., and Leung, D. Y. K., 1967, The synthesis and immunological activity of a peptide related to collagen, *Biochem. Biophys. Res. Commun.* **29**:623.

Kirrane, J. A., and Glynn, L. E., 1968, Immunology of collagen, *Int. Rev. Connect. Tissue Res.* **4**:1.

Kirrane, J. A., and Robertson, W. van B., 1968, The antigenicity of native and tyrosylated neutral-salt-soluble rat collagen, *Immunology* **14**:139.

Kobayashi, Y., Sakai, R., Kakiuchi, K., and Isemura, T., 1970, Physicochemical analysis of $(\text{Pro-Pro-Gly})_n$ with defined molecular weight-temperature dependence of molecular weight in aqueous solution, *Biopolymers* **9**:415.

Kohn, L. D., Isersky, C., Zupnik, J., Lenaers, A., Lee, G., and Lapière, C. M., 1974, Calf tendon procollagen peptidase: Its purification and endopeptidase mode of action, *Proc. Natl. Acad. Sci. U.S.A.* **71**:40.

Kriegel, W., Langness, U., Jahn, P., and Müller, W., 1970, "Collagen-like protein" (CLP) und Kollagen-antikörper bei der chronischen Polyarthritis, *Z. Rheumaforsch.* **29**:173.

Kühn, K., and Eggl, M., 1966, An electron microscope investigation of acid-soluble collagen treated with proteases, *Biochem. Z.* **344**:215.

Lee, G., Tate, R., Martin, G., and Kohn, L., 1975, Purification of type III procollagen from bovine amniotic fluid and the development of a specific antiserum to type III procollagen, *Fed. Proc.* **34**:696.

Lehman, D. H., Marquardt, H., Wilson, C. B., and Dixon, F. J., 1974, Specificity of autoantibodies to tubular and glomerular basement membranes induced in guinea pigs, *J. Immunol.* **112**:241.

LeRoy, C. E., 1968, Precipitating and complement-fixing antibodies to collagen with species and collagen subunit specificity, *Proc. Soc. Exp. Biol. Med.* **128**:341.

LeRoy, C. E., 1969, Some characteristics of the antigenic moiety of calf skin collagen, *J. Immunol.* **102:**919.

Lindsley, H. B., Mannik, M., and Bornstein, P., 1969, Immunogenicity of rat and human collagen polypeptide chains, *Arthritis. Rheum.* **12:**676.

Lindsley, H., Mannik, M., and Bornstein, P., 1971, The distribution of antigenic determinants in rat skin collagen, *J. Exp. Med.* **133:**1309.

Lustig, L., 1970, Cytotoxic action of an antiserum to soluble collagen on tissue culture of fibroblasts, *Proc. Soc. Exp. Biol. Med.* **133:**207.

Lustig, L., 1971, Ultrastructural localization of a soluble collagen antigen, *J. Histochem. Cytochem.* **19:**663.

Lustig, L., Constantini, H., and Mancini, R. E., 1969, Serological and immunofluorescent study of soluble collagen in tissue culture of fibroblasts, *Proc. Soc. Exp. Biol. Med.* **130:**283.

Mahieu, P. M., Lambert, P. H., and Maghuin-Rogister, G. R., 1973, Primary structure of a small glycopeptide isolated from human glomerular basement membrane and carrying a major antigenic site, *Eur. J. Biochem.* **40:**599.

Mancini, R. E., Paz, M., Vilar, O., Davidson, O. W., and Barquet, J., 1965, Histo-immunological detection of collagen fractions, *Proc. Soc. Exp. Biol. Med.* **118:**346.

Maoz, A., and Fuchs, S., 1974, Fractionation of mouse lymphocyte populations on collagen fibers and gels, *Isr. J. Med. Sci.* **10:**1117.

Maoz, A., Fuchs, S., and Michaeli, D., 1971, Ascaris cuticle collagen: On the antigenic specificity of the subunits, *Biochim. Biophys. Acta* **243:**106.

Maoz, A., Fuchs, S., and Sela, M., 1973*a*, Immune response to the collagen-like synthetic ordered polypeptide (L-Pro-Gly-L-Pro)$_n$, *Biochemistry* **12:**4238.

Maoz, A., Fuchs, S., and Sela, M., 1973*b*, On immunological cross-reactions between the synthetic ordered polypeptide (L-Pro-Gly-L-Pro)$_n$ and several collagens, *Biochemistry* **12:**4246.

Maoz, A., Dym, H., Fuchs, S., and Sela, M., 1973*c*, Cytotoxic activity of antibodies to a collagen-like synthetic polytripeptide on cells in tissue culture, *Eur. J. Immunol.* **3:**839.

Marquardt, H., Wilson, C. B., and Dixon, F. J., 1973*a*, Human glomerular basement membrane. Selective solubilization with chaotropes and chemical and immunological characterization of its components, *Biochemistry* **12:**3260.

Marquardt, H., Wilson, C. B., and Dixon, F. J., 1973*b*, Isolation and immunological characterization of human glomerular basement membrane antigens, *Kidney Int.* **3:**57.

Martin, G. R., Byers, P. H., and Piez, K. A., 1975, Procollagen, *Adv. Enzymol.* **42:**167.

Maurer, P. H., 1954, I. Antigenicity of oxypolygelatin and gelatin in man, *J. Exp. Med.* **100:**497.

McBride, O. W., and Harrington, W. F., 1967*a*, Ascaris cuticle collagen: On the disulfide cross-linkages and the molecular properties of the subunits, *Biochemistry* **6:**1484.

McBride, O. W., and Harrington, W. F., 1967*b*, Helix–coil transition in collagen. Evidence for a single-stranded triple helix, *Biochemistry* **6:**1499.

McDevitt, H. O., 1972, H-2 region and H-linked Ir genes, *in: Genetic Control of Immune Responsiveness* (H. O. McDevitt and M. Landy, eds.), pp. 92–101, Academic Press, New York and London.

McIntosh, R. M., and Griswold, W., 1971, Antigen identification in Goodpasture's syndrome, *Arch. Pathol.* **92:**329.

Meigel, W., Pontz, B., Timpl, R., Hieber, E., Nordwig, A., Steffen, C., and Kühn, K., 1971, Comparative immunological studies on fish and invertebrate collagens, *J. Immunol.* **107:**1146.

Michaeli, D., and Epstein, E. H., 1971, Isolation and identification of the antigenic determinants of human collagen, *Isr. J. Med. Sci.* **7**:462.

Michaeli, D., and Fudenberg, H. H., 1974, The incidence and antigenic specificity of antibodies against denatured human collagen in rheumatoid arthritis, *Clin. Immunol. Immunpathol.* **2**:153.

Michaeli, D., Kamenecka, H., Benjamini, E., Kettman, J. R., Leung, D. Y. K., and Miner, R. C., 1968, Immunochemical studies on collagen. I. The binding and precipitation of guinea-pig skin collagen and gelatin with rabbit anti-collagen, *Immunochemistry* **5**:433.

Michaeli, D., Martin, G. R., Kettman, J., Benjamini, E., Leung, D. Y. K., and Blatt, B. A., 1969, Localization of antigenic determinants in the polypeptide chains of collagen, *Science* **166**:1522.

Michaeli, D., Senyk, G., Maoz, A., and Fuchs, S., 1972, Ascaris cuticle collagen and mammalian collagens: Cell mediated and humoral immunity relationships, *J. Immunol.* **109**:103.

Michaeli, D., Benjamini, E., Leung, D. Y. K., and Martin, G. R., 1971, Immunochemical studies on collagen. II. Antigenic differences between guinea pig skin collagen and gelatin, *Immunochemistry* **8**:1.

Miller, E. J., 1973, A review of biochemical studies on the genetically distinct collagens of the skeletal system, *Clin. Orthop.* **92**:260.

Miyata, T., Sohde, T., Rubin, A. L., and Stenzel, K. H., 1971, Effects of ultraviolet irradiation on native and telopeptide-poor collagen, *Biochim. Biophys. Acta* **229**:672.

Mohos, S. C., and Wagner, B. M., 1969, Damage to collagen in corneal immune injury. Observation of connective tissue structure, *Arch. Pathol.* **88**:3.

Müller, P., Lemmen, C., Gay, S., von der Mark, K., and Kühn, K., 1975*a*, Biosynthesis of collagen by chondrocytes *in vitro, in: Extracellular Matrix Influences on Gene Expression* (H. C. Slavkin and R. C. Greulich, eds.), pp. 293–302, Academic Press, New York and London.

Müller, P. K., Lemmen, C., Gay, S., and Meigel, W. N., 1975*b*, Disturbance in the regulation of the type of collagen synthesized in a form of osteogenesis imperfecta, *Eur. J. Biochem.* **59**:97.

Needleman, S. B., and Stefanovic, N., 1969, Production of antibody for rat tail tendon collagen, *Experientia* **25**:1335.

Nist, C., von der Mark, K., Hay, E. D., Olsen, B. R., Bornstein, P., Ross, R., and Dehm, P., 1975, Location of procollagen in chick corneal and tendon fibroblasts with ferritin-conjugated antibodies, *J. Cell Biol.* **65**:75.

Nordwig, A., Nowack, H., and Hieber-Rogall, E., 1973, Sea anemone collagen: Further evidence for the existence of only one α-chain type, *J. Mol. Evol.* **2**:175.

Nowack, H., Timpl, R., and Nordwig, A., 1974, Localization and structural features of an antigenic determinant in sea anemone collagen, *Eur. J. Immunol.* **4**:698.

Nowack, H., Hahn, E., David, C. S., Timpl, R., and Götze, D., 1975*a*, Immune response control to calf collagen type I in mice: A combined control of Ir-1A and non H-2 linked genes, *Immunogenetics* **2**:331.

Nowack, H., Hahn, E., and Timpl, R., 1975*b*, Specificity of the antibody response in inbred mice to bovine type I and type II collagen, *Immunology* **29**:621.

Nowack, H., Hahn, E., and Timpl, R., 1976*a*, Requirement for T cells in the antibody response of mice to calf skin collagen, *Immunology* **30**:29.

Nowack, H., Rohde, H., and Timpl, R., 1976*b*, Isolation of a procollagen peptide from amniotic fluid, *Hoppe-Seyler's Z. Physiol. Chem.* **357**:601.

Nowack, H., Gay, S., Wick, G., Becker, U., and Timpl, R., 1976*c*, Preparation and use in immunohistology of antibodies specific for type I and type III collagen and procollagen, *J. Immunol. Meth.*, in press.

O'Dell, D. S., 1968, Immunology of Collagen and related materials, *in: Treatise on Collagen* (B. S. Gould, ed.), Vol. 2A, pp. 311–322, Academic Press, London and New York.

Olsen, B. R., and Prockop, D. J., 1974, Ferritin-conjugated antibodies used for labeling of organelles involved in the cellular synthesis and transport of procollagen, *Proc. Natl. Acad. Sci. U.S.A.* **71**:2033.

Olsen, B. R., Berg, R. A., Kishida, Y., and Prockop, D. J., 1973, Collagen synthesis: Localization of prolyl hydroxylase in tendon cells detected with ferritin-labeled antibodies, *Science* **182**:825.

Park, E., Church, R. L., and Tanzer, M. L., 1975, Immunological properties of procollagens obtained from the culture medium of dermatosparactic cells, *Immunology* **28**:781.

Parakkal, P. F., 1969, Involvement of macrophages in collagen resorption, *J. Cell Biol.* **41**:345.

Pawlowski, P. J., Gilette, M. T., Martinell, J., Lukens, L. N., and Furthmayr, H., 1975, Identification and purification of collagen-synthesizing polysomes with anti-collagen antibodies. *J. Biol. Chem.* **250**:2135.

Paz, M., and Seifter, S., 1972, Immunological studies of collagens modified by reaction with hydralazine, *Am. J. Med. Sci.* **263**:281.

Peterkofsky, B., and Diegelmann, R., 1971, Use of a mixture of proteinase-free collagenases for the specific assay of radioactive collagen in the presence of other proteins, *Biochemistry* **10**:988.

Pontz, B., Meigel, W., Rauterberg, J., and Kühn, K., 1970, Localization of two species specific antigenic determinants on the peptide chains of calf skin collagen, *Eur. J. Biochem.* **16**:50.

Pope, F. M., Martin, G. R., Lichtenstein, J. R., Penttinen, R., Gerson, B., Rowe, D. W., and McKusick, V. A., 1975, Patients with Ehlers–Danlos syndrome type IV lack type III collagen, *Proc. Natl. Acad. Sci.* **72**:1314.

Rauterberg, J., and Kühn, K., 1968, The renaturation behaviour of modified collagen molecules, *Hoppe-Seyler's Z. Physiol. Chem.* **349**:611.

Rauterberg, J., Timpl, R., and Furthmayr, H., 1972, Structural characterization of N-terminal antigenic determinants in calf and human collagen, *Eur. J. Biochem.* **27**:231.

Reid, K. B. M., Lowe, D. M., and Porter, R. R., 1972, Isolation and characterization of Clq, a subcomponent of the first component of complement from human and rabbit sera, *Biochem. J.* **130**:749.

Robbins, W. C., Watson, R. F., Pappas, G. D., and Porter, K. R., 1955, Some effects of anti-collagen serum on collagen formation in tissue culture: A preliminary report, *J. Biophys. Biochem. Cytol.* **1**:381.

Rohde, H., Nowack, H., Becker, U., and Timpl, R., 1976, Radioimmunoassay for the aminoterminal peptide of procollagen pα1(I)-chain. *J. Immunol. Meth.* **11**:135.

Rothbard, S., and Watson, R. F., 1965, Immunologic relation among various animal collagens, *J. Exp. Med.* **122**:441.

Rothbard, S., and Watson, R. F., 1967, Demonstration by immunofluorescence of the fixation of perfused antibody to human collagen in human kidney, *J. Exp. Med.* **125**:595.

Rothbard, S., and Watson, R. F., 1969, Comparison of reactions of antibodies to rat collagen and to rat kidney in the basement membranes of rat renal glomeruli, *J. Exp. Med.* **129**:1145.

Rothbard, S., and Watson, R. F., 1972, Demonstration of collagen in human tissues by immunofluorescence, *Lab. Invest.* **27:**76.

Rothbard, S., and Watson, R. F., 1973, Comparison of antigenicity of rat collagen prepared by various methods, *Arch. Pathol.* **96:**203.

Schmitt, F. O., Levine, L., Drake, M. P., Rubin, A. L., Pfahl, O., and Davison, P. F., 1964, The antigenicity of tropocollagen, *Proc. Natl. Acad. Sci. U.S.A.* **51:**493.

Schofield, J. D., and Prockop, D. J., 1973, Procollagen—a precursor form of collagen, *Clin. Orthop.* **97:**175.

Schwarz, W., and Guldner, F.-H., 1967, Elektronenmikroskopische Untersuchung des Kollagenabbaus im Uterus der Ratte nach der Schwangerschaft, *Z. Zellforsch.* **83:**416.

Sela, M., and Arnon, R., 1960, Studies on the chemical basis of the antigenicity of proteins. I. Antigenicity of polypeptidyl gelatins, *Biochem. J.* **75:**91.

Sela, M., Schechter, B., Schechter, M., and Borek, F., 1967, Antibodies to sequential and conformational antigenic determinants, *Cold Spring Harbor Symp. Quant. Biol.* **32:**537.

Senyk, G., and Michaeli, D., 1973, Introduction of cell-mediated immunity and tolerance to homologous collagen in guinea pigs: Demonstration of antigen-reactive cells for a self-antigen, *J. Immunol.* **111:**1381.

Sherr, C. J., and Goldberg, B., 1973, Antibodies to a precursor of human collagen, *Science* **180:**1190.

Sherr, C. J., Taubman, M. B., and Goldberg, B., 1973, Isolation of a disulfide-stabilized, three-chain polypeptide fragment unique to the precursor of human collagen, *J. Biol. Chem.* **248:**7033.

Stark, M., and Kühn, K., 1968, The properties of molecular fragments obtained on treating calfskin collagen with collagenase from clostridium histolyticum, *Eur. J. Biochem.* **6:**534.

Steffen, C., 1970, Considerations of pathogenesis of rheumatoid arthritis as collagen autoimmunity, *Z. Immunitätsforschg.* **139:**219.

Steffen, C., and Timpl, R., 1963, Antigenicity of collagen and its application in the serological investigation of rheumatoid arthritis sera, *Int. Arch. Allergy* **22:**333.

Steffen, C., and Timpl, R., 1965, Untersuchung über die spezifischen Reaktionsbereiche von Kollagenantigenen. I. Nachweis der spezifischen Reaktion niedermolekularer Kollagenpeptide mit Antiseren gegen natives und denaturiertes lösliches Kollagen, *Z. Immunitätsforschg.* **129:**469.

Steffen, C., and Timpl, R., 1970, Immunogenicity and specificity of collagen. X. Investigations on collagen determinants by inhibition experiments with peptides obtained after multiple degradation, *Z. Immunitätsforschg.* **139:**455.

Steffen, C., and Wick, G., 1971, Delayed hypersensitivity reactions to collagen in rats with adjuvant-induced arthritis, *Z. Immunitätsforschg.* **141:**169.

Steffen, C., Timpl, R., and Wolff, I., 1967*a*, Identification of an antigenic determinant of collagen treated with hydroxylamine, *Nature* **213:**497.

Steffen, C., Timpl, R., and Wolff, I., 1967*b*, Immunogenität und Spezifität von Kollagen. III. Erzeugung und Nachweis von Anti-Kollagen-Antikörpern und serologischer Vergleich von nativem und denaturiertem löslichen Kollagen, *Z. Immunitätsforschg.* **134:**91.

Steffen, C., Timpl, R., and Wolff, I., 1967*c*, Immunogenitat und Spezifität von Kollagen. IV. Untersuchungen zur Kollagen-Spezifität der Antiseren, *Z. Immunitätsforschg.* **134:**205.

Steffen, C., Timpl, R., and Wolff, I., 1968, Immunogenicity and specificity of collagen. V. Demonstration of three different antigenic determinants on calf collagen, *Immunology* **15:**135.

Steffen, C., Timpl, R., Wolff, I., and Furthmayr, H., 1970, Immunogenicity and specificity of collagen. IX. Resistance of different antigenic determinants of calf collagen to proteolytic treatment, *Immunology* **18:**849.

Steffen, C., Dichtl, M., Knapp, W., and Brunner, H., 1971*a*, Immunogenicity and specificity of collagen. XII. Demonstration by immunofluorescence and hemagglutination of antibodies with different specificity to human collagen, *Immunology* **21:**649.

Steffen, C., Carmann, H., Schuster, F., Tausch, G., Boesch, J., and Freilinger, G., 1971*b*, Untersuchungen über die Autoantikörpereigenschaften von Kollagenantikörpern und ihre Vorkommen in der Synovia von Patienten mit rheumatoider Arthritis, *Z. Rheumaforsch.* **30:**92.

Steffen, C., Mlczoch, F., Knapp, W., and Schuster, F., 1972, Auftreten von Gewebsantikörpern bei verschiedenen Lungenerkrankungen, *Klin. Wochenschr.* **50:**1117.

Steffen, C., Ludwig, H., Thumb, N., Frank, O., Eberl, R., and Tausch, F., 1973, Nachweis von Antikörpern mit verschiedener Kollagenspezifität bei progressiv chronischer Polyarthritis und Vergleich von Humankollagen und Kalbkollagen als Testantigen, *Klin. Wochenschr.* **51:**222.

Stenzel, K. H., Miyata, T., and Rubin, A. L., 1974, Collagen as biomaterial, *Ann. Rev. Biophys. Bioeng.* **3:**231.

Stoltz, M., Timpl, R., and Kühn, K., 1972, Non-helical regions in rat collagen α1-chain, *FEBS Lett.* **26:**61.

Stoltz, M., Timpl, R., Furthmayr, H., and Kühn, K., 1973, Structural and immunogenic properties of a major antigenic determinant in neutral salt-extracted rat-skin collagen, *Eur. J. Biochem.* **37:**287.

Taubman, M. B., Goldberg, B., and Sherr, C. J., 1974, Radioimmune assay for human procollagen, *Science* **186:**1115.

Thompson, K., Harris, M., Benjamini, E., Mitchell, G. F., and Noble, M., 1972, Cellular and humoral immunity: a distinction in antigenic recognition, *Nature (London), New Biol.* **238:**20.

Timpl, R., 1966, Fraktionierung der nicht-kollagenen Komponenten des Gelenkkapselbindegewebes, *Hoppe-Seyler's Z. Physiol. Chem.* **346:**257.

Timpl, R., Furthmayr, H., Steffen, C., and Doleschel, W., 1967, Isolation of pure anticollagen antibodies using a specific immunoadsorbent technique, *Z. Immunitätsforschg.* **134:**391.

Timpl, R., Wolff, I., Furthmayr, H., and Steffen, C., 1968*a*, Immunogenicity and specificity of collagen. VI. Separation of antibody fractions with restricted specificity from anti-collagen sera using an immunoadsorbent technique, *Immunology* **15:**145.

Timpl, R., Wolff, I., Wick, G., Furthmayr, H., and Steffen, C., 1968*b*, Immunogenicity and specificity of collagen. VII. Differences between various collagens demonstrated by cross-reaction studies, *J. Immunol.* **101:**725.

Timpl, R., Fietzek, P. P., Furthmayr, H., Meigel, W., and Kühn, K., 1970*a*, Evidence for two antigenic determinants in the C-terminal region of rat skin collagen, *FEBS Lett.* **9:**11.

Timpl, R., Wolff, I., Meigel, W., Pontz, B., Steffen, C., and Kühn, K., 1970*b*, Antigenic properties of chemically modified collagen, *J. Immunol.* **105:**1131.

Timpl, R., Beil, W., Furthmayr, H., Meigel, W., and Pontz, B., 1971, Characterization of conformation independent antigenic determinants in the triple-helical part of calf and rat collagen, *Immunology* **21:**1017.

Timpl, R., Furthmayr, H., and Beil, W., 1972, Maturation of the immune response to soluble rat collagen: Late appearance of antibodies to N-terminal sites of the α2-chain, *J. Immunol.* **108:**119.

Timpl, R., Wick, G., Furthmayr, H., Lapière, C. M., and Kühn, K., 1973*a*, Immunochemical properties of procollagen from dermatosparactic calves, *Eur. J. Biochem.* **32:**584.

Timpl, R., Furthmayr, H., Hahn, E., Becker, U., and Stoltz, M., 1973*b*, Immunochemistry of collagen. Model properties of a natural protein antigen, *Behring Institute Mitt.* **53:**66.

Timpl, R., Glanville, R. W., Nowack, H., Wiedemann, H., Fietzek, P. P., and Kühn, K., 1975, Isolation, chemical and electron microscopical characterization of neutral-salt-soluble type III collagen and procollagen from fetal bovine skin, *Hoppe-Seyler's Z. Physiol. Chem.*, **356:**1783.

Traub, W., and Piez, K. A., 1971, The chemistry and structure of collagen, *Adv. Protein Chem.* **25:**243.

von der Mark, K., Click, E. M., and Bornstein, P., 1973, The immunology of procollagen. I. Development of antibodies to determinants unique to the proα1 chain, *Arch. Biochem. Biophys.* **156:**356.

von der Mark, H., von der Mark, K., and Gay, S., 1976, Study of differential collagen synthesis during development of the chick embryo by immunofluorescence. I. Preparation of collagen type I and type II specific antibodies and their application to early stages of the chick embryo, *Develop. Biol.* **48:**237.

Wahl, L. M., Wahl, S. M., Mergenhagen, S. E., and Martin, G. R., 1975, Collagenase production by lymphokine-activated macrophages, *Science* **187:** 261.

Waksman, B. H., and Mason, H. L., 1949, The antigenicity of collagen, *J. Immunol.* **63:**427.

Watson, R. F., Rothbard, S., and Vanamee, P., 1954, The antigenicity of rat collagen, *J. Exp. Med.* **99:**535.

Wells, J. V., Michaeli, D., and Fudenberg, H. H., 1973, Antibodies to human collagen in subjects with selective IgA deficiency, *Clin. Exp. Immunol.* **13:**203.

Wick, G., Rudas, B., and Timpl, R., 1968, Die Wirkung von Antiseren gegen Rattenhautkollagen auf das reparative Granulationsgewebe der Ratte, *Arch. Klin. Exp. Dermatol.* **232:**33.

Wick, G., Kokoschka, E.-M., Kraft, D., and Timpl, R., 1974, Immunofluorescence analysis of human skin biopsies by means of anti-collagen sera, *Fed. Proc.* **33:**727.

Wick, G., Furthmayr, H., and Timpl, R., 1975, Purified antibodies to collagen: an immunofluorescence study of their reaction with tissue collagen, *Int. Arch. Allergy Appl. Immunol.* **48:**664.

Wick, G., Kraft, D., Kokoschka, E.-M., and Timpl, R., 1976*a*, The diagnostic application of specific antiprocollagen sera. I. Analysis of skin biopsies, *J. Clin. Immunol. Immunpathol.*, in press.

Wick, G., Nowack, H., Hahn, E., Timpl, R., and Miller, E. J., 1976*b*, Visualization of type I and II collagen in tissue sections by immunohistological techniques, *J. Immunol.*, in press.

Wolff, I., and Timpl, R., 1968, Reaction of the human anti-gelatine factor with gelatin from different species and with collagen peptides, *Vox Sang.* **15:**459.

Wolff, I., Timpl, R., Pecker, I., and Steffen, C., 1967, A two-component system of human serum agglutinating gelatin-coated erythrocytes, *Vox Sang.* **12:**443.

Wolff, I., Wick, G., Furthmayr, H., Timpl, R., and Steffen, C., 1970, Immunogenicity and specificity of collagen. VIII. Studies on the antigenic structure of soluble fish collagen, *Immunology* **18:**843.

Wright, R., and Alp, M. H., 1971, Autoantibodies to reticulin in patients with various gastrointestinal diseases and dietary antibodies, *Gut* **12**:858.

Zimmermann, B. K., Timpl, R., and Kühn, K., 1973, Intermolecular cross-links of collagen. Participation of the carboxy-terminal non-helical region of the $\alpha 1$-chain, *Eur. J. Biochem.* **35**:216.

8
Collagen Pathology at the Molecular Level

CHARLES M. LAPIÈRE and BETTY NUSGENS

I. Introduction

A new era of pathology has started with the development of molecular biology. Much progress has been made thanks to advancing knowledge of physiological pathways which are involved in the function of cells and tissues. Conversely, the contribution of pathology to the understanding of such pathways has proved most important. Literature provides a basis for initiating a classification of diseases at the molecular level, hopefully enabling one to track specific defects of the connective tissues from its clinical expression to the altered chemical reaction(s) which produce it.

Collagen is often involved in such disorders. This large protein occurs as several distinct, genetically unique molecular species, often tissue specific, and always containing three polypeptides. Each collagen isotype undergoes a number of covalent modifications after translation, providing it with chemical information which allows the protein to perform its functions in the extracellular space. The ultimate resistance of collagen polymers to stress, and most probably the spatial organization of the molecules in the fibers, are closely related to the chemical and physical structure of the collagen monomer. Potentially, malfunction of any one of the steps required for the production of collagen polymers could lead to defective mechanical properties. Pathology of the connective tissue might become, in the future, a highly diversified illustration of molecular

---

CHARLES M. LAPIÈRE and BETTY NUSGENS · Service de Dermatologie, Hôpital de Bavière, Université de Liège, 4020 Liège, Belgium.

pathology. At the present time few of a potentially large number of alterations of the structure or the function of collagen have received a satisfactory answer at the molecular level.

All conditions in which collagen participates, clinical in humans and animals, or experimental models, will be considered in parallel with related biochemical information, to offer a current and comprehensive view of disturbances in molecular processing. The present work reflects a montage encompassing the biochemical aspect of medical problems and the medical application of basic biological information. The proposed classification of many disorders will therefore be speculative and based upon the closest approximation trying to relate a specific defect to the malfunction of one defined step responsible for synthesis, organization, maturation, or degradation of collagen.

A. *Structure–Function Relationship*

The mechanical properties of the connective tissue are to a large extent conditioned by collagen in its polymeric form. They depend on the amount of fibers opposing the traction force, on the mechanical properties of the individual polymers, and their architectural organization. The latter parameter is potentially influenced by various components of the connective tissue, the elastic fibers, and the ground substance each of which could also be involved in pathological modifications.

Collagen in its fibrous form is a ubiquitous protein of which four different but closely related molecular species have been described in mammals (for review, see Miller and Matukas, 1974). They are the product of five different genes coding for the polypeptide α chains of the following types: α1(I), α2, α1(II), α1(III), and α1(IV). All collagen molecules are rigid rods, around 3000 Å long, and consist of the helical association of three α chains. The most widely distributed collagen, type I, is the only hybrid molecule and is formed by the association of two α1(I) and one α2 chains. All other types of collagen molecules seem to be made of three identical chains: type II, α1(II) 3; type III, α1(III) 3; and type IV, α1(IV) 3. Other molecular species or subtypes might exist. The distribution of the species of collagen molecules in various tissues is listed in Table 1. Although the distribution of the collagen molecular species is tissue dependent and most probably related to mechanical properties, only pathology provides an indirect support of this hypothesis.

A precise knowledge of the biosynthetic pathways involved in the production of a collagen molecule is required to pinpoint defective steps

TABLE 1
Types of Collagen

Organ	Source	Collagen type[a] I $(\alpha 1)_2\alpha 2$	II $(\alpha 1)3$	III $(\alpha 1)3$	IV $(\alpha 1)3$	? $(\alpha 1)3$	Reference
Dermis	Adult	+++	−	+	*		Epstein, 1974
	Embryo	++	−	++	*		
Tendon		++++	−	−	−		Bornstein, 1969
Bone		++++	−	−	−		Miller *et al.*, 1966
Lung	Alveoli,	++++	−	−	*		Bradley *et al.*, 1974
	Bronchi	−	++	−	−		
Dentin		+++	−	+	−		Volpin and Veis, 1973
Eye	Lens	−	−	−	*	++++	Trelstad and Kang, 1974
	Cornea	++++	−	−	*		
	Vitreous body	−	−	−	−	++++	
	Fibrous sclera	+++	−	−	−		
	Cartilaginous sclera	+	+++	−	−		
Cartilage		+	+++	−	−		Miller and Matukas, 1969
Cardiovascular system		+	−	+++	*		Trelstad, 1974; Miller and Matukas, 1974
Basement membrane		−	−	−	++++		Kefalides, 1971

[a] ++++: only detected molecular species, +++: predominant molecular species, ++: abundant, +: traces, −: undetermined or absent, *: basement membrane(s) not included.

that may occur in pathology. Most of the available information is related to type I collagen. Although significant differences potentially may occur in the posttranslational processing of the other collagen types, little is known about the type specificity of the enzymes involved in these reactions. The edification of a collagen fibrous framework can be divided schematically into closely related steps: synthesis, assembly, and maturation. The first step is intracellular (see Chapter 5). It can be summarized as follows: using the genetic information related to the functional activity of the fibroblast,* pro-α chain polypeptides are synthesized using the classical machinery of protein synthesis. The NH_2-terminal section of the polypeptide, some 200 residues long, is made of a sequence of amino acids more rich than the rest of the molecule in dicarboxylic amino acids, serine, leucine, and isoleucine; it also contains half-cystine residues and tryptophan. This section of the collagen precursor has been called the propeptide (or registration peptide, Speakman, 1971) since it is removed later in the processing of the molecules. It is supposed to be involved in the correct positioning of the three α chains, in the initiation of the triple-helical structure, and also in the selection of the correct number and the type of the polypeptides which form the collagen molecule. The second section, called the telopeptide, some 15 residues long containing components of the cross-links, is the region of the polypeptide joining the propeptide and the α chain proper. The latter is called the crystalline portion of the molecule. It is some 1000 amino acids long and made of a repeating sequence characterized by the presence of a glycine residue at every third position, a large proportion of proline ($\simeq$25%), as well as acidic and basic amino acids clustered together in well-defined regions. A noncrystalline sequence of ill-defined length extends the α chain at its C-terminal end.

The initial polypeptide produced by the ribosomes undergoes a series of covalent modifications. These are most significant in terms of molecular configuration and in providing the molecules with chemical groups which are involved in their ultimate organization into extracellular polymers (see Chapters 1, 2, and 3). All the posttranslational reactions are mediated by enzymes. Some of them occur shortly after formation of the peptide bonds. Suitably located proline residues (around 40% of the total) are hydroxylated to hydroxyproline under the activity of prolyl hydroxylase, using atmospheric oxygen and α-ketoglutarate as cosubstrates and requiring ascorbate and ferrous ions (Fe^{2+}) as cofactors. Hydroxyproline is required to stabilize at body temperature the triple helix of the crystalline

* Fibroblast is used here as an operational term qualifying any cell capable of synthetizing at least the fibrous proteins collagen and/or elastin.

part of the molecule. Hydroxylation of lysine in specific positions, and to a varying extent in the different types of α chain [from 6 in α1(I) to around 25 in α1(IV)], is mediated by lysyl hydroxylase. It uses the same cosubstrates and cofactors as prolyl hydroxylase. Some of the hydroxylysine residues contain a sugar moiety, enzymatically attached through an *O*-glycosidic linkage. The responsible enzyme is hydroxylysyl-*O*-galactosyltransferase. The number and location of hydroxylysyl residues participating in this reaction differ markedly in the various types of polypeptides [from one in α1(I) to more than ten in α1(IV)]. In some tissues, skin for example, a second transferase (glucosylgalactosyl-*O*-hydroxylysyltransferase) adds a glucose to the galactose. Both enzymes need Mn^{2+} as cofactor, at least *in vitro*. The function of the sugars is not known, but the occurrence of the monosaccharide form in bone and the disaccharide one in skin might be related to calcification.

The substituted polypeptides organized in a triple helix, extended on each end by nonhelical peptides, represent procollagen, the protein secreted by the fibroblast (for review, see Bornstein, 1974). Further modifications occur in the extracellular space. The oxidation of well-defined lysyl and hydroxylysyl residues into the corresponding aldehyde (allysine and hydroxyallysine) is performed by oxidase(s) requiring cuprous ions (Cu^{2+}) as a cofactor. This enzyme, partly defined in terms of specificity and chemical nature, seems to operate in the extracellular space (Layman *et al.*, 1972*a*), acting better upon polymeric than monomeric collagen (Siegel, 1974). The aldehydes located at both extremities of the collagen polypeptides are involved in the formation of covalent cross-links between the collagen molecules in the fibers to ensure their resistance to stress.

The nonhelical polypeptide extensions confer upon the procollagen molecule solubility properties required for its diffusion in the extracellular space (Lapière and Nusgens, 1974). They may also be involved in the initial steps of polymerization, and they are known to prevent cross-linking through the aldehydes. The excision of the propeptides is performed by procollagen peptidase, a specific endopeptidase. Its activity in the extracellular space seems required before the collagen monomers can achieve a correct packing into fibers and the arrangement of the polymers into bundles (Lapière *et al.*, 1975).

Further processes of maturation of extracellular collagen result in a better packing of the monomers and increased stability of the polymers through hydrophobic bonds. The final coherence of the fibers is achieved by the formation of covalent bonds between molecules through reactions involving lysyl and hydroxylysyl aldehydes and other amino acid residues (see Chapter 4).

The initial step in the degradation of fibrous collagen in the extracellular space results from the action of collagenase, a specific peptidase formed by the activation of procollagenase, its zymogen, through a cascade of activating steps (see Chapter 6). Many types of cells are capable of producing this enzyme, but whatever the source, it cleaves the three polypeptides of collagen at a specific location dividing the molecule into a three quarters N-terminal fragment and a one quarter C-terminal fragment. A significant lowering of the denaturation temperature of the fragments is responsible for a loss of the tertiary structure of the polypeptides. The subsequent gelatin is susceptible to various peptidases and cathepsins, completing the dissection of the fragments into small peptides and free amino acids. A constant fraction (5–10%) of the catabolites containing free and peptide-bound hydroxyproline is excreted in the urine (for review, see Prockop and Kivirikko, 1967; Kivirikko, 1970).

Growth, development, morphogenesis, and also the maintenance of the steady state require the synthesis and the simultaneous degradation of structural macromolecules such as collagen. The intensity of the process, its spatial distribution, and its control are still poorly understood.

The spatial orientation of the fibroblasts in the tissue, mechanical forces, possible electromagnetic fields related to mechanical stress, and/or the presence of other synthetic products of cells might be involved in the ultimate organization of the fibrous framework. This architecture is an intrinsic property of each connective tissue and is most significant in the expression of its mechanical properties.

Skin, bone, tendon, and cornea, mainly composed of type I collagen, are formed of similar polymers that are associated into bundles under a different architecture, providing these tissues with specific mechanical properties.

Skin displays a large multidirectional deformability. It is a composite connective tissue made of three structural zones in which the collagen bundles have varying architectural organization: a thin layer of intertwined fine fibers adjacent to the epidermis, a thick layer of densely packed coarse fibers in the mid-dermis, and a loosely arranged layer of coarse fibers in the deep dermis (Brown, 1972). In skin, deformability produced by stress is limited by the bundles of fibers that become oriented along the axis of loading (Finlay, 1969).

In bone, the resistance to compression, the elasticity, and the resistance to traction are related to the architectural organization of the collagen framework and the closely associated mineral phase. Collagen fibers are required for ensuring the cohesion between the organic matrix and the calcium crystals (Nusgens and Lapière, 1976). Highly structured sheets of polymeric collagen in the osteons are responsible for the tensile strength and the elasticity of this calcified structure (Ascenzi and Bonucci, 1967).

In tendon, the parallel arrangement of polymeric collagen and the mechanical properties are directly related (Viidik, 1972).

In cornea, the stroma is transparent and displays a high resistance to pressure. Both properties are related to a high degree of architectural organization. Cornea is made of multiple layers of polymeric collagen in which the fibrils are all oriented in the same direction and at approximately right angles in adjacent layers. In addition, there is an angular shift of the layers in a clockwise direction proceeding from the outer to inner layers (Coulombre, 1965).

Articular cartilage containing type II collagen is also highly organized. The elastic resistance is provided by a connective tissue made of thin fibrils supporting highly hydrated proteoglycans. The architecture of the fibrous network of cartilage forming the surface of the joints is closely related to the friction to which it is exposed. The collagen fibers run in well-defined patterns (Meachim *et al.*, 1974).

Large blood vessels, like skin, are complex connective tissues displaying multifactorial mechanical properties (Vaishnav *et al.*, 1972). Nothing is known of the involvement of type III collagen to conditioning the architectural organization of the collagen framework.

Basement membranes containing type IV collagen are less structured layers of polymers interposed between epithelia or endothelia and connective tissue. They provide the cells with a supporting base and allow filtration. As seen at the dermoepidermal interface, the dermal papillae are covered by a folded structure of high plasticity (Piérard *et al.*, 1974).

B. *Clinical Expression of Defective Collagen Framework*

A defective operation in any one of the numerous chemical reactions listed in Table 2 and involved in the ultimate production of a perfectly structured collagen framework might lead to abnormal mechanical properties. The phenotypic expression of the various molecular defects is, however, restricted to only a few clinical signs: variation in amount of fibrous proteins, changes in extensibility or in elasticity of the tissue, and fragility.

Quantitative variation in the collagen fibers can be expressed in two forms: in amount related to connective tissue thickness or in amount per unit volume (density). Sclerosis and fibrosis are most often related to increased density of the fibrous components that can be accompanied by variation in tissue thickness, increase or decrease. It can affect any connective tissue and often selected regions of it. In calcified tissues, it can be accompanied by increased mineral density. Atrophy is a reduced amount of fibrous protein in soft connective tissues. It is called osteoporosis in bone. In skin, the thinness of the dermis increases its transparency, making the blood vessels very apparent. In the eye, the thinness of the sclerae leads to the visibility of the underlying pigment.

Extensibility depends to a large extent on the architecture of the collagen bundles. It is increased when the collagen bundles are loosely associated with excessive waving or defective coherence. It is expressed by stretchable skin, joint hyperlaxity, kyphoscoliosis, floppy valves, and aneurysm. Ectopia lentis is supposed to be related to such a defective function of the supporting ligaments of the lens. Reduced extensibility results from the reversed situation. Since extensibility is related to the extent to which the bundles can be stretched, any disorder inducing a change of the resting architecture will also result in a modified extensibility.

Elasticity in terms of a biological property for a connective tissue is the sum of several physical parameters including elasticity (capacity of accumulating and releasing mechanical energy), extensibility (defining the rate of accumulation of mechanical energy), and viscosity (slowing the rate of recovering the initial architecture). Provided resistance of the polymers is compatible with accumulation of mechanical energy, biological elasticity is increased when extensibility is high and/or viscosity low. It is reduced in the reversed situation.

Fragility is closely related to polymer organization as a function of its chemical and physical properties. It is often associated with other signs. It expresses itself in skin as easy wounding, in tendons, fascia, and fetal membranes by rupture, in the vascular connective tissue by dissecting aneurysm and floppy valves, and in bone by fractures. When superficial blood vessels are affected bruising or purpura occurs.

The observation, in human beings or animals, of one or several of these symptoms affecting one or several connective tissues in a somewhat reproducible pattern has led to the definition of diseases bearing a defined name. In some instances, this name is suggestive of the clinical condition (osteogenesis imperfecta, scleroderma, or dermatosparaxis). In other examples, it bears the name of the person(s) who described it in sufficient detail. The somewhat variable expression of the phenotype of these clinical entities has led to their classification as syndromes (Ehlers–Danlos syndrome, Marfan syndrome, Werner syndrome, etc.). This represents a most suitable terminology since the word has been used in Greek antiquity to mean "an association of symptoms not related to well-defined pathological conditions."

A large body of information related to the clinical expression of many of these conditions is available in the textbook *Heritable Disorders of the Connective Tissue* by V. A. McKusick (1972).

C. Technology in Collagen Pathology

Besides clinical observation, a large variety of techniques has been used to demonstrate alteration of the connective tissue at the functional, structural, cellular, or molecular level. The proposed classification of collagen pathology at the molecular level is presented in Table 2. The

TABLE 2
Cellular and Molecular Basis of Collagen Defects[a]

Step	Biochemical event	Disease: Heritable	Disease: Acquired
A INTRACELLULAR			
1	Cistron-translation	a Osteogenesis imperfecta b Ehlers–Danlos IV	c Rheumatoid arthritis d Paget e Amino acid analogs
2	Prolyl hydroxylase	b Genetic deficiency	a Scurvy
3	Lysyl hydroxylase	a Ehlers–Danlos VI b Osteogenesis imperfecta	c Rickets
4	Galactosyl and glucosyl transferase		a Diabetes
B EXTRACELLULAR			
1	Lysyl oxidase	b Ehlers–Danlos V d Menkes' kinky hair syndrome e Aneurysm prone mice	a Lathyrism c Copper deficiency
2	Procollagen peptidase	Dermatosparaxis a Calf b Sheep c Human (ED VII)	
3	Collagenase	a Osteopetrosis b Epidermolysis bullosa	c Invasive processes d Healing e Rheumatoid arthritis f Gas gangrene

TABLE 2—*Continued*

Step	Biochemical event	Disease: Heritable		Disease: Acquired	
C INTERACTION					
1	Proteoglycans	a	Marfan	b	Scleromyxedema and pretibial myxedema
2	Glycoproteins	b	Pseudoxanthoma elasticum	a	Ectopic calcification
		c	Cutis laxa		
3	Polymerization modifiers	b	Ochronosis	a	Physiological compounds
				c	Ionizing radiations
4	Aldehydes blocking agents	b	Homocystinuria	a	Penicillamine
D METABOLISM					
1	Regulation	b	Progeria and Werner syndrome	a	Aging
				c	Endocrine disturbances
2	Healing			a	Cicatricial fibrosis
				b	Cheloids
				c	Scleroderma

[a] All disorders of collagen displaying a substantiated or potential molecular basis are classified in respect to the localization of the impaired reaction (A to D), in function of the defective chemical event for each localization (1, 2, . . .) and for each defined biochemical alteration by its most common name (a, b . . .). Each heritable or acquired condition is therefore coded by three symbols (A1a, B3e . . .) as in the table of contents, in the text, and in table 3.

scope of the techniques available to investigate these pathological processes and their successful applications are listed in Table 3.

Alteration in mechanical properties, described in clinical terms, is rarely substantiated by instrumental measurements. Extensibility and elasticity of skin can be measured, even *in vivo*, using procedures based upon the relationship between the deformation of the connective tissue and the force applied to it.

In most diseases, morphological observations, light microscopy, and histochemical techniques are most often able to disclose modifications in the texture of the connective tissue and prove useful in demonstrating their location. These are complementary to the clinical diagnosis. Transmission electron microscopy can define the structure of the collagen fibers

TABLE 3
Techniques Applied to Define Collagen Pathology [a]

Mechanical properties	
B1c:	Coulson *et al.*, 1965
B1e:	Rowe *et al.*, 1974
B2a:	Piérard and Lapière, 1975
C4a:	Harkness, 1968
D1a:	Grahame, 1970
Morphology (transmission electron microscopy)	
A1a:	Riley *et al.*, 1973; Teitelbaum *et al.*, 1974
A1d:	Rebel *et al.*, 1974
A2a:	Ross and Benditt, 1964; Hashimoto *et al.*, 1970; Phillips, 1971
A3a:	Pinnell *et al.*, 1972
A4a:	Osterby, 1972; Hagg, 1974
B1c:	Waisman *et al.*, 1969
B2a:	O'Hara *et al.*, 1970; Simar and Betz, 1971; Lapière *et al.*, 1975
B2b:	Fjølstad and Helle, 1974
B2c:	Williams *et al.*, 1974
B3c:	Hashimoto *et al.*, 1973b
C1a:	Bolande, 1963
C1b:	Lapière *et al.*, 1969
C2b:	Danielsen *et al.*, 1972; Ross, 1973; Martinez-Hernandez and Huffer, 1974
D2b:	Kischer, 1974; Kischer and Shetlar, 1974
D2c:	Braun-Falco and Rupec, 1964; Fleischmajer *et al.*, 1971; Fleischmajer and Prunieras, 1972; Kobayasi and Asboe-Hansen, 1972
Morphology (scanning electron microscopy)	
A3c:	Steendijke and Boyde, 1973
B2a:	Piérard and Lapière, 1975
D2c:	Julkunen, 1971
Cell culture	
A1a:	Martin *et al.*, 1974; Meigel *et al.*, 1974; Penttinen *et al.*, 1974, 1975
A1c:	Nimni and Deshmukh, 1973
Cell culture (continued)	
A2a:	Bates *et al.*, 1972
A3a:	Pinnell *et al.*, 1972
B1a:	Levene *et al.*, 1972
B2a:	Church *et al.*, 1973, 1974
B2c:	Lichtenstein *et al.*, 1973
B3b:	Eisen, 1969; Lazarus, 1972
C1a:	Matalon and Dorfman, 1968
C4a:	Uitto, 1969
D1a:	Martin *et al.*, 1970; Lima and Maceira-Coelho, 1972
D1b:	Martin *et al.*, 1965; Nienhaus *et al.*, 1971; Epstein *et al.*, 1974; Goldstein and Singal, 1974; Holliday *et al.*, 1974
D2c:	Kovacs and Fleischmajer, 1974
Physical and chemical properties of collagen (amount and extractibility)	
A1a:	Stevenson *et al.*, 1970; Francis *et al.*, 1974
A3a:	Pinnell *et al.*, 1972; Sussman *et al.*, 1974
A4a:	Kefalides, 1974
B1a:	Levene and Gross, 1959
B1c:	Chou *et al.*, 1968
B1e:	Rowe *et al.*, 1974
B2a:	Lenaers *et al.*, 1971
B2c:	Lichtenstein *et al.*, 1973
C1a:	Laitinen *et al.*, 1968; Priest *et al.*, 1973
C4a:	Nimni and Bavetta, 1965; Harris and Sjoerdsma, 1966a
C4b:	Harris and Sjoerdsma, 1966a; Kang and Trelstad, 1973
D2c:	Laitinen *et al.*, 1966
Physical and chemical properties of collagen (molecular type)	
A1a:	Penttinen *et al.*, 1974, 1975
A1b:	Martin *et al.*, 1974
A1c:	Nimni and Deshmukh, 1973

TABLE 3—*Continued*

Physical and chemical properties of collagen (molecular type) (*cont'd*)	Prolyl hydroxylase (*Cont'd*)
A3c: Barnes *et al.*, 1972; Toole *et al.*, 1972	D2b: Cohen *et al.*, 1971
B2a: Lenaers *et al.*, 1971	D2c: Uitto *et al.*, 1970*a*; Keiser *et al.*, 1971
B2c: Lichtenstein *et al.*, 1973	Lysyl hydroxylase
D2a: Shuttleworth and Forrest, 1974; Shuttleworth *et al.*, 1975	A3a: Krane *et al.*, 1972; Sussman *et al.*, 1974
Chemical composition, hydroxylation, glycosylation	D1a: Anttinen *et al.*, 1973
A2a: Barnes *et al.*, 1970; Wilson and Poe, 1973	Glycosyltransferase
A3a: Pinnell *et al.*, 1972; Sussman *et al.*, 1974	A4a: Spiro and Spiro, 1971
A3b: Eastoe *et al.*, 1973; Gross *et al.*, 1974; Martin *et al.*, 1974	D1a: Spiro and Spiro, 1971
A3c: Barnes *et al.*, 1972; Toole *et al.*, 1972; Gross *et al.*, 1974	Procollagen peptidase
A4a: Beisswenger, 1973; Beisswenger and Spiro, 1973; Spiro, 1973; Westberg and Michael, 1973; Kefalides, 1974	B2a: Lapière *et al.*, 1971; Kohn *et al.*, 1974; Lapière and Piérard, 1974
D2a: Cintron, 1974	B2c: Lichtenstein *et al.*, 1973
Cross-Links	B3c: Lapière and Piérard, 1974
A1c: Herbert *et al.*, 1973	D1a: Lapière and Piérard, 1974
A3a: Eyre and Glimcher, 1972	D2b: Lapière and Piérard, 1974
A3c: Mechanic *et al.*, 1972	Lysyl oxidase
B1a: Martin *et al.*, 1961	B1a: Pinnell and Martin, 1968
B1e: Rowe *et al.*, 1974	B1b: Di Ferrante *et al.*, 1975
B2a: Bailey and Lapière, 1973	B1c: Rucker and Goettlich-Riemann, 1972; Harris and O'Dell, 1974
C4a: Deshmukh *et al.*, 1973	C1a: Layman *et al.*, 1972
C4b: Kang and Trelstad, 1973	C4a: Deshmukh *et al.*, 1971; Nimni *et al.*, 1972
D1a: Bailey and Shimokomaki, 1971; Fuji and Tanzer, 1974	Collagenase
D2a: Forrest and Jackson, 1971	B3a: Walker, 1966
D2c: Herbert *et al.*, 1974	B3b: Abramson, 1969; Eisen, 1969; Lazarus, 1972
Prolyl hydroxylase	B3c: Harris *et al.*, 1972; Hashimoto *et al.*, 1973*b*
A1c: Uitto *et al.*, 1970*b*; Nigra *et al.*, 1972	B3e: Evanson *et al.*, 1967; Harris *et al.*, 1969*a,b*; Nigra *et al.*, 1972; Abe and Nagai, 1973; Harris and McCroskery, 1974
A2a: Mussini *et al.*, 1967; Stassen *et al.*, 1974	D1c: Walker, 1964; Harper and Toole, 1973; Koob *et al.*, 1974
C4a: Uitto, 1969	D2b: Milsom and Craig, 1973
D1c: Cutroneo *et al.*, 1971; Cutroneo *et al.*, 1975	Collagen metabolism (radiotracer *in vitro*)
	A1c: Uitto *et al.*, 1972*b*; Nimni and Deshmukh, 1973
	A2a: Golub, 1973

TABLE 3—*Continued*

Collagen metabolism (radiotracer *in vitro*) (*Cont'd*)		Collagen metabolism (urinary catabolites) (*Cont'd*)	
B3a:	Marks, 1974	A2b:	Nusgens and Lapière, 1973
C2b:	Blumenkrantz *et al.*, 1973	A3a:	Pinnell *et al.*, 1972
D1a:	Uitto, 1970	B1b:	Di Ferrante *et al.*, 1975
D2b:	Craig *et al.*, 1975	B2a:	Ansay *et al.*, 1968
D2c:	LeRoy, 1974	B3b:	Lazarus, 1972
Collagen metabolism (radiotracer *in vivo*)		B3c:	Kivirikko, 1970
A2a:	Barnes *et al.*, 1970; Barnes and Kodicek, 1972	C1a:	Berenson and Serra, 1959; Jones *et al.*, 1964
B2a:	Marks, 1973	C1b:	Kivirikko *et al.*, 1965*b*; Lapière *et al.*, 1969; Winand, 1968; Askenasi, 1974
C4a:	Klein and Nowacek, 1969	D1a:	Morrow *et al.*, 1967 Askenasi, 1975
D1c:	Kivirikko *et al.*, 1965*a*; Avioli and Prockop, 1967	D1b:	Nusgens and Lapière, 1973
D2a:	Madden and Peacock, 1971	D1c:	Kivirikko *et al.*, 1964; Prockop and Kivirikko, 1967; Kivirikko, 1970; Krane *et al.*, 1973
Collagen metabolism (urinary catabolites)		D2a:	Klein *et al.*, 1962; Smith *et al.*, 1974
A1a:	Mitoma *et al.*, 1959; Langness and Behnke, 1971; Riley *et al.*, 1973; Lancaster *et al.*, 1975	D2c:	Hardy *et al.*, 1971
A1d:	Krane *et al.*, 1967		
A2a:	Barnes and Kodicek, 1972		

[a] See table 2 for the code used to classify the diseases.

and the packing of the polymers in the fibers, as well as cellular morphology. Scanning electron microscopy is required to define the three-dimensional architecture of the fibrous framework and the spatial relationship between the various types of polymers in the tissue.

Chemical or physical investigation of collagen in tissue samples provides information directly or indirectly related to specific molecular defects of the collagen. The amount of collagen (hydroxyproline determined by colorimetric reaction) per unit volume (wet weight), or its concentration (amount per unit dry weight), has been measured in several tissues and in many diseases. The solubility of native collagen in salt solutions or the extractibility of the gelatin (denatured collagen) in denaturing solvents offer the possibility of indirectly assessing the state of polymerization and the effectiveness of inter- and intramolecular bonds between molecules in the fibers. The analysis of extracted molecular species by acrylamide-gel electrophoresis at acid pH or in the presence of sodium dodecyl sulfate provides more direct information about the occurrence of such covalent cross-links. Measurement of reducible cross-links by radiochemical techniques provides the required support to demonstrate their alteration.

Salt fractionation of extracted native collagen has been used for isolating various molecular species. It can be performed on collagen extracted after limited proteolysis. Acrylamide-gel electrophoresis or ion-exchange chromatography are used to characterize the type of α chains contained in the molecules. Confirmation of the chemical nature of molecules requires amino acid analysis of purified polypeptides or their fragments obtained after proteolysis or cyanogen bromide cleavage. The amino acid composition also provides indirect information about the function of the various enzymes involved in the posttranscriptional modifications of the polypeptides. It has been used to assess the extent of proline and lysine hydroxylation. The concentration of the glycosides of hydroxylysine can be measured after alkaline hydrolysis.

Techniques are available for measuring the activity of all the enzymes involved in the posttranscriptional modifications of the collagen polypeptides. Most of these techniques are suitable and have been used for screening pathology in humans using small amounts of tissue or cells grown *in vitro.* The most reliable information is related to the properties of the enzymes rather than to the amount that can be produced.

Fibroblast culture (cell culture) has been used extensively in the study of many connective tissue disorders. The information collected is related to the rate and the capacity of cellular multiplication and to the mode of expression of their genetic information.

Metabolic activity of the connective tissues can be measured *in vitro* (tissue culture) or *in vivo,* by incorporation of radiotracers in selected biosynthetic products. For collagen, incorporation of proline to be converted to hydroxyproline or lysine converted to hydroxylysine represents the most reliable parameter. *In vivo* measurement is often only usable for animal studies due to the large amount of radioisotope to be injected. Assessing metabolic activity *in vitro* using tissue samples or cell cultures is subject to discrepancies related to a possible change in factors controlling the rate of various reactions.

An indirect approach for assessing collagen metabolic activity is the measurement of its degradation products which appear in urine. Urinary hydroxyproline has been used most extensively (for review, see Prockop and Kivirikko, 1967; Kivirikko, 1970). The measurement of urinary hydroxylysine and its glycosylated derivatives (mono- and disaccharides) seems more informative since they could provide information about the tissue origin of the catabolites, the monosaccharide form arising from bone, the disaccharide from skin. Analysis of the nature and sequence of urinary peptides, as well as the relationship between peptide-bound proline and hydroxyproline, have also proved useful in some disorders.

II. Classification of Collagen Disorders at the Molecular Level

The proposed classification of collagen pathology at the molecular level is presented in Table 2.

Alteration in the chemical composition of the collagen framework is a structural defect. Modification in amount of polymeric collagen in the extracellular space most often results from a control defect. Both types of pathology could be genetically defined (often heritable disorders) or result from the abnormal function of a correct gene product (acquired diseases).

A *structural defect* is an alteration in the structure of the collagen molecule and/or in the type of collagen forming the fibers. Most often, it results from the abnormal functioning of one intracellular step of polypeptide synthesis. Structural alterations too extensive to allow the formation of a stable molecule might result in a reduced amount of fibrous collagen. In the absence of extensive modification of its tertiary structure, the collagen monomer could be used for building the fibrous framework, but its ultimate stabilization or its architecture might be modified. This will result in the formation of fibers displaying altered mechanical properties, structure, and/or architecture.

A *control defect* results from an alteration in the mechanism(s) regulating the rate of synthesis, degradation, and the balance between the two. Atrophy results from degradation being greater than synthesis; sclerosis is the reverse. If the balance is preserved, changes in the rate of remodeling will not be expressed by visible signs although chemical investigation could demonstrate it. In most instances, several biosynthetic products of the fibroblasts will be affected.

A *genetic defect* as observed in the heritable diseases is responsible for the synthesis of defective polypeptides or the nonproduction of a required protein. Impaired control can also be genetically determined.

An *acquired defect* results from the abnormal activity of an otherwise normal gene or gene products. It could depend on the lack of a required cofactor or the introduction of an inhibitor. The function of one or several enzymes or the overall synthetic activity of the cell could be modified. It could also depend on the interaction between endogenous or exogenous compounds and extracellular collagen.

A. Pathology Related to Intracellular Processes

1. Procollagen Cistron and Translation

Precise information about control in the use of the different collagen cistrons is not yet available, although its occurrence is demonstrated by compositional changes occurring in skin during embryonic life. The proportion of type III collagen diminishes progressively with development of the embryo while that of type I collagen increases (Epstein, 1974). This pattern is supported by *in vitro* observation; in culture, early embryonic skin produces both type III and type I collagen while in later stages of development it synthesizes mainly type I molecules (Vinson and Seyer, 1974). Even cancer tissue was found to synthesize the type of collagen specific to the strain of cells from which it derives, as for example, the production of type II collagen by a transplantable chondrosarcoma (Smith *et al.*, 1975*a*). This retention in the use of genetic information validates the utilization of tissue culture for investigating similar processes in pathology.

Three diseases in the human manifested as structural defects could be attributed to an abnormal expression of the genetic information. Two of these are heritable disorders, osteogenesis imperfecta and the Ehlers–Danlos syndrome, type IV. The same process might participate in rheumatoid arthritis, a condition to be considered as an acquired disorder.

a. Osteogenesis Imperfecta (A1a). This is a human heritable disorder of most connective tissues containing type I collagen as a predominant form (bone, skin, tendon, eye, ear, teeth). It is known to occur in different clinical and heritable forms with variable expressivity. The main clinical signs are bone fragility, blue sclera, otosclerosis, and thin skin (McKusick, 1972). In some patients, the defect might be related to an abnormally persistent synthesis of type III collagen after fetal and postnatal development. Fibroblasts isolated from the skin of patients exhibiting this disorder were found to synthesize *in vitro* a type III collagen in much larger proportion than normal fibroblasts from age-matched controls (Penttinen *et al.*, 1974, 1975; Martin *et al.*, 1974; Meigel *et al.*, 1974). Penttinen *et al.* (1974), found that the proportion of type III collagen produced in culture is related to the severity of the defect. These observations are supported by indirect findings speculating about the presence of "immature" collagen in several connective tissues, thinner than normal collagen fibers in the cornea (Riley *et al.*, 1973), an excess of the so-called "argyrophilic fibers or reticulin fibers" in skin (Follis, 1952), and an abnormal architecture of the collagen bundles and thin fibers in the bone of three patients (Teitelbaum *et al.*, 1974). There is, however, no evidence for an excess of type III collagen in skin, and it remains difficult to understand how the above-mentioned defect would result in bone alteration. Calcified connective tissues are supposed to contain little or no type III collagen even during development (Linsenmayer *et al.*, 1973), except perhaps in the vascular system. In some forms of the disease, the fragility of bone is reduced in adult life. This improvement with age suggests that the pathological process might be a developmental defect, to be considered in parallel with the known progressive replacement of collagen type III by type I in skin during fetal development.

The turnover of collagen does not seem strikingly modified since the values for urinary hydroxyproline were reported as normal (Lancaster *et al.*, 1975), as reduced (Mitoma *et al.*, 1959), or as increased (Langness and Behnke, 1971). Other observations point to the reduced thickness of cortical bone (Riley *et al.*, 1971), the diminished amount of collagen in skin (Stevenson *et al.*, 1970), altered polymerization of skin collagen (Francis *et al.*, 1974), and an abnormal morphology of fibroblasts in culture (Lancaster *et al.*, 1975). Another type of molecular pathology has been described and will be considered elsewhere.

b. Ehlers–Danlos Type IV (A1b). Collagen type III is a normal component of growing and adult skin and the vascular system (Trelstad, 1974). A defect in its synthesis has been claimed (Martin *et al.*, 1974) to be the cause of Ehlers–Danlos type IV (ecchymotic type), displaying clinical manifestation in the connective tissues rich in this type of molecule, skin, and the vascular system. While skin fibroblasts from normal individuals synthesize both type I and type III collagen in culture, similar cells from the patients produce only type I collagen. Skin in five patients presenting this syndrome and skin, aorta, gut, and lungs studied in one patient were found not to contain type III collagen as in normal individuals (Pope *et al.*, 1975).

Bleeding in Ehlers–Danlos type IV is explainable on the grounds of the missing type III collagen. This collagen is indeed located in the subendothelial layer and the only one found in fresh thrombi (Gay *et al.*, 1975*a*). Type III collagen is a most potent aggregating agent of platelets (Balleisen *et al.*, 1975) and if not the only, at least the main type of collagen capable of inducing platelet aggregation (Hugues *et al.*, 1976).

c. Rheumatoid Arthritis (A1c). In osteoarthritic cartilage, a shift in synthesis of collagen from type II to type I has been observed in tissue culture (Nimni and Deshmukh, 1973). It is most probably a consequence of pathological processes responsible for degradation of the cartilage that is discussed elsewhere. It is perhaps related to the presence of proteolytic enzymes. When cultured in the presence of such enzymes, cartilage cells stop producing type II collagen and reverse to a more embryonic state of differentiation (Deshmukh and Nimni, 1973). Early embryonic cartilage synthesizes up to 60% of its collagen in the form of type I, while during later embryonic development it produces less than 10% of type I (Seyer and Vinson, 1974). The shift in synthesis observed in the arthritic cartilage might be related to the loss of tissue organization, since isolated cartilage cells produce type I collagen while cartilage slices in culture produce type II molecules (Layman *et al.,* 1972*b*).

When dealing with inflamed tissues, the possibility has to be ruled out that the disturbance results from synthesis directed by cells of the granulation tissue. They are known to produce a collagen of a type different from the one synthesized in the tissue in which the inflammation develops (Bailey *et al.,* 1973).

d. Paget's Disease of Bone (A1d). It is possible that the metabolic disorder observed in this condition depends upon defective genetic information. The osteoclasts contain intranuclear inclusions (Rebel *et al.,* 1974), and the condition is known to evolve sometimes into osteosarcoma. The cells participating in the process divide rapidly, they synthesize a structurally abnormal calcifying tissue (for review, Meunier, 1975), and secrete a collagen-type polypeptide. This polypeptide or its breakdown products are found in the urine (Krane *et al.,* 1967). Its amino acid composition is similar to that observed in the urinary breakdown products of other disorders with increased bone degradation (Krane *et al.,* 1970).

e. Amino Acid Analogs (A1e). Up to now, no defect has been found that involves an alteration in the primary structure of the collagen α chain within its crystalline part or in the peptide extensions. Some degree of variation in the primary structure of the crystalline part of the α chains is most possible and will probably not modify the properties of the molecule if changes are located at positions which are known to vary with evolution (Piez *et al.,* 1968).

Genetic alterations resulting in the substitution for glycine at every third position would probably be most deleterious since it could prevent the formation of a correct tertiary structure along the total length of the polypeptides, leading to impaired stability and polymerizing properties. Detailed analysis of the primary structure of the collagen α chains has not been performed in those heritable disorders of the connective tissue still lacking a molecular basis. Chemical information about the sequence of amino acids around the cross-linking sites seems of particular interest. The introduction of a bulky or a charged amino acid in this region might alter the reaction of the aldehydes which form the cross-links.

Alterations in the primary structure of the collagen polypeptide have been produced in animals and in cell culture by introduction of analogs of proline and lysine. When *cis*-hydroxyproline is used, collagen synthesis proceeds, but the substituted molecules are not released from the cells (Rosenbloom and Prockop, 1971). Collagen synthesized in the presence of another analog, azetidine-2-carboxylic acid, is also, at least in part, retained inside the cells and probably degraded. Some substituted molecules can be used to build fibers, although their thermal stability is reduced (Lane *et al.*, 1971*a,b*). Similar observations made in cell culture allowed one to conclude that sufficient incorporation of an analog would prevent a stable tertiary structure of the molecules from forming and would result in their degradation (Uitto *et al.*, 1972*a*, Uitto and Prockop, 1974). Hydroxynorvaline also inhibits assembly and secretion of procollagen (Christner *et al.*, 1975).

Although a lysine analog, D,L-*trans*-4,5-dehydrolysine (Christner and Rosenbloom, 1971) is also incorporated in collagen during its synthesis, the impaired release of the substituted molecules from the cell is not as dramatic.

2. *Prolyl Hydroxylase (PH)*

In the absence of sufficient hydroxylation of suitable proline residues into hydroxyproline, the collagen polypeptides are unable to achieve a triple-helical structure allowing the collagen monomer to be stable at body temperature (Rosenbloom *et al.*, 1973; Berg and Prockop, 1973). In a cell-free hydroxylating system or by using the purified enzyme, the reaction mediated by prolyl hydroxylase (PH) is inefficient in the absence of cosubstrates (O_2 and α-ketoglutarate) or cofactors (Fe^{2+} and ascorbate) (for review, see Barnes and Kodicek, 1972). Conditions of reduced activity of PH can be produced *in vitro* by modifying the culture medium (Peterkofsky, 1972). In cell cultures, ascorbic acid seems to act as an activator of PH (Levene *et al.*, 1974). Inhibition of hydroxyproline formation can also be produced in organ culture by various chelating agents acting probably by complexing ferrous ions (Chvapil *et al.*, 1967). In fibroblast culture, this procedure results in the synthesis of underhydroxylated collagen which is rapidly degraded (Switzer and Summer, 1973). Prolyl hydroxylase activity in fibroblast culture is also reduced by proline analogs (Kerwar *et al.*, 1975) and the nonhelical procollagen retained in the rough endoplasmic reticulum (Uitto *et al.*, 1975).

a. Scurvy (A2a). This is the only not immediately lethal condition capable of producing defective hydroxylation *in vivo*. In the absence of ascorbate, in animal species requiring an exogenous source of this vitamin (guinea pig and human, for example), a deficient activity of hydroxylating enzymes results in the production of collagen polypeptide containing too little hydroxyproline to be stable and secreted (Barnes *et al.*, 1970). Mussini *et al.* (1967) have indeed observed reduction of PH activity in scorbutic tissue of the guinea pig. Since scurvy is observed in the human when the body pool of ascorbic acid is reduced under 300 mg (Baker *et al.*, 1971), one can assume that the mechanism resulting in reduction of PH activity perhaps involves structural alteration of the enzyme. It might be related to an activation process acting upon a pool of enzyme precursor, the existence of which has been suggested in fibroblast culture (McGee *et al.*, 1971) and in the tissues (Stassen *et al.*, 1974). The requirements for activation seem identical to those needed for the hydroxylation reaction (Kuttan *et al.*, 1975). Some analogs of ascorbic acid are effective cofactors of PH *in vitro* (Kutnink *et al.*, 1969) but not *in vivo* (Rokosova and Chvapil, 1974).

All connective tissues requiring collagen synthesis to maintain their structural integrity will display abnormal mechanical properties in scurvy. Defective blood vessels due to impaired function of basement membranes seem of importance. One of the first and most significant symptoms in human pathology is the perifollicular hemorrhage of the lower part of the legs where the capillary blood pressure is highest. Bleeding of the gums and of the eyes also occurs. Alteration of the dermal blood vessels is substantiated by ultrastructural observation (Hashimoto *et al.*, 1970). The vascular system of bone is also markedly affected (Phillips, 1971). Defective wound healing due to an impaired formation of the granulation tissue is a most dramatic feature of scurvy. It was demonstrated long ago by the observation of inefficient collagen fiber formation (Robertson and Schwartz, 1953; Gould, 1958) and distorted cell architecture (Ross and Benditt, 1964).

In the human, a deficient supply of ascorbic acid (daily need around 10 mg) produces death before the tissues are completely depleted of the vitamin. This might explain why the excretion of urinary hydroxyproline is reduced but not absent and, as observed in some patients, it is even increased. Catabolism of insufficiently hydroxylated collagen could indeed produce an excess of breakdown products (unpublished results).

Chelation of ferrous ions also inhibits hydroxylation of collagen *in vivo*. It was found effective in embryos but resulted in death; it was of little efficiency in adult animals (Chvapil *et al.*, 1974).

Little is known about the defective hydroxylation of lysine in scurvy. It has been observed in ascorbate-deficient fibroblast culture (Bates *et al.*, 1972) although ascorbate is required for enzyme function but not activation since lysyl hydroxylase is not synthesized under an inactive form (Miller, 1975).

b. Genetically Determined Prolyl Hydroxylase Deficiency (A2b). This was suspected in one patient (Lapière and Nusgens, 1969) on the basis of a greatly increased urinary excretion of peptidyl hydroxyproline and peptidyl proline contained in short peptides whose amino acid sequence was related to collagen. A major component of these peptides was the tripeptide, Gly-Pro-Pro, in which proline in the third position should have been hydroxylated under normal conditions (Nusgens and Lapière, 1973). The administration of the known cofactors of PH was ineffective. The patient presented a long-standing history of spontaneous ulceration of the skin and retarded healing since early childhood. Her first pregnancy resulted in the birth of an anencephalic child whose fibroblasts were not defective for PH (unpublished results). A second patient who presented a somewhat similar clinical and metabolic pathology was recently described (Ellis, 1972). Two other patients presenting similar metabolic disorders were also reported (Goodman *et al.*, 1968; Jackson *et al.*, 1975). In the latter, Jackson *et al.*, 1975, relate the defect to a reduction in prolidase resulting in a lack of recycling collagen breakdown products and impaired collagen synthesis.

3. Lysyl Hydroxylase (LH)

a. Ehlers–Danlos Type VI (A3a). The deficiency in lysyl hydroxylase in this disease is the first demonstration in the human of collagen molecular pathology in a heritable disorder of the connective tissue (Pinnel *et al.*, 1972). The symptoms, all related to abnormal mechanical properties of the collagen framework, are those of the Ehlers–Danlos syndrome: marked kyphoscoliosis, joint hyperlaxity, velvety and hyperelastic skin, and easy bruising. Fragility of various connective tissues of the eye can cause grave ocular problems such as retinal detachment. The enzymatic defect, first described in two sisters (Pinnel *et al.*, 1972), has recently been observed in one other patient (Sussman *et al.*, 1974). A unique biochemical marker of this condition is the extremely low level of hydroxylysine in skin collagen. In the first two described patients, a reduced lysyl hydroxylation has also been found in fascia and bone. The urinary excretion of hydroxylysine is significantly lower than that of age-matched controls but still considerable when considered in relation to the very low hydroxylysine content of skin. The relative proportion of the glycoside of hydroxylysine to total hydroxylysine is unchanged. Fibroblasts from these patients produce, in culture, little lysyl hydroxylase and a normal prolyl hydroxylase activity (Krane *et al.*, 1972; Sussman *et al.*, 1974). These observations provide a most reasonable explanation for the defect, i.e., that it results from a defective enzyme. It is unlikely that the deficiency is related to a limiting concentration of specific cofactors or to the presence of an inhibitor. The inefficient use of cofactor(s) should, however, not be ruled out (Krane, personnal communication). The observation of reduced activity of lysyl hydroxylase in the mother of the affected children supports the heritable basis of the disease and suggests its autosomal recessive inheritance (Krane *et al.*, 1972).

It is most interesting to observe that all the connective tissues are not affected to the same extent; skin most, fascia and bone much less. It is also worth noting that the ultrastructure of the collagen fibers in the dermis is not altered and that the morphology of fibroblasts is normal (Pinnell *et al.*, 1972). Human skin at the age of the patients, contains around 25% of type III collagen (Epstein, 1974), a species of molecule in which, under normal conditions, lysyl hydroxylation is larger than in type I collagen. The presence of less than one residue of hydroxylysine per collagen molecule in the dermis therefore indicates that impaired hydroxylation also affects type III collagen. Nothing is known about type IV collagen which is much more hydroxylated. It must be noted that easy bruising has been observed. The lysyl hydroxylation of type II collagen does not seem impaired.

Information collected during the study of this condition has largely contributed to the understanding of the physiological role of hydroxylysine. Production of a collagen containing less than one hydroxylysine per molecule and its efficient use in building fibers demonstrate that this hydroxylated amino acid is not required for secretion and polymerization. Impaired cross-linking (Eyre and Glimcher, 1972) and increased solubility of the fibrous protein point to its involvement in the physiological formation of the chemical bond ensuring fiber stability. This hypothesis is supported by the known participation of specific hydroxylysine residues in this process (see Chapter 4).

b. Osteogenesis Imperfecta (A3b). Increased hydroxylation of lysine has been observed in osteogenesis imperfecta (Eastoe *et al.*, 1973; Gross *et al.*, 1974). In the patient described by Gross *et al.* (1974), collagen in skin (+70%) and cartilage (+40%) was also more extensively hydroxylated than that of normal humans. Although hydroxylation of lysine was increased in both skin and bone, glycosylation was only enhanced in bone (Gross *et al.*, 1974).

One has to question the significance of these observations in relationship with the observed disturbance previously discussed, the excessive synthesis of type III collagen in tissue culture. Under normal conditions, the concentration of hydroxylysine is 6/1000 in type I collagen, 8/1000 in type III, and 14/1000 in type II. As long as additional chemical analyses of purified collagen accounting for the whole framework of these tissues is unavailable, it will not be possible to distinguish between increased lysine hydroxylation and a modification in the type of molecules forming their fibers.

c. Rickets, Hypocalcemia, and Hypoparathyroidism (A3c). In rickets, an acquired disorder induced by vitamin D deficiency (Toole *et al.*, 1972; Barnes *et al.*, 1973*a*; Gross *et al.*, 1974), in hypocalcemia, and in hypoparathyroidism (Barnes *et al.*, 1973*b*), the hydroxylation of lysine in bone collagen is also increased but not that of skin and cartilage (Gross *et al.*, 1974). In all three conditions, an excessive rate of collagen degradation might be related to hypocalcemia (Aymard, 1971). Interestingly, in rickets the pattern of the cross-links is also modified. The proportion of compounds derived from hydroxylysine and its aldehyde is increased (Mechanic *et al.*, 1972). This pattern of cross-links is similar to that observed in fetal bone.

4. *Sugar Transferases*

a. Diabetes (A4a). Up to now, there exists no demonstration in man of pathological changes specifically related to the impaired function of either one of the two enzymes involved in the glycosylation of hydroxylysine. The observation of an increased activity of sugar transferases in the renal cortex of alloxane diabetic rats has been correlated by Spiro and Spiro (1971) to an enhanced basement membrane synthesis. The kidney glomerular basement membrane of alloxane diabetic rats is indeed increased in thickness (Hagg, 1974). In the diabetic human, the thickening of the vascular basement membrane is well substantiated and has been shown to represent an early event in juvenile diabetes (Osterby, 1972). The question, however, remains unsettled whether such an excess of basement membrane results from the accumulation of normal or abnormal proteins. An excessive deposition of collagen was initially proposed by Beisswenger and Spiro (1970). It is supported by significant changes in amino acid composition such as increased proportion of hydroxylysine (Beisswenger and Spiro, 1973; Spiro, 1973; Beisswenger, 1973). The alternate hypothesis is supported by others who did not observe these modifications (Westberg and Michael, 1973; Kefalides, 1974). Most evidence is against an isolated increase in the activity of the sugar transferases in human diabetes. Not enough attention has been paid to a possible disturbance in the turnover of basement membrane in diabetes. It is known to be an active process under physiological conditions (Walker, 1973*b*).

B. *Pathology Related to Extracellular Enzymes*

1. *Lysine and Hydroxylysine Oxidase*

a. Lathyrism (B1a). The most striking and extensively documented defect in this enzyme(s) activity occurs in lathyrism. This experimental disease has been a major tool in advancing the basic knowledge of collagen structure and function. Lathyrism is produced in most animal species by administration of aminonitrile and related compounds that are specific inhibitors of the oxidase *in vivo* and in cell culture (for review see Tanzer, 1965). They were most significant in allowing the demonstration of lysyl oxidase (Pinnell and Martin, 1968) and basic features related to cross-linking of fibrous proteins, i.e., the accumulation of cold extractable collagen (Levene and Gross, 1959), the significance of interchain binding between collagen α chains (Martin *et al.,* 1961), and elastin (Sykes and Partridge, 1972).

The lathyrogen β-aminopropionitrile is an irreversible inhibitor of lysyl oxidase *in vitro* and *in vivo* (Siegel *et al.,* 1970). Suppression of aldehyde formation and the absence of the cross-links are responsible for decreased mechanical resistance of the collagen framework. It will allow a progressive distention of the connective tissue, causing aneurysms, osteophytosis, skin fragility, etc. (for review see Barrow *et al.,* 1974). Interestingly, the administration of the drug during fetal development at just the 15th day of gestation in the rat causes cleft palate (Pratt and King, 1972), pointing to the importance of adequate mechanical properties of the connective tissue in embryogenesis and development.

The pathological changes in lathyrism occur in all tissues in which collagen or elastin is being synthesized, thus explaining the tissue specificity of the defect and its relationship to growth (Gross and Levene, 1959), development, and remodeling.

As proposed by Bentley *et al.* (1970), modifications of proteoglycan metabolism induced by lathyrogens are indirect and secondary to tissue injury brought about by long-standing impairment of collagen fiber integrity.

b. Ehlers–Danlos Type V (B1b). This sex-linked form of the syndrome has been attributed to a deficient activity of lysyl oxidase(s) as observed in one patient (Di Ferrante *et al.,* 1975). Cultured skin fibroblasts synthesized excessively soluble collagen and had a low lysyl oxidase activity. The patient, and his maternal cousin presenting the same symptoms, excreted in the urine increased amounts of collagen and elastin breakdown products. The degradation of dermatan sulfate appeared to be inadequate.

c. Copper Deficiency (B1c). Lysyl oxidase(s) is inactive in the absence of cuprous ion (Pinnell and Martin, 1968). A deficient copper supply will produce alterations similar to lathyrism (Carnes, 1971). In miniature pigs the most striking lesions are cardiovascular, and the main features are related to impaired elastogenesis, although collagen is also altered (Waisman *et al.,* 1969). Similar clinical and chemical observations were made in copper-deficient chicks (Rucker and Goettlich-Riemann, 1972). It has to be stressed that a major step in the identification of the elastin monomer was achieved by its purification from aorta of copper-deficient swine (Sandberg *et al.*, 1971). In culture, copper-deficient pig aorta also produces soluble elastin devoid of aldehydes (Smith *et al.*, 1975*b*).

Although penicillamine seems to have a direct effect on collagen and elastin by combining with the lysyl aldehyde, its possible chelating activity on copper (as sought in the treatment of Wilson disease) might alter the activity of the lysyl oxidase(s). Nimni *et al.* (1972) have indeed observed in the rat that this compound blocks the aldehydes and inhibits their formation.

An enhanced collagen cross-linking and an increased aldehyde content of extractable collagen have been observed in zinc-deficient rats (McClain *et al.*, 1973), suggesting that this metal could play a role, perhaps by competing with copper in the tissues and depressing the activity of lysyl oxidase. Inhibition of cross-linking of elastin by iproniazid and its counteraction by pyridoxal phosphate have also been reported (Rucker and O'Dell, 1970), indicating a possible involvement of this cofactor in the enzymatic reaction.

The two following genetic disorders seem to be related to a defective supply of cuprous ions and impaired function of lysyl oxidase(s).

d. Menkes' Kinky-Hair Syndrome (B1d). In this X-linked inherited disorder in the human, the fragility of the connective tissue (tortuous arteries and hemorrhages, retarded growth, and bone alteration), as well as other pathological alterations (pili torti, hypothermia, muscle deficiency), might be explainable on the basis of a genetically determined deficient copper supply due to defective intestinal absorption (Danks *et al.*, 1972). The defect seems correctable by administration of exogenous copper (Dekaban and Steusing, 1974), although normal serum-copper level does not always improve the clinical state (Wehinger *et al.*, 1975).

e. Aneurysm-Prone Mice (B1e). With alleles at the X-chromosomal locus, "mottled," these mice display abnormalities that are, in many respects, similar to those found in lathyrism: aneurysms, reduced tensile strength of skin, and bone abnormalities. This pathology is related to impaired formation of the cross-links in collagen and elastin because of a failure to generate the lysine-derived aldehydes (Rowe *et al.*, 1974). In the most affected mutants, the increased extractibility of collagen, the decreased breaking strength of the skin, the reduced proportion of the cross-linked components (β chains), and the reduced amount of lysine-derived aldehyde strongly support a defect in the cross-linking of collagen. In aortic elastin, although desmosine and isodesmosine levels are normal, the concentration of the reduced aldol condensation product is markedly diminished. Although no striking differences in serum copper levels were noted between mutant and control mice, an abnormal copper transport was suggested to explain this cross-linking deficiency (Hunt, 1974).

2. *Procollagen Peptidase*

Heritable deficiency of procollagen peptidase underlies dermatosparaxis. This heritable disorder is transmitted as an autosomal recessive character in three affected species: calf (B2a) (Hanset and Ansay, 1967; Hanset and Lapière, 1974), sheep (B2b) (Helle and Ness, 1972), and humans (Ehlers–Danlos type VII, B2c) (McKusick, 1974).

Dermatosparaxis in the calf is the first heritable disorder of the connective tissue that has received a satisfactory answer at the molecular level. The defective activity of procollagen peptidase is responsible for the accumulation in the extracellular space of procollagen in the form of polymers. Information concerning this condition in sheep, still incomplete, suggests that the defect is similar to that observed in calves. The Ehlers–Danlos type VII in the human also seems to be related to the same enzyme deficiency, although many differences clearly separate it from the other two.

In the calf (Hanset and Ansay, 1967) and sheep (Helle and Ness, 1972), the most striking clinical feature is the extreme fragility of skin, resulting in easy wounding and ultimately in death due to wound infection and septicemia. The defect is present in other connective tissues containing type I collagen (Lenaers *et al.*, 1971), although their functional and structural integrity are better preserved. In the human (Lichtenstein *et al.*, 1973), the skin is hyperextensible, velvety, and somewhat fragile. The main pathology is related to tendon and fascia, resulting in hypermobility of the joints and multiple dislocations.

The ultrastructure of the collagen framework in the calf (O'Hara *et al.*, 1970; Simar and Betz, 1971) and sheep (Fjølstad and Helle, 1974) are similar but different from what is observed in the human. In animals, instead of the classical cylindrical collagen fibers, the fibrous framework of the dermis is made of ribbons of procollagen polymers joined together on their lateral sides to form elongated pseudofibers whose cross-sections show a starlike aspect. These pseudofibers are irregularly dispersed within an excessive amount of amorphous material, depending on an excess of proteoglycans (Ansay *et al.*, 1968; Winand and Nusgens, 1971). As observed by scanning electron microscopy in the calf (Piérard and Lapière, 1976) the procollagen polymers are associated to form sheets in a honeycomb arrangement. Upon traction they separate easily. Rupture occurs in the mid-dermis, a part of the connective tissue devoid of adnexae and poor in elastic fibers. The breakage location seems identical in the sheep (Fjølstad and Helle, 1974). In Ehlers–Danlos type VII, the ultrastructure of the fibers is not modified, but their packing is impaired. The fibers are frequently found in sheets instead of bundles, with an excess of ground substance (Williams *et al.*, 1974).

Much has been learned about procollagen by investigating the molecular processes responsible for dermatosparaxis in the calf. The tissues of these animals provided the tool for the demonstration of procollagen peptidase (Lapière *et al.*, 1971) and were the source of biological material for establishing the basic chemical structure of all procollagen extensions (Lenaers *et al.*, 1971; Furthmayr *et al.*, 1972). It is obvious that the collagen extracted from dermatosparactic skin is a precursor form of collagen, i.e., α chains extended by polypeptide moieties strikingly different from the crystalline part of the molecule and mostly located at the —NH_2-terminal extremity of the molecules (Stark *et al.*, 1971). The relationship between dermatosparactic collagen and normal collagen precursors is further supported by immunological similarities (Timpl *et al.*, 1973) and by the observation that dermatosparactic collagen is a substrate for procollagen peptidase from various tissues in several animal species (Lapière and Piérard, 1974). The size of the dermatosparactic collagen extension is, however, smaller than that of most reported collagen precursors (for review see Bornstein, 1974), suggesting that it represents a partly processed collagen precursor. Its accumulation in the dermatosparactic tissues results from a deficient activity of procollagen peptidase (Lapière *et al.*, 1971). This endopeptidase (Kohn *et al.*, 1974) ensures the ultimate processing of procollagen, allowing proper fibrogenesis and bundle organization to proceed. All processes related to maturation of extracellular collagen are suppressed in its absence since the cross-linking of the molecules is also impaired, although the lysyl aldehydes are present in normal or even increased concentration (Bailey and Lapière, 1973).

Normal procollagen peptidase is active on dermatosparactic collagen both in its monomeric and polymeric form. It is obvious that polymeric procollagen is an intermediate step in the architectural organization of the collagen fibers (Lapière *et al.*, 1975). The organization of procollagen fibers in the dermatosparactic dermis can be obtained by grafting fragments of dermis within a diffusion chamber implanted under the skin of a normal calf (Shoshan *et al.*, 1974).

In the dermatosparactic calf, the procollagen peptidase deficiency is probably incomplete. The dermis of these animals indeed contains some normally processed type I collagen, and other tissues, tendons, and blood vessels are mainly composed of this type of molecule. The proportion of collagen *vs.* procollagen is inversely related to the rate of turnover of the molecules; it increases with aging of the animals and increases with the "age" of the molecule in connective tissue as determined by sequential extraction.

Although no measurable procollagen peptidase activity can be extracted from skin or tendon of dermatosparactic calves, a protein can be collected from these tissues which has chemical and immunological similarities with the purified enzyme of normal calf (unpublished results). The presence of normal collagen molecules might therefore result from the activity of a mutated enzyme of low efficiency. It is most unlikely that the proteolytic activity of a nonspecific endopeptidase would simulate its effect because of the specificity of the peptide bond to be cleaved, as determined by the sequence of amino acids in the region forming the junction between the telopeptide and the propeptide (Fietzek, personnal communication). The low rate of procollagen peptidase activity in the mutant calf perhaps explains how such a defective animal can still achieve complete development.

As described in the dermatosparactic calves (Lapière *et al.,* 1975), collagen fibers in the papillary dermis and around adnexae display a more classical architecture. In cross-section these fibers display only notches, their parallel packing is enhanced, and the amount of interfibrillar substance is reduced. Intermediary structures make a progressive transition between the two types of fibers. This observation also suggests that some collagen is processed more extensively when associated with cells of a different strain than the dermal fibroblasts.

Not enough is known about the participation of type III collagen in the edification of the connective tissue, particularly in the dermatosparactic animals. In culture, dermatosparactic skin fibroblasts secrete large amount of collagen under the form of at least two soluble precursors, probably type I and type III (Church *et al.,* 1973; Church *et al.,* 1974).

Little information is available about procollagen peptidase deficiency in sheep and in humans (Ehlers–Danlos type VII). In sheep the small amounts of collagen that can be extracted seem to represent a larger precursor of type I than the molecule collected from the skin in the calf. Perhaps the mutation responsible for the enzyme deficiency is different and results in its complete absence. In humans the collagen extracted from skin contains polypeptide species representing partly processed precursors (Lichtenstein *et al.,* 1973) of a molecular size comparable to those observed in the calf. Fibroblasts cultured from the dermis of three such patients were found to produce procollagen peptidase in extensively reduced but still measurable amounts. When tested under similar conditions, dermatosparactic fibroblasts in culture also produced a small but measurable amount of procollagen peptidase activity (unpublished results).

The rate of collagen metabolism in organisms presenting this enzyme deficiency is not well known. As measured in the calf by the urinary excretion of hydroxyproline (Ansay *et al.*, 1968), it has been found similar to age-matched controls. *In vitro,* skin fibroblasts from Ehlers–Danlos type VII (Lichtenstein *et al.*, 1973) produce a larger-than-normal amount of collagen and its precursor.

3. Collagenase

This endopeptidase is produced by many types of cells and arises, in some of them, by the activation of a precursor after limited proteolysis (Harper and Gross, 1972; Vaes, 1972; Oronsky *et al.*, 1973; Robertson *et al.*, 1974). Any possible defect in its activity could therefore depend on a structural alteration of the active enzyme or its proenzyme, an alteration of the activation process, or defective activity of inhibitors (Eisen *et al.*, 1970; Abe and Nagai, 1972; Bauer *et al.*, 1972; Werb *et al.*, 1974). No existing pathology has been actually related to one of these specific mechanisms, although unsuitable collagenase activity exists in several conditions. Two types of heritable disorders could be related to a defective activity of collagenase: a decreased activity in osteopetrosis and an increase in epidermolysis bullosa dystrophica.

a. Osteopetrosis (B3a). This is a systemic disorder of bone in which resorption fails to keep pace with accretion. This heritable disorder has been recognized in many animal species, including human, rabbit, rat, and mouse. In the mouse it is present in three forms, variable in intensity and clinical expression with increased bone density, lack of tooth eruption, and retardation of growth. In the human and some animals the defect is spontaneously reversible with aging. In the human it is transmitted as an autosomal recessive condition in the malignant form and as an autosomal dominant disease in the benign form (McKusick, 1972). In contrast to the failure of bone resorption *in vivo* in the mouse, increased collagenase production occurs in bone *in vitro* (Walker, 1966) perhaps from the activation or the release of a stock of proenzyme. The observation of a lack of release of lysosomal enzymes under the influence of parathormone has led Marks (1974) to postulate that osteopetrosis is caused by a defective activation in the release mechanism of proteases necessary for bone resorption. This thesis is supported by experiments of Walker (1973*a*) showing that parabiosis and recruitment of competent osteolytic cells correct the defect in the mouse.

b. Epidermolysis Bullosa Dystrophica (B3b). This heritable disorder in the human expressed by blistering of the skin after minor trauma followed by healing with the formation of a scar has been shown to depend in part upon an increased amount of epidermal collagenase activity induced by mechanical trauma (Eisen, 1969). The destruction of the papillary dermis may be produced by an increased synthesis of the enzyme or may depend upon an associated deficiency in plasma inhibitor. The latter hypothesis is not supported by Lazarus (1972) who feels that the local increased amount of collagenase activity is a secondary reaction to tissue injury.

c. Invasive Processes (B3c). Collagenase is needed to initiate degradation of polymeric collagen. This enzyme may occur in increased amounts in all pathological conditions in which connective tissue breakdown is required. This has been demonstrated in invading cancers (Harris *et al.*, 1972; Hashimoto *et al.*, 1973*b*; Yamanishi *et al.*, 1972, 1973), cholesteatoma of the ear (Abramson, 1969), and the rheumatoid nodule (Hashimoto *et al.*, 1973*a*). Collagen degradation in metastatic cancers invading bone is further supported by many observations of an increased amount of urinary hydroxyproline (see Kivirikko, 1970, for review).

d. Healing (B3d). Collagenase has been isolated from the granulation tissue of healing wounds (Grillo and Gross, 1967). In skin burns this enzyme is probably involved in the extensive collagen degradation that is reflected in the increased urinary excretion of peptidic hydroxyproline (Klein *et al.*, 1962; Smith *et al.*, 1974). The amount of skin collagenase activity produced *in vitro* by cheloids (Milsom and Craig, 1973) is not significantly different from that produced by normal skin.

A more specific involvement of collagenase related to trauma and healing is observed in the alkali-burned cornea. A release of collagenase, perhaps in excess, is claimed to be responsible for perforating ulcers (Brown, 1971).

e. Rheumatoid Arthritis and Inflammation (B3e). Collagen degradation and cartilage alteration in rheumatoid arthritis have received much attention (for review, see Harris and Krane, 1974*a–c*). A collagenase displaying well-defined properties is secreted by the inflamed synovium (Evanson *et al.*, 1967; Harris *et al.*, 1969*a,b*). It has been purified and its mode of action upon fibrous collagen has provided interesting information in terms of control mechanism to understand the protection of the articular surfaces under normal conditions. Cartilage collagen (type II), is indeed degraded by rheumatoid synovial collagenase at a much lower rate than type I collagen (Woolley *et al.*, 1975*a*). Further, when comparing activity at or near the temperature of a normal joint (around 33°) to that observed at the temperature of an inflamed joint as in rheumatoid arthritis (around 36°), Harris and McCroskery (1974) observed a fourfold increase in the rate of degradation. In the arthritic joint, a large proportion of collagenase is present under an inactive form by combination with serum protein inhibitors (Abe *et al.*, 1973; Abe and Nagai, 1973). It suggests that some control of this enzyme activity is modulated by circulating factors. A specific inhibitor (Wooley *et al.*, 1975*b*) might exert this function.

It seems unlikely that the enhanced collagenase activity of the inflamed synovium results from the accumulation of leukocytes (Harris *et al.*, 1969*a,b*). These cells that participate in most inflammatory processes are known to contain collagenase (Lazarus *et al.*, 1968), and rheumatoid synovial fluid is able to activate a proenzyme of this origin (Kruze and Wojtecka, 1972). Specific antibodies to collagen (Wick *et al.*, 1975) are able to trigger the inflammatory mechanism as in the collagen-induced acute synovitis of the collagen-immunized rabbits (Steffen *et al.*, 1975). Such an activation might be mediated by prostaglandins and cyclic nucleotides (Castor *et al.*, 1975).

Granulocyte collagenase and/or an enzyme produced by dermal cells (Lazarus and Fullmer, 1969) are involved in degradation of collagen in several types of inflammation (Nigra *et al.*, 1972) as well as in resorption of granulation tissue that is produced by carrageenin (Pardo and Tamayo, 1974). The stimulated production or the release of active enzyme is inducible by toxin (Wahl *et al.*, 1974) or phagocytosis even of inert material (Werb and Reynolds, 1974). Lysosomal enzymes might be involved in the activation process (Vaes and Eeckhout, 1975).

Periodontal pathology is associated with tissue degradation and the production of a collagenase (Fullmer and Gibson, 1966; Fullmer *et al.*, 1972) which has been isolated from crevice fluid (Golub *et al.*, 1974). The proenzyme is activated by products issued from the microbial plaque (Robertson *et al.*, 1974). Its activity has also been related to a mast cell factor blocking the action of serum inhibitors (Simpson and Taylor, 1974).

f. Gas Gangrene (B3f). The rapid spreading of gas gangrene, a clostridial infection, is probably facilitated by the secretion of a most potent collagenase (Mandl *et al.*, 1958) and a large variety of neutral proteases involved in the degradation of most tissue components.

4. *Cathepsins*

The involvement of lysosomal hydrolases in the extracellular events of collagen degradation under physiological conditions is still debated. Cathepsin B has been found to degrade collagen at acid pH (Burleigh *et al.,* 1974), but its significance in pathological conditions is unknown. Cathepsins and proteases of lysosomal origin participate in the degradation of various components of the connective tissues such as proteoglycans and structural proteins that are involved in the organization of the connective tissue. They might be required to make the collagen framework accessible to collagenase as, for example, in bone in which degradation is a multistep process (Nusgens *et al.,* 1972). Collagen in calcified tissue is protected from the action of collagenase by the presence of crystals (Lapière and Nusgens, 1970*b*). Cathepsins might be involved in releasing the crystals, a process that is documented by morphological studies (Bonucci, 1974).

An indirect effect of lysosomal cathepsins in connective tissue degradation has been proposed by Lazarus and Barret (1974). Neutral proteinases display a chemotactic activity responsible for the accumulation of granulocytes in the inflammatory tissue, and degradation of fibrous collagen might result in the activity of the collagenase present in these cells.

Lysosomal hydrolases are responsible for the degradation of proteoglycans. Many heritable disorders result from their genetic alteration (for review, see Hers and Van Hoff, 1973). Although these storage diseases affect most connective tissues, they have little effect on the metabolism of collagen.

C. *Extracellular Interaction between Collagen and Other Compounds*

1. *Proteoglycans*

a. Marfan's Syndrome (C1a). This is a heritable disease affecting various connective tissues: the skeletal system (excessive growth of bone), the eye (ectopia lentis), and the cardiovascular system (floppy valves and aneurysm) (for details see McKusick, 1972). The nature of the defect is unknown. The dominant pattern of the gene responsible for Marfan's syndrome makes it more likely that it depends upon an alteration in a structural protein or a control system rather than in an enzyme operating for determining protein structure or metabolism. To account for the three main locations of the defect, a pathology related to elastin alone is most unlikely. The increased urinary excretion of collagen breakdown products in a large proportion of the patients presenting Marfan's syndrome (Jones *et al.,* 1964) has pointed to the possibility of a defective collagen framework in this disorder. Physical and chemical investigation disclose some anomalies, i.e., increased extractibility of skin collagen (Laitinen *et al.,*

1968) and increased solubility of collagen produced by fibroblasts in culture (Priest *et al.,* 1973). No conclusive evidence, however, supports an alteration in the structure of this molecule.

Modifications in polysaccharides have been observed that might be either the cause or the result of the defective mechanical properties of the connective tissues. Some information supports an alteration in the regulation of hyaluronic acid production. Bolande (1963) observed an increased deposition of metachromatic substances in the aorta and the endocardium of a "Marfan" patient. Berenson and Serra (1959) observed markedly increased urinary excretion of polysaccharides which had the electrophoretic mobility of hyaluronic acid and chondroitin sulfate. In some patients, Winand (1975) observed an increased urinary excretion of peptidic hydroxyproline and glycosaminoglycans composed of a high proportion of unsulfated and low-sulfated chondroitin. The alteration in polysaccharides metabolism is further suggested by the observation of an accumulation of hyaluronic acid in fibroblast culture (Matalon and Dorfman, 1968). This does not, however, represent a convincing argument since similar anomalies can be induced in normal fibroblasts by slight modifications of the culture conditions (Lie, 1974).

It is conceivable that a disturbed metabolism of proteoglycans is responsible for an enhanced rate in turnover of collagen. There is an obvious relationship between various components of the connective tissue and particularly between collagen and proteoglycans as demonstrated in articular cartilage (Simunek and Muir, 1972; Rosenberg *et al.,* 1973). It is further supported by their known interaction in the polymerization of collagen *in vitro* (Wood, 1960; Gelman and Blackwell, 1973; Nemeth-Csoka, 1974). *In vivo* the relationship is also known to exist. The formation of fibrous collagen is induced by the injection of carrageenin (Jackson, 1957), a polymeric glycan which is slowly resorbed. The deposited collagen is removed after the disappearance of the injected polymer. It is also known that impaired fibrogenesis leads to abnormal concentrations of proteoglycans as in lathyrism (Bentley *et al.,* 1973) and in dermatosparaxis (Ansay *et al.,* 1968; Winand and Nusgens, 1971).

Other pathological conditions also support the relationship between proteoglycans and collagen. An increase in the urinary excretion of both collagen and proteoglycan breakdown products was observed in our patient presenting genetically impaired hydroxylation of proline (Nusgens and Lapière, 1973; Lapière and Winand, 1971). In Down's syndrome, both hydroxyproline and glycosaminoglycan urinary excretion is reduced (Klujber and Méhes, 1973). The following two disorders also represent examples of a similar pathology associating collagen and proteoglycans.

b. Scleromyxedema (C1b). This is a rare skin disease related to the presence of localized superficial deposits of proteoglycans and fibrous collagen. The defect, perhaps depending on a pathological immunoglobulin (James *et al.*, 1967) coincides in two of our patients with an abnormal metabolism of proteoglycans. Skin fibroblasts are loaded with amorphous material that is also found in the extracellular space and displays metachromasia (Lapière *et al.*, 1969). The urine of these patients contains an excess of peptidic hydroxyproline and low-sulfated chondroitin (Winand, 1975). Both return to normal upon therapy (Lapière and Winand, 1971).

Pretibial Myxedema. In Grave's disease this also seems related to a disturbed metabolism of proteoglycans perhaps induced by an immunological process. Immunological suppressive therapy is known to prevent the connective tissue alteration (Winand and Mahieu, 1973). Glycosaminoglycans accumulate in skin (Watson and Pearce, 1947). These patients excrete in the urine an excess of proteoglycans catabolites and peptidic hydroxyproline that simultaneously return to normal upon treatment (Lapière and Winand, 1971).

2. *Glycoproteins*

Structural glycoproteins are associated with collagen in all connective tissues (for review, see Robert *et al.*, 1972). Acidic polypeptides of related amino acid composition are constituents of the basement membranes (for review see Kefalides, 1975), part of the reticulin fibers (Pras and Glynn, 1973; Pras *et al.*, 1974), and participate in the composition of the elastic fibers to form the microfibrils (Ross and Bornstein, 1969). Similar polypeptides have also been isolated from bone (Ashton *et al.*, 1974).

a. Ectopic calcification (C2a). Acidic glycoproteins are involved in calcification of bone (Lapière and Nusgens, 1970*b*). Some of the peptides isolated from bone (Shuttleworth and Veis, 1972) are similar to the phosphoprotein isolated from dentin (Veis *et al.*, 1972; Carmichael and Dodd, 1973). They might be contained in those vesicles which migrate from the cells to the calcifying front in dentin (Weinstock *et al.*, 1972; Weinstock and Leblond, 1973) and in cartilage (Simon *et al.*, 1973). A protein of that class might act as an inducing factor, promoting ectopic osteogenesis (Urist *et al.*, 1972). Under pathological conditions, such glycoproteins participate in calcification of soft connective tissues as in tumoral calcinosis (Lapière *et al.*, 1969), paraosteoarthropathy (Nusgens *et al.*, 1972), and in scleroderma (unpublished results).

b. Pseudoxanthoma Elasticum (C2b). This is a heritable disorder primarily affecting the elastic fibers. In view of the above-mentioned disorders, it is reasonable to speculate about the involvement of the microfilaments associated with elastin in the pathogenesis of calcification in pseudoxanthoma elasticum (Martinez-Hernandez and Huffer, 1974). Abnormal structural properties of the elastic fibers (Ross, 1973) coincide with calcium deposition in skin and precocious medial calcification in peripheral arteries. Accumulation of proteoglycans in the extracellular space and collagen fibers altered in structural organization have been reported by Danielsen *et al.* (1970).

c. Cutis Laxa (C2c). This is a rare condition characterized by a widely spread disorder of many connective tissues attributed to the degeneration of elastic fibers. The occurrence of this syndrome in siblings and the evidence of parental consanguinity in several pedigrees suggest an autosomal recessive inheritance (Goltz *et al.,* 1965). Skin is abnormally loose and enlarged. Piérard (personnal communication) observed by scanning electronmicroscopy the presence of thin whirling collagen bundles loosely packed together. The dermis had lost some of its elasticity (McCarthy *et al.,* 1965), although when under tension its mechanical properties were comparable to that of normal skin (Grahame and Beighton, 1971). The abnormal architecture and the laxity of the dermis point to the function of elastic fibers in the organization of the collagen framework.

3. Polymerization-Modifying Agents

a. Miscellaneous Physiological Compounds (C3a). Capable of altering collagen polymerization *in vitro,* these compounds such as urea, arginine, (Gross and Kirk, 1958) and glucose (Hayashi and Nagai, 1972) are probably ineffective *in vivo*. Even under extreme pathological conditions they would never reach suitable concentrations. Sulfate ions are known to retard polymerization of procollagen *in vitro* (Lapière and Nusgens, 1974). The lowest effective concentration of the ion is still several orders of magnitude greater than its physiological molarity. If delayed polymerization occurred *in vivo*, the persistence of collagen in an oligomeric form in the extracellular fluid for an extended period of time and at physiological temperature could result in its degradation by proteases of little or no specificity (Lapière, 1967). Suppression of sulfate might result in the reversed situation. In the rat, a diet deficient in sulfate has been shown to enhance collagen deposition (Brown and Liddy, 1971).

b. Ochronosis (C3b). Homogentisic acid, resulting from the inherited deficiency of homogentisic acid oxidase, is excreted in the urine and also accumulates in the cells and the extracellular space. Fibrous collagen in cartilage, tendon, and ligaments is coated with a deep brown deposit, dense to electron transmission. This condition is known as ochronosis, and degenerative changes affect joints, tendons, the cardiovascular system, and the eye (McKusick, 1972). Homogentisic acid *in vitro,* at higher concentrations than those observed in ochronosis, accelerates polymerization of collagen but not that of a collagen precursor (unpublished results).

c. Ionizing Radiations (C3c). This could be considered as a possible pathogen in the range used in clinical therapy, i.e., up to a few krads. In such instances, X-rays are capable of modifying the collagen molecule through at least two different mechanisms. At low doses *in vitro* they are capable of introducing covalent modifications that enhance the polymerizing properties of the molecules (Van Caneghem and Lapière, 1970). At high doses they produce ruptures in the polypeptide chains and modify the ultrastructure of the molecules and the fibers (Filisko *et al.,* 1972).

4. *Aldehyde Blocking Agents*

Compounds known to react with the aldehydes involved in cross-linking produce alterations somewhat similar to lathyrism.

a. Penicillamine (C4a), injected into animals induces skin fragility, depresses wound healing, and causes an accumulation of collagen that can be extracted in the cold by salt solutions of neutral pH (Nimni and Bavetta, 1965). This extractable collagen is newly synthesized (Klein and Nowacek, 1969; Deshmukh *et al.,* 1973). Penicillamine blocks the aldehydes involved in the cross-linking reaction *in vivo* (Deshmukh and Nimni, 1969) as well as *in vitro* (Deshmukh *et al.,* 1971). At high concentration, this compound is also capable of inhibiting lysyl oxidase (Nimni *et al.,* 1972), perhaps by chelating copper. Copper chelation by penicillamine has been used in human clinical pathology to treat Wilson's disease (Walshe, 1956), rheumatoid arthritis (Jaffe, 1964), and scleroderma (Fulghum and Katz, 1968). In the skin of most patients treated with this drug, the proportion of extractable collagen is increased (Harris and Sjoerdsma, 1966*b*).

b. Homocystinuria (C4b). This recessive inborn error of metabolism cystathionine synthetase deficiency, results in accumulation of homocysteine, leading to a condition comparable to that produced by penicillamine. The reaction of homocysteine with aldehydes to form stable compounds is known to occur at least *in vitro* (Jackson, 1973). Increased extractibility of skin collagen was observed in two homocystinuric patients (Harris and Sjoerdsma, 1966*a*). Defective cross-linking *in vivo* and *in vitro,* best substantiated by the observations of Kang and Trelstad (1975), could be responsible for the clinical condition ressembling Marfan's syndrome: altered mechanical properties of the connective tissues of the skeleton, the eye (ectopia lentis), and the vascular system (McKusick, 1972).

D. *Pathology Related to Metabolism and Turnover*

1. *Regulation*

Constant and adequate mechanical properties of all connective tissues depend upon a precise balance between the cellular functions ensuring synthesis and degradation of their constituents. Such a process is probably under control of the genome and *in vivo* seems to be modulated by hormones, nutritional status, and other mediators, such as chalones (Houck *et al.*, 1973), prostaglandins (Blumenkrantz and Søndergaard, 1972), and serotonin (Boucek *et al.*, 1972). Suppression of growth by dietary restriction reduces the proportion of salt-extractable collagen in guinea pig skin (Gross, 1958) and of newly formed bone and urinary hydroxyproline in rats (Lapière and Nusgens, 1970*a*). Inadequate food supply in children also induces a reduction of urinary hydroxyproline (Whitehead, 1965). Fasting and weight loss in the obese adult have the reverse effect, an increased urinary hydroxyproline and a high calcium excretion suggesting a loss of bone collagen (Ball *et al.*, 1972).

Much information related to connective tissue metabolism and its control has been obtained by measuring the daily excretion of urinary hydroxyproline (for review, see Prockop and Kivirikko, 1967; Kivirikko, 1970). In most conditions the proportion of free hydroxyproline contributing to total hydroxyproline is around 5%. In some physiological and pathological conditions, free urinary hydroxyproline is largely increased as in the heritable deficiency of hydroxyproline oxidase (Efron *et al.*, 1965), in newborn children and premature babies, probably due to the slow appearance of this degradative enzyme (Morrow *et al.*, 1967), and in type II hyperprolinemia (Goodman *et al.*, 1974; Applegarth *et al.*, 1974).

a. Aging (D1a). Physiological aging should not be separated from growth since both are the continuation of the same process, perhaps related to a reduced mitotic potential of the fibroblast (Hayflick, 1965, Martin *et al.*, 1970) and a progressive modification of the genome (Petes *et al.*, 1974), accompanied by changes in the endocrine status. In terms of collagen framework, all modifications related to aging support a progressive reduction in the rate of turnover of its constitutive molecules. Urinary hydroxyproline excretion in the human increases with age from birth to puberty and decreases to stable and lower values in the adult. This age-related variation in collagen turnover can be correlated with chemical observations in selected connective tissues. In the dermis, the proportion of extractable collagen progressively diminishes with the reduced rate of growth, as observed in several animal species and the human (Legrand *et al.,* 1969). In bone, the proportion of recently synthesized noncalcified matrix also diminishes with age (Nusgens *et al.,* 1972). Lung polysomes isolated from fetus induce, *in vitro*, collagen synthesis at twice the rate of the polysomes collected from adult lung (Collins and Crystal, 1975). These changes are related to a reduced synthesis of the collagen polypeptides accompanied by a reduced concentration of the enzymes involved in their covalent modifications, such as peptidyl hydroxylases (Uitto *et al.,* 1969; Anttinen *et al.,* 1973), procollagen peptidase (Lapière and Piérard, 1974), and the products of reaction of the aldehydes (Bailey and Shimokomaki, 1974; Fujii and Tanzer, 1974). This reduced metabolic activity ultimately results in modified mechanical properties of the connective tissue described by Grahame (1970) as a "stiffening up" of the soft connective tissues. It is most often accompanied by a reduction in the mass of the fibrous components and/or other compounds interacting to form the connective tissue.

Pathological alterations that are associated with aging could result from an imbalance between synthesis and degradation and a progressive loss of selected tissue components. Senile skin (also called senile elastosis) could result from a loss of collagen and proteoglycans, while elastin, more inert in terms of turnover, would be preserved. The mechanism responsible for senile osteoporosis is difficult to investigate due to the slow rate at which osteopenia occurs (Wu and Frost, 1969). The dynamic aspect of the process is better substantiated in the rapidly developing osteoporosis occurring distal to the level of a spinal cord section. Urinary hydroxyproline excretion is increased (Chantraine, 1971), and chemical analyses of bone demonstrate the presence of a high proportion of newly synthesized matrix (Nusgens *et al.,* 1972). Both observations suggest that the reduction of bone mass is due to a degradation rate surpassing the synthesis rate in a tissue undergoing active remodeling. The process is localized suggesting that local factors of regulation are responsible for bone matrix synthesis and degradation.

Increased bone resorption has been noted in patients receiving large doses of heparin (Nichols *et al.,* 1965). This observation is consistent with the findings that heparin not only increases the amount of collagenase released by bone in culture, but also enhances the activity of the enzyme when tested on polymeric collagen (Sakamoto *et al.,* 1975).

b. Premature Aging Syndromes (D1b). Four syndromes, autosomal recessive, have been described in the human: progeria (Hutchinson-Gilford), metageria, acrogeria (Gottron), and pangeria (Werner) (Gilkes *et al.,* 1974). The main difference between what is called physiological aging and the premature aging syndromes is mainly the chronology of the clinical signs of aging and their intensity. Progeria is a real caricature of aging as suggested by Rosenbloom and DeBusk (1971). In progeria and pangeria, the life span of cultured fibroblasts is often diminished (Martin *et al.*, 1965; Goldstein, 1969), the growing capacity of such fibroblasts is reduced (Nienhaus *et al.*, 1971), their functions are perturbed (Epstein *et al.*, 1974), and some of their biosynthetic products are abnormal (Holliday *et al.*, 1974; Goldstein and Singal, 1974; Goldstein and Moerman, 1975).

c. Endocrine Disturbances (D1c). Little is known of the exact mechanism through which hormones influence the metabolism of collagen. It is most likely that they act by modulating the biosynthetic activity of the cells through the pathway involving the second messenger. Calcitonin receptors exist in kidney and bone that activate adenylate cyclase (Marx, 1972). Cyclic nucleotides stimulate collagenase and hyaluronidase activity in organ culture of tadpole tail fins (Harper and Toole, 1973) while dibutyryl cAMP markedly increases the ability of progesterone to prevent the expression of collagenase activity in cultures of postpartum rat uterus (Koob and Jeffrey, 1974). Hormone involvement in collagen metabolism has been extensively investigated in the human, mainly by measurement of the urinary excretion of hydroxyproline. Little has been added since the review of Kivirikko (1970).

Glucocorticosteroids, even by topical application on the skin, induce a loss of tissue proteins. Atrophic skin and osteoporosis are most common symptoms observed in patients on long-term steroid therapy. Reduced synthesis of collagen (Kivirikko *et al.*, 1968) is supported by a decreased activity of prolylhydroxylase (Cutroneo *et al.*, 1971, 1975). Reduced collagenase production and degradation of collagen have also been observed in skin culture (Koob *et al.*, 1974).

In the absence of growth hormone (as in pituitary dwarfs), the urinary excretion of hydroxyproline is low and parallels the retardation of growth; levels comparable to normal children and acceleration of growth can be recovered by injecting the hormone (Teller *et al.*, 1973).

Estrogens have specific effects on some target tissues. They prevent collagen degradation in the involuting uterus (Ryan and Woesner, 1972) and the skin (Skosey and Damgaard, 1973). The suppressive activity of progesterone on the release of collagenase in the uterus seems to be of physiological significance (Jeffrey *et al.*, 1971*a*,*b*).

Thyroxin is obviously involved in processes regulating the metabolism of connective tissue proteins. Hyperthyroidism is always accompanied by an increased urinary excretion of peptide-bound hydroxyproline (Kivirikko *et al.*, 1964), while in hypothyroidism the excretion of collagen catabolites is reduced (Kivirikko *et al.*, 1965*b*).

Parathormone and calcitonin are polypeptide hormones mainly involved in the homeostasis of calcium and phosphorus. Their action is in part mediated by the activity of the cells of the calcified connective tissues. Parathormone enhances the catabolism of bone *in vivo* (Avioli and Prockop, 1967); in bone culture it stimulates collagenase activity (Walker *et al.*, 1964) and the release of lysosomal enzymes (Vaes, 1968). The physiological effect of calcitonin on bone is still debated. It suppresses bone resorption induced by parathormone (Robinson *et al.*, 1972), although its direct activity on bone cell metabolism is minimal (Marks, 1972). The urinary excretion of peptide-bound hydroxyproline is reduced by calcitonin in normal subjects and in patients exhibiting Paget's disease of bone (Krane *et al.*, 1973).

2. *Healing*

Healing should be considered as a homeostatic function. It is performed by the connective tissues and is required to maintain vital physical properties of the frame supporting the cells and ensuring the rigidity of the skeleton as well as the continuity of the envelope. Any parenchymal defect resulting from trauma or necrosis is usually filled by the connective tissue.

Healing and scar formation proceed through several successive steps (for review, see Shoshan and Gross, 1974): the proteolytic separation of the necrotic from the living tissues, the proliferation of blood vessels and associated cells responsible for the synthesis of the macromolecules forming the granulation tissue, and finally the remodeling of the newly formed tissue to obtain the final scar. In most healing or fibrotic processes, embryonic (type III) collagen is observed in the early phases of tissue repair as in skin wound healing (Bailey *et al.*, 1975), in the remodeling of inflamed synovium in rheumatoid arthritis (Weiss *et al.*, 1975) and in liver cirrhosis (Gay *et al.*, 1975*b*).

The newly formed connective tissue differs from that in adjacent tissues in terms of chemical composition, structure, and physical properties. In skin, scar collagen is claimed to be only type I molecules (Shuttleworth and Forrest, 1974; Shuttleworth *et al.*, 1975), instead of the normal mixed population of molecules: two thirds of type I and one third of type III (Epstein, 1974). The first fibers deposited under a random pattern evolve into large irregular masses in the mature scar (Forrester *et al.*, 1970). The individual fibers seem to be "glued" together, at least in the hypertrophic scar (Kischer, 1974). Mechanical factors are most significant in conditioning the architecture of the fibrous framework in scar tissue. They have received much attention in plastic and reconstructive surgery (Gibson, 1965; Kenedi *et al.*, 1965).

In the absence of specific metabolic disturbances affecting the connective tissue and provided a correct vascular supply of the tissue is ensured, healing is rarely defective. Even in those disorders resulting from enzyme deficiencies (dermatosparaxis and Ehlers–Danlos type VI), correct skin-wound healing has been observed.

a. Cicatricial Fibrosis (D2a). Healing and remodeling of scar tissue can lead to retraction, stenosis, and impaired function through contracture. It represents a main sequel of thermal injury (Larson *et al.*, 1974). It can also be a life-threatening or crippling consequence of many diseases. It is the end result of an activation of connective tissue metabolism in conditions such as viral hepatitis or carbon tetrachloride intoxication (Risteli and Kivirikko, 1974) which can produce cirrhosis, pyelonephritis that can lead to nephrosclerosis, rheumatic fever to heart valve sclerosis, and other diseases such as rheumatoid arthritis (Uitto *et al.*, 1972), silicosis of the lung, fibrosis in sarcoidosis, and arteriosclerosis (Langner and Fuller, 1973). Even posttraumatic healing could be a potential hazard mainly when it occurs in the walls of a hollow organ, such as the esophagus, gut, or blood vessels.

In the hypertensive rat, the synthesis of collagen is increased in blood vessels, prolyl hydroxylase activity is enhanced (Ooshima *et al.*, 1975) and all manifestations are suppressed by antihypertensive agents (Ooshima *et al.*, 1974). In the human atherosclerotic plaque a larger than normal proportion of type I and a reduced proportion of type III collagen have been found by McCullagh and Balian (1975). Type I collagen synthesis is the mode of expression of the genome of smooth muscle cells *in vitro* (Layman and Titus, 1975). The trigger mechanism of the modifications observed in the hypertensive disease might be related to the platelet release reaction perhaps induced by collagen. Ross (1975) has indeed isolated a serum factor responsible for the proliferation of fibroblasts and smooth muscle cells in culture that is derived from thrombocytes.

b. Cheloids (D2b). The interaction between connective tissue and epidermis is well substantiated during embryogenesis (for review, see Slavkin, 1971). An epidermal regulation of connective tissue function and architecture (Gillman, 1964) is also substantiated by the common observation of the healing of the full thickness skin wound and the active remodeling of the scar occurring after epidermal coverage (Madden and Peacock, 1971). In human pathology, conditions are known to exist which might be related to alteration of this control. Pyogenic granuloma can be considered as a proliferation of heavily vascularized granulation tissue covered by a poorly organized and defective epidermis. Cheloids might be related to a similar kind of defective control of granulation tissue metabolism. This condition prevails in genetically defined patients. After coverage by epidermis, the newly formed connective tissue does not undergo a normal remodeling and the scar maintains hyperactive aspects, such as a larger than normal volume and several metabolic characteristics of newly formed connective tissue, i.e., a higher than normal amount of cells (Hoopes, 1971), a higher activity of enzymes involved in collagen synthesis and assembly (Cohen *et al.,* 1971; Lapière and Piérard, 1974; Craig *et al.,* 1975), and a higher content of hydrated proteoglycans (Kischer and Shetlar, 1974). The production of collagenase has been found comparable to normal skin (Milsom and Craig, 1973).

c. Scleroderma (D2c). This could result from an abnormal control of fibroblast function and the deposition of scar tissue in several organs. Cartilage, mainly composed of type II collagen, seems unaffected. In skin, sclerosis depends upon the laying down of dense sheets of collagen fibers in the subcutis (Fleischmajer *et al.,* 1971). The wide range of diameters displayed by the collagen fibers (Braun-Falco and Rupec, 1964; Fleischmajer and Prunieras, 1972, Kobayashi and Asboe-Hansen, 1972) suggests that the process depends upon a long-persisting synthesis. This hypothesis is supported by the observation of increased prolyl hydroxylase activity in the skin of some patients (Uitto *et al.,* 1970; Keiser *et al.,* 1971) and the increased proportion of reducible aldimine bonds related to the deposition of newly synthesized collagen in active lesions (Herbert *et al.,* 1974). The ultrastructure of the fibroblasts shows signs of hyperactivity, such as a well developed rough endoplasmic reticulum (Fleischmajer and Prunieras, 1972). The increased synthetic activity of the fibroblast is retained in tissue culture (LeRoy, 1974), although an increased doubling time and a loss of contact inhibition were observed (Kovacs and Fleischmajer, 1974). It has to be recalled that scleroderma is a long-standing disease that might have alternations of active and quiescent periods. This could explain the wide variations in the urinary excretion of hydroxyproline (unpublished results and contradictory reports) and proteoglycans (Hardy *et al.,* 1971). The trigger mechanism of the abnormal control of cell function is not known (Winkelman, 1971). The presence of antinuclear antibodies (Jordon *et al.,* 1971) could be the cause or the result of the defect. An increased proportion of chromosome breakage has also been reported (Emerit and Housset, 1973).

III. Relationship between Molecular Defects and Impaired Mechanical Properties

As it leaves the cell the collagen precursor has to be considered as a molecule marked with information allowing it to perform defined functions in the extracellular space. Even though four types of collagen and probably subclasses of these molecules have been identified that participate in building the various connective tissues, additional information, related to local conditions, is required to specify the architecture of the tissue and its ultimate mechanical properties.

The pathways resulting in functional deficiencies are probably as numerous as are the steps in the processing of the molecules and the factors influencing their behavior in the extracellular space. This principle was used to analyze the clinical and experimental pathology of the connective tissue, in which collagen is involved, in a search for disturbances in each individual step of molecular processing. It was observed that diseases with obviously different clinical expressions were found related to a comparable basic mechanism, while in many syndromes individual types of disease displaying similar symptoms were found to be caused by different mechanisms. It is obvious that the structural and the mechanical expression of the defects and/or the clinical phenotype of the diseases are much more limited in variety than the chemical processes responsible for their development. Each of the major clinical symptoms (modification in amount, extensibility, elasticity, and fragility) can indeed represent a similar end result of different molecular alterations.

A reduced amount of collagen, expressed by transparent skin in osteogenesis imperfecta, seems related to a defective differentiation of the fibroblasts which synthesize an excessive proportion of type III collagen, while in the Ehlers–Danlos type IV it could be due to a reduced synthesis of the same type III collagen. The density of the collagen framework is strikingly reduced in dermatosparaxis, although the thickness of the skin is normal or even increased. Atrophy of skin and bone is observed in many endocrine disturbances, resulting in a defective control in turnover of collagen, in hyperthyroidism or in iatrogenic hypercorticism. A loss of fibrous collagen is also present in aging and various syndromes of precocious aging. Although the collagen synthesized under conditions of defective hydroxylation never achieve a structure allowing it to be used in building fibers, atrophy is not observed in scurvy. Under such conditions, basement membranes are mainly altered and the extensive collagen production required in healing is also impaired. It is obvious that besides the nature of the molecular alteration, the timing and the duration of the defect, as well as the rate at which it interferes with metabolism, are major factors in conditioning its clinical expression.

Excess of fibrous collagen is more limited in terms of the mechanism capable of producing it. It is genetically determined in bone of osteopetrotic animals and is possibly related to a defective activity of degradative enzymes. Sclerosis is also observed in several conditions resulting from an activation of connective tissue metabolism by exogenous or endogenous factors inducing healing and scar formation, scleromyxedema, and scleroderma.

Extensibility and architectural organization of the collagen framework are most closely related to one another. In many disorders affecting collagen, polymerization appears indeed normal up to fiber formation and pathological in further steps of organization. Dermatosparaxis in the calf is the only defect for which pathology displays its expression at the level of the molecules, the fibers, the bundles, and the mechanical properties.

Reduced extensibility of human skin with increasing age is related to a thickening of the collagen bundles and their association into sheetlike structures (Millington and Wilkinson, 1974). Scleroderma displays somewhat similar features that are related to deposition in the deep dermis of sheets of polymeric collagen tightly packed. McNeal (1973) also reported an increased adherence of the polymers in the sheets. A similar cohesion between collagen fibers in bundles whirling into nodules seems responsible for reduced extensibility and sclerosis in the hypertrophic scar (Linares *et al.*, 1972).

Increased extensibility is a cardinal sign observed in several connective tissues of patients presenting the Ehlers–Danlos syndrome. In this disorder no report points to alteration in the ultrastructure of the collagen fibers, but their spatial association into bundles is disturbed (Julkunen *et al.*, 1970). Four molecular defects have been described that can produce hyperextensibility: defect in type III synthesis (Ehlers–Danlos type IV), in lysyl oxidase (Ehlers–Danlos type V), in lysyl hydroxylase (Ehlers–Danlos type VI), and in procollagen peptidase (Ehlers–Danlos type VII). It must also be pointed out that defective collagen and normal elastic network in the Ehlers–Danlos syndrome (Varadi and Hall, 1965) or normal collagen and defective elastic framework in Cutis laxa result in a somewhat comparable disturbance in collagen fiber architecture. The stretchability of skin is increased in both disorders but elasticity is normal in Ehlers–Danlos and reduced in Cutis laxa.

Defective resistance to stress and tissue fragility are the functional characters that are most closely related to the molecular structure of polymeric collagen. These abnormal properties always depend upon a defective binding of the molecules within the fibers. They can be generated by several mechanisms that will directly or indirectly result in the absence of interpolypeptide cross-linking. All mechanisms that could produce it can be deduced from the sequence of events directly or indirectly involved in cross-linking. Fragility is observed when hydroxylation of lysine is deficient (Ehlers–Danlos type VI), when oxidation of lysine is impaired (Ehlers–Danlos type V, lathyrism, and copper deficiency), when the aldehydes are blocked by exogenous or endogenous compounds (penicillamine or homocysteine), and also when steric hindrance is introduced by the uncleaved propeptide extension in procollagen peptidase deficiency (dermatosparaxis and Ehlers–Danlos type VII).

Comparable phenotypes for different molecular defects, as well as differing phenotypes for similar molecular defects, indicate that additional factors are involved in the expression of defective mechanical performances.

Beside properties inherent to the structure of the molecules, such factors might participate in the architectural organization of fibers into bundles. Motion can orient fibers and bundles (Lapière *et al.,* 1975), pressure stimulates healing (Hassler *et al.,* 1974) and modulates remodeling (Larson *et al.,* 1974), mechanical activity stimulates collagen deposition and depresses cross-linking in bovine muscle (Kruggel and Field, 1974), electrical stimulation induces collagenase production in fibroblast culture (Kamrin, 1974) and enhances healing in bone (Lavine *et al.,* 1974). Polysaccharides *in vitro* modify polymerization, and *in vivo* many pathological conditions seem related to abnormal proteoglycans and collagen metabolism.

Fibroblast is an operational name given to any type of cell capable of producing compounds participating in the structure of the connective tissue. They range from ectodermal or endodermal cells producing basement membranes, to fibroblasts in soft connective tissues, osteoblasts, or odontoblasts in calcifying tissues, and myofibroblasts in elastic blood vessels. It is not yet known whether or not the part of the genome used to perform a similar function is identical in all these stems of cells. Even though the cistron could be identical, the rate at which it is operating and the local condition of activity of the ribosomal products are so varied that there is not one single rule to predict what alteration will produce its deficiency. It has indeed been observed that a large variation exists at the level of the tissues in the degree of expression of the defect. Observations in lathyrism support the hypothesis that a molecular defect will display a most dramatic clinical expression in those connective tissues having an active metabolism. The intensity of the defect expressed by osteophytosis, skin fragility, and increased extractibility of collagen is directly related to the rate of growth, i.e., it is highest in skin and bone; it is important in young animals undergoing rapid remodeling and much reduced in the adults. A similar type of reasoning could also be applied to other types of molecular alteration even when inherited. Although recessivity for most of the enzyme defects supposes that only the homozygous animal will display the clinical phenotype, some enzyme activity has to be present in Ehlers–Danlos type VI, to account for the small amount of hydroxylysine, as well as in dermatosparaxis, to explain the presence of some normally processed collagen molecules. In both disorders, the most defective tissues are again those displaying the highest rate of turnover. As observed in dermatosparaxis, skin is most fragile and wounding occurs most readily in the early postnatal period; bone is also disturbed while tendon and blood vessels display normal mechanical properties. Everything occurs as if a limited amount of activity is large enough to allow the edification of a mechanically suited framework in those tissues displaying a slow rate of collagen synthesis and insufficient for those tissues which display the highest rate of metabolism.

Several other characteristics of collagen are tissue specific and might also influence the expression of a defect. Besides variation in the type of collagen used to build the fibers, the nature or the extent of some covalent reactions after translation display known differences, i.e., the amount and location of the aldehydes are not identical in skin and tendon (Deshmukh, 1973), and the pattern of the cross-links is different in skin, bone, and tendon (see Chapter 4), the hydroxylation of lysine is higher in bone as compared to skin (Stoltz *et al.,* 1973), the hydroxylation of proline is larger in skin as compared to tendon (Bornstein, 1967), the disaccharide glycosylated hydroxylysine is mainly found in skin and the monosaccharide in bone (Pinnell *et al.,* 1971), and the telopeptide of skin $\alpha 1$ chain is two amino acids shorter than that in tendon (Bornstein, 1969). Other still unknown details in chemical structure probably exist that are perhaps involved in function.

Up to now, pathology has offered most of the models allowing us to foresee the relationship between the molecular structure of collagen and its function. Many more models remain available. Four out of the seven clinical forms of Ehlers–Danlos syndrome in man (McKusick, 1974) are reaching the stage of chemical investigation and little is known for a similar type of pathology described in dogs and mink (Hegreberg *et al.,* 1970) as well as cat (Butler, 1975). The various other syndromes add to the collection of altered molecular processes. They will provide a large variety of additional models as suspected by their differing phenotypes and/or heredity. Finally, a large group of rare conditions also exist (see McKusick, 1972), and most likely many of them will also prove helpful in unraveling the processes that Nature uses in holding our cells together.

IV. Therapy

Successful therapy is a most rewarding end result of any investigation in clinical pathology. Chemists, biologists, clinical investigators, and practitioners all contribute to its realization.

A. *Genetic Defects*

In many polysaccharide storage diseases, the intracellular defective hydrolases and their substrates have been identified (for review see Neufeld and Fratantoni, 1970; McKusick, 1972), and promising results have been observed in their management by the introduction of enzyme replacement therapy (Di Ferrante *et al.*, 1971). The cells are able to absorb the injected enzyme by pinocytosis (von Figura and Kresse, 1974) and concentrate it in contact with its substrate inside the lysosomes.

Little progress has been made in restoring enzyme activities involved in the anabolic pathway. On theoretical grounds, it seems feasible to introduce a missing gene or replace a defective gene inside the affected genome (Osterman *et al.*, 1971; Friedmann and Roblin, 1972). This approach still needs much work before being applied for those diseases that have received an answer at the molecular level.

Perhaps more realistic at the present time is a therapy directed towards correcting the defective function of enzymes active in the extracellular space: procollagen peptidase, the oxidase(s), and collagenase.

In vitro it has been shown that polymeric procollagen is a suitable substrate for procollagen peptidase (Lapière *et al.*, 1971). *In vivo* dermatosparactic dermis within a diffusion chamber implanted under the skin of a normal calf is processed by a diffusing enzyme. Polymeric procollagen is converted into collagen and the architecture of the implanted dermis proceeds towards a better organization of the fibers into bundles (Shoshan *et al.*, 1974). Work is presently in progress to assess the possibility of correcting the defect of dermatosparactic calves by injecting active procollagen peptidase.

Restoring the activity of lysyl oxidase(s) might also become feasible. Collagen in its polymeric form is indeed the substrate of the enzyme (Siegel, 1974).

Collagenase of bacterial origin has been used in human therapy for debridement of cutaneous wounds (Moserova *et al.*, 1974). It seems helpful in suppressing the necrotic tissue. It could not, however, replace the endogenous enzyme responsible for the precise cleavage performed in the granulation tissue.

It is quite obvious that for all three enzymes of potential use in therapy, the timing of their activity and the precise location of their action will be difficult, if not impossible, to monitor in order to simulate their physiological activity so accurately modulated by the cells.

B. *Acquired Defects*

Many defects can be corrected that result from abnormal conditions in the function of otherwise normal gene products. Such therapy did not wait for the definition of collagen pathology at the molecular levels to be introduced, e.g., the treatment of gas gangrene by killing the microorganisms producing collagenase, the management of scurvy by providing a correct dietary supply of vitamin C, and the therapy for rickets by exogenous administration of vitamin D or exposure to the sun.

Abnormal function of the fibroblast related to hormonal imbalance can also be improved by suppressing excess production of the hormone, by introducing an inactive analog of the hormone, or by administration of a hormone displaying an opposite effect. Pituitary deficiency, hyper- and hypothyroidism, Cushing's syndrome, hyper- and hypoparathyroidism are correctable disorders. Although the pathogenesis of Paget's disease of bone is not clear, the disturbed activity of the pathological bone cells seems corrected by calcitonin (Bijvoet and Jansen, 1967). Bone resorption can be reduced by administration of diphosphonates (for review, see Fleisch and Russell, 1972). Clinical and metabolic improvement in Marfan's syndrome has been induced by an anabolic androgenic steroid (Jones *et al.*, 1969).

Several other alterations of connective tissue related to pathological conditions in the function of otherwise normal enzymes can be corrected. The suppression of an inhibitor such as a lathyrogen will allow a progressive recovery of oxidase(s) activity. It will not be immediate (Pinto and Bentley, 1974) since the inhibitor is irreversibly bound to the enzyme and *de novo* synthesis seems to be required. Improvement has been effected in homocystinuria by a diet poor in cystine, thereby depleting the metabolic pathway leading to the formation of the compound blocking the aldehyde. Restoring the supply of copper to correct the defective cross-linking in copper deficiency is easily done. A similar corrective therapy is not as operative in a related disorder such as Menkes' kinky-hair disease (Dekaban and Steusing, 1974). Not enough is known concerning the reason for the abnormal transport of copper.

Suppressing the enhanced activity of connective tissue cells or the pathological effect of their biosynthetic products is an everyday problem in medical practice. A most elegant application of basic biological information to therapy is the management of corneal ulcer following alkali burn. The observed overproduction of collagenase and the known inhibition of this enzyme by sulfhydryl compounds has led to their use, by topical application, with successful results (Brown and Weller, 1970). Antirheumatic drugs are perhaps effective by inhibiting the activity of mediators responsible for cell activation during the inflammatory process (Boucek *et al.*, 1972; Blumenkrantz and Søndergaard, 1972; Castor, 1973). These drugs are known to depress the formation of induced granulation tissue (Kulonen and Potilla, 1975), to inhibit leukocyte collagenase (Wojtecka-Lukasik and Dancewicz, 1974), and even to interfere with normal connective tissue cell function. Aspirin inhibits *in vitro* osteolysis induced by parathormone or prostaglandin E_1 (Powles *et al.*, 1973).

Excessive healing and fibrosis resulting from trauma or inflammation are most serious conditions in human pathology and too few possibilities exist for controlling their course. A most logical approach has been used by several groups of researchers to reduce the formation of scar tissue. It is based upon the introduction of compounds capable of interfering with one of the steps required in collagen biosynthesis or its extracellular organization. A potential inhibitor exists for each one of the enzymes involved in processing the collagen polypeptide and thus ensuring the structural characteristics required to establish stress-resistant polymers. Some of these inhibitors have been studied. Depressing proline hydroxylation is known to diminish the secretion of operational collagen molecules *in vitro*. The introduction of proline analogs could achieve this goal *in vivo*. The formation of scars with reduced mechanical properties has been reported by Daly *et al.*, (1973). The efficiency of this procedure is still debated. Chronic administration of 3,4-dehydroproline in mice proved to be toxic and did not affect collagen synthesis in normal or repaired tissues (Madden *et al.*, 1973*a*). *In vivo* chelation of iron is not feasible, although it is most effective *in vitro* (Chvapil *et al.*, 1974).

Inhibiting the formation of cross-links is the second most logical approach in suppressing sclerosis and retraction of the scar after healing. It would allow the deposition of structurally normal collagen fibers but prevent the development of their mechanical properties. This can be achieved by two routes: by inhibiting the oxidase(s), therefore preventing the formation of the aldehydes, or by blocking the lysyl and hydroxylysyl aldehydes. Both are of potential use *in vivo*. Inhibition of the oxidase(s) by lathyrogens has been largely used in animals (see Barrow *et al.*, 1974) and has also been applied to human therapy (Keiser and Sjoerdsma, 1967). Penicillamine, recommended in rheumatoid arthritis, has also been tested for treating scleroderma. It seems to interfere with the formation of the cross-links in newly formed tissue in the active form of the disease (Herbert *et al.*, 1974). Improvement in the mechanical properties of skin has been observed in some patients (Bluestone *et al.*, 1970). Most stimulating results have been obtained in the control of the experimental esophageal stenosis in the dog by treatment using β aminopropionitrile (Madden *et al.*, 1973*b*; Davis *et al.*, 1972).

Physical factors, normally involved in the organization of the collagen framework, can be used for modeling the architecture of the scar tissue and improve its mechanical performances. Prevention of contracture in the burned patient has been obtained by pressure (Larson *et al.*, 1974).

Many other possibilities exist that could be deduced from a close analysis of the disease process, the function and the nature of the cells that participate in it, and the defined mechanisms allowing them to express these functions.

ACKNOWLEDGMENTS

We wish to thank M. L. Tanzer and G. E. Piérard for their help in the preparation of the manuscript and their useful suggestions for its improvement. Mrs. N. Noelen expertly typed the manuscript and kindly verified the correctness of the bibliography. K. Zupnik and L. Hudson kindly helped in organizing the manuscript.

References

Abe, S., and Nagai, Y., 1972, Interaction between tadpole collagenase and human α2-macroglobulin, *Biochim. Biophys. Acta* **278:**125.

Abe, S., and Nagai, Y., 1973, Evidence for the presence of a complex of collagenase with α2-macroglobulin in human rheumatoid synovial fluid: A possible regulatory mechanism of collagenase activity *in vivo*, *J. Biochem.* **73:**897.

Abe, S., Shinmei, M., and Nagai, Y., 1973, Synovial collagenase and joint diseases: The significance of latent collagenase with special reference to rheumatoid arthritis, *J. Biochem.* **73:**1007.

Abramson, M., 1969, Collagenolytic activity in middle ear cholesteatoma, *Ann. Otol., Rhinol. Laryngol.* **78:**112.

Ansay, M., Gillet, A., and Hanset, R., 1968, La dermatosparaxie héréditaire des bovidés: Observations complémentaires sur le collagène et les mucopolysaccharides acides, *Ann. Med. Vet.* **6:**465.

Anttinen, H., Orava, S., Ryhanen, L., and Kivirikko, K. I., 1973, Assay of protocollagen lysyl hydroxylase activity in the skin of human subjects and changes in the activity with age, *Clin. Chim. Acta* **47:**289.

Applegarth, D. A., Ingram, P., Hingston, J., and Hardwick, D. F., 1974, Hyperprolinemia type II, *Clin. Biochem.* **7:**14.

Ascenzi, A., and Bonucci, E., 1967, The tensile properties of single osteons, *Anat. Rec.* **158:**375.

Ashton, B. A., Triffitt, J. T., and Herring, G. M., 1974, Isolation and partial characterization of a glycoprotein from bovine cortical bone, *Eur. J. Biochem.* **45:**525.

Askenasi, R., 1974, Urinary hydroxylysine and hydroxylysyl glycoside excretions in normal and pathologic states, *J. Lab. Clin. Med.* **83:**673.

Askenasi, R., 1975, Urinary excretion of free hydroxylysine, peptide-bound hydroxylysine and hydroxylysyl glycosides in physiological conditions, *Clin. Chim. Acta* **59:**87.

Avioli, L. V., and Prockop, D. J., 1967, Collagen degradation and the response to parathyroid extract in the intact rhesus monkey, *J. Clin. Invest.* **46:**217.

Aymard, P., 1971, L'hydroxyproline dans le rachitisme du rat, *C.R. Soc. Biol.* **165:**1573.

Bailey, A. J., and Lapière, C. M., 1973, Effect of an additional peptide extension of the N-terminus of collagen from dermatosparactic calves on the crosslinking of the collagen fibres, *Eur. J. Biochem.* **34**:91.

Bailey, A. J., and Shimokomaki, M. S., 1971, Age related changes in the reducible cross-links of collagen, *FEBS Lett.* **16**:86.

Bailey, A. J., Bazin, S., and Delaunay, A., 1973, Changes in the nature of the collagen during development and resorption of granulation tissue, *Biochim. Biophys. Acta* **328**:383.

Bailey, A. J., Bazin, S., Sims, T. J., Le Lous, M., Nicoletis, C., and Delaunay, A., 1975, Characterization of the collagen of human hypertrophic and normal scars, *Biochim. Biophys. Acta* **405**:412.

Baker, E. M., Hodges, R. E., Hood, J., Sauberlich, H. E., March, S. C., and Canham, J. E., 1971, Metabolism of 14C- and 3H-labeled L-ascorbic acid in human scurvy, *Am. J. Clin. Nutr.* **24**:444.

Ball, M. F., Canary, J. J., and Houck, J. C., 1972, Studies of the hydroxyprolinuria of fasting, *J. Clin. Endocrinol. Metab.* **35**:416.

Balleisen, L., Marx, R., and Kuhn, K., 1975, Uber die stimulierende Wirkung von Kollagen und Kollagenderivaten auf die Ausbreitung und Folienadhäsion von Thrombozyten in defibrinogenisiertem Menschencitratplasma und in tierischen Citratplasmen, *Blut* **31**:95.

Barnes, M. J., Constable, B. J., Morton, L. F., and Kodicek, E., 1970, Studies *in vivo* on the biosynthesis of collagen and elastin in ascorbic acid-deficient guinea pigs. Evidence for the formation and degradation of a partially hydroxylated collagen, *Biochem. J.* **119**:575.

Barnes, M. J., Constable, B. J., Morton, L. F., and Kodicek, E., 1973*a*, Bone collagen metabolism in vitamin D deficiency, *Biochem. J.* **132**:113.

Barnes, M. J., Constable, B. J., Morton, L. F., and Kodicek, E., 1973*b*, The influence of dietary calcium deficiency and parathyroidectomy on bone collagen structure, *Biochim. Biophys. Acta* **328**:373.

Barnes, M. J., and Kodicek, F., 1972, Biological hydroxylations and ascorbic acid with special regard to collagen metabolism, *Vitam. Horm.* **30**:1–43.

Barrow, M. V., Simpson, C. F., and Miller, E. J., 1974, Lathyrism: A review, *Q. Rev. Biol.* **49**:101.

Bates, C. J., Prynne, C. J., and Levene, C. I., 1972, Ascorbate-dependent differences in the hydroxylation of proline and lysine in collagen synthesized by 3T6 fibroblasts in culture, *Biochim. Biophys. Acta* **278**:610.

Bauer, E. A., Eisen, A. Z., and Jeffrey, J. J., 1972, Regulation of vertebrate collagenase activity *in vivo* and *in vitro*, *J. Invest. Dermatol.* **59**:50.

Beisswenger, P. J., 1973, Specificity of the chemical alteration in the diabetic glomerular basement membrane, *Diabetes* **22**:744.

Beisswenger, P. J., and Spiro, R. G., 1970, Human glomerular basement membrane: Chemical alteration in diabetes mellitus, *Science* **168**:596.

Beisswenger, P. J., and Spiro, R. G., 1973, Studies on the human glomerular basement membrane. Composition, nature of the carbohydrate units and chemical changes in diabetes mellitus, *Diabetes* **22**:180.

Bentley, J. P., Wuthrich, R. C., and Van Bueren, A. M., 1970, Lathyrism and mucopolysaccharide metabolism in aorta, skin and cartilage, *Atherosclerosis* **12**:159.

Berenson, G. S., and Serra, M. T., 1959, Mucopolysaccharides in urine from patients with Marfan's syndrome, *Fed. Proc.* **18**:190.

Berg, R. A., and Prockop, D. J., 1973, The thermal transition of a nonhydroxylated form of collagen. Evidence for a role for hydroxyproline in stabilizing the triple-helix of collagen, *Biochem. Biophys. Res. Commun.* **52:**115.

Bijvoet, O. L. M., and Jansen, A. P., 1967, Thyrocalcitonin in Paget's disease, *Lancet* **2:**471.

Bluestone, R., Grahame, R., Holloway, V., and Holt, P. J. L., 1970, Treatment of systemic sclerosis with D-penicillamine. A new method of observing the effects of treatment, *Ann Rheum. Dis.* **29:**153.

Blumenkrantz, N., and Søndergaard, J., 1972, Effect of prostaglandins E1 and F1α on biosynthesis of collagen, *Nature* **239:**246.

Bolande, R. P., 1963, The nature of the connective tissue abiotrophy in the marfan syndrome, *Lab. Invest.* **12:**1087.

Bonucci, E., 1974, The organic-inorganic relationships in bone matrix undergoing osteoclastic resorption, *Calcif. Tissue Res.* **16:**13.

Bornstein, P., 1967, The incomplete hydroxylation of individual prolyl residues in collagen, *J. Biol. Chem.* **242:**2572.

Bornstein, P., 1969, Comparative sequence studies of rat skin and tendon collagen. II. The absence of a short sequence at the amino terminus of the skin α1 chain, *Biochemistry* **8:**63.

Bornstein, P., 1974, The biosynthesis of collagen, *Annu. Rev. Biochem.* **43:**567.

Boucek, R. J., Speropoulos, A. J., and Noble, N. L., 1972, Serotonin and ribonucleic acid and collagen metabolism of fibroblasts *in vitro*, *Proc. Soc. Exp. Biol. Med.* **140:**599.

Bradley, K., McConnell-Breul, S., and Crystal, R. G., 1974, Lung collagen heterogenity, *Proc. Natl. Acad. Sci. U.S.A.* **71:**2828.

Braun-Falco, O., and Rupec, M., 1964, Collagen fibrils of the scleroderma in ultra-thin skin sections, *Nature* **202:**708.

Brown, S. I., 1971, Collagenase and corneal ulcers, *Invest. Ophthalmol.* **10:**203.

Brown, I. A., 1972, Scanning electron microscopy of human dermal fibrous tissue, *J. Anat.* **113:**159.

Brown, R. G., and Liddy, E. P., 1971, Collagen metabolism in bone and carrageenin granuloma by sulfate-deprived and sulfate-supplemented rats, *Can. J. Physiol. Pharmacol.* **49:**1008.

Brown, S. I., and Weller, C. A., 1970, Collagenase inhibitors in prevention of ulcers of alkali-burned cornea, *Arch. Ophthalmol.* **83:**352.

Burleigh, M. C., Barrett, A. J., and Lazarus, G. S., 1974, Cathepsin B1. A lysosomal enzyme that degrades native collagen, *Biochem. J.* **137:**387.

Butler, W. F., 1975, Fragility of the skin in a cat, *Res. Vet. Sci.* **19:**213.

Carmichael, D. J., and Dodd, C. M., 1973, An investigation of the phosphoprotein of the bovine dentin matrix, *Biochim. Biophys. Acta* **317:**187.

Carnes, W. H., 1971, Role of copper in connective tissue metabolism, *Fed. Proc.* **30:**995.

Castor, C. W., 1973, Connective tissue activation. V. The flux of connective tissue activating peptide during acute inflammation, *J. Lab. Clin. Med.* **81:**95.

Castor, C. W., Harnsberger, S. C., Scott, M. E., and Ritchie, J. C., 1975, Connective tissue activation. VII. Evidence supporting a role for prostaglandins and cyclic nucleotides, *J. Lab. Clin. Med.* **85:**392.

Chantraine, A., 1971, Clinical investigation of bone metabolism in spinal cord lesions, *Paraplegia* **8:**253.

Chou, W. S., Savage, J. E., and O'Dell, B. L., 1968, Relation of monoamine oxidase activity and collagen crosslinking in copper-deficient and control tissues, *Proc. Soc. Exp. Biol. Med.* **128:**948.

Christner, P. J., and Rosenbloom, J., 1971, Effects of incorporation of trans-4,5-dehydrolysine on collagen biosynthesis and extrusion in embryonic chick tibiae, *J. Biol. Chem.* **246:**7551.

Christner, P., Carpousis, A., Harsch, M., and Rosenbloom, J., 1975, Inhibition of the assembly and secretion of procollagen by incorporation of a threonine analogue, hydroxynorvaline, *J. Biol. Chem.* **250:**7623.

Church, R. L., Tanzer, M. L., and Lapière, Ch. M., 1973, Identification of two distinct species of procollagen synthesized by a clonal line of calf dermatosparactic cells, *Nature* **244:**188.

Church, R. L., Yaeger, J. A., and Tanzer, M. L., 1974, Isolation and partial characterization of procollagen fractions produced by a clonal strain of calf dermatosparactic cells, *J. Mol. Biol.* **86:**785.

Chvapil, M., Hurych, J., Ehrlichova, E., and Cmuchalova, B., 1967, Effects of various chelating agents, quinones, diazoheterocyclic compounds and other substances on proline hydroxylation and synthesis of collagenous and non-collagenous proteins, *Biochim. Biophys. Acta* **140:**339.

Chvapil, M., McCarthy, D., Madden, J. W., and Peacock, E. E., 1974, Effect of 1,10-phenanthroline and desferrioxamine *in vivo* on prolyl hydroxylase and hydroxylation of collagen in various tissues of rats, *Biochem. Pharmacol.* **23:**2165.

Cintron, C., 1974, Hydroxylysine glycosides in the collagen of normal and scarred rabbit corneas, *Biochem. Biophys. Res. Commun.* **60:**288.

Cohen, I. K., Keiser, H. R., and Sjoerdsma, A., 1971, Collagen synthesis in human keloid and hypertrophic scar, *Surg. Forum* **22:**488.

Collins, J. F., and Crystal, R. G., 1975, Characterization of cell-free synthesis of collagen by lung polysomes in a heterologous system, *J. Biol. Chem.* **250:**7332.

Coulombre, A. J., 1965, Problems in corneal morphogenesis, *Adv. Morphog.* **4:**81.

Coulson, W. F., Weissman, N., and Carnes, W. H., 1965, Cardiovascular studies on copper-deficient swine VII. Mechanical properties of aortic and dermal collagen, *Lab. Invest.* **14:**303.

Craig, R. D. P., Schofield, J. D., and Jackson, D. S., 1975, Collagen biosynthesis in normal human skin, normal and hypertrophic scar and keloid, *Eur. J. Clin. Invest.* **5:**69.

Cutroneo, K. R., Costello, D., and Fuller, G. C., 1971, Alteration of proline hydroxylase activity by gluco-corticoids, *Biochem. Pharmacol.* **20:**2797.

Cutroneo, K. R., Stassen, F. L. H., and Cardinale, G. J., 1975, Antiinflammatory steroids and collagen metabolism: Glucocorticoid-mediated decrease of prolyl hydroxylase, *Mol. Pharmacol.* **11:**44.

Daly, J. M., Steiger, E., Prockop, D. J., and Dubrick, S. J., 1973, Inhibition of collagen synthesis by the proline analogue *cis*-4-hydroxyproline, *J. Surg. Res.* **14:**551.

Danielsen, L., Kobayasi, T., Larsen, H. W., Midtgaard, K., and Christensen, H. E., 1970, Pseudoxanthoma elasticum. A clinico-pathological study. *Acta Derm. Venereol.* **50:**355.

Danks, D. M., Campbell, P. E., Stevens, B. J., Mayne, V., and Cartwright, E., 1972, Menkes's kinky hair syndrome: An inherited defect in copper absorption with widespread effects, *Pediatrics* **50:**188.

Davis, W. M., Madden, J. W., and Peacock, E. E., 1972, A new approach to the control of esophageal stenosis, *Ann. Surg.* **176:**469.

Dekaban, A. S., and Steusing, J. K., 1974, Menkes' kinky hair disease treated with subcutaneous copper sulphate, *Lancet* **2:**1523.

Deshmukh, K., 1974, A comparative study of the crosslinking precursors present in rat skin and tail tendon collagen, *Connect. Tissue Res.* **2:**100.

Deshmukh, K., and Nimni, M. E., 1969, A defect in the intramolecular and intermolecular cross-linking of collagen caused by penicillamine. II. Functional groups involved in the interaction process, *J. Biol. Chem.* **244:**1787.

Deshmukh, K., and Nimni, M. E., 1973, Effects of lysosomal enzymes on the type of collagen synthesized by bovine articular cartilage. *Biochem. Biophys. Res. Commun.* **53:**424.

Deshmukh, A., Deshmukh, K., and Nimni, M. E., 1971, Synthesis of aldehydes and their interactions during the *in vitro* aging of collagen, *Biochemistry* **10:**2337.

Deshmukh, K., Just, M., and Nimni, M. E., 1973, A defect in the intramolecular and intermolecular crosslinking of collagen caused by penicillamine. III. Accumulation of acid soluble collagen with a high hydroxylysine content in bone, *Clin. Orthop.* **91:**186.

Dhem, A., Piret, N., Nicaise, M., and Nusgens, B., 1976, Bone in dermatosparaxis. I. Morphological analysis, *Calcif. Tissue Res.* (in press).

Di Ferrante, N., Nichols, B. L., Donnelly, P. V., Neri, G., Hrgovcic, R., and Berglund, R., 1971, Induced degradation of glycosaminoglycans in Hurler's and Hunter's syndrome by plasma infusion, *Proc. Natl. Acad. Sci. U.S.A.* **68:**303.

Di Ferrante, N., Leachman, R. D., Angelini, P., Donnelly, P. V., Francis, G., Almazan, A., and Segni, G., 1975, Lysyl oxidase deficiency in Ehlers-Danlos syndrome type V, *Connect. Tissue Res.* **3:**49.

Eastoe, J. E., Martens, P., Thomas, N. R., 1973, The amino-acid composition of human hard tissue collagens in osteogenesis imperfecta and dentinogenesis imperfecta, *Calcif. Tissue Res.* **12:**91.

Efron, M. L., Bixby, E. M., and Pryles, C. V., 1965, Hydroxyprolinemia. II. A rare metabolic disease due to a deficiency of the enzyme "hydroxyproline oxidase," *New Engl. J. Med.* **272:**1299.

Eisen, A. Z., 1969, Human skin collagenase: Relationship to the pathogenesis of epidermolysis bullosa dystrophica, *J. Invest. Dermatol.* **52:**449.

Eisen, A. Z., Bloch, K. J., and Sakai, T., 1970, Inhibition of human skin collagenase by human serum, *J. Lab. Clin. Med.* **75:**258.

Ellis, J. P., 1972, Chronic leg ulceration? Connective tissue defect. *Proc. R. Soc. Med.* **65:**1078.

Emerit, I., and Housset, E., 1973, Chromosome studies on bone marrow from patients with systemic sclerosis evidence for chromosomal breakage *in vivo*, *Biomedicine* **19:**550.

Epstein, E. H., 1974, $\alpha 1$(III)3 human skin collagen. Release by pepsin digestion and preponderance in fetal life, *J. Biol. Chem.* **249:**3225.

Epstein, J., Williams, J. R., and Little, J. B., 1974, Rate of DNA repair in progeric and normal human fibroblasts, *Biochem. Biophys. Res. Commun.* **59:**850.

Evanson, J. M., Jeffrey, J. J., and Krane, S. M., 1967, Human collagenase: Identification and characterization of an enzyme from rheumatoid synovium in culture, *Science* **158:**499.

Eyre, D. R., and Glimcher, M. J., 1972, Reducible crosslinks in hydroxylysine deficient collagens of a heritable disorder of connective tissue, *Proc. Natl. Acad. Sci. U.S.A.* **69:**2594.

Filisko, F., Novak, P., and Geil, P. H., 1972, Electron irradiation-induced splitting of native collagen fibrils, *J. Ultrastruct. Res.* **38:**102.

Finlay, B., 1969, Scanning electron microscopy of the human dermis under uni-axial strain, *BioMed. Eng.* **4:**322.

Fjølstad, M., and Helle, O., 1974, A hereditary dysplasia of collagen tissues in sheep, *J. Pathol.* **112:**183.

Fleisch, H., and Russell, R. G. G., 1972, A review of the physiological and pharmacological effects of pyrophosphate and diphosphonates on bones and teeth, *J. Dent. Res.* **51**:324.
Fleischmajer, R., and Prunieras, M., 1972, Generalized morphea. II. Electron microscopy of collagen, cells, and the subcutaneous tissue, *Arch. Dermatol.* **106**:515.
Fleischmajer, R., Damiano, V., and Nedwich, A., 1971, Scleroderma and the subcutaneous tissue, *Science* **171**:1019.
Follis, R. J., Jr., 1952, Osteogenesis imperfecta congenita; A connective tissue diathesis, *J. Pediat.* **41**:713.
Forrest, L., and Jackson, D. S., 1971, Intermolecular cross-linking of collagen in human and guinea pig scar tissue, *Biochim. Biophys. Acta* **229**:681.
Forrester, J. C., Zederfeldt, B. H., Hayes, T. L., and Hunt, T. K., 1970, Tape closed and sutured wounds: A comparison by tensiometry and scanning electron microscopy, *Brit. J. Surg.* **57**:729.
Francis, M. J. O., and Smith, R., 1974, Evidence of a generalized connective-tissue defect in Paget's disease of bone, *Lancet* **1**:841.
Francis, M. J. O., Smith, R., and Bauze, R. J., 1974, Instability of polymeric skin collagen in osteogenesis imperfecta, *Br. Med. J.* **1**:421.
Friedmann, T., and Roblin, R., 1972, Gene therapy for human genetic disease, *Science* **175**:949.
Fujii, K., and Tanzer, M. L., 1974, Age-related changes in the reducible crosslinks of human tendon collagen, *FEBS Lett.* **43**:300.
Fulghum, D. D., and Katz, R., 1968, Penicillamine for scleroderma, *Arch. Dermatol.* **98**:51.
Fullmer, H. M., and Gibson, W., 1966, Collagenolytic activity in gingivae of man, *Nature* **209**:728.
Fullmer, H. M., Taylor, R. E., and Guthrie, R. W., 1972, Human gingival collagenase: Purification, molecular weight and inhibitor studies, *J. Dent. Res.* **51**:349.
Furthmayr, H., Timpl, R., Stark, M., Lapiere, Ch. M., and Kuhn, K., 1972, Chemical properties of the peptide extension in the p α_1 chain of dermatosparactic skin procollagen, *FEBS Lett.* **28**:247.
Gay, S., Balleisen, L., Remberger, K., Fietzek, P. P., Adelmann, B. C., and Kuhn, K., 1975*a*, Immunohistochemical evidence for the presence of collagen type III in human arterial walls, arterial thrombi, and in leukocytes, incubated with collagen *in vitro, Klin. Wschr.* **53**:899.
Gay, S., Fietzek, P. P., Remberger, K., Eder, M., and Kuhn, K., 1975*b*, Liver cirrhosis: immunofluorescence and biochemical studies demonstrate two types of collagen, *Klin. Wschr.* **53**:205.
Gelman, R. A., and Blackwell, J., 1973, Interaction between collagen and chondroitin-6-sulfate, *Connect. Tissue Res.* **2**:31.
Gibson, T., 1965, Biomechanics in plastic surgery, *in: Biomechanics and Related Bioengineering Topics* (R. M. Kenedi, ed.), pp. 129–134, Pergamon Press, New York.
Gilkes, J. J. H., Sharvill, D. E., and Wells, R. S., 1974, The premature ageing syndromes. Report of eight cases and description of a new entity named metageria, *Br. J. Dermatol.* **91**:243.
Gillman, T., 1964, Possible importance of dermal–epidermal interactions in the pathogenesis of human and experimental wound healing and skin cancers, *in: Progress in the Biological Sciences in Relation to Dermatology*, Vol. 2, pp. 113–134, University Press, Cambridge.
Goldstein, S., 1969, Lifespan of cultured cells in progeria, *Lancet* **1**:424.

Goldstein, S., and Moerman, E. J., 1975, Heat-labile enzymes in Werner's syndrome fibroblasts, *Nature* **255:**159.

Goldstein, S., and Singal, D. P., 1974, Alteration of fibroblast gene products in vitro from a subject with Werner's syndrome, *Nature* **251:**719.

Goltz, R. W., Hult, A. M., Goldfarb, M., and Gorlin, R. J., 1965, Cutis laxa. A manifestation of generalized elastolysis. *Arch. Dermatol.* **92:**373.

Golub, L. M., 1973, The effect of ascorbic acid on the turnover of collagen in bone in tissue culture, *J. Periodont. Res.* **8:**71.

Golub, L. M., Stakiw, J. E., and Singer, D. L., 1974, Collagenolytic activity of human gingival crevice fluid, *J. Dent. Res.* **53:**1501.

Goodman, S. I., Solomons, C. C., Muschenheim, F., McIntyre, C. A., Miles, B., and O'Brien, D., 1968, A syndrome resembling lathyrism associated with iminodipeptiduria, *Amer. J. Med.* **45:**152.

Goodman, S. I., Mace, J. W., Miles, B. S., Teng, C. C., and Brown, S. B., 1974, Defective hydroxyproline metabolism in type II hyperprolinemia, *Biochem. Med.* **10:**329.

Gould, B. S., 1958, Biosynthesis of collagen. III. The direct action of ascorbic acid on hydroxyproline and collagen formation in subcutaneous polivynil sponge implants in guinea pigs, *J. Biol. Chem.* **232:**637.

Grahame, R., 1970, A method for measuring human skin elasticity in vivo with observations on the effects of age, sex and pregnancy, *Clin. Sci.* **39:**223.

Grahame, R., and Beighton, P., 1971, The physical properties of skin in cutis laxa, *Br. J. Dermatol.* **84:**326.

Grillo, H. C., and Gross, J., 1967, Collagenolytic activity during mammalian wound repair, *Dev. Biol.* **15:**300.

Gross, J., 1958, Studies on the formation of collagen. II. The influence of growth rate on neutral salt extracts of guinea pig dermis, *J. Exp. Med.* **107:**265.

Gross, J., and Kirk, D., 1958, The heat precipitation of collagen from neutral salt solution: Some rate regulating factors, *J. Biol. Chem.* **233:**355.

Gross, J., and Levene, C. I., 1959, Effect of beta-aminoproprionitrile on extractability of collagen from skin of mature guinea pigs, *Am. J. Pathol.* **35:**687.

Gross, J., Toole, B. P., and Trelstad, R. L., 1974, Bone cell biology: The emerging role of hydroxylysine. *International Santa Catalina Island Colloquium.* Extracellular matrix influences on gene expression (Sept. 18–24).

Hagg, E., 1974, Glomerular basement membrane thickening in rats with long-term alloxan dibetes. A quantitative electron microscopic study, *Acta Pathol. Microbiol. Scand.* **82:**211.

Hanset, R., and Ansay, M., 1967, Dermatosparaxie (peau déchirée) chez le veau: Un défaut général du tissu conjonctif, de nature héréditaire, *Ann. Med. Vet.* **7:**451.

Hanset, R., and Lapière, Ch. M., 1974, Inheritance of dermatosparaxis in the calf. A genetic defect of connective tissues, *J. Hered.* **65:**356.

Hardy, K. H., Rosevear, J. W., Sams, W. M., and Winkelmann, R. K., 1971, Scleroderma and urinary excretion of acidic glycosaminoglycans, *Mayo Clin. Proc.* **46:**119.

Harkness, R. D., 1968, *In vitro* and *in vivo* observations on the effect of penicillamine on collagen, *Postgrad. Med. J.* **1:**34.

Harper, E., and Gross, J., 1972, Collagenase, procollagenase and activator relationships in tadpole tissue cultures, *Biochem. Biophys. Res. Commun.* **48:**1147.

Harper, E., and Toole, B. T., 1973, Collagenase and hyaluronidase stimulation by dibutyryl adenosine cyclic 3′,5′-monophosphate, *J. Biol. Chem.* **248:**2625.

Harris, E. D., and Krane, S. M., 1974*a*, Collagenase (first of three parts), *N. Engl. J. Med.* **291:**557.

Harris, E. D., and Krane, S. M., 1974*b*, Collagenase (second of three parts), *N. Engl. J. Med.* **291**:605.

Harris, E. D., and Krane, S. M., 1974*c*, Collagenase (third of three parts), *N. Engl. J. Med.* **291**:652.

Harris, E. D., and McCroskery, P. A., 1974, The influence of temperature and fibril stability on degradation of cartilage collagen by rheumatoid synovial collagenase, *N. Engl. J. Med.* **290**:1.

Harris, E. D., and O'Dell, B. L., 1974, Copper and amine oxidases in connective tissue metabolism, *in; Advances in Experimental Medicine and Biology, Protein–Metal Interactions,* (Mendel Friedman, ed.), pp. 267–284, Plenum Press, New York.

Harris, E. D., and Sjoerdsma, A., 1966*a*, Collagen profile in various clinical conditions, *Lancet* **2**:707.

Harris, E. D., and Sjoerdsma, A., 1966*b*, Effect of penicillamine on human collagen and its possible application to treatment of scleroderma, *Lancet* **2**:996.

Harris, E. D., Cohen, G. L., and Krane, S. M., 1969*a*, Synovial collagenase: Its presence in culture from joint disease of diverse etiology, *Arthritis Rheum.* **12**:92.

Harris, E. D., Dibona, D. R., and Krane, S. M., 1969*b*, Collagenase in human synovial fluid, *J. Clin. Invest.* **48**:2104.

Harris, E. D., Faulkner, C. S., and Wood, S., 1972, Collagenase in carcinoma cells, *Biochem. Biophys. Res. Commun.* **48**:1247.

Hashimoto, K., Kitabchi, A. E., Duckworth, W. C., and Robinson, N., 1970, Ultrastructure of scorbutic human skin, *Acta Derm. Venereol.* **50**:9.

Hashimoto, K., Yamanishi, Y., Dabbous, M. K., and Maeyens, E., 1973*a*, Collagenase activity in rheumatoid nodules, *Acta Derm. Venerol.* **53**:439.

Hashimoto, K., Yamanishi, Y., Maeyens, E., Dabbous, M. K., and Kanzaki, T., 1973*b*, Collagenolytic activities of squamous cell carcinoma of the skin, *Cancer Res.* **33**:2790.

Hassler, C. R., Rybicki, E. F., Simonen, F. A., and Weis, E. B., 1974, Measurements of healing at an osteotomy in a rabbit calvarium: the influence of applied compressive stress on collagen synthesis and calcification, *J. Biomech.* **7**:545.

Hayashi, T., and Nagai, Y., 1972, Factors affecting the interactions of collagen molecules as observed by in vitro fibril formation. I. Effects of small molecules, especially saccharides, *J. Biochem.* **72**:749.

Hayflick, L., 1965, The limited in vitro lifetime of human diploid cell strains, *Exp. Cell Res.* **37**:614.

Hegreberg, G. A., Padgett, G. A., Ott, R. L., and Henson, J. B., 1970, A heritable connective tissue disease of dogs and mink resembling Ehlers–Danlos syndrome of man, *J. Invest. Dermatol.* **54**:377.

Helle, O., and Ness, N. N., 1972, A hereditary skin defect in sheep, *Acta Vet. Scand.* **13**:443.

Herbert, C. H., Jayson, M. I. V., and Bailey, A. J., 1973, Joint capsule collagen in osteoarthrosis, *Ann. Rheum. Dis.* **32**:510.

Herbert, C. M., Lindberg, K. A., Jayson, M. I. V., and Bailey, A. J., 1974, Biosynthesis and maturation of skin collagen in scleroderma and effect of D-penacillamine, *Lancet* **1**:187.

Hers, H. G., and Van Hoff, F., 1973, *Lysosomes and Storage Diseases,* Academic Press, New York.

Holliday, R., Porterfield, J. S., and Gibbs, D. D., 1974, Premature ageing and occurrence of altered enzyme in Werner's syndrome fibroblasts, *Nature* **248**:762.

Hoopes, J. E., Su, C. T., and Im, J. C., 1971, Enzyme activities in hypertrophic scars and keloids, *Plast. Reconstr. Surg.* **47**:132.

Houck, J. C., Sharma, V. K., and Cheng, R. F., 1973, Fibroblast chalone and serum mitogen (anti-chalone), *Nature (London), New Biol.* **246**:111.

Hugues, J., Herion, F., Nusgens, B., and Lapière, Ch. M., 1976, Type III collagen and probably not type I collagen aggregates platelets, *Thromb. Res.* (in press).

Hunt, D. M., 1974, Primary defect in copper transport underlies mottled mutants in the mouse, *Nature* **249**:852.

Jackson, D. S., 1957, Connective tissue growth stimulated by carrageenin. I. The formation and removal of collagen, *Biochem. J.* **65**:277.

Jackson, S. H., 1973, The reaction of homocysteine with aldehyde: An explanation of the collagen defects in homocystinuria, *Clin. Chim. Acta* **45**:215.

Jackson, S. H., Dennis, A. W., and Greenberg, M., 1975, Iminodipeptiduria: A genetic defect in recycling collagen; A method for determining prolidase in erythrocytes, *Canad. Med. Ass. J.* **113**:759.

Jaffe, I. A., 1964, Rheumatoid arthritis with arteritis—report of a case treated with penicillamine, *Ann. Intern. Med.* **61**:556.

James, K., Fundenberg, H., Epstein, W. L., and Shuster, J., 1967, Studies on a unique diagnostic serum globulin in papular mucinosis *(Lichen myxedematosus), Clin. Exp. Immunol.* **2**:153.

Jeffrey, J. J., Coffey, R. J., and Eisen, A. Z., 1971*a*, Studies on uterine collagenase in tissue culture. I. Relationship of enzyme production to collagen metabolism, *Biochim. Biophys. Acta* **252**:136.

Jeffrey, J. J., Coffey, R. J., and Eisen, A. Z., 1971*b*, Studies on uterine collagen in tissue culture. II. Effect of steroid hormones on enzyme production, *Biochim. Biophys. Acta* **252**:143.

Jones, C. R., Bergman, M. W., Kittner, P. J., and Pigman, W., 1964, Urinary hydroxyproline excretion in Marfan's syndrome as compared with age matched controls, *Proc. Soc. Exp. Biol. Med.* **116**:931.

Jones, C. R., van Itallie, T. B., Stevenson, S. S., and Pigman, W. W., 1969, Decrease in hydroxyproline excretion in a patient with Marfan's syndrome after methandrostenolone therapy, *JAMA* **207**:2085.

Jordan, R. E., Deheer, D., Schroeter, A., and Winkelmann, R. K., 1971, Antinuclear antibodies: Their significance in scleroderma, *Mayo Clin. Proc.* **46**:111.

Julkunen, H., 1971, Scanning electron microscopic study in scleroderma, *Ann. Med. Exp. Fenn.* **49**:180.

Julkunen, H., Rokkanen, P., and Inoue, H., 1970, Scanning electron microscopic study of the collagen bundles of the skin in the Ehlers-Danlos syndrome, *Ann. Med. Exp. Fenn.* **48**:201.

Kamrin, B. B., 1974, Induced collagenolytic activity by electrical stimulation of embryonic fibroblasts in tissue culture, *J. Dent. Res.* **53**:1475.

Kang, A. H., and Trelstad, R. L., 1973, A collagen defect in homocystinuria, *J. Clin. Invest.* **52**:2571.

Kefalides, N. A., 1971, Isolation of a collagen from basement membranes containing three identical α-chains, *Biochem. Biophys. Res. Commun.* **45**:226.

Kefalides, N. A., 1974, Biochemical properties of human glomerular basement membrane in normal and diabetic kidneys, *J. Clin. Invest.* **53**:403.

Kefalides, N. A., 1975, Basement membranes: Current concept of structure and synthesis. *Dermatologica* **150**:4.

Keiser, H. R., and Sjoerdsma, A., 1967, Studies on beta-aminopropionitrile in patients with scleroderma, *Clin. Pharmacol. Ther.* **8**:593.

Keiser, H. R., Stein, H. D., and Sjoerdsma, A., 1971, Increased protocollagen proline hydroxylase activity in sclerodermatous skin, *Arch. Dermatol.* **104**:57.

Kenedi, R. M., Gibson, R., and Daly, C. H., 1965, Bio-engineering studies of the human skin II. *in: Biomechanics and Related Bio-engineering Topics* (R. M. Kenedi, ed.), pp. 147–158, Pergamon Press, New York.

Kerwar, S. S., Marcel, R. J., and Salvador, R. A., 1975, Reduction of prolyl hydroxylase activity in L-929 fibroblasts by proline analogs, *Biochem. Biophys. Res. Commun.* **66:**1275.

Kischer, C. W., 1974, Collagen and dermal patterns in the hypertrophic scar, *Anat. Rec.* **179:**137.

Kischer, C. W., and Shetlar, M. R., 1974, Collagen and mucopolysaccharides in the hypertrophic scar, *Conn. Tissue Res.* **2:**205.

Kivirikko, K. I., 1970, Urinary excretion of hydroxyproline in health and disease, *in: International Review of Connective Tissue Research* (D. A. Hall and D. S. Jackson, eds.), pp. 93–163, Academic Press, New York.

Kivirikko, K. I., Koivusalo, M., Laitinen, O., and Lamberg, B. A., 1964, Hydroxyproline in the serum and urine of patients with hyperthyroidism, *J. Clin. Endocrinol. Metab.* **24:**222.

Kivirikko, K. I., Laitinen, O., Aer, J., and Halme, J., 1965*a*, Studies with ^{14}C-proline on the action of cortisone on the metabolism of collagen in the rat, *Biochem. Pharmacol.* **14:**1445.

Kivirikko, K. I., Laitinen, O., and Lamberg, B. A., 1965*b*, Value of urine and serum hydroxyproline in the diagnosis of thyroid disease, *J. Clin. Endocrinol Metab.* **25:**1347.

Klein, L., and Nowacek, C. J., 1969, Effect of penicillamine on new and pre-existing [^{3}H]collagen in vivo, *Biochim. Biophys. Acta* **194:**504.

Klein, L., Curtiss, P. H., Jr., and Davis, J. H. D., 1962, Collagen breakdown in thermal burns, *Surg. Forum* **13:**459.

Klujber, L., and Méhes, K., 1973, Urinary hydroxyproline and glycosaminoglycan excretion in Down's syndrome, *Z. Kinderheilk.* **114:**205.

Kobayasi, T., and Asboe-Hansen, G., 1972, Ultrastructure of generalized scleroderma, *Acta Derm. Venerol.* **52:**81.

Kohn, L. D., Isersky, C., Zupnik, J., Lenaers, A., Lee, G., and Lapière, Ch. M., 1974, Calf tendon procollagen peptidase: Its purification and endopeptidase mode of action, *Proc. Natl. Acad. Sci. U.S.A.* **71:**40.

Koob, T. J., and Jeffrey, J. J., 1974, Hormonal regulation of collagen degradation in the uterus: Inhibition of collagenase expression by progesterone and cyclic AMP, *Biochim. Biophys. Acta* **354:**61.

Koob, T. J., Jeffrey, J. J., and Eisen, A. Z., 1974, Regulation of human skin collagenase activity by hydrocortisone and dexamethasone in organ culture, *Biochem. Biophys. Res. Commun.* **61:**1083.

Kovacs, C. J., and Fleischmajer, R., 1974, Properties of scleroderma fibroblasts in culture, *J. Invest. Dermatol.* **63:**456.

Krane, S. M., Munoz, A. J., and Harris, E. D., 1967, Collagen-like fragments: Excretion in urine of patients with Paget's disease of bone, *Science* **157:**713.

Krane, S. M., Munoz, A. J., and Harris, E. D., 1970, Urinary polypeptides related to collagen synthesis, *J. Clin. Invest.* **49:**716.

Krane, S. M., Pinnell, S. R., and Erbe, R. W., 1972, Lysyl-protocollagen hydroxylase deficiency in fibroblasts from siblings with hydroxylysine-deficient collagen, *Proc. Natl. Acad. Sci. U.S.A.* **69:**2899.

Krane, S. M., Harris, E. D., Singer, F. R., and Potts, J. T., 1973, Acute effects of calcitonin on bone formation in man, *Metabolism* **22:**51.

Kruggel, W. G., and Field, R. A., 1974, Cross-linking of collagen in active and quiescent bovine muscle, *Growth* **38:**495.

Kruze, D., and Wojtecka, E., 1972, Activation of leucocyte collagenase proenzyme by rheumatoid synovial fluid, *Biochim. Biophys. Acta* **285:**436.

Kulonen, E., and Potilla, M., 1975, Effect of the administration of antirheumatic drugs on experimental granuloma in rats, *Biochem. Pharmacol.* **24:**219.

Kuttan, R., Cardinale, G. J., and Udenfriend, S., 1975, An activatable form of prolyl hydroxylase in fibroblast extracts, *Biochem. Biophys. Res. Commun.* **64:**947.

Kutnink, M. A., Tolbert, B. M., Richmond, V. L., and Baker, E. M., 1969, Efficacy of the ascorbic acid stereoisomers in proline hydroxylation in vitro, *Proc. Exp. Biol. Med.* **132:**440.

Laitinen, O., Uitto, J., Hannuksela, M., and Mustakallio, K. K., 1966, Increased soluble collagen content of affected and normal looking skin in dermatomyositis, lupus erythematosus and scleroderma, *Ann. Med. Exp. Fenn.* **44:**507.

Laitinen, O., Uitto, J., Iivanainen, M., Hannuksela, M., and Kivirikko, K. I., 1968, Collagen metabolism of the skin in Marfan's syndrome, *Clin. Chim. Acta* **21:**321.

Lancaster, G., Goldman, H., Scriver, C. R., Gold, R. J. M., and Wong, I., 1975, Dominantly inherited osteogenesis imperfecta in man: An examination of collagen biosynthesis, *Pediat. Res.* **9:**83.

Lane, J. M., and Miller, E. J., 1969, Isolation and characterization of the peptides derived from the α2 chain of chick bone collagen after cyanogen bromide cleavage, *Biochemistry* **8:**2134.

Lane, J. M., Dehm, P., and Prockop, D. J., 1971*a*, Effect of the proline analogue azetidine-2-carboxylic acid on collagen synthesis in vivo. I. Arrest of collagen accumulation in growing chick embryos, *Biochim. Biophys. Acta* **236:**517.

Lane, J. M., Parkes, L. J., and Prockop, D. J., 1971*b*, Effect of the proline analogue azetidine-2-carboxylic acid on collagen synthesis in vivo. II. Morphological and physical properties of collagen containing the analogue, *Biochim. Biophys. Acta* **236:**528.

Langner, R. O., and Fuller, G. C., 1973, Collagen synthesis in thoracic aortas of rabbits with epinephrine-thyroxine induced arteriosclerosis, *Atherosclerosis* **17:**463.

Langness, U., and Behnke, H., 1971, Collagen metabolites in plasma and urine in osteogenesis imperfecta, *Metabolism* **20:**456.

Lapière, Ch. M., 1967, Structural change in native collagen below melting temperature, *VII Int. Cong. of Biochem. Tokyo* (Aug. 19–25) A.40.

Lapière, Ch. M., and Nusgens, B., 1969, Plaies cutanées torpides et trouble du métabolisme du collagène, *Arch. Belg. Dermatol. Syphiligr.* **25:**353.

Lapière, Ch. M., and Nusgens, B., 1970*a*, Effets du jeûne sur le métabolisme du collagène de la peau et de l'os chez le rat, *Arch. Belg. Dermatol. Syphiligr.* **26:**383.

Lapière, Ch. M., and Nusgens, B., 1970*b*, Maturation related changes of the protein matrix of bone, *in: Chemistry and Molecular Biology of the Intercellular Matrix, Collagen, Basal Laminae, Elastin* (E. A. Balazs, ed.), Vol. I, pp. 55–81, Academic Press, New York.

Lapière, Ch. M., and Nusgens, B., 1974, Polymerization of procollagen in vitro, *Biochim. Biophys. Acta* **342:**237.

Lapière, Ch. M., and Piérard, G., 1974, Skin procollagen peptidase in normal and pathologic conditions, *J. Invest. Dermatol.* **62:**582.

Lapière, Ch. M., and Winand, R., 1971, Simultaneous urinary excretion of breakdown products of collagen and glycosaminoglycans in human diseases involving the skin, First annual meeting, *European Society for Dermatological Research,* Noordwijk-aan-Zee, Holland, April, 20–21 (abstract).

Lapière, Ch. M., Nusgens, B., Quinaux, N., and Declercq, A., 1968, Calcinose tumorale par hyperparathyroidisme secondaire (étude physique et chimique des produits calcifiés), *Arch. Belg. Dermatol. Syphiligr.* **24:**153.

Lapière, Ch. M., Winand, R., and Simar, L. J., 1969, Quelques observations concernant les troubles métaboliques d'un patient souffrant de scléromyxoedème, *Arch. Belg. Dermatol. Syphiligr.* **25:**349.

Lapière, Ch. M., Lenaers, A., and Kohn, L., 1971, Procollagen peptidase: An enzyme excising the coordination peptides of collagen, *Proc. Natl. Acad. Sci. U.S.A.* **68:**3054.

Lapière, Ch. M., Nusgens, B., Piérard, G., and Hermanns, J. F., 1975, The involvement of procollagen in spatially oriented fibrogenesis, *in: Dynamics of Connective Tissues Macromolecules* (M. Burleigh and R. Poole, eds.), pp. 30–50, North Holland, Amsterdam.

Larson, D. L., Abston, S., Willis, B., Linares, H., Dobrkovsky, M., Evans, E. B., and Lewis, S. R., 1974, Contracture and scar formation in the burn patient, *Clin. Plast. Surg.* **1:**653.

Lavine, L., Lustrin, I., Rinaldi, R., and Shamos, M., 1974, Clinical and ultrastructural investigations of electrical enhancement of bone healing, *Ann. N.Y. Acad. Sci.* **238:**552.

Layman, D. L., and Titus, J. L., 1975, Synthesis of type I collagen by human smooth muscle cells *in vitro*, *Lab. Invest.* **33:**103.

Layman, D. L., Narayanan, A. S., and Martin, G. R., 1972*a*, The production of lysyl oxidase by human fibroblasts in culture, *Arch. Biochem. Biophys.* **149:**97.

Layman, D. L., Sokoloff, L., and Miller, E. J., 1972*b*, Collagen synthesis by articular chondrocytes in monolayer culture, *Exp. Cell Res.* **73:**107.

Lazarus, G. S., 1972, Collagenase and connective tissue metabolism in epidermolysis bullosa, *J. Invest Dermatol.* **58:**242.

Lazarus, G. S., and Barrett, A. J., 1974, Neutral proteinase of rabbit skin: An enzyme capable of degrading skin protein and inducing an inflammatory response, *Biochim. Biophys. Acta* **350:**1.

Lazarus, G. S., and Fullmer, H. M., 1969, Collagenase production by human dermis in vitro, *J. Invest. Dermatol.* **52:**545.

Lazarus, G. S., Brown, R. S., Daniels, J. R., and Fullmer, H. M., 1968, Human granulocyte collagenase, *Science* **159:**1483.

Legrand, Y., Lapière, Ch. M., Pignaud, G., and Caen, J., 1969, Microméthode d'extraction du collagène à partir de biopsie de peau humaine. Application à l'étude des maladies du saignement, *Pathol. Biol.* **17:**991.

Lenaers, A., Ansay, M., Nusgens, B., and Lapière, Ch. M., 1971, Collagen made of extended α-chains, procollagen, in genetically defective dermatosparaxic calves, *Eur. J. Biochem.* **23:**533.

LeRoy, E. C., 1974, Increased collagen synthesis by scleroderma skin fibroblasts in vitro: A possible defect in the regulation or activation of the scleroderma fibroblast, *J. Clin. Invest.* **54:**880.

Levene, C. I., and Gross, J., 1959, Alteration in state of molecular aggregation of collagen induced in chick embryos by β-aminopropionitrile (Lathyrus factor), *J. Exp. Med.* **110:**771.

Levene, C. I., Bates, C. J., and Bailey, A. J., 1972, Biosynthesis of collagen cross-links in cultured 3T6 fibroblasts; Effect of lathyrogens and ascorbic acid, *Biochim. Biophys. Acta* **263:**574.

Levene, C. I., Aleo, J. J., Prynne, C. J., and Bates, C. J., 1974, The activation of protocollagen proline hydroxylase by ascorbic acid in cultured 3T6 fibroblasts, *Biochim. Biophys. Acta* **338:**29.

Lichtenstein, J. R., Martin, G. R., Kohn, L. D., Byers, P. H., and McKusick, V., 1973, Defect in conversion of procollagen to collagen in a form of Ehlers–Danlos syndrome, *Science* **182**:298.

Lie, S. O., 1974, Indication of "storage disease" in normal human fibroblasts, *in: Skeletal Dysplasias* (D. Bergsma, ed.), pp. 203–211, The National Foundation-March of Dimes, Excerpta Medica, Amsterdam.

Lima, L., and Macieira-Coelho, A., 1972, Parameters of ageing in chicken embryo fibroblasts cultivated in vitro, *Exp. Cell Res.* **70**:279.

Linares, H. A., Kischer, C. W., Dobrkovsky, M., and Larson, D. L., 1972, The histiotypic organization of the hypertrophic scar in humans, *J. Invest. Dermatol.* **59**:323.

Linsenmayer, T. F., Toole, B. P., and Trelstad, R. L., 1973, Temporal and spatial transitions in collagen types during embryonic chick limb development, *Dev. Biol.* **35**:232.

Madden, J. W., and Peacock, E. E., 1971, Studies on the biology of collagen during wound healing: III. Dynamic metabolism of scar collagen and remodeling of dermal wounds, *Ann. Surg.* **174**:511.

Madden, J. W., Chvapil, M., Carlson, E. C., and Ryan, J. N., 1973*a*, Toxicity and metabolic effects of 3,4-dehydroproline in mice, *Toxicol. Appl. Pharmacol.* **26**:426.

Madden, J. W., Davis, W. M., Butler, C., and Peacock, E. E., 1973*b*, Experimental esophageal lye burns. II. Correcting established strictures with beta-aminopropionitrile and bougienage, *Ann. Surg.* **178**:277.

Mandl, I., Zipper, H., and Fergusson, L. T., 1958, *Clostridium histolyticum* collagenase. Its purification and properties, *Arch. Biochem. Biophys.* **74**:465.

Marks, S. C., 1972, Lack of effect of thyrocalcitonin on formation of bone matrix in mice and rats, *Horm. Metab. Res.* **4**:296.

Marks, S. C., 1973, Pathogenesis of osteopetrosis in the rat reduced bone resorption due to reduced osteoclast function, *Am. J. Anat.* **138**:165.

Marks, S. C., 1974, A discrepancy between measurements of bone resorption in vivo and in vitro in newborn osteopetrotic rats, *Am. J. Anat.* **141**:329.

Martin, G. M., Gartler, S. M., Epstein, C. J., and Motulsky, A. G., 1965, Diminished lifespan of cultured cells in Werner's syndrome, *Fed. Proc.* **24**:678.

Martin, G. M., Sprague, C. A., and Epstein, C. J., 1970, Replicative lifespan of cultivated human cells. *Lab. Invest.* **23**:86.

Martin, G. R., Gross, J., Piez, K. A., and Lewis, M. S., 1961, On the intramolecular cross-linking of collagen in lathyritic rats. *Biochem. Biophys. Acta* **53**:599.

Martin, G. R., Orkin, R. W., Penttinen, R., and Lichtenstein, J. R., 1974, The synthesis of genetically distinct collagens in certain disorders, *International Santa Catalina Island colloquium.* Extracellular matrix influence on gene expression.

Martinez-Hernandez, A., and Huffer, W. E., 1974, Pseudoxanthoma elasticum: Dermal polyanions and the mineralization of elastic fibres, *Lab. Invest.* **31**:181.

Marx, S. J., Woodard, C. J., and Aurbach, G. D., 1972, Calcitonin receptors of kidney and bone, *Science* **178**:999.

Matalon, R., and Dorfman, A., 1968, The accumulation of hyaluronic acid in cultured fibroblasts of the Marfan syndrome, *Biochem. Biophys. Res. Commun.* **32**:150.

McCarthy, C. F., Warin, R. P., and Read, A. E. A., 1965, Loose skin (Cutis laxa) associated with systemic abnormalities, *Arch. Intern. Med.* **115**:62.

McClain, P. E., Wiley, E. R., Beecher, G. R., Anthony, W. L., and Hsu, J. M., 1973, Influence of zinc deficiency on synthesis and cross-linking of rat skin collagen, *Biochim. Biophys. Acta* **304**:457.

McCullagh, K. A., and Balian, G., 1975, Collagen characterization and cell transformation in human atherosclerosis, *Nature* **258**:73.

McGee, J. O. D., Langness, U., and Udenfriend, S., 1971, Immunological evidence for an inactive precursor of collagen proline hydroxylase in cultured fibroblasts, *Proc. Natl. Acad. Sci. U.S.A.* **68:**1585.

McKusick, V. A., 1972, *Heritable Disorders of Connective Tissue,* 4th ed., C. V. Mosby, St. Louis.

McKusick, V. A., 1974, Multiple forms of the Ehlers–Danlos syndrome, *Arch. Surg.* **109:**475.

McNeal, J. E., 1973, Scleroderma and the structural basis of skin compliance, *Arch. Dermatol.* **107:**699.

Meachim, G., Denham, D., Emery, I. H., and Wilkinson, P. H., 1974, Collagen alignments and artificial splits at the surface of human articular cartilage, *J. Anat.* **118:**101.

Mechanic, G. L., Toverud, S. U., and Ramp, W. K., 1972, Quantitative changes of bone collagen crosslinks and precursors in vitamin-D deficiency, *Biochem. Biophys. Res. Commun.* **47:**760.

Meigel, W. N., Müller, P. K., Pontz, B. F., Sorensen, N., and Spranger, J., 1974, A constitutional disorder of connective tissue suggesting a defect in collagen biosynthesis, *Klin. Wochenschr.* **52:**906.

Menkes, J. H., Alter, M., Steigleder, G. K., Weakley, D. R., and Sung, J. H., 1962, A sex-linked recessive disorder with retardation of growth, peculiar hair and focal cerebral and cerebellar degeneration, *Pediatrics* **29:**764.

Meunier, P., 1975, La maladie osseuse de Paget. Histologie quantitative, histopathogénie et perspectives thérapeutiques, *Lyon Méd.* **233:**839.

Miller, E. J., and Matukas, V. J., 1969, Chick cartilage collagen: A new type of α1 chain not present in bone or skin of the species, *Proc. Natl. Acad. Sci. U.S.A.* **64:**1264.

Miller, E. J., and Matukas, V. J., 1974, Biosynthesis of collagen: The biochemist's view, *Fed. Proc.* **33:**1197.

Miller, E. J., Martin, G. R., Piez, K. A., and Powers, M. J., 1967, Characterization of chick bone collagen and compositional changes associated with maturation, *J. Biol. Chem.* **242:**5481.

Miller, R. L., 1975, The effect of ascorbic acid on lysyl and prolyl hydroxylase activity of cultured fibroblasts, *Arch. Biochem. Biophys.* **170:**341.

Millington, P. F., and Wilkinson, R., 1974, Changes in skin with age. *Scand. J. Clin. Lab. Invest.* **34:**52.

Milsom, J. P., and Craig, R. D. P., 1973, Collagen degradation in cultured keloid and hypertrophic scar tissue, *Br. J. Dermatol.* **89:**635.

Mitoma, C., Smith, T. E., Davidson, J. D., Udenfriend, S., Dacosta, F. M., and Sjoerdsma, A., 1959, Improvement in methods for measuring hydroxyproline: Application to human urine, *J. Lab. Clin. Med.* **53:**970.

Morrow, G., Kivirikko, K. I., and Prockop, D. J., 1967, Catabolism and excretion of free hydroxyproline in infancy, *J. Clin. Endocrinol. Metab.* **27:**1365.

Moserova, J., Konickova, Z., Behounkova, E., and Janecek, J., 1974, Experimental use of collagenase in the debridement of thermally damaged skin, *J. Hyg. Epidem. Microbiol. Immunol.* **18:**494.

Mussini, E., Hutton, J. J., and Udenfriend, S., 1967, Collagen proline hydroxylase in wound healing, granuloma formation scurvy and growth, *Science* **157:**927.

Nemeth-Csoka, M., 1974, The effect of acid mucopolysaccharides on the activation energy of collagen fibril-formation, *Exp. Pathol.* **9:**256.

Neufeld, E. F., and Fratantoni, J. C., 1970, Inborn errors of mucopolysaccharide metabolism. Faulty degradative mechanisms are implicated in this group of human diseases, *Science* **169:**141.

Nichols, G., Griffith, G. C., and Asher, J. D., 1965, Heparin osteoporosis, *J. Clin. Invest.* **44:**1080.

Nienhaus, A. J., de Jong, B., ten Kate, L. P., and Oswald, F. H., 1971, Fibroblast culture in Werner's syndrome, *Humangenetik* **13:**244.

Nigra, T. P., Friedland, M., and Martin, G. R., 1972, Controls of connective tissue synthesis: Collagen metabolism, *J. Invest Dermatol.* **59:**44.

Nimni, M. E., and Bavetta, L. A., 1965, Collagen defect induced by penicillamine, *Science* **150:**905.

Nimni, M. E., and Deshmukh, K., 1973, Differences in collagen metabolism between normal and osteoarthritic human articular cartilage, *Science* **181:**751.

Nimni, M. E., Deshmukh, K., and Gerth, N., 1972, Collagen defect induced by penicillamine, *Nature* **240:**220.

Nusgens, B., and Lapière, Ch. M., 1973, The relationship between proline and hydroxyproline urinary excretion in human as an index of collagen catabolism. *Clin. Chim. Acta* **48:**203.

Nusgens, B., and Lapière, Ch. M., 1976, Bone in dermatosparaxis II. chemical analysis, *Calcif. Tissue Res.* (in press).

Nusgens, B., Chantraine, A., and Lapière, Ch. M., 1972, The protein in the matrix of bone, *Clin. Orthop. Relat. Res.* **88:**252.

O'Hara, P. J., Read, W. K., Romane, W. M., and Bridges, C. H., 1970, A collagenous tissue dysplasia of calves, *Lab. Invest.* **23:**307.

Ooshima, A., Fuller, G. C., Cardinale, G. J., Spector, S., and Udenfriend, S., 1974, Increased collagen synthesis in blood vessels of hypertensive rats and its reversal by antihypertensive agents, *Proc. Natl. Acad. Sci. U.S.A.* **71:**3019.

Ooshima, A., Fuller, G., Cardinale, G., Spector, S., and Udenfriend, S., 1975, Collagen biosynthesis in blood vessels of brain and other tissues of the hypertensive rat, *Science* **190:**898.

Oronsky, A. L., Perper, R. J., and Schroder, H. C., 1973, Phagocytic release and activation of human leukocyte procollagenase, *Nature* **246:**417.

Osterby, R., 1972, Morphometric studies of the peripheral glomerular basement membrane in early juvenile diabetes. I. Development of initial basement membrane thickening, *Diabetologia* **8:**84.

Osterman, J. V., Waddell, A., and Aposhian, H. V., 1971, Gene therapy systems: The need, experimental approach and implications, *Ann. N.Y. Acad. Sci.* **179:**514.

Pardo, A., and Tamayo, R. P., 1974, The collagenase of carrageenin granuloma, *Connect. Tissue Res.* **2:**243.

Penttinen, R. P., Lichtenstein, J., Byers, P. H., Sussman, M. D., Rowe, D. W., McKusick, V. A., and Martin, G. R., 1974, Disorders identified as inherited defects in collagen: Osteogenesis imperfecta, *Isr. J. Med. Sci.* **10:**1469.

Penttinen, R. P., Lichtenstein, J. R., Martin, G. R., and McKusick, V. A., 1975, Abnormal collagen metabolism in cultured cells in osteogenesis imperfecta, *Proc. Natl. Acad. Sci. U.S.A.* **72:**586.

Peterkofsky, B., 1972, The effect of ascorbic acid on collagen polypeptide synthesis and proline hydroxylation during the growth of cultured fibroblasts, *Arch. Biochem. Biophys.* **152:**318.

Petes, T. D., Farber, R. A., Tarrant, G. M., and Holliday, R., 1974, Altered rate of DNA replication in ageing human fibroblast cultures, *Nature* **251:**434.

Phillips, H. J., 1971, The vascularization of scorbutic bone. An experimental study in the guinea pig, *J. Anat.* **108**:347.

Piérard, G. E., and Lapière, Ch. M., 1976, Skin in dermatosparaxis. Dermal microarchitecture and biomechanical properties, *J. Invest. Derm.* **66**:2.

Piérard, G. E., Hermanns, J. F., and Lapière, Ch. M., 1974, Stéréologie de l'interface dermo-épidermique. Observation de la plasticité de la membrane basale au microscope électronique à balayage, *Dermatologica* **149**:266.

Piez, K. A., Bladen, H. A., Lane, J. M., Miller, E. J., Bornstein, P., Butler, W. T., and Kang, A. H., 1968, Comparative studies on the chemistry of collagen utilizing cyanogen bromide cleavage. Structure, function and evolution in proteins, *Brookhaven Symposium of Biology,* **21**:345.

Pinnell, S. R., and Martin, G. R., 1968, The cross linking of collagen and elastin: Enzymatic conversion of lysine in peptide linkage to α-aminoadipic-δ-semialdehyde (allysine) by an extract from bone, *Proc. Natl. Acad. Sci. U.S.A.* **61**:708.

Pinnell, S. R., Fox, R., and Krane, S. M., 1971, Human collagens: Differences in glycosylated hydroxylysines in skin and bone. *Biochim. Biophys. Acta* **229**:119.

Pinnell, S. R., Krane, S. M., Kenzora, J. E., and Glimcher, M. J., 1972, A heritable disorder of connective tissue: Hydroxylysine deficient collagen disease, *N. Engl. J. Med.* **286**:1013.

Pinto, J. S., and Bentley, J. P., 1974, The timecourse of collagen crosslinking, *Biochim. Biophys. Acta* **354**:254.

Pope, F. M., Martin, G. R., Lichtenstein, J. R., Penttinen, R., Gerson, B., Rowe, D. W., and McKusick, V. A., 1975, Patients with Ehlers–Danlos syndrome type IV lack type III collagen, *Proc. Natl. Acad. Sci. U.S.A.* **72**:1314.

Powles, T. J., Easty, D. M., Easty, G. C., Bondy, P. K., and Munro-Neville, A., 1973, Aspirin inhibition of in vitro osteolysis stimulated by parathyroid hormone and PGE, *Nature* **245**:83.

Pras, M., and Glynn, L. E., 1973, Isolation of a non-collagenous reticulin component and its primary characterization, *Br. J. Exp. Pathol.* **54**:449.

Pras, M., Johnson, G. D., Holborow, E. J., and Glynn, L. E., 1974, Antigenic properties of a non-collagenous reticulin component of normal connective tissue, *Immunology* **27**:469.

Pratt, R. M., and King, C. T. G., 1972, Inhibition of collagen cross-linking associated with β-aminopropionitrile-induced cleft palate in the rat, *Dev. Biol.* **27**:332.

Priest, R. E., Moinuddin, J. F., and Priest, J. H., 1973, Collagen of Marfan's syndrome is abnormally soluble, *Nature* **245**:264.

Prockop, D. J., and Kivirikko, K. I., 1967, Relationship of hydroxyproline excretion in urine to collagen metabolism, *Ann. Intern. Med.* **66**:1243.

Rebel, A., Malkani, K., and Basle, M., 1974, Anomalies nucléaires des osteoclastes de la maladie osseuse de Paget. *Nouv. Presse Med.* **3**:1299.

Riley, F. C., Jowsey, J., and Brown, D., 1971, Connective tissue ultrastructure and bone remodeling in osteogenesis imperfecta, *J. Bone Surg.* **53A**:801.

Riley, F. C., Brown, D., and Jowsey, J., 1973, Osteogenesis imperfecta: Morphologic and biochemical studies of connective tissue, *Pediat. Res.* **7**:757.

Risteli, J., and Kivirikko, K. I., 1974, Activities of prolyl hydroxylase, lysyl hydroxylase, collagen galactosyltransferase and collagen glucosyltransferase in the liver of rats with hepatic injury, *Biochem. J.* **144**:115.

Robert, L., Robert, B., Moczar, E., and Moczar, M., 1972, Les glycoprotéines de structure du tissu conjonctif, *Pathol. Biol.* **20**:1001.

Robertson, P. B., Cobb, C. M., Taylor, R. E., and Fullmer, H. M., 1974, Activation of latent collagenase by microbial plaque, *J. Periodontal Res.* **9:**81.

Robertson, W. van B., and Schwartz, B., 1953, Ascorbic acid and the formation of collagen. *J. Biol. Chem.* **201:**689.

Robinson, C. J., Rafferty, B., and Parsons, J. A., 1972, Calcium shift into bone: A calcitonin-resistant primary action of parathyroid hormone, studied in rats, *Clin. Sci.* **42:**235.

Rokosova, B., and Chvapil, M., 1974, Relationship between the dose of ascorbic acid and its structural analogs and proline hydroxylation in various biological systems, *Connect. Tissue Res.* **2:**215.

Rosenberg, L. C., Pal, S., and Beale, R. J., 1973, Proteoglycans from bovine proximal humeral articular cartilage, *J. Biol. Chem.* **248:**3681.

Rosenbloom, A. L., and De Busk, F. L., 1971, Progeria of Hutchinson–Gilford: A caricature of ageing. *Am. Heart J.* **82:**287.

Rosenbloom, J., and Prockop, D. J., 1971, Incorporation of *cis*-hydroxyproline into protocollagen and collagen. Collagen containing *cis*-hydroxyproline in place of proline and transhydroxyproline is not extruded at a normal rate, *J. Biol. Chem.* **246:**1549.

Rosenbloom, J., Harsch, M., and Jimenez, S., 1973, Hydroxyproline content determines the denaturation temperature of chick tendon collagen, *Arch. Biochem. Biophys.* **158:**478.

Ross, R., 1973, The elastic fibre. A review, *J. Histochem. Cytochem.* **21:**199.

Ross, R., 1975, Connective tissue cells, cell proliferation and synthesis of extracellular matrix—a review, *Phil. Trans. R. Soc. Lond.* **271:**247.

Ross, R., and Benditt, E. P., 1964, Wound healing and collagen formation. IV. Distortion of ribosomal patterns of fibroblasts in scurvy, *J. Cell Biol.* **22:**365.

Ross, R., and Bornstein, P., 1969, The elastic fiber. I. The separation and partial characterization of its macromolecular components, *Cell Biol.* **40:**366.

Rowe, D. W., McGoodwin, E. B., Martin, G. R., Sussman, M. D., Grahn, D., Faris, B., and Franzblau, C., 1974, A sex-linked defect in the cross linking of collagen and elastin associated with the mottled locus in mice, *J. Exp. Med.* **139:**180.

Rucker, R. B., and Goettlich-Riemann, W., 1972, Isolation and properties of soluble elastin from copper deficient chicks, *J. Nutr.* **102:**563.

Rucker, R. B., and O'Dell, B. L., 1970, Inhibition of elastin cross linking by iproniazid and its counteraction by pyridoxal phosphate, *Biochim. Biophys. Acta* **222:**527.

Ryan, J. N., and Woessner, J. F., 1972, Oestradiol inhibits collagen breakdown in the involuting rat uterus. *Biochem. J.* **127:**705.

Sakamoto, S., Sakamoto, M., Goldhaber, P., and Glimcher, M. J., 1975, Studies on the interaction between heparin and mouse bone collagenase, *Biochim. Biophys. Acta* **385:**41.

Sandberg, L. B., Zeikus, R. D., and Coltrain, I. M., 1971, Tropoelastin purification from copper deficient swine: A simplified method, *Biochim. Biophys. Acta* **236:**542.

Seyer, J. M., and Vinson, W. C., 1974, Synthesis of type I and type II collagen by embryonic chick cartilage, *Biochem. Biophys. Res. Commun.* **58:**272.

Shoshan, S., and Gross, J., 1974, Biosynthesis and metabolism of collagen and its role in tissue repair processes, *Isr. J. Med. Sci.* **10:**537.

Shoshan, S., Segal, N., Traub, W., Salem, G., Kuhn, K., and Lapière, Ch. M., 1974, Normal characteristics of dermosparactic calf skin collagen fibers following their subcutaneous implantation within a diffusion chamber into a normal calf, *FEBS Lett.* **41:**269.

Shuttleworth, C. A., and Forrest, L., 1974, Pepsin-solubilized collagens of guinea-pig dermis and dermal scar, *Biochim. Biophys. Acta* **365:**454.

Shuttleworth, C. A., and Veis, A., 1972, The isolation of anionic phosphoproteins from bovine cortical bone via the periodate solubilization of bone collagen, *Biochim. Biophys. Acta* **257:**414.

Shuttleworth, C. A., Forrest, L., and Jackson, D. S., 1975, Comparison of the cyanogen bromide peptides of insoluble guinea-pig skin and scar collagen, *Biochim. Biophys. Acta* **379:**207.

Siegel, R. C., 1974, Biosynthesis of collagen crosslinks: Increased activity of purified lysyl oxidase with reconstituted collagen fibrils, *Proc. Natl. Acad. Sci. U.S.A.* **71:**4826.

Siegel, R. C., Pinnell, S. R., and Martin, G. R., 1970, Cross-linking of collagen and elastin. Properties of lysyl oxidase, *Biochemistry* **9:**4486.

Simar, L. J., and Betz, E. H., 1971, Dermatosparaxis of the calf, a genetic defect of the connective tissue. 2. Ultrastructural study of the skin, *Hoppe-Seyler's Z. Physiol. Chem.* **352:**13.

Simon, D. R., Berman, I., and Howell, D. S., 1973, Relationship of extracellular matrix vesicles to calcification in normal and healing rachitic epiphyseal cartilage, *Anat. Rec.* **176:**167.

Simpson, J. W., and Taylor, A. C., 1974, Regulation of gingival collagenase: A possible role for a mast cell factor, *Proc. Soc. Exp. Biol. Med.* **145:**42.

Simunek, Z., and Muir, H., 1972, Changes in the protein-polysaccharides of pig articular cartilage during prenatal life, development and old age, *Biochem. J.* **126:**515.

Skosey, J. L., and Damgaard, E., 1973, Effect of estradiol benzoate on the degradation of insoluble collagen of rat skin, *Endocrinology* **93:**311.

Slavkin, H. C., 1971, The dynamics of extracellular and cell surface protein interactions, *in: Cellular and Molecular Renewal in the Mammalian Body* pp. 221–276, Academic Press, New York.

Smith, B. D., Martin, G. R., Miller, E. J., Dorfman, A., and Swarm, R., 1975*a*, Nature of the collagen synthesized by a transplanted chondrosarcoma, *Arch. Biochem. Biophys.* **166:**181.

Smith, B. D., Abraham, P. A., and Carnes, W. H., 1975*b*, Crosslinkage of salt-soluble elastin *in vitro, Biochem. Biophys. Res. Commun.* **66:**893.

Smith, J. G., Wehr, R. F., Badger, N. L., and Pirkle, D. E., 1974, Urinary hydroxyproline: Source of increase after thermal burns, *Proc. Soc. Exp. Biol. Med.* **145:**897.

Speakman, P. T., 1971, Proposed mechanism for the biological assembly of collagen triple helix, *Nature* **229:**241.

Spiro, R. G., 1973, Biochemistry of the renal glomerular basement membrane and its alterations in diabetes mellitus, *N. Engl. J. Med.* **288:**1337.

Spiro, R. G., and Spiro, M. J., 1971, Studies on the biosynthesis of the hydroxylysine-linked disaccharide unit of basement membrane and collagens. III. Tissue and subcellular distribution of glycosyl-transferases and the effect of various conditions on the enzyme levels, *J. Biol. Chem.* **246:**4919.

Stark, M., Lenaers, A., Lapière, Ch. M., and Kuhn, K., 1971, Electronoptical studies of procollagen from the skin of dermatosparaxic calves. *FEBS Lett.* **18:**225.

Stassen, F. L. H., Cardinale, G. J., and Udenfriend, S., 1973, Activation of prolylhydroxylase in L-929 fibroblasts by ascorbic acid, *Proc. Natl. Acad. Sci. U.S.A.* **70:**1090.

Stassen, F. L. H., Cardinale, G. J., McGee, J., O'D., and Udenfriend, S., 1974, Prolyl hydroxylase and an immunologically related protein in mammalian tissues, *Arch. Biochem. Biophys.* **160:**340.

Steendijk, R., and Boyde, A., 1973, Scanning electron microscopic observations on bone from patients with hypophosphataemic (vitamin D resistant) rickets, *Calcif. Tissue Res.* **11**:242.

Steffen, C., Ludwig, H., Kovac, W., and Menzel, J., 1975, Collagen-induced acute synovitis in collagen-immunized rabbits, *Z. Immun. Forsch.* **150**:432.

Stevenson, C. J., Bottoms, E., and Shuster, S., 1970, Skin collagen in osteogenesis imperfecta, *Lancet* **1**:860.

Stoltz, M., Furthmayr, H., and Timpl, R., 1973, Increased lysine hydroxylation in rat bone and tendon collagen and localization of the additional residues, *Biochim. Biophys. Acta* **310**:461.

Sussman, M., Lichtenstein, J. R., Nigra, T. P., Martin, G. R., and McKusick, V. A., 1974, Hydroxylysine-deficient skin collagen in a patient with a form of the Ehlers–Danlos syndrome, *J. Bone Jt. Surg.* **56A**:1228.

Switzer, B. R., and Summer, G. K., 1973, Inhibition of collagen synthesis by α,α'-dipyridyl in human skin fibroblasts in culture, *In Vitro* **9**:160.

Sykes, B. C., and Partridge, S. M., 1972, Isolation of a soluble elastin from lathyritic chicks, *Biochem. J.* **130**:1171.

Tanzer, M. L., 1965, Experimental lathyrism, *in: International Review of Connective Tissue Research*, Vol. 3, pp. 91–112, Academic Press, New York.

Teitelbaum, S. L., Kraft, W. J., Lang, R., and Avioli, L. V., 1974, Bone collagen aggregation abnormalities in osteogenesis imperfecta, *Calcif. Tissue Res.* **17**:75.

Teller, W. M., Genscher, U., Burkhardt, H., and Rommel, K., 1973, Hydroxyproline excretion in various forms of growth failure, *Arch. Dis. Child.* **48**:127.

Timpl, R., Wick, G., Furthmayr, H., Lapière, Ch. M., and Kuhn, K., 1973, Immunochemical properties of procollagen from dermosparactic calves, *Eur. J. Biochem.* **32**:584.

Toole, B. P., Kang, A. H., Trelstad, R. L., and Gross, J., 1972, Collagen heterogeneity within different growth regions of long bones of rachitic and non-rachitic chicks, *Biochem. J.* **127**:715.

Trelstad, R. L., 1974, Human aorta collagens: Evidence for three distinct species, *Biochem. Biophys. Res. Commun.* **57**:717.

Trelstad, R. L., and Kang, A. H., 1974, Collagen heterogeneity in the avian eye: Lens, vitreous body, cornea and sclera, *Exp. Eye Res.* **18**:395.

Uitto, J., 1969, Effect of D-penicillamine on collagen biosynthesis in organ culture, *Biochim. Biophys. Acta* **194**:498.

Uitto, J., 1970, A method for studying collagen biosynthesis in human skin biopsies in vitro, *Biochim. Biophys. Acta* **201**:438.

Uitto, J., and Prockop, D. J., 1974, Incorporation of proline analogues into collagen polypeptides. Effects on the production of extracellular procollagen and on the stability of the triple-helical structure of the molecule, *Biochim. Biophys. Acta* **336**:234.

Uitto, J., Halme, J., Hannuksela, M., Peltokallio, P., and Kivirikko, K. I., 1969, Protocollagen proline hydroxylase activity in the skin of normal human subjects and of patient with scleroderma, *Scand. J. Clin. Lab. Invest.* **23**:241.

Uitto, J., Hannuksela, M., and Ramussen, O. G., 1970*a*, Protocollagen proline hydroxylase activity in scleroderma and other connective tissue disorders, *Ann. Clin. Res.* **2**:235.

Uitto, J., Lindy, S., Rokkanen, P., and Vainio, K., 1970*b*, Increased protocollagen proline hydroxylase activity in synovial tissue in rheumatoid arthritis, *Clin. Chim. Acta* **30**:741.

Uitto, J., Dehm, P., and Prockop, D. J., 1972*a*, Incorporation of *cis*-hydroxyproline into collagen by tendon cells. Failure of the intracellular collagen to assume a triple-helical conformation, *Biochim. Biophys. Acta* **278**:601.

Uitto, J., Lindy, S., Turto, H., and Vainio, K., 1972*b*, Collagen biosynthesis in rheumatoid synovial tissue, *J. Lab. Clin. Med.* **79:**960.

Uitto, J., Hoffmann, H. P., and Prockop, D. J., 1975, Retention of nonhelical procollagen containing *cis*-hydroxyproline in rough endoplasmic reticulum, *Science* **190:**1202.

Urist, M. R., Iwata, H., and Strates, B. S., 1972, Bone morphogenetic protein and proteinase in the guinea pig, *Clin. Orthop.* **85:**275.

Vaes, G., 1968, On the mechanisms of bone resorption. The action of parathyroid hormone on the excretion and synthesis of lysosomal enzymes and on the extracellular release of acid by bone cells, *J. Cell Biol.* **39:**676.

Vaes, G., 1972, The release of collagenase as an inactive proenzyme by bone explants in culture, *Biochem. J.* **126:**275.

Vaes, G., and Eeckhout, Y., 1975, The precursor of bone collagenase and its activation, *in: Protides of Biological Fluids,* (H. Peeters, ed.), pp. 391–397, Pergamon Press, Oxford.

Vaishnav, R. N., Young, J. T., Janicki, J. S., and Patel, D. J., 1972, Nonlinear anisotropic elastic properties of the canine aorta, *Biophys. J.* **12:**1008.

Van Caneghem, P., and Lapière, Ch. M., 1970, Etudes concernant le mode d'action des radiations ionisantes sur l'interrelation des molécules de collagène dans les fibrilles, *Int. J. Radiat. Biol.* **18:**401.

Varadi, D. P., and Hall, D. A., 1965, Cutaneous elastin in Ehlers–Danlos syndrome, *Nature* **208:**1224.

Veis, A., Spector, A. R., and Zamoscianyk, H., 1972, The isolation of an EDTA-soluble phosphoprotein from mineralizing bovine dentin, *Biochim. Biophys. Acta* **257:**404.

Viidik, A., 1972, Simultaneous mechanical and light microscopic studies of collagen fibers, *Z. Anat. Entwickl. Gesch.* **136:**204.

Vinson, W. C., and Seyer, J. M., 1974, Synthesis of type III collagen by embryonic chick skin, *Biochem. Biophys. Res. Commun.* **58:**58.

Volpin, D., and Veis, A., 1973, Cynanogen bromide peptides from insoluble skin and dentin bovine collagens, *Biochemistry* **12:**1452.

von Figura, K., and Kresse, H., 1974, Quantitative aspects of pinocytosis and the intracellular fate of *N*-acetyl-α-D-glycosaminidase in Sanfilippo B fibroblasts, *J. Clin. Invest.* **53:**85.

Wahl, L. M., Wahl, S. M., Mergenhagen, S. E., and Martin, G. R., 1974, Collagenase production by endotoxin-activated macrophages, *Proc. Natl. Acad. Sci. U.S.A.* **71:**3598.

Waisman, J., Concilla, P. A., and Coulson, W. F., 1969, Cardiovascular studies on copper-deficient swine. XIII. The effect of chronic copper deficiency on the cardiovascular system of miniature pigs, *Lab. Invest.* **21:**548.

Walker, D. G., 1966, Elevated bone collagenolytic activity and hyperplasia of parafollicular light cells of the thyroid gland in parathormone-treated grey-lethal mice, *Z. Zellforsch. Mikrosk. Anat.* **72:**100.

Walker, D. G., 1973, Osteopetrosis cured by temporary parabiosis, *Science* **180:**875.

Walker, D. G., Lapière, Ch. M., and Gross, J., 1964, A collagenolytic factor in rat bone promoted by parathyroid extract, *Biochem. Biophys. Res. Commun.* **15:**397.

Walker, F., 1973, The origin, turnover and removal of glomerular basement membrane, *J. Pathol.* **110:**233.

Walshe, J. M., 1956, Penicillamine, a new oral therapy for Wilson's disease, *Am. J. Med.* **21:**487.

Watson, A. M., and Pearce, R. H., 1947, The mucopolysaccharide content of the skin in localized pretibial myxoedema, *Am. J. Clin. Pathol.* **17:**507.

Wehinger, H., Witt, I., Losel, I., Denz-Seibert, G., and Sander, C., 1975, Intravenous copper in Menkes kinky-hair syndrome, *Lancet* **I:**1143.

Weinstock, A., Weinstock, M., and Leblond, C. P., 1972, Autoradiographic detection of H^3-fucose incorporation in glycoprotein by odontoblasts and its deposition at the site of the calcification front in dentin, *Calcif. Tissue Res.* **8:**181.

Weinstock, M., and Leblond, C. P., 1973, Radioautographic visualization of the deposition of a phosphoprotein at the mineralization front in the dentin of the rat incisor, *J. Cell Biol.* **56:**838.

Weiss, J. B., Shuttleworth, C. A., Brown, R., Sedowfia, K., Baildam, A., and Hunter, J. A., 1975, Occurrence of type III collagen in inflamed synovial membranes: a comparison between nonrheumatoid, rheumatoid, and normal synovial collagens, *Biochem. Biophys. Res. Commun.* **65:**907.

Werb, Z., and Reynolds, J. J., 1974, Stimulation of endocytosis of the secretion of collagenase and neutral proteinase from rabbit synovial fibroblasts, *J. Exp. Med.* **140:**1482.

Werb, Z., Burleigh, M. C., Barrett, A. J., and Starkey, P. M., 1974, The interaction of α_2-macroglobulin with proteinases: Binding and inhibition of mammalian collagenases and other metal proteinases, *Biochem. J.* **139:**359.

Westberg, N. G., and Michael, A. F., 1973, Human glomerular basement membrane: Chemical composition in diabetes mellitus, *Acta Med. Scand.* **193:**39.

Whitehead, R. G., 1965, Hydroxyproline:creatinine ratio as an index of nutritional status and rate of growth, *Lancet* **2:**567.

Wick, G., Furthmayr, H., and Timpl, R., 1975, Purified antibodies to collagen: an immunofluorescence study of their reaction with tissue collagen, *Int. Archs. Allergy Appl. Immun.* **48:**664.

Williams, B. R., Cranley, R. E., Doty, S. B., Lichtenstein, J., and McKusick, V. A., 1974, Disorganization of collagen in individuals with procollagen peptidase deficiency (Ehlers–Danlos type VII). *Isr. J. Med. Sci.* **10:**1470.

Wilson, R. P., and Poe, W. E., 1973, Impaired collagen formation in the scorbutic channel catfish, *J. Nutr.* **103:**1359.

Winand, R. J., 1968, Increased urinary excretion of acidic mucopolysaccharides in exophthalmos, *J. Clin. Invest.* **47:**2563.

Winand, R. J., 1975, Collagen-proteoglycans relationship in pathology, *Ital. J. Biochem.* **24:**96.

Winand, R. J., and Mahieu, P., 1973, Prevention of malignant exophthalmos after treatment of thyrotoxicosis, *Lancet* **1:**1196.

Winand, R. J., and Nusgens, B., 1971, Dermatosparaxis, a genetic defect of the connective tissues. 4. The modified glycosaminoglycans and proteoglycans, *Hoppe-Seyler's Z. Physiol. Chem.* **352:**14.

Winkelmann, R. K., 1971, Classification and pathogenesis of scleroderma, *Mayo Clin. Proc.* **46:**83.

Wojtecka-Lukasik, E., and Dancewicz, A. M., 1974, Inhibition of human leucocyte collagenase by some drugs used in the therapy of rheumatic diseases, *Biochem. Pharmacol.* **23:**2077.

Wood, G. C., 1960, The formation of fibrils from collagen solutions. 3. Effect of chondroitin sulphate and some other naturally occurring polyanions on the rate of formation, *Biochem. J.* **75:**605.

Woolley, D. E., Lindberg, K. A., Glanville, R. W., and Evanson, J. M., 1975*a*, Action of rheumatoid synovial collagenase on cartilage collagen. Different susceptibilities of cartilage and tendon collagen to collagenase attack, *Eur. J. Biochem.* **50:**437.

Woolley, D. E., Roberts, D. R., and Evanson, J. M., 1975*b*, Inhibition of human collagenase activity by a small molecular weight serum protein, *Biochem. Biophys. Res. Commun.* **66**:747.

Wu, K., and Frost, H. M., 1969, Bone formation in osteoporosis, *Arch. Pathol.* **88**:508.

Yamanishi, Y., Dabbous, M. K., and Hashimoto, K., 1972, Effect of collagenolytic activity in basal cell epithelioma of the skin on reconstituted collagen and physical properties and kinetics of the crude enzyme, *Cancer Res.* **32**:2551.

Yamanishi, Y., Maeyens, E., Dabbous, M. K., Ohyama, H., and Hashimoto, K., 1973, Collagenolytic activity in malignant melanoma: physiochemical studies, *Cancer Res.* **33**:2507.

9
Collagen and Cell Differentiation

A. H. REDDI

I. Introduction

Cell differentiation is one of the central problems of contemporary biology and medicine. How does the fertilized egg develop into an adult organism with a myriad of cell types? It is common knowledge that all cells in a multicellular organism possess the same genetic information, hence an identical genotype. What is the molecular basis of the progressive appearance of diverse phenotypes from cells with the same genotype? The alteration of gene expression during development is of cardinal importance in cell differentiation.

A detailed analysis of the mechanisms of differential gene expression will improve our knowledge of differentiation, with attendant implications for cancer and developmental anomalies. Ample attention is currently focused on intracellular molecules involved in transcription and translation of genetic information and its regulation. The origin and evolution of multicellular animals was accompanied by extracellular matrix macromolecules such as collagen and proteoglycans. There has been a relative paucity of information about the developmental appearance and possible role of extracellular constituents in cell differentiation. The purpose of this chapter is to marshall recent experimental evidences implicating collagen in cell differentiation and to present the concept that collagenous

---

A. H. REDDI · The Ben May Laboratory for Cancer Research, University of Chicago, Chicago, Illinois 60637. This article is dedicated with admiration to Dr. Charles Huggins on the occasion of his seventy-fourth birthday.

matrix imparts positional information locally at critical phases of morphogenesis. This idea is mainly based on the fact collagen is predominantly in the solid state, and the requisite specificity may be conferred by the molecular heterogeneity (see Chapter 1) of collagens. Furthermore, important physical characteristics of collagen, especially piezoelectricity, may play a crucial role in altering cell-surface behavior. Considerable emphasis will be laid on the work done in our laboratory on the influence of collagenous matrix on gene expression in responding fibroblasts, as many of the ideas discussed arose out of these experiments.

II. Collagen in Early Embryogenesis

When does collagen appear first in an embryo? Collagen synthesis was detected in sea urchin, *Paracentrotus lividus* at cleavage stage, underwent a several-fold increase during gastrulation, and also was correlated with stages of spicule formation (Pucci-Minafra *et al.*, 1972). Similar results were obtained in two other genera of sea urchin embryos by Golob *et al.* (1974). Edds (1958) detected collagen in early stage-22 frog embryo. In other studies on frog embryos, collagen synthesis has been observed in ectoderm–mesodern and endoderm cells (Klose and Flickinger, 1971). Employing the incorporation of radioactive proline into macromolecular hydroxyproline as an index of collagen synthesis, several-fold increase was detected during neurulation in *Xenopus laevis* (Green *et al.*, 1968). The aforementioned investigations relied mostly on incorporation of radioactive proline into collagen. The possible heterogeneity of collagen molecules in early embryos appears not to have been investigated thus far. Using electron microscopic techniques, collagen was first detected in chick embryos as the incomplete basal lamina on the undersurface of epiblast (Trelstad *et al.*, 1967). It is noteworthy that collagen first appears in the embryo before the appearance of specialized connective tissue cells.

III. Tissue Interactions and Organogenesis

A. Epithelial–Mesenchymal Interactions

During early phases of organogenesis the epithelial and mesenchymal components interact and influence each other. These epithelial–mesenchymal interactions are crucial for the morphogenesis of diverse tissues (Grobstein, 1967; Fleischmajer and Billingham, 1968; Jackson, 1968; Kratochwil, 1972). The incisive studies of Grobstein and his co-workers (Grobstein, 1967, for a review) have delineated the principles underlying these tissue interactions. The technique consists of dissociating the epithelium and mesenchyme of embryonic rudiments by mild trypsin digestion and thereafter simply reassociating the tissues or interposing a Millipore filter between the two components. The entire assembly is then maintained in organ culture. When the dissociated epithelium and mesenchyme were cultured separately, there was no epithelial morphogenesis. However, in the recombinants, including the ones with a Millipore filter (20 μm thick; pore size, 1.0 μm), the morphogenesis was complete. During the course of these investigations the remarkable specificity of salivary mesenchyme for salivary morphogenesis was demonstrated (Grobstein, 1954). However, the specificity for tubulogenesis in metanephric kidney was not absolute, as the dorsal spinal cord also served as an inductor in addition to the normal ureteric bud epithelium.

Based on these transfilter epitheliomesenchymal interactions and the histological appearance of interfacial material, Grobstein (1954) recognized the importance of extracellular matrix in morphogenesis. These early studies were then extended to other organ systems, including endodermal rudiments such as pancreas and lung. Kallman and Grobstein (1964, 1965), using ultrastructural and autoradiographic techniques, identified collagen at the epitheliomesenchymal interface of mouse pancreas and salivary glands. Mesenchyme appears to be the source of collagen (Bernfield, 1970). The branching morphogenesis of salivary epithelium is abolished by treatment with collagenase (Grobstein and Cohen, 1965). Despite the fact that epithelial–mesenchymal interactions have been studied in numerous organ systems, we still do not understand the precise mechanisms involved. However, the increasing knowledge gained about collagen and other extracellular molecules should aid this task.

It was generally accepted until recently that in view of the fact that transfilter inductions occurred across Millipore filters, the putative morphogenetic molecules diffused short distances. However, using Nucleopore filters (produced by bombardment of polycarbonate by charged particles) with uniform straight pores, it was demonstrated that tubulogenesis induced by spinal cord required a close cell apposition (Wartiovaara *et al.*, 1974). The Millipore filters have tortuous pores and early studies may have overlooked the presence of cell processes therein. This example illustrates how technical advances can radically alter the firmly established ideas in experimental biology. In view of this finding, many of the early studies on epithelial–mesenchymal interactions need to be reassessed with reference to the possible role of cell–cell contact in organogenesis.

B. *Epithelial Collagens*

Mesenchyme is generally considered to be the source of collagen (Bernfield and Wessells, 1970). There is a growing body of evidence that epithelia synthesize collagen. The elegant studies of Hay and her colleagues demonstrated the secretion of collagen by neuroepithelium (Cohen and Hay, 1971; Trelstad *et al.*, 1973) and isolated corneal epithelium (Hay and Dodson, 1973). Notochord, an epithelial tissue, also synthesizes collagen, and it cochromatographed with cartilage collagen chains (Linsenmayer *et al.*, 1973). Notochordal collagen from the sturgeon, *Scaphirhynchus platorhynchus* has been characterized and is identical to the cartilage collagen of the same species (Miller and Mathews, 1974). It would appear that collagens are synthesized by both mesenchymal and epithelial cells.

C. *Molecular Heterogeneity of Collagens*

The early work of Piez and co-workers (see Chapter 1) established that most vertebrate collagens consisted of two $\alpha 1$ chains and one $\alpha 2$ chain. Miller and Matukas (1969), during the course of characterization of chick sternal cartilage, observed an excess of $\alpha 1$ chains relative to $\alpha 2$ than would be accounted by a classical $\alpha 1$ ratio of 2:1. Additional studies with cyanogen bromide peptides revealed that cartilage was composed of a new collagen, and they designated it as type II, $[\alpha 1(\mathrm{II})]_3$. The already well-characterized collagen in adult bone and skin was termed type I and was symbolized $[\alpha 1(\mathrm{I})]_2\alpha 2$ (see Miller, 1973 for a review). This important discovery by Miller and Matukas (1969) of a collagen distinct in primary structure, hence genetically different, was almost immediately confirmed by Trelstad *et al.* (1970). A new era in collagen biochemistry ensued, and newer types of collagen differing in primary sequence are being added. For example, type III, $[\alpha 1(\mathrm{III})]_3$ in embryonic skin, with cysteine, an amino acid hitherto unsuspected in collagen, was described (Chung and Miller, 1974). The basement membrane collagens consist of three identical $\alpha 1$ chains and were termed type IV, $[\alpha 1(\mathrm{IV})]_3$ (Kefalides, 1973). At present we know of four types of collagen with four different $\alpha 1$ chains and one $\alpha 2$ chain. It would appear that at least five different structural genes code for collagens. This rapidly expanding knowledge about primary structure heterogeneity has important implications for organogenesis in particular (see Section III-D) and in general for possible specific roles in developmental biology.

There exists, in addition to primary structure differences among collagens, an enormous potential for posttranslational modifications (see Chapter 5). These are: the hydroxylation of proline and lysine, galactosylation and glucosylation of hydroxylysine, formation of cross-link precursors (see Chapter 4) by oxidative deamination of lysine and hydroxylysine and, finally, proteolytic cleavage and processing of procollagen to collagen. The possible permutations and combinations of chemically distinct collagens that can arise by the degree and site of various posttranslational modifications is staggering. The collagen in both skin and bone is type I; however, galactosylhydroxylysine was predominant in bone and the disaccharide derivative, glucosylgalactosylhydroxylysine in human skin (Pinnell *et al.*, 1971; Segrest and Cunningham, 1970). This example illustrates the chemical differences in the collagens of identical primary structure. In relation to the morphogenetic significance of collagens these potential postsynthetic modifications assume added importance. With this as background, we shall now consider specific examples of organogenesis.

D. Organogenesis

1. Eye

The eye is a remarkable structure precisely engineered to relay visual images to the brain. A detailed description of the equally precise development is beyond the scope of this article, and the reader is directed to the excellent work of Coulombre (1965). The optical function of the eye requires a precise geometrical structure, and this is achieved by numerous tissue interactions during the morphogenesis of component structures and their final assembly. For example, cornea, lens, and vitreous body must be transparent; the musculoskeletal structures develop in exact relationship to lens to aid in the reception and transmission of light energy to brain as nervous impulse. The outer sclera maintains the shape of the eye, and the intraocular pressure in the developing eye determines the geometry of the cornea. Truly, the embryology of eye challenges even today's modern electronic technology.

Collagen constitutes an integral molecule in several parts of eye: cornea, lens, vitreous body, and sclera. The collagenous cornea is transparent and consists of collagen fibrils oriented at approximately right angles to each other in orthogonal layers (Trelstad and Coulombre, 1971). This epithelial collagen on extrusion into subepithelial space self-assembles into a spiral, orthogonal matrix. The self-assembly of collagen in a morphogenetic system is best illustrated by the avian corneal morphogenesis (Trelstad *et al.*, 1974). Using *in vitro* techniques, Dodson and Hay (1974) demonstrated that frozen-killed lens promoted collagen synthesis by corneal epithelium. A variety of collagenous substrata promoted collagen synthesis by isolated corneal epithelium. A threefold stimulation of [^{3}H]proline incorporation was elicited by lens capsule basal lamina (basement membrane), rat-tail tendon collagen gels, and type II collagen from a chondrosarcoma (Meier and Hay, 1974). Collagen fibers were detected on the third day of development (stage 17½) in the presumptive cornea (Conrad, 1970). Dische (1970) has suggested the possible importance of carbohydrates associated with collagen in the developing corneal stroma.

The molecular heterogeneity of collagens (see Section III-C) is perhaps best illustrated by the detection of five different collagens in the eye of 3-week-old chicks (Trelstad and Kang, 1974). In addition to the established type I and type II collagens in cornea and cartilaginous sclera, two new $\alpha 1$ chains were detected in lens and vitreous body and a novel $\alpha 2$-like molecule in scleral cartilage. Further details about these new collagens is awaited with interest. The lens capsule basement membrane consists of type IV collagen (Kefalides, 1973). What developmental roles, if any, do each of these different collagen molecules play in early development and morphogenesis of eye?

2. *Heart*

Collagen was detected in developing heart of chick embryos (Woessner *et al.,* 1967; Johnson *et al.,* 1974). Collagen is present in cardiac jelly of early stages of developing heart. Studies on molecular heterogeneity in adult bovine cardiac muscle revealed both type I collagen and another molecule with three identical $\alpha 1$ chains unique to heart (McClain, 1974). Type I-like collagen was identified in early chick embryo heart (Johnson *et al.,* 1974).

3. *Lung*

The developing lung exhibits typical epithelial–mesenchymal interactions. Bronchial mesenchyme directs the branching morphogenesis of bronchial epithelial buds (Alescio and Cassini, 1962; Spooner and Wessells, 1970). The stabilizing influence of collagen on mouse lung morphogenesis was shown by the use of collagenase (Wessells and Cohen, 1968). In later studies, Wessells (1970) observed the inhibitory effects of ordered collagen fibers of tracheal mesoderm on bronchial morphogenesis; the disoriented collagen of bronchial mesoderm is stimulatory to branching of bronchial buds. The collagen synthesis in rabbit lungs is high in late gestation and declines to a lower level in postnatal life (Bradley *et al.*, 1974*a*). Pulmonary alveoli synthesize type I collagen. The tracheobronchial tree predominantly synthesizes type II collagen (Bradley *et al.*, 1974*b*). Lung morphogenesis *in vitro* was inhibited by a proline analog L-azetidine-2-carboxylic acid (Alescio, 1973).

4. *Skin*

Sengel (1971) has reviewed the tissue interactions in the development of cutaneous structures in birds. The 12-day-old chick embryonic skin was employed to investigate the epidermal–dermal interactions. The functional development of epidermis is dependent on dermis; however, live dermal cells are not essential as frozen-killed dermis will suffice. Collagen gels can substitute for dermis (Dodson, 1963). The development of feathers in embryonic chick skin is dependent on a birefringent, collagenous lattice in the dermis (Stuart *et al.,* 1972). Further, treatment of this lattice with collagenase abolishes the feather morphogenesis. The collagenous lattice determines the pattern of distribution of the dermal papillae. Goetinck and Sekellick (1972), using scaleless mutant chick skin which lacks feathers, noted the absence of fibrous lattice in the dermis. However, the rates of synthesis of collagen in the mutant skin was unaltered. It is likely that there might be an enzymatic deficiency in the postsynthetic processing of procollagen and fibrillogenesis connected with the dermal lattice formation.

The human skin consists of both type I and type III collagens (Epstein, 1974). In fetal skin, the type III, $[\alpha 1(III)]_3$, is the predominant species (Chung and Miller, 1974; Epstein, 1974). Synthesis of type III collagen was demonstrated in early embryonic chick skin (Vinson and Seyer, 1974). It is pertinent, therefore, to enquire whether the type III collagens play a crucial role in early skin morphogenesis and merit further investigation.

5. *Skeletal System*

The vertebral cartilage developmentally is formed by somite sclerotome. Somite chondrogenesis is stimulated by notochord and spinal cord (Lash, 1968) and is indicated by appearance of collagen fibrils (Minor, 1973). The induction of chondrogenesis by spinal cord in chick embryo somites in chorioallantoic cultures is prevented by collagenase and hyaluronidase in combination, but not singly (O'Hare, 1972). Addition of 5-bromodeoxyuridine to chondroblast cultures shifts the α chain composition from type II to type I collagen (Schiltz *et al.*, 1973). Similar differences in collagen biosynthesis were observed in limb bud cells exposed to bromodeoxyuridine (Levitt *et al.*, 1974). A detailed analysis of the influence of bromodeoxyuridine on chondrocytes in culture revealed a new collagen, type I trimer, in addition to type I (Mayne *et al.*, 1975).

The changes in embryonic limb during endochondral ossification consist of sequential changes from mesenchymal cells to cartilage and subsequent replacement with new bone. The ontogeny of collagens during these biological transitions was examined by Linsenmeyer and colleagues (1973). Type I collagen was detected in mesenchymal cells; the formation of cartilaginous core at stage 25–26 correlated with appearance of type-II-like collagen. The peripheral cells continued to make type I collagen. The shaft continued to synthesize type II collagen until stage 33; with the onset of ossification of tibial diaphysis, bone collagen, type I, was synthesized (Gross, 1974). These observations were corroborated by specific immunofluorescent localization of types I and II collagens (von der Mark *et al.*, 1976).

6. *Tooth*

Slavkin (1972) has reviewed the developmental aspects of tooth formation. The tooth cervical extracellular matrix is morphogenetically active in inducing odontogenesis. RNA-containing matrix vesicles have been implicated in tissue interactions (Slavkin, 1972). Kollar and Baird (1969) demonstrated the cardinal influence of dental papilla in determining the shape of the mouse embryonic incisor or molar tooth. Chemical heterogeneity of collagens has been investigated in the 26-day embryonic rabbit molar tooth rudiments. The isolated epithelium appeared to synthesize types I and IV collagens; the mesenchyme made only type I (Trelstad and Slavkin, 1974). The *in vitro* differentiation of ameloblasts is inhibited by the proline analog L-azetidine-2-carboxylic acid, hinting at the possible role of collagen in this phenomenon (Ruch *et al.*, 1974).

In conclusion, collagen appears to play an important role in organogenesis, although the mode of action is unclear. The experiments with collagenase and inhibitors of collagen synthesis constitute suggestive proof for the role of collagen. For more recent information about the influence of extracellular matrices on gene expression, the compendium by Slavkin and Greulich (1975) may be consulted.

IV. Collagen as "Permissive" Substratum

The growth of cells in tissue culture was extensive on collagen gels (Ehrmann and Gey, 1956). The differentiation of chick myoblasts to myotubes is generally promoted by conditioned medium, that is, tissue culture medium of muscle fibroblasts. The studies of Hauschka and Konigsberg (1966) revealed that collagen can substitute for conditioned medium. It is likely the muscle fibroblastic cells may provide the requisite collagen for myotube differentiation. Detailed investigation of the molecular specificity revealed that constituent $\alpha 1$ and $\alpha 2$ chains of rat skin collagen and cyanogen bromide fragments $\alpha 1$-CB7 and $\alpha 1$-CB8 support muscle differentiation; however, $\alpha 1$-CB3 and $\alpha 1$-CB6 are ineffective in this regard (Hauschka and White, 1972). These authors, in addition, observed that a component in horse serum may aid cell attachment to collagen. Klebe (1974) purified a collagen-dependent cell-attachment factor from calf serum and described a two-step process: binding of serum factor to collagen and a divalent cation-dependent reaction for cells to attach to collagen–serum factor complex. The relevance of these *in vitro* observations to phenomena in living animals remains to be demonstrated. There may be species differences. Hauschka (1974) observed that human muscle colonies differentiate in the absence of collagen. It is noteworthy that chick-conditioned medium permits differentiation of human clonal myoblasts.

Elsdale and Bard (1972) have described methods for the preparation of hydrated collagen lattices and the growth of virus-transformed rat cells therein. A microcrystalline form of collagen (Avitene) is an effective substratum for cell growth (Rucker *et al.*, 1972). In conclusion, collagenous substrata are effective in promoting cell growth in culture conditions for a variety of cells and cell differentiation in muscle.

V. Collagenous Matrix in the Solid State and Differentiation

Our discussion thus far was focused on early embryogenesis and organogenesis and collagens associated with these phases of development and morphogenesis. Considerations of the role of collagen as substratum *in vitro* revealed the important influence on myoblast differentiation. However, we must obtain direct evidence for the influence of collagen on cells *in vivo*. For technical reasons these studies are best done using tissues abundant in collagen, such as bone, cartilage, tendon, and teeth.

Autologous transplantation of canine incisor tooth to intramuscular sites resulted in new bone formation locally (Huggins *et al.*, 1936). Despite the importance of this finding, the progress in this area was slow until Urist (1965) made the discovery that demineralized, lyophilized segments of tooth and bone induced bone formation. We have developed simple techniques for the preparation of collagenous matrix from diaphyseal bone in the form of coarse powder of discrete particle size (Reddi and Huggins, 1972*a*). Transplantation of these powders results in bone formation at the site of implantation by a typical sequential endochondral ossification. Hemopoietic bone marrow develops *in situ* in the newly formed ossicle (Reddi and Huggins, 1975). This section will describe the results of these studies in relation to possible mode of action (Reddi, 1974, 1975*a*).

A. Preparation of the Matrix and Bioassay

Diaphyseal shafts from long bones of rat were cleaned of the medullary bone marrow and periosteum by repeated washing by stirring in water. These shafts were dehydrated in ethanol and ether and stored. Powders of the diaphyseal bone were obtained by pulverization in a micromill and sieved to discrete particle size. Demineralization was carried out routinely in 0.5 M HCl and washed in water, dehydrated in ethanol and ether, and dried overnight at 37°C (Reddi and Huggins, 1972*a*). A similar procedure was employed for teeth. However, when comparisons of other "soft" tissue matrices were made, in order to prevent swelling after demineralization, the samples were treated with 0.25 M $NaHCO_3$ briefly following the acid extraction. The remainder of the procedure was identical to that described for bone matrix. Demineralization of the bone matrix can also be accomplished by 0.5 M EDTA, pH 7.4 and 0.2 M sodium citrate buffer, pH3.5 (Reddi, 1974). Powders of demineralized rat diaphyseal matrix were devoid of calcium and inorganic phosphate; the organic phosphorus content was 12.9 ± 0.8 μmoles/g dry weight. The total anthrone-reactive sugars range between 0.7 and 1.0% by weight. The amino acid composition revealed 33% glycine and was generally similar to collagens (Table 1). This nonliving collagenous diaphyseal bone matrix has the remarkable property of inducing new bone formation in subcutaneous space of rat. Since the visible and biochemical characteristics of the responding fibroblasts were drastically changed, we refer to this phenomenon as transformation. The alteration of the phenotype of subcutaneous

TABLE 1
Amino Acid Composition of the Rat Diaphyseal Bone Matrix

Amino acid	Residues/1000
3-Hydroxyproline	1
4-Hydroxyproline	94
Aspartic acid	50
Threonine	21
Serine	42
Glutamic acid	79
Proline	106
Glycine	326
Alanine	105
Valine	21.4
Methionine	6.6
Isoleucine	8.7
Leucine	28
Tyrosine	9
Phenylalanine	16
Hydroxylysine	12
Lysine	21
Histidine	5.3
Arginine	48

fibroblasts in response to collagenous matrix is reminiscent of embryonic differentiation.

The bioassay of collagenous matrices was routinely performed *in vivo* in young rats. Under sterile conditions, the matrices were transplanted to a subcutaneous pocket prepared by blunt dissection. On designated days the implants, which now were discrete button-like transformation plaques, were obtained for histological and biochemical investigations (see Section V-B). The quantitative assay of alkaline phosphatase and ^{45}Ca incorporation 9–10 days after transplantation represented a standard bioassay and was performed routinely.

B. *Sequential Histological and Biochemical Changes*

Transplantation of collagenous diaphyseal bone matrix elicits sequential changes in the responding fibroblasts (Reddi and Huggins, 1972*a*, 1975).

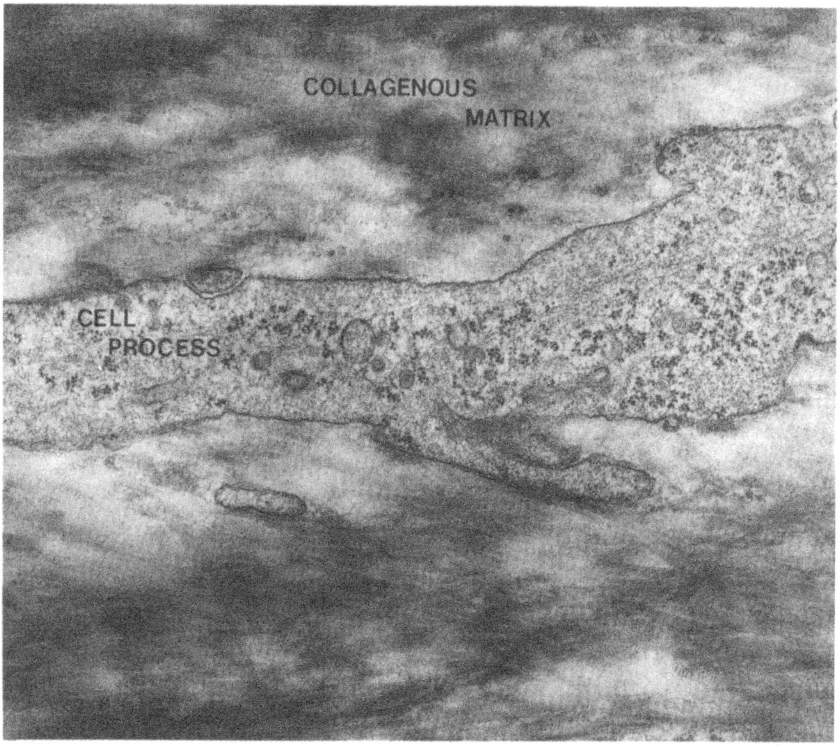

FIGURE 1. An electronmicrograph depicting a fibroblastic cell process in contiguity with the implanted collagenous diaphyseal bone matrix on day 5 (× 20,000). Obtained in collaboration with Dr. W. A. Anderson.

On day 1, a button-like, planoconvex transformation plaque was observed. Histologically, it consisted of implanted matrix and several polymorphonuclear leukocytes enmeshed in an irregular fibrin network. The leukocytes in the plaque were transient. On days 3 and 4, the transformation plaque was a conglomerate of fibroblasts and matrix, with an occasional leukocyte. The transient occurrence of polymorphonuclear leukocytes correlated with alkaline phosphatase, a characteristic enzyme in leukocytes (Reddi, 1975*b*).

Recent electron microscopic investigations of the initial interactions between matrix and fibroblasts revealed the close contiguity of fibroblast cell membrane to the collagenous matrix (Figure 1). There was no evidence of phagocytosis of collagenous matrix by fibroblasts (Reddi and Anderson, 1976).

The first chondroblasts were evident on day 5 and were abundant on days 7 and 8. The incorporation of ^{35}S into chondromucoprotein (Figure 2) correlated with chondrogenesis (Reddi and Huggins, 1972*a*; Reddi, 1974). With the appearance of invading capillaries on day 9, chondrolytic foci were seen in close proximity to blood vessels. Cell-free spaces were seen in regions of chondrolysis and new osteoblasts with characteristic basophilia were distinguished. On day 10, the cell-free spaces had enlarged and coalesced, forming an irregular honeycomb-like structure. New bone formed by appositional growth on the surface of newly formed calcified cartilage matrix and the implanted nonliving matrix. Alkaline phosphatase increased several-fold with new bone formation (Huggins and Urist, 1970). The incorporation of ^{32}P into acid-soluble bone mineral (Figure 2) exhibited excellent correlation with increases in alkaline phosphatase activity (Reddi, 1974). Further, the osteogenesis in response to implanted matrix also correlated with decline in lactic dehydrogenase (Reddi and Huggins, 1971), and increase in citrate (Reddi and Huggins, 1972*b*).

By day 12, the transformation plaque was an osseous conglomerate with cavernous sinuses. The first signs of extravascular colonies of hemopoiesis were evident; it was composed of: basophilic hemocytoblasts

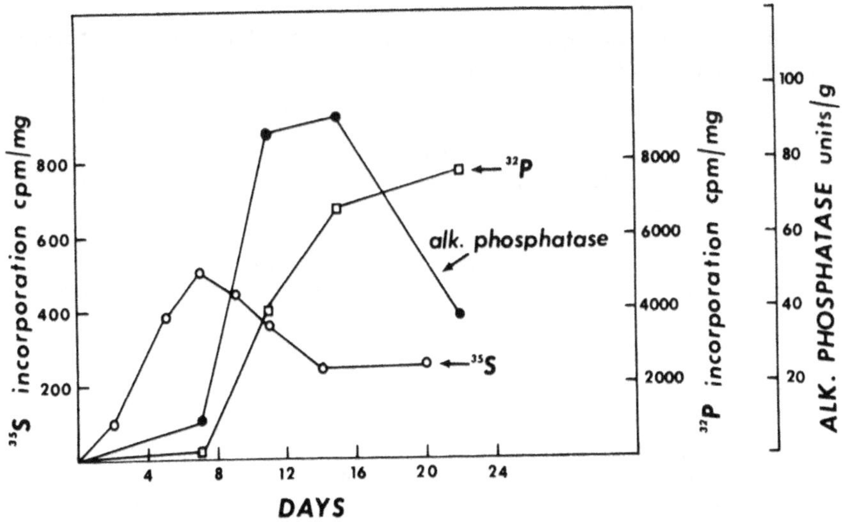

FIGURE 2. Sequential changes in ^{35}S incorporation into "total" chondromucoprotein, alkaline phosphatase activity, and ^{32}P incorporation into bone mineral during the transformation of fibroblasts (from Reddi, 1974).

and early stages of erythrocytic and granulocyte cells. The transformation ossicle was pinkish in the gross on day 14. There was progressive vascular confluence, and the vascular sinuses were engorged. On day 21, megakaryocytes were observed among the hemopoietic cells. Erythropoiesis was complete by day 24 as indicated by a plateau in values for incorporation of ^{59}Fe into heme (Reddi and Huggins, 1975; Reddi and Anderson, 1976). In addition, there was a difference in hemopoiesis that was dependent on anatomical site of the transformation ossicle. The ossicles were red and hemopoiesis extensive in the thoracic location; however, their abdominal counterparts were generally white and hemopoiesis was sparse (Reddi and Huggins, 1975).

C. *Specificity*

Mineralized matrix was ineffective in transformation of fibroblasts. The mineral phase has to be removed for the collagenous matrix to express its potential (Reddi, 1974). The transforming potency is equal in bone and tooth. Only certain collagenous matrices are active. It would appear there is a broad spectrum of activities among various collagenous matrices from rat. At one end of the spectrum is compact diaphyseal bone matrix and tooth, followed in decreasing order by: "spongy" bone from scapula and pelvis, membranous bone from calvarium; finally, tail tendon, Achilles tendon, cartilage, and skin matrices are devoid of transforming potency (Table 2) (Reddi, 1975*a*).

Table 2
Comparison of Transformation Potency of Matrices from Different Bones[a]

Bone matrix	Alkaline phosphatase (units/g)	^{45}Ca incorporation (cpm/mg)	Calcified cartilage
Humerus	56.0 ± 7.1[b]	7459 ± 1131	++++
Femur	80.2 ± 9.3	3749 ± 822	++++
Tibia	50.1 ± 4.5	4170 ± 1074	++++
Fibula	25.1 ± 3.3	2141 ± 422	++++
Pooled diaphysis	66.8 ± 6.3	5634 ± 1057	++++
Calvarium	5.3 ± 0.9	530 ± 376	+
Scapula	11.2 ± 3.2	1715 ± 1110	+
Pelvis	8.3 ± 2.1	689 ± 333	+

[a] From Reddi (1975*a*).
[b] ± standard error of mean of 8 observations.

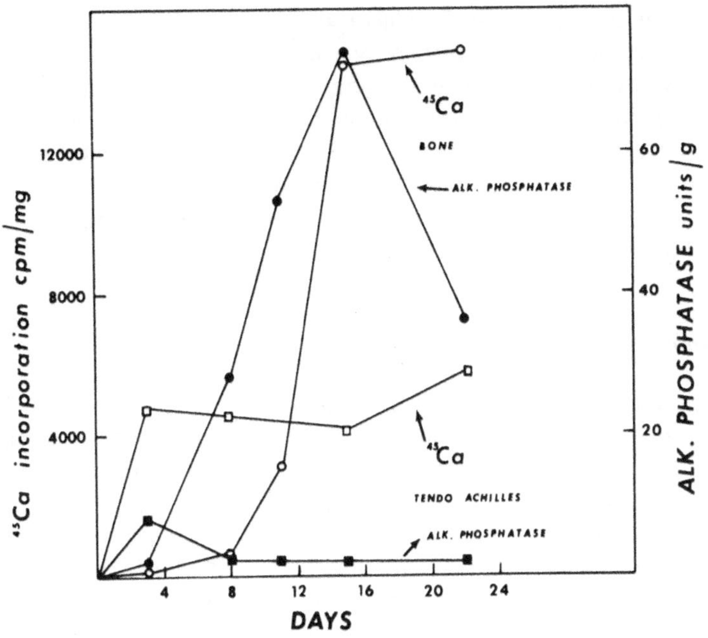

FIGURE 3. Dichotomy between calcification of tendon matrix and ossification of bone matrix. The large increase in alkaline phosphatase preceding ossification is not seen with calcification.

When matrices from Achilles tendon and diaphyseal bone matrix were transplanted in rats, the former calcified rapidly; the latter induced bone and ossified as indicated by ^{45}Ca incorporation into bone mineral (Figure 3). Also seen was the large increase in alkaline phosphatase preceding ossification. Thus, there is a dichotomy between calcification and ossification. Also, it is noteworthy under normal *in situ* conditions tendon does not calcify and bone matrix calcifies; however, when 0.5 M HCl-treated matrix was transplanted subcutaneously the reverse was observed. There was species specifity among the matrices tested in allogeneic and xenogeneic transplants (Reddi, 1974). In conclusion, the transforming property is restricted to only bone and tooth among the matrices tested thus far.

D. Humoral and Nutritional Influences

The transformation of fibroblasts by collagenous matrix is hormone-dependent (Reddi, 1974). Hypophysectomy has profound influence on osteogenesis as assessed by ^{32}P incorporation into bone mineral. Bovine growth hormone administration alleviates the inhibitory influence of hypophysectomy (Reddi, 1974). Thompson and Urist (1970), observed the stimulating influence of calcitonin on bone formation. Experiments designed to investigate the influence of severe dietary deprivation by rats fed exclusively with sucrose ration revealed an obligatory transformation of fibroblasts to bone. However, hemopoiesis was markedly impaired (Reddi and Huggins, 1974*a*).

E. Geometry

There is a profound influence of geometry of the collagenous matrix on the sequential transformation of fibroblasts (Reddi and Huggins, 1973; Reddi, 1974, 1975*a*). The temporal sequence is altered by the shape of the collagenous matrix. Using incisor teeth of conoid tips and cylindrical tubes, the cardinal influence of shape and oxygen tension in differentiation of cartilage and bone was demonstrated (Reddi, 1974).

The size of the collagenous matrix has important bearing on the quantitative yield of transformation products, as assessed by alkaline phosphatase activity and ^{45}Ca incorporation, 9 days after transplantation of collagenous diaphyseal bone matrix. There is an optimal size requirement in the range of 74–840 μm (Figure 4) and is most likely due to the optimum surface area-to-volume ratio of the implanted matrix. Fine powders (44–74 μm) are rather weak in transformation potency and yielded sporadic and sparse islands of calcified cartilage and bone. These observations deserve additional comments. Animal cells in tissue culture must attach to a solid surface such as glass or plastic in order to spread and divide, and this has been termed "anchorage dependence" (Stoker *et al.*, 1968). Using BHK 21 clone C13 cells, they observed that glass fibrils 250–750 μm in length, and glass beads ranging in diameter between 50 and 90 μm had attached colonies of cells; however, there were no colonies in the presence of silica fragments 5 μm in size. Maroudas (1972) found the optimal size for cell growth to be 50-μ-diameter glass beads. Maroudas *et al.* (1973) observed a correlation between the length of asbestos fibers and induction of lung mesotheliomas in rats. Fibers less than 20 μm in length did not induce mesotheliomas. From these seemingly disparate observations, one conclusion is almost inevitable: the size of the various materials must be at least 44 μm to elicit a fibroblastic change. In other words, the size of the collagenous matrix and other fibers must be larger than the responding cells. After all, collagen is the extracellular fibrillar matrix with which living cells interact. It is, therefore, noteworthy that

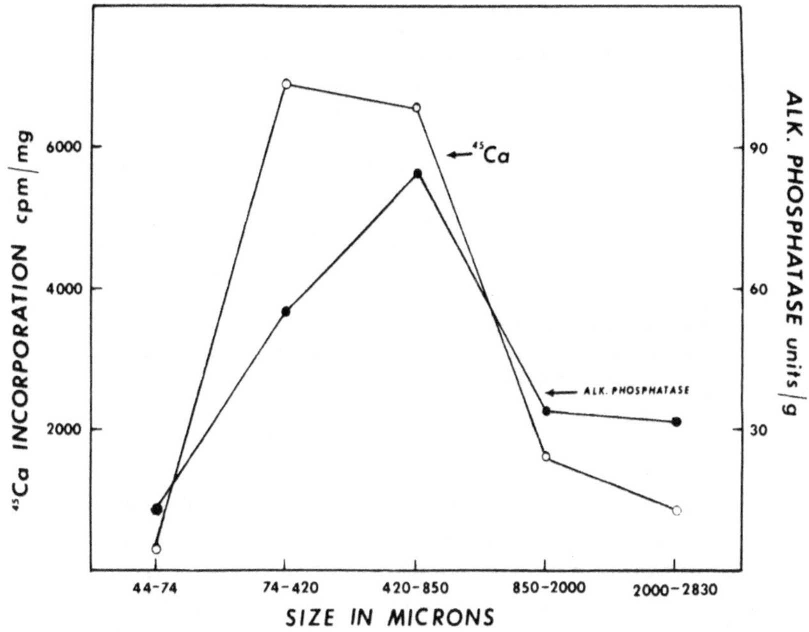

FIGURE 4. Influence of geometry of the collagenous matrix on transformation. The optimal size is in the range of 74–850 μm for transformation.

there are interesting parallels between behavior of cells *in vitro* and *in vivo* and that our experiment with collagenous matrices is an example of anchorage dependence of cells in the living organism.

F. Surface Charge Characteristics

During the course of investigations on the influence of pretreatment of matrix by endogenous plasma enzymes, we observed that transforming potency was abolished by heparinized blood plasma (Huggins and Reddi, 1973). In addition, the blood plasma coagulated. Other studies revealed that collagenous bone matrix bound heparin from solution and could explain the coagulative potency of matrix. This property was also shared by "soft" tissue collagenous matrices. The inhibitory influence of various polyanions was tested, and among the compounds tested, heparin, polyvinyl sulfonate, and dextran sulfate were inhibitory (Table 3). Among other polyanions tested,the following were not inhibitory: DNA, RNA, chondroitin sulfate, dermatan sulfate, keratan sulfate, and hyaluronic acid. There

TABLE 3
Inhibitory Effects of Polyanions on Bone Matrix-Induced Differentiation of Fibroblasts[a]

	Alkaline phosphatase (units/g)	^{45}Ca incorporation (cpm/mg)
Control	49.1 ± 6.6[b]	5243 ± 1147
Heparin 0.1%	6.1 ± 1.0	69 ± 10
Dextran sulfate 1.0%	5.8 ± 0.8	82 ± 17
Polyvinylsulfonate 1.0%	5.0 ± 0.9	83 ± 15

[a] From Reddi (1975*a*).
[b] ± standard error of mean of 8 observations.

is an excellent correlation between the degree of sulfation of various heparan sulfates and their inhibitory effects on collagenous bone matrix-induced differentiation of fibroblasts (Table 4). These experiments suggest the possible interactions between collagenous matrix and anionic components on cell membrane (Kojima and Yamagata, 1971; Kraemer, 1971). It is conceivable that alteration of surface charge characteristics on the matrix is inhibitory to transformation (Reddi and Huggins, 1974*b*; Reddi, 1975*a*). It was also observed that there is a pH-dependent inhibition of fibroblast transformation by an anionic azo dye, Evans blue.

TABLE 4
Correlation between the Sulfate Content and Inhibitory Effects of Heparan Sulfates on Diaphyseal Bone Matrix-Induced Differentiation of Fibroblasts[a]

	Molar ratio sulfate/hexosamine	Alkaline phosphatase (units/g)	^{45}Ca incorporation (cpm/mg)
Control	—	47.5 ± 6.5[b]	5027 ± 1413
Heparan sulfate (low sulfate)	0.54	56.5 ± 6.5	4126 ± 941
Heparan sulfate (medium sulfate)	1.00	30.9 ± 8.5	2725 ± 1398
Heparan sulfate (high sulfate)	1.25	18.0 ± 3.5	2128 ± 688
Heparin	2.33	4.2 ± 1.2[c]	80 ± 27[c]

[a] From Reddi (1975*a*).
[b] ± standard error of mean of 9 observations.
[c] Significant difference, $P < 0.01$.

In addition, under the brief time periods examined, the dye was restricted to the surface of the powdered matrix (Reddi and Huggins, 1974*b*).

The interaction between proteoglycans and collagen is well known (Mathews, 1970). In our studies collagenous bone matrix in solid state interacted with and bound acid mucopolysaccharides in solution. Experiments with [^{35}S]heparin revealed the electrostatic nature of binding of glycosaminoglycans to collagen. Whether steric factors were involved in the observed inhibitory influence of heparin, in addition to electrostatic interactions, is not known. It would appear that one of the early steps in bone matrix-induced differentiation of fibroblasts is an initial electrostatic interaction between collagenous matrix and fibroblast membrane. The inhibitory effects of heparin on matrix-induced bone formation are not without clinical implications. Patients with cardiovascular problems on long-term heparin therapy exhibited spontaneous fractures of ribs and vertebrae, possibly due to diminished osteogenesis, among other reasons (Griffith *et al.*, 1965).

G. Mechanism of Action: Matrix–Membrane Interactions

What is the mode of action of collagenous diaphyseal bone matrix on alteration of gene expression in responding fibroblasts? Although the precise molecular mechanisms underlying this phenomenon are not known, the various characteristics of the process may be discussed.

The geometry of the collagenous matrix is crucial. The dimensions of the matrix must be larger than the responding fibroblasts. It would appear that the matrix may have to interact over large "domains" in the cell surface. This process may involve multiple interactions with several "receptors" on the cell surface and/or membrane. The surface charge is critical, and perturbation by pretreatment with polyanions is deleterious to transformation of fibroblasts. The optimal positive charge and charge distribution on the matrix results in initial electrostatic interaction with cell surface. This may aid in the proper orientation of membrane structures to the matrix prerequisite for more specific and stronger interaction with cellular receptors. The surface charge is a necessary, but not sufficient, condition for transformation, as other collagenous matrices with identical dye-binding properties were ineffective in transformation. The specific matrix–membrane interaction sets into motion a train of interconnected cellular reactions culminating in the appearance of new phenotypes. How are the events at cell-surface membrane transduced to the nucleus? It should be recalled that there was no evidence of phagocytosis of the matrix by responding fibroblasts (Reddi and Anderson, 1976).

TABLE 5

Effect of Treatment of Bone Matrix with Buffers on Its Competence to Transform Fibroblasts[a]

		Plaque		
Buffer	pH	Alkaline phosphatase (units/g)	^{45}Ca (cpm/mg)	Calcified cartilage
Acetic acid–acetate	3.6	43.2 ± 4.3[b]	8496 ± 2864	++++
Tris-HCl	7.4	35.2 ± 5.3	8574 ± 2074	++++
Gly-NaOH	10.8	34.0 ± 2.2	6190 ± 1456	++++
Gly-NaOH	11.4	26.7 ± 4.6	2301 ± 633	++++
Gly-NaOH	12.0	7.5 ± 1.7	542 ± 144	++
Gly-NaOH	12.8	1.32 ± 0.33	90 ± 7	0

[a] From Reddi and Huggins, (1974*a*).
[b] ± standard error of mean of 8 observations.

TABLE 6

Influence of Pretreatment of Bone Matrix by Various Enzymes on Its Transforming Potency

Enzyme pretreatment[a]	Alkaline phosphatase (units/g)	^{45}Ca incorporation (cpm/mg)
Control	64.8 ± 5.5	6478 ± 730
DNAase	40.7 ± 4.3	5575 ± 1140
RNAase	61.6 ± 7.5	10820 ± 2070
Trypsin	0.5 ± 0.1	not detectable
Papain	1.1 ± 0.1	not detectable
Pronase	0.8 ± 0.1	not detectable
Collagenase[b]	—	—

[a] Collagenous bone matrix (74–420 μm) was pretreated with the specified enzymes at pH 7.4 in the presence of appropriate metal ions at 25° for 3 hr (enzyme to matrix ratio 1:100 w/w). Thereafter the matrix was washed thrice with respective buffers. In order to prevent possible electrostatic binding of enzyme protein to the matrix in the solid state an additional wash with 1.5 M NaCl was instituted. Then the matrix was dehydrated in ethanol and ether and bioassayed *in vivo* in the routine manner.
[b] Pretreatment of the collagenous rat bone matrix by collagenase from *Clostridium histolyticum* rendered the matrix soluble on transplantation. All the transplants underwent dissolution *in vivo*.

TABLE 7
Influence of Periodate Oxidation of the Collagenous Diaphyseal Bone Matrix on Transformation of Fibroblasts

Treatment	Alkaline phosphatase (units/g)	^{45}Ca incorporation (cpm/mg)
Control	27.1 ± 5.4	4461 ± 1066
Sodium periodate, 0.1 M[a]	3.7 ± 0.3	165 ± 23

[a] The diaphyseal bone matrix (74–420 μm) was treated with 0.1 M sodium periodate in 0.1 M sodium phosphate buffer, pH 6.0, for 1 hr at 25°C, with constant stirring. Thereafter the matrix was washed thrice with the sodium phosphate buffer, dehydrated in ethanol and ether, and bioassayed in routine manner.

What is the chemical basis of matrix specificity? Both bone and tendon collagens are type I, yet only the former transforms. Tendon swells in dilute acid; bone does not (Glimcher and Krane, 1968). The transforming property is acid stable but alkali labile (Table 5) (Reddi and Huggins, 1974*b*). A systematic study of the influence of various exogenous enzymes revealed that DNA and RNA are probably not involved (Table 6). The importance of the proteinaceous component is revealed by the inhibitory action of various proteolytic enzymes. It would appear that both helical and nonhelical regions of the collagen are essential for biological response. Periodate oxidation of short duration (1 hr) abolishes the transforming potency (Table 7). This hints at the possible role of carbohydrates in this phenomenon. Recent experiments with pretreatment of the matrix with galactose oxidase revealed the importance of galactose in the matrix (A. H. Reddi, unpublished observations). In view of this, there is an analogy with the collagen-mediated platelet interaction (Jamieson *et al.*, 1971; Chesney *et al.*, 1972; Kang *et al.*, 1974). The importance of cell-surface glycosyltransferases in cell recognition is well known (Roth, 1973). Other experiments designed to investigate the influence of sulfhydryl reducing agents on the matrix revealed the importance of disulfide-bonded regions (Table 8). It is likely that either molecules covalently bound to the collagenous matrix are released or the conformation of a critical region is perturbed. Studies currently in progress will help distinguish between these possibilities.

Among other enzymes tested, acid and alkaline phosphatases, hyaluronidase, and neuraminidase have no effect on the matrix-induced fibroblast transformation. It is therefore unlikely that matrix phospho- and sialoproteins (Herring, 1972) are involved in this process. Urist and

TABLE 8
Influence of Sulfhydryl Reducing Agents on the Collagenous Diaphyseal Bone Matrix-Induced Differentiation of Fibroblasts

Treatment[b]	Concentration	Alkaline phosphatase units/g[a]	^{45}Ca incorporation cpm/mg
Control	—	48.4 ± 7.4	6740 ± 2060
Dithiothreitol	1 mM	23.1 ± 4.3	2800 ± 694
Dithiothreitol	5 mM	14.4 ± 2.8	1400 ± 620
Dithiothreitol	10 mM	8.7 ± 1.8	340 ± 97
Dithiothreitol	25 mM	2.6 ± 0.3	76 ± 15
2-Mercaptoethanol	100 mM	2.4 ± 0.3	58 ± 12

[a] Standard error of 8 observations.
[b] Collagenous diaphyseal bone matrix (74–420 μm) was pretreated with the specified reagent in 0.1 M tris-HCl buffer, pH 7.4 for 1 hr at 25°C with stirring. The solutions were flushed with nitrogen prior to use. The matrix was washed thrice with the buffer and dehydrated in ethanol and ether and bioassayed.

Iwata (1973) proposed that a bone morphogenetic protein is involved in inducing bone, but provided no compelling evidence. Clearly, future investigations will focus on defining the precise matrix–membrane interactions involved prior to alteration of gene expression.

VI. Collagen and Cancer

If collagen is implicated in normal cell differentiation and morphogenesis, it is possible that it has significance for tumors. After all, cancer is a disease of differentiation. In metastases, the extracellular matrix architecture is intimately involved. It is conceivable that epithelial carcinoma cells loose their "moorings" to basement membrane and drift. It is noteworthy that the new niche of a metastatic cell is often a bony skeletal site. The osteolytic and osteosclerotic lesions are well known. Is there a propensity for tumor cells for collagenous bone matrix?

Smith *et al.* (1975) identified type II collagen in a transplantable chondrosarcoma. In tissue culture randomly oriented Chinese hamster cells were converted to fibroblast-like cells by dibutyryl cAMP with accompanying increase in collagen synthesis (Hsie *et al.*, 1971). Addition of dibutyryl cAMP to Kirsten sarcoma virus-transformed 3T3 cells, increased collagen synthesis and altered the morphology (Peterkofsky and Prather,

1974). Are there any transitions in the type of collagen synthesized during such alterations?

VII. *Biological Implications and Conclusions*

A. *Specification of Positional Information*

Wolpert (1969) has contributed much to the concept of spatial pattern formation, broadly referred to as specification of positional information. While considerations of gradients of diffusible molecules aiding interpretation of positional information have been made, it is important to bear in mind the other alternatives. It is likely that insoluble molecules, such as collagen, may play a crucial role in specifying positional information. Collagen may function in the form of a supramolecular assembly in conjunction with certain associated molecules. In order for the collagenous matrix to possess the requisite specificity for diverse process, it must have a potential for modification of structure. As pointed out before (Section III-C), the observed molecular heterogeneity confers the needed properties. Also, physical properties of collagen lend another dimension to control of morphogenetic processes. Collagen is piezoelectric and pyroelectric in structure (Athenstaedt, 1970; Bassett, 1971; Liboff and Shamos, 1973; Fukada, 1974). In view of this, subtle electrical charge alterations may be influencing the cell surface and ultimately the nucleus. Taken together, collagen is an ideal molecule for imparting positional information. The importance of extracellular matrix in differentiation has been pointed out before several times (Baitsell, 1925; Weiss, 1933; Grobstein, 1954; Gross, 1956). Now we are on the threshold of some direct experiments demonstrating matrix–membrane interaction.

B. *Fracture Healing*

One of the marvels of medical science is the precise healing of fractures. Despite a long history of study of this phenomenon, there is no basic mechanism to explain the local nature of the healing process. The sequential events in collagenous matrix-induced bone formation (see Section V) is similar to fracture healing: the early inflammation-like response, clot formation, cartilage callus, and finally ossification and remodeling of new bone (Ham and Harris, 1971). It is likely that the collagenous matrix has the necessary positional information for fracture healing. This hypothesis deserves further study.

C. *Developmental Anomalies*

The emerging knowledge concerning the role of collagen in normal cell differentiation may provide a rational basis for prevention of developmental anomalies. The horrendous problems of craniofacial anomalies is well known. What is the possible role of collagen in craniofacial morphogenesis? Is collagen involved in migration of neural crest cells? For example, administration of β-aminopropionitrile (a collagen cross-linking inhibitor) at a critical stage of gestation in the rat resulted in cleft palate in the offspring (Pratt and King, 1972). Future studies are expected to answer many of these questions.

ACKNOWLEDGMENTS

I wish to thank Mrs. L. B. Altschul and Mr. Ken Anderson for their assistance in the preparation of this article. Our experiments have been supported by Grants from The National Foundation–March of Dimes, American Cancer Society, The Jane Coffin Childs Memorial Fund for Medical Research, and The National Institutes of Health, U.S. Public Health Service.

References

Alescio, T., 1973, Effect of a proline analogue, azetidine-2-carboxylic acid, on the morphogenesis *in vitro* of mouse embryonic lung, *J. Embryol. Exp. Morphol.* **29:**439.

Alescio, T., and Cassini, A., 1962, Induction *in vitro* of tracheal buds by pulmonary mesenchyme grafted on tracheal epithelium, *J. Exp. Zool.* **150:**83.

Athenstaedt, H., 1970, Permanent longitudinal electric polarization and pyroelectric behaviour of collagenous structures and nervous tissues in man and other vertebrates, *Nature* **228:**830.

Baitsell, G. A., 1925, On the origin of the connective-tissue ground substance in the chick embryo, *Q. J. Microsc. Sci.* **69:**571.

Bassett, C. A. L., 1971, Biophysical principles affecting bone structure, *in: The Biochemistry and Physiology of Bone* (G. H. Bourne, ed.), Vol. 3, pp. 1–76, Academic Press, New York.

Bernfield, M. R., 1970, Collagen synthesis during epetheliomesenchymal interactions, *Dev. Biol.* **22:**213.

Bernfield, M. R., and Wessells, N. K., 1970, Intra- and extracellular control of epithelial morphogenesis. *Dev. Biol., Suppl.* **4:**195.

Bradley, K. H., McConnel, S. D., and Crystal, R. G., 1974*a*, Lung collagen composition and synthesis, *J. Biol. Chem.* **249**:2674.

Bradley, K., McConnell-Breul, S., and Crystal, R. G., 1974*b*, Lung collagen heterogeneity, *Proc. Natl. Acad. Sci.* **71**:2828.

Chesney, C. M., Harper, E., and Colman, R. W., 1972, Critical role of the carbohydrate side chains of collagen in platelet aggregation, *J. Clin. Invest.* **51**:2693.

Chung, E., and Miller, E. J., 1974, Collagen polymorphism: Characterization of molecules with the chain composition $[\alpha 1(\mathrm{III})]_3$ in human tissues, *Science* **183**:1200.

Cohen, A. M., and Hay, E. D., 1971, Secretion of collagen by embryonic neuroepithelium at the time of spinal cord-somite interaction, *Dev. Biol.* **26**:578.

Conrad, G. W., 1970, Collagen and mucopolysaccharide biosynthesis in the developing chick cornea, *Dev. Biol.* **21**:292.

Coulombre, A. J., 1965, Eye, *in: Organogenesis* (R. L. DeHaan and H. Ursprung, eds.), pp. 219–251, Holt, Rinehart and Winston, New York.

Dische, Z., 1970, Collagen of embryonic type in the vertebrate eye and its relation to carbohydrates and subunit structure of tropocollagen, *Dev. Biol., Suppl.* **4**:164.

Dodson, J. W., 1963, On the nature of tissue interactions in embryonic skin, *Exp. Cell Res.* **31**:233.

Dodson, J. W., and Hay, E. D., 1974, Secretion of collagen by corneal epithelium. Effect of the underlying substratum on secretion and polymerization of epithelial products, *J. Exp. Zool.* **189**:51.

Edds, M. V., Jr., 1958, Development of collagen in the frog embryo. *Proc. Natl. Acad. Sci. U.S.A.* **44**:296.

Ehrmann, R. L., and Gey, G. O., 1956, The growth of cells on a transparent gel of reconstituted rat-tail collagen, *J. Natl. Cancer Inst.* **16**:1375.

Elsdale, T., and Bard, J., 1972, Collagen substrata for studies on cell behavior, *J. Cell Biol.* **54**:626.

Epstein, E. H., Jr., 1974, $[\alpha 1(\mathrm{III})]_3$ Human skin collagen. Release by pepsin digestion and preponderance in fetal life, *J. Biol. Chem.* **249**:3225.

Fleischmajer, R., and Billingham, R. E., 1968, Epithelial-Mesenchymal Interactions, Williams and Wilkins Co., Baltimore.

Fukada, E., 1974, Piezoelectric properties of biological macromolecules, *Adv. Biophys.* **6**:121.

Glimcher, M. J., and Krane, S. M., 1968, The organization and structure of bone and the mechanism of calcification, *in: Treatise on Collagen* (B. S. Gould, ed.), Vol. 2, Part B, pp. 67–251, Academic Press, New York.

Goetinck, P. F., and Sekellick, P. F., 1972, Observations on collagen synthesis, lattice formation, and morphology of scaleless and normal embryonic skin, *Dev. Biol.* **28**:636.

Golob, R., Chetsanga, C. J., and Doty, P., 1974, The onset of collagen synthesis in sea urchin embryos, *Biochim. Biophys. Acta* **349**:135.

Green, H., Goldberg, B., Schwartz, M., and Brown, D. D., 1968, The synthesis of collagen during the development of *Xenopus laevis*, *Dev. Biol.* **18**:391.

Griffith, G. C., Nichols, G., Asher, J. D., and Flanagan, B., 1965, Heparin osteoporosis, *J. Am. Med. Assoc.* **193**:85.

Grobstein, C., 1954, Tissue interaction in the morphogenesis of mouse embryonic rudiments *in vitro*, *in: Aspects of Synthesis and Order in Growth* (D. Rudnick, ed.), pp. 233–256, Princeton University Press, Princeton, New Jersey.

Grobstein, C., 1967, Mechanisms of organogenetic tissue interaction, *Natl. Cancer Inst. Monogr.* **26**:279.

Grobstein, C., and Cohen, J., 1965, Collagenase: Effect on the morphogenesis of embryonic salivary epithelium *in vitro, Science* **150:**626.

Gross, J., 1956, The behavior of collagen units as a model in morphogenesis, *J. Biophys. Biochem. Cytol.* **2**(Suppl):261.

Gross, J., 1974, Collagen Biology: Structure, degradation and disease, *Harvey Lect.* **68:**351.

Ham, A. W., and Harris, W. R., 1971, Repair and transplantation of bone, *in: The Biochemistry and Physiology of Bone* (G. H. Bourne, ed.), Vol. 3, pp. 337–399, Academic Press, New York.

Hauschka, S. D., 1974, Clonal analysis of vertebrate myogenesis II. Environmental influences upon human muscle differentiation, *Dev. Biol.* **37:**329.

Hauschka, S. D., and Konigsberg, I. R., 1966, The influence of collagen on the development of muscle clones, *Proc. Natl. Acad. Sci. U.S.A.* **55:**119.

Hauschka, S. D., and White, N. K., 1972, Studies of myogenesis *in vitro*, *in: Research in Muscle Development and the Muscle Spindle* (B. Q. Banker, R. J. Przbylski, J. P. van Der Meulen, and M. Victor, eds.), pp. 53–71, Excerpta Medica, Amsterdam.

Hay, E. D., and Dodson, J. W., 1973, Secretion of collagen by corneal epithelium I. Morphology of the collagenous products produced by isolated epithelia grown on frozen-killed lens, *J. Cell Biol.* **57:**190.

Herring, G. M., 1972, The organic matrix of bone, *in: The Biochemistry and Physiology of Bone* (G. H. Bourne, ed.), Vol. 1, pp. 127–189, Academic Press, New York.

Hsie, A. W., Jones, C., and Puck, T. T., 1971, Further changes in differentiation state accompanying the conversion of chinese hamster cells to fibroblastic form by dibutyryl adenosine cyclic 3′:5′-monophosphate and hormones, *Proc. Natl. Acad. Sci. U.S.A.* **68:**1648.

Huggins, C. B., and Reddi, A. H., 1973, Coagulation of blood plasma of guinea pig by the bone matrix, *Proc. Natl. Acad. Sci. U.S.A.* **70:**929.

Huggins, C. B., and Urist, M. R., 1970, Dentin matrix transformation: Rapid induction of alkaline phosphatase and cartilage, *Science* **167:**896.

Huggins, C. B., McCarroll, H. R., and Blocksom, B. H., 1936, Experiments on the theory of osteogenesis. The influence of local calcium deposits on ossification; the osteogenic stimulus of epithelium, *Arch. Surg.* **32:**915.

Jackson, S. F., 1968, The morphogenesis of collagen, *in: Treatise on Collagen* (B. S. Gould, ed). Vol. 2, Part B, pp. 1–66, Academic Press, New York.

Jamieson, G. A., Urban, C. L., and Barber, A. J., 1971, Enzymatic basis for platelet:collagen adhesion as the primary step in haemostasis, *Nature (London), New Biol.* **234:**5.

Johnson, R. C., Manasek, F. J., Vinson, W. C., and Seyer, J. M., 1974, The biochemical and ultrastructural demonstration of collagen during early heart development, *Dev. Biol.* **36:**252.

Kallman, F., and Grobstein, C., 1964, Fine structure of differentiating mouse pancreatic exocrine cells in transfilter culture, *J. Cell Biol.* **20:**399.

Kallman, F., and Grobstein, C., 1965, Source of collagen at epitheliomesenchymal interfaces during inductive interactions, *Dev. Biol.* **11:**169.

Kang, A. H., Beachey, E. H., and Katzman, R. L., 1974, Interaction of an active glycopeptide from chick skin collagen (α1-CB5) with human platelets, *J. Biol. Chem.* **249:**1054.

Kefalides, N. A., 1973, Structure and biosynthesis of basement membranes, *Int. Rev. Connect. Tissue Res.* **6:**63.

Klebe, R. J., 1974, Isolation of a collagen-dependent cell attachment factor, *Nature* **250**:248.

Klose, J., and Flickinger, R. A., 1971, Collagen synthesis in frog embryo endoderm cells, *Biochim. Biophys. Acta* **232**:207.

Kojima, K., and Yamagata, T., 1971, Glycosaminoglycans and electrokinetic behavior of rat ascites hepatoma cells, *Exp. Cell Res.* **67**:142.

Kollar, E. J., and Baird, G. R., 1969, The influence of the dental papilla on the development of tooth shape in embryonic mouse tooth germs, *J. Embryol. Exp. Morphol.* **21**:131.

Kraemer, P. M., 1971, Heparan sulfates of cultured cells I. Membrane-associated and cell-sap species in chinese hamster cells, *Biochemistry* **10**:1437.

Kratochwil, K., 1972, Tissue interactions during embryonic development: General properties, *in: Tissue Interactions in Carcinogenesis* (D. Tarin, ed.), pp. 1–47, Academic Press, New York.

Lash, J. W., 1968, Chondrogenesis: Genotypic and phenotypic expression, *J. Cell. Physiol.* **72**:35.

Levitt, D., Ho, P. L., and Dorfman, A., 1974, Differentiation of cartilage, *in: Cell Surface in Development* (A. A. Moscona, ed.), pp. 101–125, John Wiley, New York.

Liboff, A. R., and M. H. Shamos, 1973, Solid state physics of bone, *in: Biological Mineralization* (I. Zipkin, ed.), pp. 335–395, John Wiley, New York.

Linsenmeyer, T. F., Trelstad, R. L., and Gross, J., 1973*a*, The collagen of chick embryonic notochord, *Biochem. Biophys. Res. Commun.* **53**:39.

Linsenmeyer, T. F., Toole, B. P., and Trelstad, R. L., 1973*b*, Temporal and spatial transitions in collagen types during embryonic chick limb development, *Dev. Biol.* **35**:232.

Maroudas, N. G., 1972, Anchorage dependence: Correlation between amount of growth and diameter of bead for single cells grown on individual beads, *Exp. Cell Res.* **74**:337.

Maroudas, N. G., O'Neill, C. H., and Stanton, M. F., 1973, Fibroblast anchorage in carcinogenesis by fibres, *Lancet* **1973**:807.

Mathews, M. B., 1970, The interactions of proteoglycans and collagen—model systems, *in: Chemistry and Molecular Biology of the Intercellular Matrix* (E. A. Balazs, ed.), pp. 1155–1169, Academic Press, New York.

Mayne, R., Vail, M. S., and Miller, E. J., 1975, Analysis of changes in collagen biosynthesis that occur when chick chondrocytes are grown in 5-bromo-2′-deoxyuridine, *Proc. Natl. Acad. Sci. U.S.A.* **72**:4511.

McClain, P. E., 1974, Characterization of cardiac muscle collagen. Molecular heterogeneity, *J. Biol. Chem.* **249**:2303.

Meier, S., and Hay, E. D., 1974, Control of corneal differentiation by extracellular materials. Collagen as promoter and stabilizer of epithelial stroma production, *Dev. Biol.* **38**:249.

Miller, E. J., 1973, A review of biochemical studies on the genetically distinct collagens of the skeletal system, *Clin. Orthop. Relat. Res.* **92**:260.

Miller, E. J., and Mathews, M. B., 1974, Characterization of notochord collagen as a cartilage-type collagen, *Biochem. Biophys. Res. Commun.* **60**:424.

Miller, E. J., and Matukas, V. J., 1969, Chick cartilage collagen: A new type of α-1 chain not present in bone or skin of the species, *Proc. Natl. Acad. Sci. U.S.A.* **64**:1264.

Minor, R. R., 1973, Somite chondrogenesis: A structural analysis, *J. Cell Biol.* **56**:27.

O'Hare, M. J., 1972, Aspects of spinal cord induction of chondrogenesis in chick embryo somites, *J. Embryol. Exp. Morphol.* **27**:235.

Peterkofsky, B., and Prather, W. B., 1974, Increased collagen synthesis in Kirsten Sarcoma virus-transformed BALB 3T3 cells grown in the presence of dibutyryl cyclic AMP, *Cell* **3**:291.

Pinnell, S. R., Fox, R., and Krane, S. M., 1971, Human collagens: Differences in glycosylated hydroxylysines in skin and bone, *Biochim. Biophys. Acta* **229**:119.

Pratt, R. M., Jr., and King, C. T. G., 1972, Inhibition of collagen cross-linking associated with β-aminopropionitrile-induced cleft palate in the rat, *Develop. Biol.* **27**:322.

Pucci-Minafra, I., Casana, C., and LaRosa, C., 1972, Collagen synthesis and spicule formation in sea urchin embryos, *Cell Differ.* **1**:157.

Reddi, A. H., 1974, Bone matrix in the solid state: Geometric influence on differentiation of fibroblasts, *Adv. Biol. Med. Phys.* **15**:1.

Reddi, A. H., 1975*a*, Collagenous bone matrix and gene expression in fibroblasts, *in: Extracellular Matrix Influences on Gene Expression* (H. Slavkin and R. C. Greulich, eds.), pp. 619–625, Academic Press, New York.

Reddi, A. H., 1975*b*, The matrix of rat calvarium as transformant of fibroblasts, *Proc. Soc. Exp. Biol. Med.* **150**:324.

Reddi, A. H., and Anderson, W. A., 1976, Collagenous bone matrix-induced endochondral ossification and hemopoiesis, *J. Cell. Biol.* **69**:557.

Reddi, A. H., and Huggins, C. B., 1971, Lactic/malic dehydrogenase quotients during transformation of fibroblasts into cartilage and bone, *Proc. Soc. Exp. Biol. Med.* **137**:127.

Reddi, A. H., and Huggins, C. B., 1972*a*, Biochemical sequences in the transformation of normal fibroblasts in adolescent rats, *Proc. Natl. Acad. Sci. U.S.A.* **69**:1601.

Reddi, A. H., and Huggins, C. B., 1972*b*, Citrate and alkaline phosphatase during transformation of fibroblasts by the matrix and minerals of bone, *Proc. Soc. Exp. Biol. Med.* **140**:807.

Reddi, A. H., and Huggins, C. B., 1973, Influence of geometry of transplanted tooth and bone on transformation of fibroblasts, *Proc. Soc. Exp. Biol. Med.* **143**:634.

Reddi, A. H., and Huggins, C. B., 1974*a*, Obligatory transformation of fibroblasts by bone matrix in rats fed sucrose ration, *Proc. Soc. Exp. Biol. Med.* **145**:475.

Reddi, A. H., and Huggins, C. B., 1974*b*, Cyclic electrochemical inactivation and restoration of competence of bone matrix to transform fibroblasts, *Proc. Natl. Acad. Sci. U.S.A.* **71**:1648.

Reddi, A. H., and Huggins, C. B., 1975, Formation of bone marrow in fibroblast transformation ossicles, *Proc. Natl. Acad. Sci. U.S.A.* **72**:2212.

Roth, S., 1973, A molecular model for cell interactions, *Q. Rev. Biol.* **48**:541.

Ruch, J. V., Fabre, M., Karcher-Djuricic, V., and Stäublei, A., 1974, The effects of L-azetidine-2-carboxylic acid (analogue of proline) on dental cytodifferentiations *in vitro, Differentiation* **2**:211.

Rucker, I., Kettrey, R., and Zeleznick, L. D., 1972, Microcrystalline collagen (Avitene) film as a substrate for outgrowth of primary rabbit kidney fibroblasts, *Proc. Soc. Exp. Biol. Med.* **139**:749.

Schiltz, J. R., Mayne, R., and Holtzer, H., 1973, The synthesis of collagen and glycosaminoglycans by dedifferentiated chondroblasts in culture, *Differentiation* **1**:97.

Segrest, J. P., and Cunningham, L. W., 1970, Variations in human urinary *O*-hydroxylysylglycoside levels and their relationship to collagen metabolism. *J. Clin. Invest.* **49**:1497.

Sengel, P., 1971, The organogenesis and arrangement of cutaneous appendages in birds, *Adv. Morphog.* **9:**181.

Slavkin, H. C., 1972, Intercellular communication during odontogenesis, *in: Developmental Aspects of Oral Biology* (H. C. Slavkin and L. A. Bavetta, eds.), pp. 165–199, Academic Press, New York.

Slavkin, H. C., and Greulich, R. C., eds., 1975, *Extracellular Matrix Influences on Gene Expression,* Academic Press, New York.

Smith, B. D., Martin, G. R., Miller, E. J., Dorfman, A., and Swarm, R., 1975, Nature of the collagen synthesized by a transplanted chondrosarcoma, *Arch. Biochem. Biophys.* **166:**181.

Spooner, B. S., and Wessels, N. K., 1970, Mammalian lung development: Interactions in primordium formation and bronchial morphogenesis, *J. Exp. Zool.* **175:**445.

Stoker, M., O'Neill, C., Berryman, S., and Waxman, V., 1968, Anchorage and growth regulation in normal and virus transformed cells, *Int. J. Cancer* **3:**683.

Stuart, E. S., Garber, B., and Moscona, A. A., 1972, An analysis of feather germ formation in the embryo and *in vitro*, in normal development and in skin treated with hydrocortisone, *J. Exp. Zool.* **179:**97.

Thompson, J., and Urist, M. R., 1970, Influence of cortisone and calcitonin on bone morphogenesis, *Clin. Orthop. Relat. Res.* **71:**253.

Trelstad, R. L., and Coulombre, A. J., 1971, Morphogenesis of the collagenous stroma in the chick cornea, *J. Cell Biol.* **50:**840.

Trelstad, R. L., and Kang, A. H., 1974, Collagen heterogeneity in the avian eye: Lens, vitreous body, cornea and sclera, *Exp. Eye Res.* **18:**395.

Trelstad, R. L., and Slavkin, H. C., 1974, Collagen synthesis by the epithelial enamel organ of the embryonic rabbit tooth, *Biochem. Biophys. Res. Commun.* **59:**443.

Trelstad, R. L., Hay, E. D., and Revel, J. P., 1967, Cell contact during early embryogenesis in the chick embryo, *Dev. Biol.* **16:**78.

Trelstad, R. L., Kang, A. H., Igarashi, S., and Gross, J., 1970, Isolation of two distinct collagens from chick cartilage, *Biochemistry* **9:**4993.

Trelstad, R. L., Kang, A. H., Cohen, A. M., and Hay, E. D., 1973, Collagen synthesis *in vitro* by embryonic spinal cord epithelium, *Science* **179:**295.

Trelstad, R. L., Hayashi, K., and Toole, B. P., 1974, Epithelial collagens and glycosaminoglycans in the embryonic cornea. Macromolecular order and morphogenesis in the basement membrane, *J. Cell Biol.* **62:**815.

Urist, M. R., 1965, Bone: Formation by autoinduction, *Science* **150:**893.

Urist, M. R., and Iwata, H., 1973, Preservation and biodegradation of the mophogenetic property of bone matrix, *J. Theor. Biol.* **38:**155.

Vinson, W. C., and Seyer, J. M., 1974, Synthesis of type III collagen by embryonic chick skin, *Biochem. Biophys. Res. Commun.* **58:**58.

von der Mark, H., von der Mark, K., and Gay, S., 1976, Study of differential collagen synthesis during development of the chick embryo by immunofluorescence, *Develop. Biol.* **48:**237.

Wartiovaara, J., Nordling, S., Lehtonen, E., and Saxen, L., 1974, Transfilter induction of kidney tubules: Correlation with cytoplasmic penetration into nucleopore filters, *J. Embryol. Exp. Morphol.* **31:**667.

Weiss, P., 1933, Functional adaptation and the role of ground substances in development, *Am. Nat.* **67:**322.

Wessells, N. K., 1970, Mammalian lung development: Interactions in formation and morphogenesis of tracheal buds, *J. Exp. Zool.* **175:**455.

Wessells, N. K., and Cohen, J. H., 1968, Effects of collagenase on developing epithelia *in vitro*: Lung, ureteric bud, and pancreas, *Dev. Biol.* **18**:294.

Woessner, J. F., Bashey, R. I., and Boucek, R. J., 1967, Collagen development in heart and skin of the chick, *Biochim. Biophys. Acta* **140**:329.

Wolpert, L., 1969, Positional information and the spatial pattern of cellular differentiation, *J. Theor. Biol.* **25**:1.

10

Synthetic Polypeptide Models of Collagen: Synthesis and Applications

RAJENDRA S. BHATNAGAR and RAO S. RAPAKA

I. *Introduction*

Collagen, the major skeletal protein in the vertebrate body, possesses many unique features which contribute directly to its role as a structural material. Collagen is characterized by a left-handed conformation which is very similar to polyproline II or polyglycine II. Since these amino acids occur more frequently in collagen than in any other vertebrate protein, considerable interest has arisen in the use of polymers containing glycine and proline as models for collagen. The relationship between the primary structure and function of a protein is not always easy to follow; however, the unusual repeating structure of collagen with glycine occurring in every third position makes it ideally suited for model building. Model building based on synthetic polypeptide analogs permits the examination of isolated features of the natural protein molecule in a manner uncomplicated by the influence of neighboring components of differing structure and properties. Synthetic polyproline models of collagen have provided invaluable information concerning the structure of collagen and have been of considerable help in elucidating the mechanism of some of the synthetic steps in the elaboration of collagen.

RAJENDRA S. BHATNAGAR and RAO S. RAPAKA · Laboratory of Connective Tissue Biochemistry, School of Dentistry, University of California, San Francisco, California 94143.

In this chapter we have summarized the current technology of synthesis of polypeptide models of collagen and their application in elucidating several biochemical aspects of collagen synthesis and structure. Details of the structures of several of the more important collagen models and the theoretical and experimental considerations used in the elucidation of these structures have been discussed in Chapter 2.

In the first section of this chapter, we discuss the salient features of the collagen molecule which must be considered in designing models for the protein. The second section is concerned with the chemical synthesis of polypeptide models of collagen and includes a comprehensive list of models which have been synthesized. In the third section some physicochemical properties of the model peptides, relevant to the subsequent discussion of their applications, have been considered. Finally we have discussed some of the interesting applications of the polypeptide models of collagen in resolving biochemical mechanisms related to collagen.

II. Considerations in Model Building: Structural Features of the Collagen Molecule

Despite the superficially apparent repeating order in its structure, the collagen molecule is a complex entity whose properties seem to vary with the tissue of its origin. The variations in the function-related physicochemical properties may have their origin in the somewhat subtle alterations in the primary structure, introduced during synthesis at the level of the gene, as in case of the tissue-specific collagens, or at steps subsequent to the assembly of the polypeptide chain. The lack of regularity in the primary structure complicates a clear interpretation of the molecular structure from data obtainable by physicochemical techniques such as X-ray diffraction, since all parts of the molecule apparently do not conform to the overall picture. Thus, while the conformation of collagen is related to polyproline II, there are prolonged segments in the molecule with very few or no proline or hydroxyproline residues. These segments also seem to be relatively rich in polar residues and by themselves would not acquire the polyproline-II-type helix. However, the polyproline II-type helix appears to be superimposed on the molecule by several regions which are especially rich in imino acids.

A fruitful approach in investigating the role of the diverse structural features in a protein molecule is to synthesize polymers approximating the composition of the region under examination and examine the properties of the polymer. This task is rendered much easier for the collagen chemist by the fact that collagen may be considered as being made up of triplets led by glycine, and polymers of glycine-containing triplets can be used to investigate individual regions in an amplified state. Polymeric models prepared in this manner offer many advantages for structural studies; for instance, polymers made up of identical repeating monomers give rise to sharply detailed X-ray diffraction patterns from which intensities can be determined more precisely and allow comparisons with stereochemical models. Model building allows investigation of the roles of individual amino acid residues in determining the conformation of localized regions in collagen. The interaction between the collagen peptide backbone and enzymes which act on it, for instance proline and lysine hydroxylases, is conformation specific, and studies on hydroxylation of polymers containing various residues have provided insight into some of the factors which may regulate the formation of hydroxyproline and hydroxylysine.

The following description of the characteristic features of the collagen molecule is meant to emphasize some of the structural considerations used in synthesizing polypeptide models of the protein.

A. *Collagen as a Polymer of Tripeptide Units: The Distribution of Imino Acid Residues*

Glycine occurs in every third position in the collagen molecule, and imino acid residues proline and hydroxyproline occur together in many triplets with other residues in sequence such as Gly-Pro-X and Gly-X-Hyp. Hydroxyproline invariably occurs in the last position in the triplet. The high degree of regularity with which glycine and proline occur in a large number of triplets derived from partial hydrolysis of collagen led early workers to suggest that collagen had a highly regular primary structure consisting of certain periodically repeated amino acid sequences (Astbury, 1934; Bergman, 1935). Astbury's collagen model consisted of recurring triplets containing Pro-Gly-X where X was any other amino acid. Subsequent studies indicated, however, that glycine was often linked to the carboxyl group of amino acids other than proline (Schroeder *et al.* 1953, 1954; Kroner *et al.* 1953, 1955). The presence of a large proportion of proline and hydroxyproline in the sequence -Gly-Pro-Hyp-Gly- led Schroeder and colleagues (1954) to suggest that collagen was made up of units, each of which contained four amino acids, the most important of these quadruplets being Gly-Pro-Hyp-Gly. In their model, periodicity existed only in certain regions of the collagen molecule. More recent studies on the primary structure of collagen (discussed in Chapter 1) indicate that the principal manifestation of regularity in the collagen molecule is in the distribution of glycine in every third position except in the first sixteen residues at the N-terminal end and approximately the last 20 residues at the C-terminal end. Combinations of enzymatic and chemical hydrolytic techniques give rise to tripeptides, most of which contain glycine in the N-terminal position. Traub and Piez (1971) determined the frequency of occurrence of the imino acids proline and hydroxyproline in such triplets derived from the helical region of the collagen molecule which constitutes over 95% of the entire molecule. Tripeptides of composition (Gly-X-Y) where X and Y are residues other than proline or hydroxyproline comprised 44% of the total triplets. Triplets of composition (Gly-X-I), where I is proline or hydroxyproline, accounted for 20% and (Gly-I-Y) provided 27% of all triplets. (Gly-I-I) accounted for 9% of all triplets. These numbers are very close to the proportions of these triplets that would arise from a random distribution of imino acids between positions X and Y. Distribution of individual amino X and Y acid residues in the second and third positions appears, however, not to be completely random (Balian *et al.*, 1972; Fietzek *et al.*, 1973). Glutamic acid, leucine, and phenylalanine residues occur mostly in the second position, whereas most of threonine, glutamine, methionine, arginine, and lysine residues occur in the third position, with a frequency much greater than that could be accounted for by random distribution. The absence of cysteine, tryptophan, and tyrosine from the helical region indicates another example of restriction in the distribution of amino acids in collagen. Examination of the sequence of an $\alpha 1$ chain, made up as a composite from residues 1–402 of rat and 403–1011 from cow skin collagens (Hulmes *et al.,* 1973) shows that the N- and C-terminal regions of collagen are especially rich in imino acid residues. The first 18 N-terminal triplets in the helical region* contain 18 imino residues, and the 25 C-terminal triplets in the helical region contain 27 imino residues. This enrichment is even more remarkable in the six terminal triplets of the helical region at the C-terminal end with six proline and five hydroxyproline residues. The concentration of imino residues near the two terminals of the molecule may be related to the greater need for stabilization of the polyproline II helix in these regions in order to maintain the overall integrity of the collagen molecule. One manifestation of this increased helix stability is the somewhat greater resistance of the C-terminal helical

* The term "helical region" is used here to denote the part of the molecule capable of existing in the characteristic triple helical conformation of collagen. The short sequences at the N and C terminals do not have the repeating sequence characteristic of the large body of the molecule, and their conformation in the intact native molecule is not known.

region to proteolysis. Another interesting feature of the primary structure of collagen is the high frequency with which proline occurs in a triplet with hydroxyproline or with alanine. In the composite α1, of the 116 triplets which contain proline in the second position, hydroxyproline occupies the third position in 37 triplets and alanine in 31. Serine appears in nine triplets. Interestingly, the basic residues arginine, in eight triplets, and lysine + hydroxylysine, in another eight, seem to be the next most favored neighbors of proline. Examination of the same composite α1 sequence shows that nearly half of the hydroxyproline residues share triplets with proline. Alanine is the next most frequent neighbor of hydroxyproline, appearing with it in 20 triplets. Hydroxyproline also shares 11 triplets each with glutamic acid and leucine. On the basis of probability analysis, Balian *et al.* (1972) showed that the frequency of hydroxyproline sharing a triplet with leucine is almost twice that expected from chance, even if leucine were to be restricted to the second position.

B. *Clustering of Hydrophobic and Polar Residues*

While no short range regularity is apparent in the distribution of various amino acid residues in the individual collagen chain, electron microscopic and X-ray diffraction techniques show that there is a distinct segregation between residues with polar and nonpolar side chains. The charged side chains are intensely stained by both cationic and anionic electron-dense materials and, because of the clustering of these residues along the length of the molecule, a distinct banded pattern emerges in electron micrographs. The regions between the bands of the polar residues are made up largely of residues with hydrophobic side chains, and these interband regions do not stain strongly. There is no sharp limitation of distribution of residues by charge since the bands may contain both positively and negatively charged residues and polar regions may be interspersed with nonpolar residues and *vice versa.* With the advent of high-resolution electron microscopes, it has become possible to obtain good separation between bands, and correlations have been made between the bands observed in the electron micrographs with the primary structure (von der Mark *et al.*, 1970; Balian *et al.*, 1971, 1972).

C. *Variations in the Composition of Collagen in Regard to Residues Which Are Modified after Chain Assembly*

It has become clear in recent years that the term collagen refers to a variety of closely related proteins, products of different genes. The variations in primary structure of collagen are probably related to the wide variety of structural organization of this protein in different tissues, and they may have bearing on the differences in the mechanical properties of these tissues. Differentiation between these collagens has been made on the basis of the differences in their affinities for negatively charged chromatographic media, and these differences are reflected in their overall amino acid compositions, which are close enough to have prevented their identification until quite recently. In the absence of more detailed sequence data, the gathering of which is an awesome task, it is difficult to determine if tissue-specific variability of collagens differing in their overall composition has arisen from changes in the primary sequences throughout the length of the individual chains or from local changes in certain specific, function-related regions, separated by large segments of sequences which are homologous in all collagens.

In addition to the genetic variability in the collagen molecule, microheterogeneity appears in collagen chains of the same genetic origin in regard to biochemical changes occurring subsequent to the translation step in synthesis. The major intracellular posttranslational reactions result in the hydroxylation of specific proline and lysine residues by a somewhat complex, sequence-specific dioxygenase mechanism and in the two-step glycosylation of some hydroxylysine hydroxyl groups. An extracellular enzymatic reaction results in the oxidative deamination of lysine and hydroxylysine residues in preparation for cross-linking. Modification of collagen chains in this manner contributes to changes in their physicochemical properties and in the mechanical properties of their aggregates. The real contributions of these posttranslational modifications to the tissue function of collagen cannot be fully defined in the absence of complete sequence data on collagens derived from different tissues. However, studies on natural collagens and their precursors, and more extensively on synthetic polypeptide models of collagen, have provided information concerning the role of hydroxyproline, hydroxylysine, and the derivatives of the latter. Studies with synthetic polypeptide models have contributed to the elucidation of mechanisms of hydroxylation of proline and lysine.

With the availability of more sophisticated analytical techniques and the development of CNBr fragmentation procedures considerable progress has been made in elucidating the primary structure of collagen (see Chapter 1). Although less-sensitive techniques had indicated the presence of some Pro-Gly sequences in collagen, Bornstein (1967) presented the first definitive evidence for reduced hydroxylation of certain proline residues in position 3. Bornstein sequenced two homologous 36-residue peptide segments (α1-CB2) from rat skin and rat tendon. Each contained six proline residues susceptible to hydroxylation, but there were only 5.1

hydroxyproline residues in the peptide derived from skin and 2.9 residues in the tendon peptide. The same analysis showed that two of the susceptible positions in the tendon peptide contained only negligible amounts of hydroxyproline. Since then, similar deficiencies in hydroxyproline content have been reported at many other sites in the collagen backbone (Balian *et al.*, 1971, 1972; Butler and Ponds, 1971; Wendt *et al.*, 1972; Fietzek *et al.*, 1972, 1973). The possible role of neighboring residues in dictating the extent of hydroxylation is discussed in Section IV.

In addition to the heterogeneity observed in the contents of hydroxyproline, several observations indicate that incomplete hydroxylation of lysine also occurs in many sequences in rat skin (Bornstein, 1969, 1970; Kang *et al.*, 1969; Butler, 1970), chick skin (Kang *et al.*, 1969), calf skin (Wendt *et al.*, 1972), and in chick bone (Miller *et al.*, 1969; Lane and Miller, 1969). Some of the hydroxylysine residues are glycosylated with the formation of 2-*O*-α-D-glucopyranosyl-*O*-β-D-galactopyranosyl, bound to the δ-hydroxyl group (Butler and Cunningham, 1966; Spiro, 1969; Butler, 1970). The amounts of carbohydrate bound thus to collagen varies with the tissue of origin, and may have a regulatory role in determining the arrangement of the collagen molecules into fibers and in regulating the extent of cross-linking.

III. *Synthesis of Polypeptide Models of Collagen**

The large concentration of glycine (33%) and imino residues (25%) in collagen implies that polymers of compositions rich in these residues should bear a structural resemblance to collagen. The simplest models for collagen consist of homopolymers of glycine, proline, or hydroxyproline and copolymers of these in various combinations. Since glycine is present in every third position in collagen, a large number of synthetic polypeptide models of collagen have been prepared by polymerizing tripeptides containing a glycine residue, with the other two residues being any amino acid including proline and hydroxyproline. In view of the growing interest in the use of these polymeric models in studies on the structure and biochemical properties of collagen, we have prepared a summary of the general procedures used in the synthesis of polypeptides, with special reference to polypeptides related to collagen which, in many cases, present unusual problems.

* Unless specifically indicated, all amino acid residues are of the L-configuration.

A. Synthesis of Polypeptides

1. General Procedures in Peptide Synthesis

a. Conventional Methods. A large number of polypeptide models have been synthesized for structural and biological studies on collagen. The synthetic procedures used have ranged from conventional techniques to the solid-phase methods and combinations of both. The synthesis of collagen models often presents special problems, and the purpose of this section is to elaborate on these. For a more complete description of techniques and procedures involved in peptide synthesis the reader is referred to several excellent reviews which have appeared recently (Johnson, 1974; Kapoor, 1970; Meienhofer, 1973).

Poly α-amino acids are generally synthesized by the polymerization of *N*-benzyloxy carbonyl amino acid anhydrides, the polymerization being initiated by a base such as triethylamine (Katchalski *et al.*, 1964). Since the most widely used collagen models are sequential polypeptides, the initial steps in the synthesis involve the assembly of a di-, tri-, or tetrapeptide which is subsequently polymerized. In the early synthetic procedures such as those employed by Fisher (1906) alkyl esters of the monomer were used as the starting material and the reaction required heating. This procedure produced low-molecular-weight products. The introduction of thiophenyl esters by Wieland and Bernhard (1953) and *p*-nitrophenyl esters by DeTar *et al.* (1963) provided significant advances in the synthesis of sequential of copolymers. Kovacs and co-workers (1966) also introduced the pentachlorophenyl ester technique in sequential polypeptide synthesis which was effectively employed by Shibnev and co-workers (Shibnev and Debabov, 1964; Shibnev and Lazareva, 1969; Shibnev and Lisvenko, 1966; Shibnev *et al.*, 1966*a,b*; 1967*a,b*; 1968*a–d*, 1969*a–c*, 1971*a,b*; 1973) for the synthesis of a large number of sequential copolymer models of collagen. Jones (1969) utilized *o*-hydroxyphenyl esters for sequential polypeptide synthesis. These esters undergo coupling through intramolecular base catalysis which accelerates aminolysis but not oxazolone formation. The special advantage in these esters is that they can be initially used for carboxyl protection and then transformed to an active ester on acidolysis or hydrogenolysis. Fairweather and Jones (1972*b*) synthesized collagen models using this procedure and it was claimed to be racemization-free (Cowell and Jones, 1971*a,b*; Jones, 1969).

A promising approach to synthesis of collagen-like polypeptides is the use of *N*-hydroxysuccinimide esters (Anderson *et al.*, 1964; Segal, 1969). These esters are usually crystalline, highly reactive compounds and with their use the synthesis of the peptide bond occurs in aqueous media. This technique has been used for the synthesis of several collagen models including polyhexapeptides (Segal, 1969).

Kitoaka *et al.* (1958) in Sakakibara's laboratory polymerized Gly-Pro-Leu utilizing tetraethyl pyrophosphite, and later several other collagenlike polymers were synthesized by this method (Shibnev and Lazareva, 1969). However, even though Kitaoka *et al.* (1958) found less than 0.01% phosphorus contamination in their preparation, Shibnev and Debabov (1964) reported up to 3% phosphorus in their polymers prepared by this method. This seems to be a definite drawback in this procedure, and the optical purity of the products has also been questioned (Anderson *et al.*, 1952). These disadvantages may be weighed against the fact that a tripeptide may be polymerized without first synthesizing active esters.

b. Solid-Phase Synthesis. The solid-phase method of peptide synthesis was originally introduced by Merrifield (1963), and its potential is well demonstrated by the wide employment of this method in peptide synthesis of angiotensin (Marshall and Merrifield, 1965), bradykinin (Merrifield, 1964), and insulin chains, as well as by the synthesis of ribonucleases (Denkewalter and Hirschmann, 1969; Gutte and Merrifield, 1969).

The basic idea of this method is to assemble the peptide chain in a step-wise manner, while it is attached at one end to an insoluble, easily filtered, solid support. After the assembly of the desired amino acid sequence, the peptide is cleaved from the support and isolated. This makes it possible to separate the peptide from starting materials and by-products by simple filtration. The number of steps, the difficult manipulations, and the time taken for the peptide synthesis is decreased considerably. The availability of automatic peptide synthesizers has furthered the usefulness of this method (Meienhofer, 1973). Solid-phase methods have been useful in synthesizing collagen-like polypeptides of differing chain lengths (Sakakibara *et al.*, 1968; Suzuki and Koyama, 1969) with differing compositions (Kivirikko *et al.*, 1972; Sakakibara *et al.*, 1973) and with differing triplets labeled in proline (Kivirikko *et al.*, 1971). In each of these studies the synthetic approach involved a two-step assembly of the first triplet on the resin, followed by coupling of a tripeptide in each subsequent step, until the desired molecular-weight peptide was obtained. Different amino acids could be added wherever needed (Kivirikko *et al.*, 1972). The versatility of the method made it increasingly useful in biological studies related to collagen immunology and biosynthesis.

The usefulness of the solid-phase technique was increased with the introduction of new protecting groups for amino acids (Erickson and Merrifield, 1972), different coupling reagents and some novel amino protecting groups (Verber *et al.*, 1972; Yamashiro *et al.*, 1972), and different active esters (Bodansky *et al.*, 1972). The major drawback of solid-phase peptide synthesis is that it is not as flexible as the classical methods, and purification of the intermediates is laborious. The final purification of the peptides is very tedious and, if the coupling steps do not proceed with a yield of 100%, a mixture of peptide chains differing in composition as well as in chain length can be anticipated. Thus problems associated with solid phase cannot be underestimated (Bayer *et al.,* 1970; Stewart and Matsueda, 1972).

2. *Strategy and Problems in the Synthesis of Collagen Models*

Since collagen is considered to be a polymer with a repeating tripeptide sequence, most of the polypeptide models of collagen are based on polymers of tripeptide units. In considering the synthesis of a sequence, such as (Gly-Pro-Ala)$_n$, it becomes obvious that this polypeptide sequence can be obtained by the polymerization of either of three monomeric tripeptides, Pro-Ala-Gly, Ala-Gly-Pro, or Gly-Pro-Ala. In selecting the starting tripeptide unit, the important criteria must involve consideration of the following: (1) synthesis of the tripeptide should be racemization-free, (2) the nature of the N-terminal residue should be such that it should not sterically interfere with polymerization, and (3) the C-terminal residue should be one which does not racemize during polymerization.

A major concern in studies on polypeptides is their optical purity (Kovacs *et al.,* 1972). Even if only 1% of the condensation steps occur with concomitant racemization, after 100% polymerization, approximately 39% of the resulting chains will have an inverted configuration, at least at one residue (DeTar, 1967). Racemization is always a hazard whenever the carboxyl end of an amino acid or peptide is being activated. Either the coupling procedures should be such as to avoid racemization, or the C-terminal moiety to be activated should be either glycine or proline which are resistant to racemization. Goodman and Stueben (1959) have reported polypeptide synthesis by the "backing-off" procedure which is a valuable addition to the synthesis of racemization-free peptides. This procedure involves the direct coupling of an amino acid active ester.

The N-terminal residue should be selected so as not to interfere sterically in polycondensation. Fairweather and Jones (1972*a*) obtained unsatisfactory yields when polymerizing Pro-Ala-Gly active esters because of steric interference by N-terminal proline, although Brown *et al.* (1972) did not experience similar difficulties with this procedure. Even a small steric difference can have a profound influence on polymerization. Polymerization of Pro-Sar-Gly pentachlorophenyl ester hydrobromide (Rapaka and Bhatnagar, 1974) resulted in molecular weights up to 22,000, whereas polymerization of Sar-Pro-Gly pentachlorophenyl ester hydrobromide, resulted in much lower molecular weights (approximately 6000).

In synthesizing the monomeric triplet, as well as in the polymerization of the triplet, if the C-terminal residue is glycine or proline, racemization during activation is not a problem. When the C-terminal amino acid residue is phenylalanine or leucine, formation of oxazolones which are intermediates in racemization occurs (Goodman and Glaser, 1970). DeTar *et al.* (1966) observed racemization via oxazolone formation during the synthesis of *p*-nitrophenyl ester of peptides containing phenylalanine as the C-terminal residue. However, Kovacs *et al.* (1966) did not find any evidence for racemization during the synthesis of (Gly-Gly-Phe)$_n$. In our synthesis of (Pro-Gly-Leu)$_n$ via the pentachlorophenyl ester and (Pro-Gly-Phe)$_n$ via the *p*-nitrophenyl ester very little racemization was detected in the polymer (Rapaka and Bhatnagar, 1974). However, when Pro-Gly-Val pentachlorophenyl ester was polymerized, approximately 25% racemization was detected (Rapaka and Bhatnagar, 1974). It is interesting to note that Gly-Val-Pro pentachlorophenyl ester showed 2–4% D-valine, and on polymerization the polymer showed no racemization at all. The failure to detect racemized polymers may be explained by the probable formation of low-molecular-weight products with D-valine which would be eliminated during the dialysis. Sometimes a previously successful approach may not work for similar tripeptide sequences. This is exemplified by the observation that while Ala-Gly-Pro pentachlorophenyl ester HBr could be polymerized satisfactorily, Ala-Gly piperidine-2 carboxylic acid pentachlorophenyl ester HBr gave no measurable products (Fairweather and Jones, 1972*b*).

Purification of the intermediates is very important. In the synthesis of (Pro-Ala-Gly)$_n$ by Fairweather and Jones (1972*a*), the intermediate Hbr-H-Pro-Ala-Gly-ONp could not be purified, and the resulting hygroscopic material was polymerized to yield a polymer of low molecular weight and with only 22% yield. However, Brown *et al.* (1972), following the same procedure, were able to purify the intermediate, and subsequent polymerization resulted in a higher-molecular-weight polymer, with better yields (81% yield; 7000–14,000 mol. wt.).

In the synthesis of collagen models, if glycine is one of the amino acid residues and is next to an N-terminal benzyloxycarbonyl protected amino acid residue, saponification of the C-terminal esters with alkali results in the elimination of benzyl alcohol to form hydantoin derivatives, followed by a rearrangement to the urea derivative (Bodanszky and Ondetti, 1966; Bodanszky *et al.*, 1963). However, if only one equivalent of alkali is used for saponification, the side reaction may be avoided. Benzyloxycarbonyl Pro-Gly-Leu methyl ester was saponified with one equivalent of 1 N sodium hydroxide, the yield of the benzyloxycarboxyl-protected acid was satisfactory (60%), and no appreciable amount of urea was obtained. However, where benzyoxycarboxyl Pro-Gly-Val methyl ester was hydrolyzed under similar conditions with one equivalent of alkali, only 30% of the acid was obtained (Rapaka and Bhatnagar 1974).

The polymerization reaction is carried out in a polar solvent in the presence of a base, and the monomer is employed in as high a concentration as possible, which favors the polymerization reaction and minimizes the formation of cyclic peptides. It is useful to understand the factors that favor the formation of cyclic peptides, which is a competing reaction. The probability of cyclization increases with configurational and conformational stability of the ring to be formed, and decreases with increases in the loss of internal freedom that results from ring formation. 3-, 5- and 6-membered alicyclic rings are more easily formed relative to rings of other sizes. At higher concentrations cyclization of a tripeptide is unlikely. However, condensation of two tripeptide units which are not sterically inhibited may be possible. Cyclization of the so-formed hexapeptide is probably facilitated by chain folding of a 10-membered ring. The sequence of L-residue-D-residue or its enantiomer greatly facilitates the formation of cyclic peptides, as does the presence of (1) a D-amino acid residue, (2) a Gly residue, (3) an imino acid, or (4) a chain consisting entirely of imino acids. Cyclization does not occur if one of the residues Val, Thr, or Pro is inverted in configuration. An LD peptide segment introduces an elbow-like bend in the peptide backbone which brings the reactive amino and active carboxyl groups into proximity. Temperatures profoundly influence the course of the reaction, with greater cyclization occurring at elevated temperatures.

In the synthesis of collagen models, proline is one of the residues invariably present. For optical purity, a good strategy would be to synthesize the monomer with either proline or glycine as a C-terminal residue. Recently the effect of the presence of proline at the C-terminal end and of alkyl substitution on the other two residues in linear tripeptide systems was studied by Conte *et al.* (1973). They observed cyclol formation as a significant side reaction in many such sequences. Rapaka *et al.* (1976*a*) investigated racemization in the synthesis of polytripeptide models of collagen using pentachlorophenyl and *p*-nitrophenyl activation of *N*-protected tripeptide acids. Racemization occurred during the activation step if the C-terminal residue was an optically active α-amino acid but not if proline was the C-terminal residue. Activation by *N*-hydroxysuccinimide, on the other hand, resulted in optically pure active esters (Rapaka *et al.*, 1976*b*). These studies (Rapaka *et al.*, 1976*a,b*) also showed that the synthesis of optically pure active fragments does not guarantee that the polymers will be optically pure and considerable racemization may occur during polymerization.

Rothe and Mazanek (1972) reported some of the possible side reactions during the solid-phase peptide synthesis. An important side reaction is the cyclization at the dipeptide stage with simultaneous liberation of hydroxymethyl groups. The cyclization tendency depends on the nature and sequence of amino acids. Diketopiperazine formation is high with *N*-alkylated amino acids, and especially high for Pro-Pro, Sar-Pro, Gly-Sar, Pro-Sar, and Sar-Sar. In a similar reaction Gisin and Merrifield (1972) observed that the bound peptide D-valyl-L-proline underwent intramolecular aminolysis, and the resulting loss of dipeptide was about 70%.

The hydroxymethyl groups formed on the resin support can react in subsequent coupling steps, thus providing starting points for new peptide chains. When proline is one of the amino acid residues in the tripeptide monomer, initially one amino acid is attached to the resin and, after removal of the protecting group, a dipeptide unit is attached. After formation of the first tripeptide unit, tripeptide units are added until the desired chain length is achieved (Sakakibara *et al.,* 1968).

In polymerization of dipeptides, formation of diketopiperazines is a competing reaction. Polymerization of HBr·Pro-Gly-ONp (DeTar, 1967; Rapaka and Bhatnagar, 1974) resulted in only 10% yield and polymerization of HBr·H-His-(H)-Gly-ONp gave completely cyclic material (DeTar, 1967). Formation of diketopeperazines of the proline-containing peptides, especially of dipeptides, is well known. When Z-Gly-Pro-ONp was hydrolyzed or aminolyzed, benzyloxycarbonylglycyl L-prolyl diketopiperazine was a side product (Goodman and Stueben, 1962). Diketopiperazine readily forms from unactivated dipeptide esters, and the ease of formation is greater if proline is in one of the residues. If sarcosine is to be incorporated into the synthesis of collagen models it should be borne in mind that it is similar to proline in the formation of diketopiperazines. Gly-Sar-O-Me cyclizes rapidly to the diketopiperazine, whereas Sar-Gly-O-Me is slow because of steric hindrance (Purdie and Benoiton, 1973).

In spite of the various available methods of detection of optical purity, simpler and more efficient methods need to be developed. Better methods of ascertaining the purity and homogeneity are also needed, as well as more reliable methods of determining the molecular weight of the polymers, since the use of different presently available methods of determining molecular weights gives gross differences in the results obtained.

B. *A List of Collagen-like Polypeptides*

Most studies on collagen structure and the biochemical steps in its synthesis have concentrated on a small number of polypeptides. A review of the literature revealed, however, that a considerable number of polymeric models of collagen have been synthesized utilizing a large variety of chemical approaches. It seemed appropriate to list these in a table (Table 1) along with a summary of methods used in their synthesis.

IV. *Applications of Synthetic Polymers in Studies on the Structure and Synthesis of Collagen*

Major contributions in elucidating the structure of collagen have been made through investigations on model compounds and have been discussed in Chapter 2. This section is not intended to be an exhaustive review of the properties and structures of the models being discussed here but is designed to emphasize how certain localized features in the collagen molecules may have somewhat different physicochemical properties, and by inference, have different biological properties. One can only speculate on the nature of interactions between different segments of the collagen polypeptide chain, such as the imposition of the polyproline II helix on imino acid-poor regions, and possible interruptions in the continuity of the helix by residues resistant to this conformation. Such speculations have to be based on the properties of synthetic models resembling different regions of the collagen molecule.

A. *Properties of Synthetic Polypeptide Models: Some Physicochemical Considerations*

Polypeptide models of collagen have been used extensively in a wide variety of research on collagen and unavoidably certain terms have been used in more than one sense in the recent literature. In discussing the properties of polymer models of collagen it is important to clarify certain commonly misused terms. In referring to collagen and collagen-like synthetic polypeptides the terms "triple helix" or "triple-helical structure" have often been used without consideration of the actual configuration of the three chains with respect to one another. The individual helices of polyglycine II, polyproline II, and several ordered polymers containing

TABLE 1
Synthesis of Polypeptide Models of Collagen[a]

Polypeptide	Method	Reference
1. Copolymers		
$(Pro^x, Gly^y)_n$	*N*-Benzyloxycarbonyl amino acid anhydride	Borek *et al.*, 1969; Fuchs *et al.*, 1974; Maoz *et al.*, 1973*a,b*
$(Pro^x Sar^y)_n$	*N*-Benzyloxycarbonyl amino acid anhydride	Fasman and Blout, 1963
2. Proline Containing Peptides		
$(Pro\text{-}Gly)_n$	ONp, hydrazide, TEPP 2,4-ODNp; 2,5-ODNp	DeTar *et al.*, 1972; Huggins *et al.*, 1968
$(Ala\text{-}Gly\text{-}Pro)_n$	TEPP, OPcp, ONp, ONSu	DeTar *et al.*, 1972; Fairweather and Jones, 1972*a*, 1973; Lorenzi *et al.*, 1971; Shibnev and Lazareva, 1969; Shibnev *et al.*, 1968*a–d*
$(DAla\text{-}Gly\text{-}DPro)_n$	OPcp	Fairweather and Jones, 1972*a*
$(DAla\text{-}Gly\text{-}Pro)_n$		
$(Ala\text{-}Gly\text{-}DPro)_n$		
$(\beta\text{-}Ala\text{-}Pro\text{-}Gly)_n$	ONp	Fairweather and Jones, 1973
$(Ala\text{-}Pro\text{-}Gly)_n$	TEPP	Huggins *et al.*, 1968
$(Ala\text{-}Pro\text{-}Pro)_n$	TEPP	Shibnev and Lazareva, 1969
$(Ala\text{-}Pro\text{-}Pro)_n$	OPcp	Rapaka and Bhatnagar, 1976; Rapaka *et al.*, 1976*a*
$(Gly\text{-}Ala\text{-}Pro)_n$		
$(Gly\text{-}Gly\text{-}Pro)_n$		
$[Gly\text{-}Glu\text{-}(OCH_3)\text{-}Pro]_n$	OTcp	Shibnev *et al.*, 1973
$(Gly\text{-}Pro\text{-}Ala)_n$ also $n = 1, 2,$ and 4	ONSU, Otcp, TEPP, OHp, ONp	Bloom *et al.*, 1966; Cowell and Jones, 1971*b*; Jones, 1969; Shibnev and Lazareva, 1969; Shibnev *et al.*, 1968*c*, 1969*b*

TABLE 1—*Continued*

Polypeptide	Method	Reference
$(Gly\text{-}Pro\text{-}Gly)_n$	OHp, ONp, OQu, OTcp	Bloom *et al.*, 1966; Shibnev *et al.*, 1968*c,d*, 1969*a*, 1970*a*
$(Gly\text{-}Pro\text{-}Leu)_n$	TEPP, ONSu	Kitaoka *et al.*, 1958; Shibnev and Lazareva, 1969
$[Gly\text{-}Pro\text{-}Lys\text{-}(Tos)]_n$	ONp, OTcp	Shibnev *et al.*, 1966*b*
$[Gly\text{-}Pro\text{-}Lys\text{-}(Z)]_n$	ONp	Shibnev *et al.*, 1967*b*
$(Gly\text{-}Pro\text{-}Pro)_n$ and $n = 4, 5, 6$	OPcp, ONp, TEPP	Shibnev and Lisvenko, 1966; Shibnev and Lazareva, 1969; Shibnev *et al.*, 1971*a,b*
$(Gly\text{-}Pro\text{-}Ser)_n$	TEPP, BPP	Heidemann and Nill, 1969
$(Gly\text{-}Pro\text{-}Tyr)_n$	TEPP	Shibnev and Lazareva, 1969
$(Gly\text{-}Val\text{-}Pro)_n$	Opcp	Rapaka and Bhatnagar, 1975*a*
$(Gly\text{-}Ser\text{-}Pro)_n$	BPP, TEPP, OPcp	Heidemann and Nill, 1969; Shibnev *et al.*, 1970*b*
$(Glu\text{-}Pro\text{-}Gly)_3$, $(Glu\text{-}Pro\text{-}Gly)_4$, and other peptides having glutamine and proline	ONp, DCCI	Okada *et al.*, 1972
$(Leu\text{-}Pro\text{-}Gly)_n$	TEPP	Shibnev and Lazareva, 1969
$(Phe\text{-}Pro\text{-}Gly)_n$	Onp	Tamburro *et al.*, 1968
$(Pro\text{-}Gly\text{-}Gly)_n$	ONp, ONcp, BPP, ONSu, TEPP	Brown *et al.*, 1972; Fairweather and Jones, 1972*a*; Heidemann and Bernhardt, 1968; Huggins *et al.*, 1968
$(Pro\text{-}Gly\text{-}Gly)_n$	Hydrazide, ONp, Pcp	Stewart, 1965; Wolman *et al.*, 1961
$(Pro\text{-}Gly\text{-}Leu)_n$	OPcp	Rapaka and Bhatnagar, 1975*a*
$(Pro\text{-}Gly\text{-}Phe)_n$	OPcp	Rapaka and Bhatnagar, 1975*a*
$(Pro\text{-}Gly\text{-}Pro)_n$	TEPP	Engel *et al.*, 1966
$([^{14}C]Pro\text{-}Gly\text{-}Pro)_n$		
$(Pro\text{-}Gly\text{-}Ser)_n$	TEPP, BPP	Heidemann and Nill, 1969
$[Pro\text{-}His\text{-}(Z)\text{-}Gly]_n$	OTcp	Shibnev *et al.*, 1973
$(Pro\text{-}Leu\text{-}Gly)_n$	DCCI, TEPP	Kitaoka *et al.*, 1958
$(Pro\text{-}Phe\text{-}Gly)_n$	ONp	Tamburro *et al.*, 1968

$(\text{Pro-Pro-}\beta\text{-Ala})_n$	OPcp	Bhatnagar and Rapaka, 1975*b*
$(\text{Pro-Sar-Gly})_n$	OPcp	Rapaka and Bhatnagar, 1975*a*
$(\text{Pro-Ser-Gly})_n$	ONp, BPP, ONSu, OPcp	Heidemann and Nill, 1969; Shibnev *et al.*, 1970*c*
$(\text{Ser-Gly-Pro})_n$	BPP, TEPP	Heidemann and Nill, 1969
$(\text{Ser-Pro-Gly})_n$		
$(\text{Val-Pro-Gly})_n$	Hydrazide; 2,4-ODnp, 2,5-ODnp	Huggins *et al.*, 1968
$(\text{Val-Pro-Pro})_n$	OPcp	Rapaka and Bhatnagar, 1976; Rapaka *et al.*, 1976*a*
$(\text{Ala-Ala-Gly-Pro-Pro-Gly})_n$	ONSu	Segal, 1969
$(\text{Ala-Pro-Gly-Pro-Ala-Gly})_n$		
$(\text{Ala-Pro-Gly-Pro-Pro-Gly})_n$		
$(\text{Pro-Ala-Gly-Pro-Pro-Gly})_n$		
3. Polypeptides containing hydroxyproline		
$(\text{Hyp-Gly})_n$	ONp	DeTar, 1967; DeTar *et al.*, 1972
$(\text{Ala-Hyp-Hyp})_n$	TEPP	Shibnev and Lazareva, 1969
$(\text{Gly-Ala-Hyp})_n$	OPep, TEPP	Heidemann and Bernhardt, 1968; Shibnev and Lazareva, 1969; Shibnev *et al.*, 1970*a*
$(\text{Gly-Hyp-Gly})_n$	OPyr	Shibnev *et al.*, 1967*a*
$(\text{Gly-Hyp-Hyp})_n$	TEPP, OTcp, OPcp	Shibnev *et al.*, 1966*a*, 1968*a,b*, 1969*a*, 1970*a*
$(\text{Gly-Pro-Hyp})_n$	OTcp, TEPP, 2,6-ODNp, ethoxyactylene, OPfp	Andreeva *et al.*, 1961; Huggins *et al.*, 1968; Shibnev, 1964; Shibnev and Debabov, 1964; Shibnev and Lazareva, 1969; Shibnev *et al.*, 1969*a*, 1970*a*
$(\text{Gly-Ser-Hyp})_n$	OPcp	Shibnev *et al.*, 1970*b*
$(\text{Hyp-Glu-Gly})_n$	OTcp	Khalikov *et al.*, 1968
$(\text{Hyp-Glu-(}O\text{Bzl)-Gly})_n$	OTcp	Shibnev *et al.*, 1970*d*
$(\text{Hyp-Ser-Gly})_n$	OTcp	Khalikov *et al.*, 1968; Shibnev *et al.*, 1970*d*

Table 1—*Continued*

Polypeptide	Method	Reference
4. Collagen models containing proline analogs and other unnatural compounds		
(Ala-Gly-Pipec)$_n$	OPcp, OHp	Fairweather and Jones, 1972*b*
(Ala-Gly-ThZ)$_n$		
(Gly-Azet-Ala)$_n$		
(Gly-Pipec-Ala)$_n$		
(Pipec-Ala-Gly)$_n$		
(Pro-Ala-Glycol)$_n$	ONp	Stewart, 1969
5. Peptides made by solid-phase method		
(Pro-Gly-FPro)$_n$Pro, where n = 2, 3, and 4		Hutton *et al.*, 1968; Kikuchi *et al.*, 1969; Kivirikko *et al.*, 1969, 1971; Sakakibara *et al.*, 1968, 1973; Suzuki and Koyama, 1969
(Pro-Gly-MePro)$_n$ where n = 2, 3, and 4		
(3,4-^{3}H-Pro-Gly-Pro)$_n$		
[Gly-Pro-(3,4-^{3}H-Pro)]$_3$-Gly-Pro-Pro-OH		
(Pro-Pro-Gly)$_n$ where n = 1, 3, 5, 10, 20		
(Pro-Hyp-Gly)$_n$ n = 5 and 10		Sakakibara *et al.*, 1973
Z-Gly-Pro-Hyp-Gly-(Pro-Pro-Gly)$_5$-OH		Kikuchi *et al.*, 1971
AOC-(Pro-Pro-Gly)$_6$-OH		
Arg-Gly-(Pro-Pro-Gly)$_5$		Kivirikko *et al.*, 1972
Glu-Gly-(Pro-Pro-Gly)$_5$		
Arg-Gly-(Leu-Pro-Gly)$_5$		
Ala-Arg-Gly-Met-Lys-Gly-His-Arg-Gly-(Pro-Pro-Gly)$_4$		
(Pro-Pro-Gly)$_4$-Ala-Arg-Gly-Met-Lys-Gly-His-Arg-Gly-(Pro-Pro-Gly)$_4$		

[a] The nomenclature used is as suggested by the IUPAC-IUB Commission on Biochemical Nomenclature: Abbreviated Nomenclature of Synthetic Polypeptides (Polymerized Amino Acids) Revised Recommendations (1972) *J. Biol. Chem.* **247:**323. Unless otherwise stated, all amino acids are of L-configuration. We have used the form (Pro-Pro-Gly)$_n$ instead of the recommended (Pro$_2$-Gly)$_n$ to facilitate comparison between collagen-like peptides. In case of unconventional amino acids, the nomenclature used is that of the original authors. Some of the abbreviations used are AZet, azetidine carboxylic acid; BPP, bisphenylene pyrophosphite; DCCI, dicyclohexyl carbodiimide; FPro, *trans*-4-fluoroproline; MePro, *trans*-3-methyl proline; ODNp, dinitrophenyl; OHp, hydroxyphenyl; OPcp, pentachlorophenyl; OPfp, pentafluorophenyl; OPyr, 3-hydroxypyridyl; OTcp, trichlorophenyl; Pipec, pipendinyl; TEpp, tetraethylpyrophosphite; Thz, thiazohdyl.

these residues exist in the same helical conformation as do the individual helices of collagen. In many cases, however, the model polypeptides seem to form orthogonal structures in which the individual helices run parallel but do not form the "coiled-coil" triple-helical structure characteristic of collagen. Although some studies on biochemical properties of such polypeptides discuss "collagen-like triple helices," such statements are not always warranted since the analytical techniques applicable to solutions usually do not permit evaluation of properties characteristically observable in solid-state structures. The solid-state structures of many polypeptide models differ in the mode of packing from the triple-stranded, "coiled-coil" or "superhelical" structure of collagen, and it is not correct to assume that the solution structures of these polymers are similar to collagen structures.

The term "denaturation" in solutions should be applied more cautiously. Denaturation of collagen is not a one-step process. Engel (1962) examined the denaturation of collagen using light-scattering techniques which permit the simultaneous measurement of molecular weight and the estimation of the shape of the macromolecular species in solution. Engel observed a two-step process in the first of which the helical structure appeared to collapse, followed by a much slower second step resulting in strand separation. Flory and Weaver (1960) observed that the fractional change in the parameters most often used for measuring the denaturation of collagen, such as optical rotation and viscosity, was not necessarily proportional to the fractional change from native collagen to gelatin. Von Hippel and Harrington (1960) suggested the use of the more accurate parameter T_m, the temperature at which the most ordered segments of the crystalline structure melt. Unfortunately T_m is much harder to determine than the so-called denaturation temperature T_d, which represents the midpoint of thermal transition when optical rotation or viscosity are used as a combined measure of helical content and the tertiary structure. These methods are not usually sensitive enough to permit conclusions regarding interchain interactions, which should be elucidated by procedures permitting evaluation of molecular weights and major changes in particle dimensions.

In considering the mechanism of melting of the helix in collagen or collagen-like peptides one must take into account the total absence of rotational freedom about the N—C^α and restricted rotation about the C^α—C bond of proline. These stereochemical factors severely restrict rotation about the pyrrolidine peptide bonds during helix $\rightleftharpoons$ coil transition, and the conformational entropy change per imino residue is close to zero (Von Hippel and Harrington, 1959) since the polyproline II helix does not appear to undergo a conformational change in aqueous solution at temperatures up to 90°C (Harrington and Sela, 1958). The implication of this is apparent in increased T_m as the pyrrolidine content of the polymer goes up. The significance of imino residues in stabilizing the individual helix is further emphasized by Josse and Harrington (1964) who suggested that only those triplets or regions which do not contain proline or hydroxyproline are involved in the melting of the helix. Hopfinger (1973) has calculated the conformational entropy of various glycine-led triplets from theoretical considerations, and his analysis supports the concept that the presence as well as the positioning of proline residues in the triplet regulate the stability of the helical parameters of the triplet.

The presence of pyrrolidine rings in the polypeptide chain reduces the number of interchain H bonds. Apparent structural stability in collagen and collagen-like polymers as represented by the temperature dependence of viscosity or optical rotation therefore represents mainly the stability of the individual helix rather than the tertiary structure. Interchain interactions and bonding may not contribute greatly to the T_d as demonstrated by the very small increase (1.4°) in T_d of ichthyocol on the introduction of 10 additional covalent cross-links with formaldehyde (Veis and Drake, 1963). The significance of this is that the thermal transition temperature may not necessarily reflect interchain interactions and "triple-helix" formation and stability as discussed in some articles. These concepts also suggest caution in the interpretation of biochemical experiments based on helix $\rightleftharpoons$ random-coil transitions, especially those with synthetic polymers containing large concentrations of proline.

As mentioned earlier in this chapter, the conformation of collagen is best described as intertwined polyproline helices, the supercoiling being made possible by the presence of a glycine residue in every third position. For this reason polyproline and several polytripeptides containing glycine in each triplet have been used extensively in studies on the biochemical reactions of collagen. Before proceeding with a discussion of the biochemical properties of these polymers, it is of interest to examine some of their characteristic physicochemical properties.

Elucidation of solution properties of polyproline has important implications in explaining the properties and behavior of proline-rich regions of collagen. Polyproline occurs in two predominant helical forms with different optical rotation properties in solution—polyproline I and polyproline II—which differ in that in form I, all peptide bonds are *cis* whereas in polyproline II, the peptide bonds are in a *trans* configuration (Harrington and Sela, 1958). The two forms are interconvertible and perturbation of the solvent leading to increased hydrophobicity initiates mutarotation, resulting in the conversion of *trans*-bonded polyproline II to *cis*-bonded polyproline I in a highly cooperative transition (Winklmair *et al.*, 1971). In aqueous solutions polyproline II appears to have a structure similar to that in the solid state. Because the N—C^{α} bond of the proline peptide bond is part of the pyrrolidine ring and the C^{α}—C bond is restricted by stereochemical factors, the conformational possibilities for polyproline II are severely limited (Balasubramanian *et al.*, 1971).

When a nonpyrrolidine residue is introduced into the polypyrrolidine polymer, a sort of a conformational "weak spot" is introduced into the chain, permitting limited conformational variability at the peptide bond involving the nonpyrrolidine residue. An excellent structural analog of collagen is $(\text{Gly-Pro-Pro})_n$, containing glycine in every third position and possessing all the characteristics of polyproline II. Extensive structural and biochemical investigations have been made with $(\text{Gly-Pro-Pro})_n$ which exhibits properties very closely resembling collagen. $(\text{Gly-Pro-Pro})_n$, in contrast to polyproline II, undergoes well-defined conformational transitions when heated (Engel *et al.*, 1966; Berg *et al.*, 1970; Kobayashi *et al.*, 1970) and in the presence of hydrogen-bond disrupting agents (Engel *et al.*, 1966). The polymer is considerably more stable than collagen. However, in the collagen molecule extended regions containing repeating -Gly-Pro-I- sequences, where I is hydroxyproline and sometimes proline, are located only near the two terminals, presumably to impart greater conformational stability and therefore, greater resistance to proteolysis at the two ends since native collagen is not susceptible to proteases. $(\text{Gly-Pro-Pro})_n$ apparently remains a triple-helical conformational in solution since the studies of Kobayashi *et al.* (1970) indicated a threefold decrease in molecular weight on denaturation. The helix stability of $(\text{Gly-Pro-Pro})_n$ increased with the degree of polymerization (Kobayashi *et al.*, 1970).

Although $(\text{Gly-Pro-Pro})_n$ has been the polymer of choice in studies on the structure of collagen, there are very few triplets in collagen which actually contain proline in both the second and the third position since the proline residue in the third position is subject to hydroxylation. Thus $(\text{Gly-Pro-Hyp})_n$ should be an even better model for collagen than $(\text{Gly-Pro-Pro})_n$. As discussed in Chapter 2, hydroxyproline provides additional stability to the collagen helix by participating in hydrogen-bond formation.

Although a considerable number of imino acid residues are clustered in collagen with many sharing the same triplets, in a large part of the collagen molecule they appear in triplets where either the second or the third position is occupied by an α-amino acid. The composition of a triplet directly influences its physicochemical properties and the steric interactions of side chains, and the possibility of hydrogen-bond formation determines the stability of the rudiments of helix in the individual triplet. The stability of the overall collagen helix is much lower than the stability of the imino-rich regions. No simple analysis, such as averaging of the helical stabilities of different regions, is possible since even the position of a single imino residue in a triplet can markedly alter the stability, as seen in the case of certain polytripeptides containing $(Gly\text{-}X\text{-}I)_n$ or $(Gly\text{-}I\text{-}X)_n$ sequences. Considerable information is available on the structure and physicochemical properties of polytripeptides such as $(Gly\text{-}Gly\text{-}Pro)_n$, $(Gly\text{-}Ala\text{-}Pro)_n$, and $(Gly\text{-}Pro\text{-}Ala)_n$. The alanine-containing peptides are of special interest since alanine appears in a very large number of triplets which contain a proline residue in the second position or a hydroxyproline residue in the third position. Since several polytripeptides of composition $(Gly\text{-}X\text{-}I)_n$ and $(Gly\text{-}I\text{-}X)_n$ have been useful in biochemical studies on collagen, we will briefly mention the physicochemical properties of some of these polymers. As mentioned above, nonpyrrolidine residues destabilize the polypyrrolidine helix, and this becomes even more apparent in polymers where only one of the residues in each triplet is a pyrrolidine residue. This is abundantly clear in the case of $(Gly\text{-}Gly\text{-}Pro)_n$, $(Gly\text{-}Ala\text{-}Pro)_n$, and $(Gly\text{-}Ser\text{-}Pro)_n$.

There are few -Gly-Gly-Pro- sequences in collagen. Even though the polymer $(Gly\text{-}Gly\text{-}Pro)_n$ seems to exist in the solid state in left-handed helical forms resembling polyproline II (Traub and Yonath, 1966; Oriel and Blout, 1966; Traub, 1969), in aqueous solutions $(Gly\text{-}Gly\text{-}Pro)_n$ does not seem to have an ordered structure. The very small negative optical rotation seen in aqueous solutions is very close to that of proline in random chains (Oriel and Blout, 1966). In 1.4 M acetic acid solutions, however, a collagen-like structure appears to be generated. The helix is very stable in this solvent, melting at 67° and easily reforming on cooling. Interchain interactions are involved in stabilizing the ordered aggregates in acetic acid.

Both $(Gly\text{-}Ala\text{-}Pro)_n$ and $(Gly\text{-}Pro\text{-}Ala)_n$ can exist in several different structural forms in the solid state and these include structures which closely resemble collagen (Schwartz *et al.,* 1970; Doyle *et al.,* 1971; Traub and Yonath, 1966). In aqueous solutions, however, $(Gly\text{-}Ala\text{-}Pro)_n$ does not exist in a collagen-like conformation, and even the exposure of collagen-like forms in the solid state to water results in the loss of structure (Doyle *et al.,* 1971). $(Gly\text{-}Pro\text{-}Ala)_n$, on the other hand, exhibits properties in aqueous solutions which are very similar to collagen, indicating an ordered collagen-like conformation (Brown *et al.,* 1972). These workers also found evidence for chain association in $(Gly\text{-}Pro\text{-}Ala)_n$ resulting in triple helix formation. Blout and colleagues (Doyle *et al.,* 1971) concluded that -Gly-Pro-Ala- and other sequences of the form -Gly-Pro-X- may have a helix-stabilizing role in collagen, whereas sequences of the type -Gly-Ala-Pro- (or -Gly-X-Pro-) may reduce the stability of the helix. These workers suggested that the conformational differences between (Gly-Ala-Pro)$_n$ and (Gly-Pro-Ala)$_n$ may arise from differences in the hydrogen-bonding properties of alanine NH in the two sequences. Our recent studies (Ananthanarayanan *et al.*, 1976) showed similar conformational differences between (Gly-Sar-Pro)$_n$, which exists in random conformations in aqueous solution, and (Gly-Pro-Sar)$_n$, which forms collagen-like ordered helices in aqueous solutions. Since sarcosine peptide bonds do not have an -NH hydrogen available for hydrogen-bonding, interactions other than hydrogen bonds appear to be involved in the different conformational properties of -Gly-X-Pro- and -Gly-Pro-X- sequences. In native collagen most sequences of the type -Gly-X-I- contain hydroxyproline rather than proline in the third position. Recent studies (Ramachandran *et al.*, 1973, 1975; Traub, 1974) have shown that hydroxyproline may stabilize the collagen triple helix by participating in hydrogen bonds. Comparative studies on the properties of polymers of composition (Gly-X-Pro)$_n$ and (Gly-X-Hyp)$_n$ should resolve this question.

Physicochemical properties have been examined for many more polypeptide models of collagen in connection with the elucidation of the structure of collagen. An extended discussion of the properties of all polymeric models of collagen is outside the scope of this chapter.

B. *Application of Collagen Models in Studies on the Hydroxylation of Proline*

Very few proteins besides collagen contain significant amounts of hydroxyproline. Hydroxylation of proline occurs after the assembly of the polypeptide chain during collagen synthesis, and it is a critical biosynthetic step since hydroxyproline contributes to the conformational stability and perhaps to the acquisition of the eventual secondary and tertiary structure of collagen. Additionally, the synthesis of hydroxyproline also somehow facilitates the secretion of collagen; unhydroxylated collagens fail to be secreted unless hydroxylation is allowed to occur. The biological and biochemical aspects of hydroxylation of proline are discussed in Chapter 5.

An important requirement for the hydroxylation of proline seems to be in the location of the susceptible residue. In vertebrate collagens hydroxyproline appears only in the third position in each glycine-led triplet, and the requirement for glycine at the C-terminal of the susceptible proline appears to be nearly absolute. At present only one exception to this is known in the case of vertebrate collagens. This involves a 3-

hydroxyproline residue in position 983 followed by a 4-hydroxyproline residue in the helical region of the $\alpha 1$ chain. Recent studies on the sequence of collagen from various vertebrate tissues indicate that not all prolines in the third position are equally good substrates for proline hydroxylase (also referred to as protocollagen proline hydroxylase, collagen proline hydroxylase, and peptidyl proline hydroxylase). An examination of the sequences surrounding the fully or partially hydroxylated proline residues suggests that the extent of hydroxylation is regulated by neighboring residues and sequences. Because of the complexity of the collagen molecule itself, the role of sequence and conformation in regulating the extent of hydroxylation can be studied best by using synthetic polypeptide models incorporating the sequence and conformational features under investigation. Although considerable data have accumulated on the interaction between polypeptide models of collagen and proline hydroxylase, no systematic evaluation has been made of these features in determining the extent of hydroxylation. In the following discussion, we will attempt to resolve these questions by discussing available data on polypeptides in enzyme interaction in terms of some of the known properties of polypeptides.

1. Interaction between Polypeptide Models and Proline Hydroxylase

A necessary condition for the transformation of a substrate analog by an enzyme is that substantial resemblance exists between the analog and the substrate. A substrate analog occupies the same site on the enzyme that complexes the substrate, and the ensuing competitive inhibition can be utilized in examining similarities between the substrate and the analog. Such investigations are often useful in elucidating enzyme mechanisms. Interaction between an enzyme and a substrate analog does not always result in the enzymatic transformation of the analog. Of the many polypeptide models of collagen which have been examined for their ability to interact with proline hydroxylase, several undergo significant hydroxylation, whereas others form tight complexes with the enzyme, resulting in the competitive inhibition of the natural substrate of the enzyme without undergoing measurable hydroxylation themselves (Table 2).

a. Role of Sequence and Conformation. Although sequence and conformation seem to play a role in polypeptide–enzyme interaction, no generalizations can be made since polymers with a high degree of order as well as some which show no organized structure in aqueous solutions interact equally well. In low-molecular-weight polymers, chain length also becomes an important factor in determining the extent of interaction. Polyproline II forms strong complexes with proline hydroxylase (Prockop and Kivirikko, 1969) whereas polyproline I does not show any interaction (Kivi-

TABLE 2

Collagen Models Which Show Significant Interaction with Proline Hydroxylase

Polypeptide	Reference
1. Peptides which do not undergo significant hydroxylation but are competitive inhibitors	
Polyproline II	Kivirikko *et al.*, 1967; Prockop and Kivirikko, 1969
$(\text{Gly-Pro})_n$	Bhatnagar, Rapaka and Ryu, unpublished
$(\text{Gly-Gly-Pro})_n$	Kivirikko *et al.*, 1969
$(\text{Gly-Sar-Pro})_n$	Bhatnagar and Rapaka, 1974
$(\text{Gly-Pro-3-MePro})_{n=1-4}$[a]	Hutton *et al.*, 1968
$(\text{Gly-Pro-4-FluoroPro})_{n=1-4}$[a]	Hutton *et al.*, 1968
2. Polypeptides which undergo significant hydroxylation[b]	
$(\text{Gly-Ala-Pro})_n$[c]	Kivirikko *et al.*, 1967, 1969
$(\text{Gly-Leu-Pro})_n$	Bhatnagar and Rapaka, 1974
$(\text{Gly-Pro-Pro})_n$	Kivirikko *et al.*, 1967
Z-Gly-Pro-Hyp-Gly-$(\text{Pro-Pro-Gly})_5$	Kikuchi *et al.*, 1971
AOC-Glu-Pro-Gly-Leu-Pro-Gly-Pro-Pro-Gly-OH	Okada *et al.*, 1971
AOC-Glu(OMe)-Pro-Gly-Leu-Pro-Gly-Pro-Pro-Gly-OMe	Okada *et al.*, 11971
Glu-Pro-Gly-Leu-Pro-Gly-Pro-Pro-Gly	Okada *et al.*, 1971
Ala-Arg-Gly-Ile-Lys-Gly-Ile-Arg-Gly-Phe-Ser-Gly[d]	Kivirikko *et al.*, 1972
Ala-Arg-Gly-Met-Lys-Gly-His-Arg-Gly-$(\text{Pro-Pro-Gly})_4$[e]	Kivirikko *et al.*, 1972
$(\text{Pro-Pro-Gly})_4$-Ala-Arg-Gly-Met-Lys-Gly-His-Arg-Gly-$(\text{Pro-Pro-Gly})_4$[e]	Kivirikko *et al.*, 1972
$(\beta\text{-Ala-Pro-Pro})_n$	Bhatnagar and Rapaka (1975*b*)

[a] Hydroxylation of these polypeptides was not discussed in the original report.
[b] Since these polypeptides interact with the enzyme, it is assumed that they also cause competitive inhibition.
[c] $(\text{Gly-Ala-Pro})_n$ oligopeptides with protective groups still attached were somewhat better substrates. (Kivirikko *et al.*, 1969).
[d] These polymers showed significant hydroxylation of the lysine residue on interaction with lysyl hydroxylase.
[e] These polypeptides also showed significant hydroxylation of proline and lysine.

rikko *et al.*, 1967). Since polyhydroxyproline and poly-*O*-acetyl hydroxyproline show many structural characteristics of polyproline II, we examined these polymers for interaction with proline hydroxylase. Neither polymer showed significant interaction. Recent investigations have suggested that small conformational differences may exist between the proline and hydroxyproline polymers. Polyproline II chains are flexible and the proline rings show considerable mobility, whereas polyhydroxyproline chains are relatively rigid and investigations also show that the γ-carbon in polyhydroxyproline is out of the ring plane and that the hydroxyl group enters into a stabilizing interaction with a backbone carbonyl (Torchia, 1972; Torchia and Lyerla, 1974). Since the hydroxyl group in poly-*O*-acetyl hydroxyproline is blocked, one may expect a more flexible chain because of the abolition of the hydroxyl–back bone interaction. The failure of poly-*O*-acetyl hydroxyproline to bind to the enzyme indicated that the substitution at the γ-carbon rather than backbone conformation abolished interaction. These studies are useful in explaining the reduced affinity of partially hydroxylated natural substrate observed by Juva and Prockop (1969). Reduced binding of partially hydroxylated substrates is also implicated by the kinetics of the hydroxylation reaction which shows a very high initial rate followed by a steady decline in the rate of hydroxylation until the plateau is reached.

Since hydroxyproline is always followed by a glycine residue in collagen and hydroxylation in synthetic polymers occurs only at proline residues which are followed by glycine residues, the most interesting and useful investigations have been made using polytripeptides of the generalized composition $(\text{Gly-X-Y})_n$, where proline is located in either the second or third position. Kikuchi *et al.*, (1969) examined the effect of the presence of hydroxyproline in the sequence on the interaction between polypeptide analogs of collagen and proline hydroxylase. They found no difference in the interaction properties of *tert*-pentyloxy-carbonyl-$(\text{Pro-Pro-Gly})_6$ and benzyloxycarbonyl-Gly-Pro-Hyp-Gly-$(\text{Pro-Pro-Gly})_5$. A possible explanation for their results is that a single hydroxyproline residue in the N-terminal triplet of a low-molecular-weight polypeptide may not have a significant effect on the overall conformation, especially since these small polypeptides would not be expected to generate ordered structures under the experimental conditions used (37°, aqueous medium). An interesting observation in this connection was made by Hutton *et al.* (1968) who observed inhibition of proline hydroxylase by oligopeptides of composition $(\text{Gly-Pro-4-fluoroproline})_n$ (n = 1–4) and $(\text{Gly-Pro-3-methylproline})_n$ (n = 1–4). These were compared with $(\text{Gly-Pro-Pro})_n$ (n = 1–4) for their ability to inhibit proline hydroxylase. $(\text{Gly-Pro-Pro})_n$ is both a good substrate and a good inhibitor of the enzyme and the 3-methylproline analog inhibited the enzyme to the same extent as did $(\text{Gly-Pro-Pro})_n$; the 4-fluoroproline-containing oligopeptide was much less inhibitory. Although no definite conclusions can be drawn in the absence of detailed conformation data on the oligopeptides containing the methyl- or fluoro-substituted prolines, the speculation can be made that these polymers would have backbone conformations similar to the (Gly-Pro-Pro)$_n$ oligopeptides. These studies suggest that while the backbone conformation may serve as a recognition mechanism, it is not the most important single regulatory factor in determining the extent of complex formation with the enzyme. Further evidence for this has come from studies on the interaction between several different sequential polypeptides and proline hydroxylase. (Gly-Pro-Pro)$_n$ and (Gly-Pro-Ala)$_n$, which exhibit highly ordered collagen-like conformations in solution (Engel *et al.*, 1966; Brown *et al.*, 1972), behave quite differently towards proline hydroxylase. (Gly-Pro-Pro)$_n$ exhibits strong interaction with the enzyme (Kivirikko and Prockop, 1967; Hutton *et al.*, 1968), whereas (Gly-Pro-Ala)$_n$ shows very little complex formation (Kivirikko *et al.*, 1969). We have recently synthesized (Gly-Pro-Sar)$_n$ as a model for collagen (Rapaka and Bhatnagar, 1974). (Gly-Pro-Sar)$_n$, which has a collagen-like conformation in solution (Ananthanarayanan *et al.*, 1976), also did not interact with proline hydroxylase (Bhatnagar and Rapaka, 1974). In contrast (Gly-Gly-Pro)$_n$ and (Gly-Pro)$_n$, both of which are unstructured in solution (Oriel and Blout, 1966; Mattice and Mandelkern, 1971), interact very strongly with proline hydroxylase leading to competitive inhibition of the enzyme without undergoing significant hydroxylation themselves. (Gly-Ala-Pro)$_n$, which is structureless in solution (Doyle *et al.*, 1971), also interacts with the enzyme, resulting in high levels of hydroxylation, and competitively inhibits the hydroxylation of the natural substrate (Kivirikko *et al.*, 1969; Bhatnagar and Rapaka, 1974).

The role of conformation in regulating the extent of hydroxylation was also studied in a more direct manner by Kikuchi *et al.* (1969) who examined the hydroxylation of (Gly-Pro-Pro)$_n$ where n was 1, 3, 5, 10, 15, or 20. They found that the pentamer was the most efficient substrate, and (Gly-Pro-Pro)$_{15}$ and (Gly-Pro-Pro)$_{20}$ were not very good substrates. Heating and quenching the polypeptides increased the efficiency of hydroxylation in (Gly-Pro-Pro)$_{15}$ whereas (Gly-Pro-Pro)$_5$ did not show any change in substrate properties. Since (Gly-Pro-Pro)$_5$ exists as a random coil under the experimental conditions and (Gly-Pro-Pro)$_{15}$ and (Gly-Pro-Pro)$_{20}$ are triple helical, they concluded that the random-coil conformation was more efficient for substrate activity. Prockop and colleagues (Kivirikko *et al.*, 1972) examined the hydroxylation of (Gly-Pro-Pro)$_{10}$ in what they considered "triple-helical" and random conformations, and they did not obtain any differences in the ability of the polymer in the two conformations to undergo hydroxylation. They concluded that the enzyme could interact equally well with the triple-helical and random-coil conformations. However, in these studies the hydroxylation was carried out under conditions in which the polypeptides could not be expected to generate triple-helical conformations; Kikuchi and colleagues examined the hydroxylation at 37°, while Prockop and his co-workers studied $(Gly\text{-}Pro\text{-}Pro)_{10}$ at 30°. Earlier Sakakibara *et al.* (1968) and Kobayashi *et al.* (1970) had shown that the denaturation temperatures of the $n = 5$ polymer was well below $-15°$ and that $n = 10$ polymer denatured at 24°. In view of this, while the initial conclusion of Kikuchi and co-workers concerning triple helix in $(Gly\text{-}Pro\text{-}Pro)_{20}$ was valid, the studies with $(Gly\text{-}Pro\text{-}Pro)_{10}$ were inconclusive. These experiments were further complicated by the fact that the $n = 15$ and $n = 20$ polymers were not very soluble in aqueous media at the pH at which the enzyme reaction was carried out. More recently Berg and Prockop (1973) presented evidence which supports the idea that the triple-helical conformation of the natural substrate reduces the efficiency of hydroxylation. These observations, coupled with the studies on the interaction of proline hydroxylase with structured and unstructured polymers in solution, suggest that if there is a conformation requirement for interaction it may be fulfilled by the induction of appropriate conformational changes in the polymer which may accompany the enzyme–polypeptide complex formation. Induced conformational changes which facilitate substrate–enzyme interaction have been proposed for many other enzymatic reactions usually involving a conformational change in the enzyme.

b. Size of the Polypeptide. Synthetic polypeptides have also been useful in establishing a minimum size requirement for a polypeptide to interact with proline hydroxylase. While free proline and the tripeptide Gly-Pro-Pro failed to undergo hydroxylation (Kivirikko and Prockop, 1967), Pro-Pro-Gly was significantly hydroxylated (Kikuchi *et al.*, 1969). These observations are consistent with the requirement of a glycine residue on the C-terminal side of a susceptible proline for hydroxylation to occur. As discussed above, the effect of the size of the polypeptide on its interaction with proline hydroxylase is directly related to conformation, and this must be considered in the examination of large polypeptides of sequences which favor triple-helix generation. A very interesting aspect of the molecular-size requirement suggested by the studies of Kikuchi *et al.* (1969, 1971) and Kivirikko *et al.* (1971) concerns the preferential hydroxylation of proline residues in the interior of the polymer, the susceptible proline residues in triplets at the two terminals being hydroxylated less efficiently. Increasing the number of triplets in the interacting peptide provides larger numbers of internally located susceptible proline residues and, as a direct consequence, results in relatively greater efficiency of hydroxylation. The length of the polypeptide chain also plays a significant role in determining the level of competitive inhibition by polyproline, since increasing chain length results in progressively increasing efficiency as an inhibitor (Prockop and Kivirikko, 1969). Taken together, these observations seem to suggest better interaction between the enzyme and regions located within the interacting polypeptide rather than at its N- or C-terminal ends. We have suggested the possibility of a conformational change induced as a consequence of enzyme polypeptide interaction earlier in this section. It is tempting to speculate that such a conformational change may stabilize the enzyme–polypeptide complex. The internal regions of the polypeptide would be more amenable to such changes than the free terminal ends, which would be more likely to be involved in initiating destruction of ordered conformations rather than serving as nuclei for generation of helical conformations. More physicochemical data are needed to establish whether such conformational changes do indeed occur and to determine the nucleation sites during random coil–helix transformation.

c. Side Chains and Enzyme–Polypeptide Interaction. An examination of the available sequence data indicates that many of the underhydroxylated hydroxyproline residues occur adjacent to either a charged residue, such as glutamic acid, asparagine, glutamine, lysine, or arginine, or they have a neighboring residue with a bulky side chain. It should be mentioned that not all hydroxyproline residues occurring next to one of these residues are underhydroxylated. A fruitful approach in investigating the role of neighboring residues in determining the extent of hydroxylation is to investigate the interaction between the enzyme and polytripeptide containing the residue in question. Okada *et al.* (1972) examined the effect of glutamic acid in the second position on the hydroxylation of proline in polytripeptides. $(\text{Gly-Pro-Glu})_3$ did not show appreciable interaction since it did not undergo significant hydroxylation and showed only weak inhibitory properties. Esterification of the side-chain carboxyl group of the glutamic acid residue did not increase the substrate activity of the polypeptides, indicating that charge was probably not the major factor in the low interaction properties of the polypeptide and suggesting that the size of the side chain may play a role. We have examined the role of the side chain in determining the interaction between polytripeptides and proline hydroxylase, as a function of the side chain, to investigate the stereochemical aspects of the interaction. The results of our study are summarized in Table 3. The interaction was examined in terms of the ability of the polypeptide to bind to the enzyme and measured as their ability to undergo hydroxylation and to competitively inhibit the enzyme. When the residue in the second position had no side chain on the α-carbon as in glycine and sarcosine, very strong competitive inhibition but no hydroxylation was observed. Alanine, with the smallest side chain, was a good substrate, whereas valine with two vicinal methyl groups did not show any interaction at all. Polypeptides containing leucine, with a

TABLE 3
Effect of Side Chain on the Hydroxylation of (Gly-X-Pro)$_n$

Polymer[a]	Fraction of susceptible proline hydroxylated (%)[b]	K_m (μg/ml)
(Gly-Gly-Pro)$_n$	Traces	Not determined
(Gly-Sar-Pro)$_n$	Traces	Not determined
(Gly-Ala-Pro)$_n$	16.0	160
(Gly-Val-Pro)$_n$	Traces	Not determined
(Gly-Leu-Pro)$_n$	10.2	380
(Gly-Pro-Pro)$_n$	30.0	118

[a] Molecular weight approximately 4000 on the basis of gel filtration techniques.
[b] Hydroxylation was carried out with 400-fold purified proline hydroxylase. The reaction mixture contained substrate, 300 μg/ml; enzyme 0.2 mg/ml; ascorbate, 0.5 mM; ferrous ammonium sulfate, 0.1 mM, and α-ketoglutarate, 0.1 mM. Hydroxylation was carried out for 1 hr at 37°C.

branched side chain and a longer "stalk" than in valine, showed good interaction leading to hydroxylation, but the polypeptides containing alanine or leucine were not as good as (Gly-Pro-Pro)$_n$. These studies indicated that small side chains facilitate interaction and hydroxylation. Since (Gly-Pro-Pro)$_n$ showed the best hydroxylation, we may assume that the presence of proline in the second position most readily permits the acquisition of the conformation favored for hydroxylation. In examining the effect of the side chains on the polypeptide–enzyme interaction, one must bear in mind the stereochemical constraints imposed by the imino peptide bond. Peptides involving an imino residue have limited conformations because of the pyrrolidine ring across the N-C$^\alpha$ bond, and the peptide bond

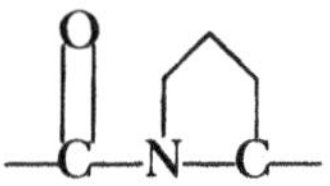

is essentially one large, relatively rigid unit which dictates the conformational parameters of the preceding residue. The side chain on the α-carbon of the preceding residue interacts with the δ-methylene of the pyrrolidine ring (Schimmel and Flory, 1968). The effect of the steric repulsions between the side chain of the preceding residue and the pyrrolidine ring is to alter the conformational energy of the prolyl residue. The polyproline II conformation is restricted to the energy minimum, and it also happens to be the preferred conformation for interaction, as seen in the strong interaction between polyproline II and proline hydroxylase.

By raising the conformational energy, an adjacent residue side chain would increase the energy requirement for the conformational transition to the preferred transition and hinder the induced conformational change. The conformational transitions are also subject to the range of "allowed conformations" for the glycyl-X peptide bond involving the second residue in the triplet. Leach *et al.* (1966) determined the effect of side chains on the sterically permitted conformation, and they determined that, while the peptide groups adjacent to glycine residues could assume only 50% of all conceivable conformations, the alanine side chain restricts these to 16% and the valine or isoleucine residues further reduce the backbone conformations to only about 5%. Leucine is less restrictive considering its side chain complexity, reducing the conformational possibilities to regions comparable to alanine peptides. Our data in Table 3 show that the side-chain complexity reduces interaction essentially in the same order as predicted by the stereochemical influences of the side chains on the backbone conformation, and they confirm the role of induced conformational transitions in enzyme–polypeptide interaction.

2. *Role of the Prolyl–Glycine Peptide Bond in the Hydroxylation of Proline*

As seen in the earlier discussion, in order for a proline residue to undergo hydroxylation, it must be present in peptide linkage with glycine at its C-terminus. This nearly absolute requirement for glycine suggests that hydroxylation of proline may involve some highly specific stereochemical properties of the prolyl–glycine peptide bond. The overall conformation of collagen-like polytripeptides is influenced not only by the stereochemical properties of the pyrrolidine residue, but they are also subject to the constraints imposed by the glycyl-X peptide bond. In the repeating sequence -Gly-X-Pro-Gly-X-Pro-, the largest degree of conformational freedom at an internal proline peptide bond is at the prolyl C^{α}—C, and there is no interaction between proline and the following residue (Schimmel and Flory, 1968). The absence of a side chain on the glycine residue contributes to the relatively large stereochemical freedom enjoyed by glycine in its peptide bonds. We have examined the possibility that the requirement for glycine may be related to the large freedom of rotation made possible by its presence next to proline. Our studies mentioned earlier (Ananthanarayanan *et al.*, 1976) indicated that sarcosine may replace glycine on the N-terminal side of proline $(\text{Gly-Gly-Pro})_n$, with the resulting polytripeptide $(\text{Gly-Sar-Pro})_n$ having properties very similar to $(\text{Gly-Gly-Pro})_n$. Comparison of the two collagen analogs $(\text{Gly-Gly-Pro})_n$ and $(\text{Gly-Sar-Pro})_n$ indicated that sarcosine in the second position acted exactly like glycine, probably because of the absence of a side-chain on the α-carbon.

Both polymers lacked structure in solution and formed strong complexes with proline hydroxylase without undergoing significant hydroxylation (Table 3). The methyl group astride the -Gly-Sar- peptide bond apparently does not introduce any stereochemical influences that would detract from the properties of $(\text{Gly-Gly-Pro})_n$. Sarcosine on the C terminal of proline, however, had completely different stereochemical properties and behaved like a residue with a side chain since $(\text{Gly-Pro-Sar})_n$, unlike $(\text{Gly-Pro-Gly})_n$ [which is the same as $(\text{Gly-Gly-Pro})_n$], showed ORD and CD spectra characteristic of collagen. Significant in the present context, $(\text{Gly-Pro-Sar})_n$ did not show interaction with the hydroxylase, a property shared with $(\text{Gly-Pro-Ala})_n$ and in contrast to $(\text{Gly-Pro-Gly})_n$. We interpret these observations as indicating at least limited stereochemical interaction between the proline residue and the sarcosine following it. The presence of the methyl group astride the prolyl-sarcosine peptide bond presumably alters the conformational properties from those of the prolyl-glycine peptide bond. In view of the postulated conformational changes necessary for polypeptide–enzyme interaction and hydroxylation, it seemed that maximal conformational flexibility at the prolyl peptide bond is required for hydroxylation. To test this hypothesis further it became necessary to synthesize a polymer in which glycine would be replaced by a larger amino acid residue which would allow maximal conformational freedom at the prolyl C^α—C bond. These criteria are met in β-alanine, which is larger than glycine but does not contain a side chain. It incorporates an additional degree of rotational freedom at the C^α—C^β methylene bond. $(\beta\text{Ala-Pro-Pro})_n$ showed solution properties very similar to those of $(\text{Gly-Pro-Pro})_n$, with almost overlapping ORD and CD spectra (Bhatnagar and Rapaka, 1975*a*). $(\beta\text{Ala-Pro-Pro})_n$ is stabilized in solution by hydrogen bonds and undergoes a conformational transition on heating, with a denaturation temperature comparable to $(\text{Gly-Pro-Pro})_n$ of the same degree of polymerization. In view of the close similarities in solution properties, it may be concluded that the role of glycine in the sequence is to provide the largest possible rotational freedom at C^α—C in addition to facilitating the closer packing of chains, since β-alanine is able to mimic glycine in the sequence. The only similarity between glycine and β-alanine is in the large rotational freedom at their peptide bonds. We compared the interactions of $(\text{Gly-Pro-Pro})_n$ and $(\beta\text{Ala-Pro-Pro})_n$ with proline hydroxylase (Bhatnagar and Rapaka, 1975*b*). Both polytripeptides underwent hydroxylation at comparable rates and exhibited similar interaction constants (K_m) (Table 4). In addition both polymers act as competitive inhibitors of the enzyme. These studies confirm our thesis that a large

TABLE 4
Comparison of (β-Ala-Pro-Pro)$_n$ and (Gly-Pro-Pro)$_n$ as Substrates and Inhibitors for Proline Hydroxylase[a]

Substrate activity	K_m (μg/ml)	V_{max} (nmole Hyp formed/ml/min)
(β-Ala-Pro-Pro)$_n$	190	1.3
(Gly-Pro-Pro)$_n$	174	1.1
Inhibitor activity	**Concentration (μg/ml)**	**Inhibition (%)**
(β-Ala-Pro-Pro)$_n$	100	29
	400	67
(Gly-Pro-Pro)$_n$	100	41
	400	73

[a] Both polymers were fractionated by gel filtration and fractions corresponding to a molecular weight of 3500 were used. Experimental conditions for hydroxylation were as described in Table 3 and the reaction was carried out for 30 min. Inhibitor activity was determined using ^{3}H-labeled substrate as described by Hutton *et al.* (1968).

freedom of rotation at prolyl C^α—C is necessary for hydroxylation. This may be related to the conformational changes accompanying interaction between a polypeptide substrate and the enzyme.

C. *Use of Polypeptide Models in Various Biological Studies on Collagen*

In addition to its involvement throughout the body as a structural material, collagen has been implicated in crucial physiological phenomena such as specific immunologic responses and the initiation of blood coagulation by the activation of the Hageman factor and by platelet aggregation and release.

As a structural material collagen is often present in the body in complexes with proteoglycans. Such complexes provide completely different mechanical properties from structures made up entirely of collagen. Interaction between proteoglycans and collagen is involved in collagen fibrillogenesis. Although such interactions have been the subject of extensive investigations, no studies have been made with collagen models. Interaction of acidic proteoglycans with homopolymers of basic amino acids have been investigated (Gelman *et al.*, 1973; Gelman and Blackwell, 1973). It would be very useful to investigate the interaction between synthetic polypeptides resembling the polar and nonpolar regions of the collagen molecule and proteoglycans. Such studies would provide valuable information concerning nucleation during fibrillogenesis and factors which regulate the diameter and strength of collagen fibers.

Immunochemical properties of collagen and collagen-like synthetic peptides have been the subject of considerable investigation and are described by Timpl in this volume (Chapter 7). The immunochemical properties of a protein are determined by conformation and sequence. Antibodies to native proteins are conformation specific whereas sequence specificity is limited to antibodies to the random-coil form of the antigenic protein. The immunogenic properties of collagen-like polypeptides must be considered within these limitations. With the current technology of peptide synthesis, it has become possible to synthesize models which mimic either the conformation, or sequence, or both, of specific regions of the collagen molecule. Sela and colleagues (Borek *et al.*, 1969; Maoz *et al.*, 1973*a,b*) have examined the immunological properties of $(\text{Gly-Pro-Pro})_n$, a polymer known to generate collagen-like conformations in solution. The lack of cross-reaction between antibodies to $(\text{Gly-Pro-Pro})_n$ and $(\text{Pro}^{66}\text{-Gly}^{34})_n$ or gelatin and the ability to cross-react with anticollagen antibodies supported the recognition of the triple-helical conformation as the antigenic marker. The role of conformation is emphasized in studies on cross-reaction with the polyhexapeptides $(\text{Gly-Pro-Ala-Gly-Pro-Pro})_n$, $(\text{Gly-Ala-Pro-Gly-Pro-Pro})_n$, $(\text{Gly-Ala-Ala-Gly-Pro-Pro})_n$, and $(\text{Gly-Ala-Pro-Gly-Pro-Ala})_n$. The cross-reaction decreased in the above order, which is also the order of stability of the polymers. The possibility that the antigenicity was due to the sequence -Gly-Pro-Pro- was ruled out by the lack of significant cross-reaction with the tripeptide Gly-Pro-Pro. The very weak cross-reaction with Gly-Pro-Pro was attributed to the terminal tripeptide unit in the polymer. The lack of significant cross-reactivity with smaller polypeptides of the same sequence is related to their inability to generate triple-helical conformations. Antigenic determinants of collagen are not confined to the helical region and are conformational as well as sequential (see Chapter 7). Since the largest triple-helix-promoting sequences are located near the two ends of the collagen molecule, the studies with $(\text{Gly-Pro-Pro})_n$ have provided only an approximation of these regions, but they have been invaluable in establishing the conformational aspect of the immunogenicity of collagen.

Collagen is known to play an important role in blood coagulation and thrombus formation. Nossel and colleagues (1969) indicated that the free carboxyl groups and triple-helical conformation of collagen are important in the activation of the Hageman factor. More recently, Walton (1974) examined the activation of the Hageman factor in the presence of (Gly-

Pro-Pro)$_n$, mol. wt. = 8000; (Gly-Pro-Lys)$_n$, mol. wt. = 20,000; (Gly-Lys-Pro)$_n$, mol. wt. = 25,000; and (Gly-Glu-Pro)$_n$, mol. wt. = 40,000. None of the polymers activated the Hageman factor. However, of the polymers used, the only one with free carboxyl groups, (Gly-Glu-Pro)$_n$, does not exist in triple-helical conformation. Walton also did not observe activation by native collagen, although a high level of activity was seen in the presence of proteoglycan. It was concluded that the activation observed in earlier studies may have been due to proteoglycan contamination in collagen. Collagen is known to be involved in initiating platelet aggregation and release. Walton did not observe aggregation with any of the above polypeptides. Platelet-release reactions occur on charged solid surfaces, and the models examined may have lacked the structural and charge distribution requirements.

V. Concluding Remarks

In spite of the superficially apparent order in the molecule, collagen is indeed a complex entity, and a fruitful approach in elucidating its structure and function is through the application of synthetic models. Indeed, the elucidation of the structure of collagen and the development of major concepts concerning its physicochemical properties would have been seriously hampered in the absence of model compounds.

We have presented a state of the art summary of procedures available for synthesis of collagen-like polypeptides, and we have also presented a complete listing of all the model polymers that have been synthesized. It became apparent to us that much of the biochemical work on collagen has been conducted with a limited number of models, and it is our intention to point out the availability of the large range of material available for systematic investigations of many facets of collagen chemistry. In compiling this chapter, it also became apparent that, although collagen has been implicated in many biological phenomena, models have been used in investigations on but a few. One of the difficulties in investigations on physiological functions of collagen is that very few reactions involving collagen occur in the intracellular milieu where it is possibly still in a soluble form. An example of a well-investigated biochemical reaction are the studies on proline hydroxylation. Models have not been used for investigating subsequent intracellular reactions; for instance, those involving the glycosylation of hydroxylysine and even the hydroxylation of lysine has not received the scrutiny that the hydroxylation of proline has received. Since the glycosylation of collagen may play a major role in the eventual functioning of collagen, it would be desirable to investigate the sequence and conformation which regulate this step. The completed collagen molecule always occurs in the extracellular milieu in the form of highly ordered arrays of fibrils and fibers, the architecture of which is ultimately traceable to the interaction properties of the specific collagen molecules comprising them. Since subtle differences in physicochemical properties may presumably be introduced during the posttranslational processing of collagen, it is of interest to elucidate the features built within the collagen polypeptide chain which may participate in regulating the posttranslational events.

Many of the biological functions attributed to collagen seem to involve highly specific interactions involving specific types of collagen aggregates. Although many of the polypeptide models of collagen are known to form aggregates resembling collagen in the solid state, the conditions required for the examination of many biological phenomena such as platelet aggregation preclude the use of any but the most insoluble collagen models. Most of the polypeptide models which have been discussed in this chapter have molecular-weight distributions which do not favor generation of ordered fiber-like structures. More homogeneous molecular-weight distributions are seen in polymers synthesized by solid-phase techniques. Unfortunately the current state of the art does not permit synthesis of very large molecular weight polymers by this technique in appreciable yields. Applications of collagen-like polymers in investigating the specific functions of collagen in the fibrous state must await further developments in the technology of polypeptide synthesis.

ACKNOWLEDGMENTS

The authors wish to express their thanks to Martha Fisher Jenkins and Jamie McManus for their assistance in various aspects of this project and to Sandra Hodess for her invaluable secretarial assistance. The studies from the authors' laboratory have been supported by USPHS grants AM-15178, HD-05812, and DE-03861 and were carried out partly during the tenure of a Research Career Development Award, DE-41311 to R.S.B.

References

Ananthanarayanan, V. S., Rapaka, R. S., Brahmachari, S. K., and Bhatnagar, R. S., 1976, Polypeptide models of collagen: Solution properties of (Gly-Pro-Sar)$_n$ and (Gly-Sar-Pro)$_n$, *Biopolymers* **15:**707.

Anderson, G. W., Blodinger, J., and Welcher, A. D., 1952, Tetraethyl pyrophosphite as a reagent for peptide synthesis, *J. Am. Chem. Soc.* **74:**5309.

Anderson, G. W., Zimmerman, J. E., and Callahan, F. M., 1964, The use of esters of N-hydroxysuccinimide in peptide synthesis, *J. Am. Chem. Soc.* **86:**1839.

Andreeva, N. S., Debanov, V. A., Millinová, M. I., Shibnev, V. A., and Chirgadze, Yu. N., 1961, A synthetic polymer isomorphous to collagen, *Biofizika* **6:**244.

Astbury, W. T., 1934, X-ray studies of protein structure, *Cold Spring Harb. Symp. Quant. Biol.* **2:**15.

Balasubramanian, R., Lakshminaraynan, A. V., Sabesan, M. N., Tegoni, G., Venkatesan, K., and Ramachandran, G. N., 1971, Conformation of amino acids. VI. Conformation of the proline ring as observed in crystal structures of amino acids and peptides, *Int. J. Protein Res.* **3:**25.

Balian, G., Click, E. M., and Bornstein, P., 1971, Structure of rat skin collagen α1-CB8. Amino acid sequence of the hydroxylamine-produced fragment HA1, *Biochemistry* **10:**4470.

Balian, G., Click, E. M., Hermodson, M. A., and Bornstein, P., 1972, Structure of rat skin collagen α1-CB8. Amino acid sequence of the hydroxylamine-produced fragment HA2, *Biochemistry*, **11:**3798.

Bayer, E., Eckstein, H., Hagele, K., Konig, W. A., Bruning, W., Hagenmaier, H., and Parr, W., 1970, Failure sequences in the solid phase synthesis of polypeptides, *J. Am. Chem. Soc.* **92:**1735.

Berg, R. A., and Prockop, D. J., 1973, Purification of [^{14}C]protocollagen and its hydroxylation by prolyl hydroxylase, *Biochemistry* **12:**3395.

Berg, R. A., Olsen, B. R., and Prockop, D. J., 1970, Titration and melting curves of the collagen-like triple helices formed from $(\text{Pro-Pro-Gly})_{10}$, *J. Biol. Chem.* **245:**5759.

Bergmann, M., 1935, Complex salts of amino acids and peptides. II. Denaturation of L-proline with the aid of rhodanilic acid. The structure of gelatin, *J. Biol. Chem.* **110:**471.

Bhatnagar, R. S., and Rapaka, R. S., 1974, Conformational aspects of proline hydroxylation in polytripeptide models of collagen, *Fed. Proc.* **33:**1596.

Bhatnagar, R. S., and Rapaka, R. S., 1975*a*, Polypeptide models of collagen: Properties of $(\text{Pro-Pro-}\beta\text{-Ala})_n$, *Biopolymers* **14:**597.

Bhatnagar, R. S., and Rapaka, R. S., 1975*b*, Hydroxylation of proline in a glycine-less polytripeptide model of collagen, *Fed. Proc.* **34:**698.

Bloom, S. M., Dasgupta, S. K., Patel, R. P., and Blout, E. R., 1966, The synthesis of glycyl-L-prolyl glycyl and glycyl-L-prolyl-L-alanyl oligopeptides and sequential polypeptides, *J. Am. Chem. Soc.* **88:**2035.

Bodanszky, M., and Ondetti, M. A., 1966, Easily removable protecting groups and their removal, *in: Peptide Synthesis* (G. A. Olah, ed.), p. 26, Interscience Publishers, New York.

Bodanszky, M., Sheehan, J. F., Ondetti, M. A., and Lande, S., 1963, Glycine analogs of bradykinin, *J. Am. Chem. Soc.* **85:**991.

Bodanszky, M., Bater, R. J., Chang, A., Fink, M. L., and Funk, K. W., 1972, Experiments with active esters in solid-phase peptide synthesis, *in: Chemistry and Biology of Peptides* (J. Meinhofer, ed.), pp. 203–207, Ann Arbor Science Publishers Inc., Ann Arbor, Michigan.

Borek, F., Kurtz, J., and Sela, M., 1969, Immunological properties of collagen-like synthetic polypeptide, *Biochim. Biophys. Acta* **188:**314.

Bornstein, P., 1967, Comparative sequence studies of rat skin and tendon collagen. I.

Evidence for incomplete hydroxylation of individual prolyl residues in the normal proteins, *Biochemistry* **6**:3082.

Bornstein, P., 1969, Comparative sequence studies of rat skin and tendon collagen. II. The absence of a short sequence at the amino terminus of the skin α1 chain, *Biochemistry* **8**:63.

Bornstein, P., 1970, Structure of α1-CB8, a large cyanogen bromide fragment from the α1 chain of rat collagen. The nature of a hydroxylamine-sensitive bond and composition of tryptic peptides, *Biochemistry* **9**:2408.

Brown, F. R., Carver, J. P., and Blout, E. R., 1969, Low temperature circular dichroism of poly(glycyl-L-prolyl-L-alanine), *J. Mol. Biol.* **39**:307.

Brown, F. R., DiCorato, A., Lorenzi, G. P., and Blout, E. R., 1972, Synthesis and structural studies of two collagen analogues: Poly(L-prolyl-L-seryl-glycyl) and poly(L-prolyl-L-alanyl-glycyl), *J. Mol. Biol.* **63**:85.

Butler, W. T., 1970, Chemical studies on the cyanogen bromide peptides of rat skin collagen. The covalent structure of α1-CB5, the major hexose-containing cyanogen bromide peptides of α1, *Biochemistry* **9**:44.

Butler, W. T., and Cunningham, L. W., 1966, Evidence for the linkage of a disaccharide to hydroxylysine in tropocollagen, *J. Biol. Chem.* **241**:3882.

Butler, W. T., and Ponds, S. L., 1971, Chemical studies on the cyanogen bromide peptides of rat skin collagen. Amino acid sequence of α1-CB4, *Biochemistry* **10**:2076.

Conte, F., Lucente, G., Romeo, A., and Zanotti, G., 1973, Cyclos-formation from tripeptide systems and structure assignment by carbon-13 nuclear magnetic resonance, *Int. J. Peptide Protein Res.* **5**:353.

Cowell, R. D., and Jones, J. H., 1971*a*, The use of monoesters of catechol in the racemization-free synthesis of sequential polypeptides with amino or carboxy side-chains, *J. Chem. Soc.* **1971**:1009.

Cowell, R. D., and Jones, J. H., 1971*b*, Sequential polypeptides. Part I. Use of monoesters of catechol in the synthesis of sequential polypeptides, *J. Chem. Soc.* **1971**:1082.

Cowell, R. D., and Jones, J. H., 1972, Sequential polypeptides. Part V. The use of monoesters of catechol in the synthesis of sequential polypeptides with amino or carboxyl side-chains, *J. Chem. Soc.* **1972**:2236.

Denkewalter, R. G., and Hirschmann, R., 1969, The synthesis of an enzyme, *Am. Sci.* **57**:389.

DeTar, D. F., 1967, The active ester synthesis of sequence peptide polymers, *in: Peptides* (H. C. Beyerman, A. Van De Linde, and W. Massen van den Brink, eds.), pp. 125–130, North-Holland Publishing Company, Amsterdam.

DeTar, D. F., Honsberg, W., Honsberg, U., Wieland, A., Gouge, M., Bach, H., Tahara, A., Brinigar, W. S., and Rogers, F. F., Jr., 1963, Synthesis of peptide polymers with repeating sequence, *J. Am. Chem. Soc.* **85**:2873.

DeTar, D. F., Silverstein, R., and Rogers, F. F., Jr., 1966, Reactions of carbodiimides. III. The reactions of carbodiimides with peptide acids, *J. Am. Chem. Soc.* **88**:1024.

DeTar, D. F., Alberts, R. F., and Gilmore, F., 1972, Synthesis of sequence peptide polymers related to collagen, *J. Org. Chem.* **37**:6377.

Doyle, B. B., Traub, W., Lorenzi, G. P., and Blout, E. R., 1971, Conformational investigations on the polypeptide and oligopeptides with the repeating sequence L-alanyl-L-prolylglycine, *Biochemistry* **10**:3052.

Engel, J., 1962, Investigation of the denaturation and renaturation of soluble collagen by light scattering, *Arch. Biochem. Biophys.* **97**:150.

Engel, J., Kurtz, J., Katchalski, E., and Berger, A., 1966, Polymers of tripeptides as collagen models. II. Conformational changes of poly(L-prolyl-glycyl-L-prolyl) in solution, *J. Mol. Biol.* **17**:255.

Erickson, B. W., and Merrifield, R. B., 1972, Improved protecting groups for solid phase synthesis, *in: Chemistry and Biology of Peptides* (J. Meinhofer, ed.), pp. 191–195, Ann Arbor Science Publishers Inc., Ann Arbor, Michigan.

Fairweather, R., and Jones, J. H., 1972*a*, Sequential polypeptides. Part IV. The synthesis of poly-(L-alanyl-L-glycyl-L-proline) and its stereoisomers, *J. Chem. Soc.* **1972**:1908.

Fairweather, R., and Jones, J. H., 1972*b*, Sequential polypeptides. Part VI. The synthesis of some sequential polypeptide collagen models containing proline analogues, *J. Chem. Soc.* **1972**:2475.

Fairweather, R., and Jones, J. H., 1973, The antigenecity of sequential polypeptides, *Immunology* **25**:241.

Fasman, G. D., and Blout, E. R., 1963, Copolymers of L-proline and sarcosine: Synthesis and physical-chemical studies, *Biopolymers* **1**:99.

Fietzek, P. P., Rexrodt, F. W., Wendt, P., Stark, M., and Kuhn, K., 1972, The covalent structure of collagen. Amino acid sequence of peptide α1-CB6-C2, *Eur. J. Biochem.* **30**:163.

Fietzek, P. P., Rexrodt, F. W., Hopper, K. E., and Kuhn, K., 1973, The covalent structure of collagen. 2. The amino acid sequence of α1-CB7 from calfskin collagen, *Eur. J. Biochem.* **38**:396.

Fischer, E., 1906, Synthese von peptiden XV, *Chem. Ber.* **39**:2893.

Flory, P. J., and Weaver, E. S., 1960, Helix ⇆ coil transitions in dilute aqueous collagen solutions, *J. Am. Chem. Soc.* **82**:4518.

Fuchs, S., Mozes, E. Maoz, A., and Sela, M., 1974, Thymus independence of a collagen-like synthetic polypeptide and of collagen and the need for thymus and bone marrow-cell cooperation in the immune response to gelatin, *J. Exp. Med.* **139**:148.

Gelman, R. A., and Blackwell, J., 1973, Interactions between mucopolysaccharides and cationic polypeptides in aqueous solution: Chondroitin 4-sulfate and dermatan sulfate, *Biopolymers* **12**:1959.

Gelman, R. A., Rippon, W. B., and Blackwell, J., 1973, Interaction between chondroitin 6-sulfate and poly-L-lysine in aqueous solution: Circular dichroism studies, *Biopolymers* **12**:541.

Gisin, B. F., and Merrifield, R. B., 1972, Carboxyl-catalyzed intramolecular aminolysis. A side chain reaction in solid phase synthesis, *J. Am. Chem. Soc.* **94**:3102.

Goodman, M., and Glaser, R. C., 1970, Racemization mechanisms in peptide synthesis, *in: Peptides: Chemistry and Biochemistry* (B. Weinstein, ed.), pp. 269–272, Marcel Dekker Inc., New York.

Goodman, M., and Stueben, K. C., 1959, Peptide synthesis via amino acid active esters, *J. Am. Chem. Soc.* **81**:3980.

Goodman, M., and Stueben, K., 1962, Peptide synthesis via amino acid active esters. II. Some abnormal reactions during peptide synthesis, *J. Am. Chem. Soc.* **84**:1279.

Gutte, G., and Merrifield, R. B., 1969, The total synthesis of an enzyme with ribonuclease A activity, *J. Am. Chem. Soc.* **91**:501.

Harrington, W. F., and Sela, M., 1958, Studies on the structure of poly-L-proline in solution, *Biochim. Biophys. Acta* **27**:24.

Heidemann, E., and Bernhardt, H. W., 1968, Synthetic polypeptides as models for collagen, *Nature* **220**:1326.

Heidemann, E., and Nill, H. W., 1969, Synthetische Polypeptide der Sequenztypen (Glycin-LProlin-L-Serin)$_n$ und (Glycin-L-Serin-L-Prolin), *Z. Naturforsch. B* **24**:843.

Hopfinger, A. J., 1973, *Conformational Properties of Macromolecules,* Academic Press, N.Y.

Huggins, M. L., Ohtsuka, K., and Morimoto, S., 1968, Synthesis of certain polypeptides and polytripeptides, *J. Polym. Sci. C* **23:**343.

Hulmes, D. J. S., Miller, A., Parry, D. A. D., Piez, K. A., and Woodhouse-Galloway, J., 1973, Analysis of the primary structure of collagen to the origins of molecular packing, *J. Mol. Biol.* **79:**137.

Hutton, J. J., Marglin, A., Witkop, B., Kurtz, J., Berger, A., and Udenfriend, S., 1968, Synthetic polypeptides as substrates and inhibitors of collagen proline hydroxylase, *Arch. Biochim. Biophys.* **125:**779.

Johnson, B. J., 1974, Synthesis, structure, and biological properties of sequential polypeptides, *J. Pharm. Sci.* **63:**313.

Jones, J. H., 1969, Racemization free polypeptide synthesis, *J. Chem. Soc.* **1969:**1436.

Josse, J., and Harrington, W. F., 1964, Role of pyrrolidine residues in the structure and stabilization of collagen, *J. Mol. Biol.* **9:**269.

Juva, K., and Prockop, D. J., 1969, Formation of enzyme–substrate complexes with protocollagen proline hydroxylase and large polypeptide substrates, *J. Biol. Chem.* **244:**6486.

Kang, A. H., Piez, K. A., and Gross, J., 1969, Characterization of the cyanogen bromide peptides from the $\alpha 1$ chain of chick skin collagen, *Biochemistry* **8:**1506.

Kapoor, A., 1970, Recent trends in the synthesis of linear peptides, *J. Pharm. Sci.* **59:**1.

Katchalski, E., Sela, M., Silman, H. I., and Berger, A., 1964, Polyamino acids as protein models, *in: The Proteins* (H. Neurath, ed.), pp. 449–452, Vol. 2, Academic Press, New York.

Kettman, J. R., Jr., Benjamini, E., Michaeli, D., and Leung, D. Y. K., 1967, The synthesis and immunological activity of a peptide related to collagen, *Biochim. Biophys. Res. Commun.* **29:**623.

Khalikov, Sh. Kh., Poroshin, K. T., Shibnev, V. A., 1968, Preparation of polypeptides with stable conformation as possible models of esterase activity, *Dokl. Akad. Nauk. Tadzh. SSR.* **11:**28.

Kikuchi, Y., Fujimoto, D., and Tamiya, N., 1969, The enzymic hydroxylation of protocollagen models, *Biochem. J.* **115:**569.

Kikuchi, Y., Fujimoto, D., and Tamiya, N., 1971, Synthesis and enzymatic hydroxylation of protocollagen model peptides containing a hydroxyproline residue, *Biochem. J.* **124:**695.

Kitaoka, H., Sakakibara, S., and Tani, H., 1958, Synthesis of poly(L-prolyl-L-leucyl-glycyl). An attempted synthesis of model collagen, *Bull. Chem. Soc. Jpn.* **31:**802.

Kivirikko, K. I., and Prockop, D. J., 1967, Hydroxylation of proline in synthetic polypeptides with purified protocollagen hydroxylase, *J. Biol. Chem.* **242:**4007.

Kivirikko, K. I., Ganser, V., Engel, J., and Prockop, D. J., 1967, Comparison of poly-L-proline I and II as inhibitors of protocollagen hydroxylase, *Hoppe-Seyler's Z. Physiol. Chem.* **348:**1341.

Kivirikko, K. I., Prockop, D. J., Lorenzi, G. P., and Blout, E. R., 1969, Oligopeptides with the sequences Ala-Pro-Gly and Gly-Pro-Gly as substrates or inhibitors for protocollagen proline hydroxylase, *J. Biol. Chem.* **244:**2755.

Kivirikko, K. I., Suga, K., Kishida, Y., Sakakibara, S., and Prockop, D. J., 1971, Asymmetry in the hydroxylation of $(\text{Pro-Pro-Gly})_5$ by protocollagen proline hydroxylase, *Biochem. Biophys. Res. Commun.* **45:**1591.

Kivirikko, K. I., Kishida, Y., Sakakibara, S., and Prockop, D. J., 1972*a*, Hydroxylation of $(\text{X-Pro-Gly})_n$ by protocollagen proline hydroxylase: Effect of chain length, helical conformation and amino acid sequence in the substrate, *Biochim. Biophys. Acta* **271**:347.

Kivirikko, K. I., Shudo, K., Sakakibara, S., and Prockop, D. J., 1972*b*, Studies on protocollagen lysine hydroxylase. Hydroxylation of synthetic peptides and the stoichiometric decarboxylation of α-ketoglutarate, *Biochemistry* **11**:122.

Kobayashi, Y., Sakai, R., Kakiuchi, K., and Isemura, T., 1970, Physicochemical analysis of $(Pro\text{-}Pro\text{-}Gly)_n$ with defined molecular weight. Temperature dependence of molecular weight in aqueous solution, *Biopolymers* **9**:415.

Kovacs, J., Gionnotti, R., and Kapoor, A., 1966, Polypeptides with known repeating sequence of amino acids. Synthesis of poly-L-glutamyl-L-alanyl-L-glutamic acid and poly glycyl glycyl-L-phenylalanine through pentachlorphenyl active ester, *J. Am. Chem. Soc.* **88**:2282.

Kovacs, J., Mayer, G. L., Johnson, R. H., and Gionotti, R., 1972, On the problem of racemization during the synthesis of sequential polypeptides, *in: Progress in Peptide Research* (S. Lande, ed.), Vol. 11, pp. 185–193, Gordon and Breach, New York.

Kroner, Th. D., Tabroff, W., and MacGarr, J. J., 1953, Peptides isolated from a partial hydrolysate of steer hide collagen, *J. Am. Chem. Soc.* **75**:4084.

Kroner, Th. D., Tabroff, W., and MacGarr, J. J., 1955, Peptides isolated from a partial hydrolysate of steer hide collagen. II. Evidence for the prolyl-hydroxyproline linkage in collagen, *J. Am. Chem. Soc.* **77**:3356.

Lane, J. M., and Miller, E. J., 1969, Isolation and characterization of the peptides derived from the α2 chain of chick bone collagen after cyanogen bromide cleavage, *Biochemistry* **8**:2134.

Leach, S. J., Nemethy, G., and Scheraga, H. A., 1966, Computation of the sterically allowed conformations of peptides, *Biopolymers* **4**:369.

Lorenzi, G. P., Doyle, B. B., and Blout, E. R., 1971, Synthesis of polypeptides and oligopeptides with repeating sequence L-alanyl-L-prolylglycine, *Biochemistry* **10**:3046.

Maoz, A., Fuchs, S., and Sela, M., 1973*a*, Immune response to the collagen-like ordered polypeptide $(L\text{-}Pro\text{-}Gly\text{-}L\text{-}Pro)_n$, *Biochemistry* **12**:4238.

Maoz, A., Fuchs, S., and Sela, M., 1973*b*, On immunological cross-reactions between the synthetic ordered polypeptide $(L\text{-}Pro\text{-}Gly\text{-}L\text{-}Pro)_n$ and several collagens, *Biochemistry* **12**:4246.

Marshall, G. R., and Merrifield, R. B., 1965, Synthesis of angiotensins by the solid phase method, *Biochemistry* **4**:2396.

Mattice, W. L., and Mandelkern, L., 1971, Development of ordered structures in sequential copolypeptides containing L-proline and γ-hydroxy-L-proline, *Biochemistry* **10**:1926.

Meienhofer, J., 1973, Peptide synthesis: A review of the solid-phase method, *In: Hormonal Proteins and Peptides* (C. H. Li, ed.), pp. 45–267, Academic Press, New York.

Merrifield, R. B., 1963, Solid phase peptide synthesis. I. The synthesis of a tetrapeptide, *J. Am. Chem. Soc.* **85**:2149.

Merrifield, R. B., 1964, Solid phase peptide synthesis. III. An improved synthesis of bradykinin, *Biochemistry* **3**:1385.

Miller, E. J., Lane, J. M., and Piez, K. A., 1969, Isolation and characterization of the peptides derived from the α1 chain of chick bone collagen after cyanogen bromide cleavage, *Biochemistry* **8**:30.

Nossel, H. L., Wilner, G. D., and LeRoy, E. C., 1969, Importance of polar groups for initiating blood coagulation and aggregating platelets, *Nature* **221**:75.

Okada, K., Kikuchi, Y., Kawashiri, Y., and Hiramoto, M., 1972, Syntheses and enzymic hydroxylation of protocollagen model peptides containing glutamyl or leucyl residue, *FEBS Lett.* **28:**226.

Oriel, P. J., and Blout, E. R., 1966, On the structure of Gly-Pro-Gly and Gly-Pro-Ala oligopeptides and sequential polypeptides, *J. Am. Chem. Soc.* **88:**2041.

Poroshin, K. T., Maryash, L. I., Grechishko, V. S., and Shibnev, V. A., 1970, Synthesis of structurally regular polypeptides containing alanine and N^{ϵ}-benzoxycarbonyl lysine, *Dokl. Akad. Nauk. Tadzh. SSSR Khim.* **13:**19.

Prockop, D. J., and Kivirikko, K. I., 1969, Effect of polymer size on the inhibition of protocollagen proline hydroxylase by polyproline II, *J. Biol. Chem.* **244:**4838.

Purdie, J. E., and Benoiton, N. L., 1973, Piperazine dione formation from esters of dipeptides containing glycine, alanine and sarcosine. Kinetics in aqueous solution, *J. Chem. Soc.* **1973:**1845.

Ramachandran, G. N., Bansal, M., and Bhatnagar, R. S., 1973, A hypothesis on the role of hydroxyproline in stabilizing collagen structure, *Biochim. Biophys. Acta* **322:**166.

Ramachandran, G. N., Bansal, M., and Ramakrishnan, C., 1975, Hydroxyproline stabilizes both intrafibrillar structure as well as inter-protofibrillar linkages in collagen, *Curr. Sci.* **44:**1.

Rapaka, R. S., and Bhatnagar, R. S., 1975*a*, Synthesis of polypeptide models of collagen, *Int. J. Peptide and Protein Res.* **7:**119.

Rapaka, R. S., and Bhatnagar, R. S., 1975*b*, Polypeptide models of collagen. Synthesis of (Pro-Pro-β-Ala)$_n$. *Int. J. Peptide and Protein Res.* **7:**475.

Rapaka, R. S., and Bhatnagar, R. S., 1976, Polypeptide models of collagen. Synthesis of (Pro-Pro-Ala)$_n$ and (Pro-Pro-Val)$_n$, *Int. J. Peptide and Protein Res.* (in press).

Rapaka, R. S., Bhatnagar, R. S., and Nitecki, D. E., 1976*a*, Racemization in the synthesis of polytripeptide models of collagen, *Biopolymers* **15:**317.

Rapaka, R. S., Bhatnagar, R. S., and Nitecki, D. E., 1976*b*, Racemization in the synthesis of sequential polypeptides using *N*-hydroxysuccinimide, *Biopolymers* (in press).

Rothe, M., and Mazanek, J., 1972, Possible side-reactions during solid-phase peptide synthesis. II. Reaction between neighbouring chains. Formation of hydroxy groups on the resin and their consequences, *in: Chemistry and Biology of Peptides* (J. Meienhofer, ed.), pp. 89–91, Ann Arbor Science Publishers Inc., Ann Arbor, Michigan.

Sakakibara, S., Kishida, Y., Kikuchi, Y., Sakai, R., and Kakiuchi, K., 1968, Synthesis of poly-(L-prolyl-L-prolyl glycyl) of defined molecular weights, *Bull. Chem. Soc. Jpn.* **41:**1273.

Sakakibara, S., Inouye, K., Shudo, K., Kishida, Y., Kobayashi, Y., and Prockop, D. J., 1973, Synthesis of (Pro-Hyp-Gly) of defined molecular weights. Evidence for the stabilization of collagen triple helix by hydroxyproline, *Biochim. Biophys. Acta* **303:**198.

Schimmel, P. L., and Flory, P. J., 1968, Conformational energies and configurational statistics of copolypeptides containing L-proline, *J. Mol. Biol.* **34:**105.

Schroeder, W. A., Honnen, L., and Green, F. C., 1953, Chromatographic separation and identification of some peptides in partial hydrolysates of gelatin, *Proc. Natl. Acad. Sci. U.S.A.* **39:**23.

Schroeder, W. A., Kay, L. M., LeGette, J., Honnen, L., and Green, F. C., 1954, The constitution of gelatin. Separation and estimation of peptides in partial hydrolysates, *J. Am. Chem. Soc.* **76:**3556.

Schwartz, A., Andries, J. C., and Walton, A. G., 1970, Structural and morphological investigations of poly(Gly-Ala-Pro), *Nature* **226:**161.

Segal, D., M., 1969, Polymers of tripeptides as collagen models. VII. Synthesis and solution properties for collagen-like polyhexapeptides, *J. Mol. Biol.* **43:**497.

Shibnev, V. A., 1964, Use of ethoxyacetylene for polymerization and cyclization of tripeptides containing amino acids, *Izv Akad. Nauk. SSSR, Ser. Khim.* **8**:1545.

Shibnev, V. A., and Debabov, V. G., 1964, Regular polypeptide with glycyl-prolyl-hydroxyprolyl sequence that is isomorphous with collagen, *Izv. Akad. Nauk. SSSR, Ser. Khim.* **6**:1043.

Shibnev, V. A., and Lazareva, A. V., 1969, Use of tetraethylpyrrophosphite for synthesizing polypeptides simulating the nonpolar region of the collagen molecule, *Izv. Akad. Nauk. SSSR, Ser. Khım.* **2**:398.

Shibnev, V. A., and Lisvenko, A. V., 1966, Synthesis of polypeptide Gly-Pro-Pro model of collagen by the method of activated esters, *Izv. Akad. Nauk. SSSR, Ser. Khim.* **7**:1287.

Shibnev, V. A., Lisvenko, A. V., Rogulenkova, V. N., Millinova, M. I., Esipova, N. G., and Chirgadze, U. N., 1966*a*, Configuration of a polypeptide chain (glycyl-L-hydroxyprolyl-L-hydroxyproline), *Biofizika* **11**:1067.

Shibnev, V. A., Poroshin, K. T., and Grechishko, V. S., 1966*b*, The effect of proline residue on formation of specific structure in a polypeptide with the sequence of Gly-Pro-Lys (ϵ-Tos). *Izv. Akad. SSSR, Ser. Khim.* **8**:1493.

Shibnev, V. A., Dyumaev, K. M., Chuvaeva, T. P., Smirnovl, D., and Poroshin, K. T., 1967*a*, Use of β-pyridol esters and their nitro derivatives in the synthesis of peptides and polypeptides with a regulated composition, *Izv. Akad. Nauk. SSSR, Ser. Khim.* **7**:1634.

Shibnev, V. A., Grechishko, V. S., and Poroshin, K. T., 1967*b*, Synthesis and polymerization of activated esters of tripeptides involving *N*-tosyl-L-lysine, *Izv. Akad. Nauk. SSSR, Ser. Khim.* **10**:2327.

Shibnev, V. A., Chuvaeva, T. P., and Poroshin, K. T., 1968*a*, Synthesis of polypeptides with regular structure having molecular weight similar to tropocollagen, *Izv. Akad. Nauk. SSSR, Ser. Khim.* **1**:225.

Shibnev, V. A., Chuvaeva, T. P., and Poroshin, K. T., 1968*b*, Application of pentachlorophenyl esters for the synthesis of collagen models, *Izv. Akad. Nauk. SSSR, Ser. Khim.* **8**:1825.

Shibnev, V. A., Lisvenko, A. V., Chuvaeva, T. P., and Poroshin, K. T., 1968*c*, Comparison of some methods used for the synthesis of regular polypeptides, *Izv. Akad. Nauk. SSSR, Ser. Khim.* **11**:2564.

Shibnev, V. A., Poroshin, K. T., Chuvaeva, T. P., and Martynova, G. A., 1968*d*, The use of 8-hydroxyquinoline peptide esters in the synthesis of polypeptides with regular structure, *Izv. Akad. Nauk. SSSR, Ser. Khim.* **5**:1144.

Shibnev, V. A., Chuvaeva, T. P., Martynova, G. A., and Poroshin, K. T., 1969*a*, 2,4,5-trichlorophenyl esters and their use in the synthesis of polypeptides of regular structure, *Izv. Akad. Nauk. SSSR, Ser. Khim.* **3**:637.

Shibnev, V. A., Chuvaeva, T. P., Martynova, G. A., and Poroshin, K. T., 1969*b*, Use of *N*-hydroxysuccinimide esters for synthesizing models of collagen structure, *Izv. Akad. Nauk. SSR, Ser. Khim.* **11**:2532.

Shibnev, V. A., Chuvaeva, T. P., and Poroshin, K. T., 1969*c*, Synthesis of various activated tripeptide esters representing monomers for preparing models of collagen structure, *Izv. Akad. Nauk. SSSR, Ser. Khim.* **11**:2527.

Shibnev, V. A., Chuvaeva, T. P., Poroshin, K. T., 1970*a*, Application of different activated esters for synthesis of polypeptides with regular structure and their comparative evaluation, *Izv. Akad. Nauk. SSSR, Ser. Khim.* **1**:121.

Shibnev, V. A., Khalikov, Sh. Kh., Finogenova, M. P., and Poroshin, K. T., 1970*b*, Synthesis of poly(glycyl-seryl-hydroxyproline) and poly(glycyl-seryl-proline) polypeptides using pentachlorophenyl esters, *Izv. Akad. Nauk. SSSR, Ser. Khim.* **2**:399.

Shibnev, V. A., Khalivkov, Sh. Kh., Finogenova, M. P., and Poroshin, K. T., 1970c, Synthesis of a polytripeptide (Pro-Ser-Gly)$_n$ simulating the collagen type structure, *Izv. Akad. Nauk. SSSR, Ser. Khim.* **12**:2822.

Shibnev, V. A., Khalivkov, Sh. Kh., Finogenova, M. P., and Poroshin, K. T., 1970d, Synthesis of polypeptides of regular structure containing serine and glutamic acid and modeling nonpolar regions of the collagen protein molecule, *Izv. Akad. Nauk. SSSR, Ser. Khim.* **4**:880.

Shibnev, V. A., Finogenova, M. P., and Poroshin, K. T., 1971a, Synthesis of high-molecular weight (Gly-Pro-Pro)$_n$ polytripeptide collagen-type structure, *Dokl. Akad. Nauk. SSSR* **198**:862.

Shibnev, V. A., Poroshin, K. T., and Grechishko, V. S., 1971b, Synthesis of protected oligopeptides with the Gly-Pro-Pro sequence and the formation of collagen-type structure in them, *Dokl. Akad. Nauk. SSSR* **198**:608.

Shibnev, V. A., Khalikov, Sh. Kh., Ismailov, M. I., and Poroshin, K. T., 1973, Synthesis of regular polypeptides containing a sequence of collagen-type Glu-Glu (OCH_3)-Pro and Pro-His-(CBO)-Gly-, *Izv. Akad. Nauk. SSSR, Ser. Khim.* **8**:1874.

Spiro, R. G., 1969, Characterization and quantitative determination of the hydroxylysine-linked carbohydrate units of several collagens, *J. Biol. Chem.* **244**:602.

Stewart, F. H. C., 1965, The synthesis and polymerization of peptide *p*-nitrophenyl esters, *Aust. J. Chem.* **18**:887.

Stewart, F. H. C., 1969, Synthesis of polydepsipeptides with regularly repeating unit sequences, *Aust. J. Chem.* **22**:1291.

Stewart, J. M., and Matsueda, G. R., 1972, Some problems in solid phase peptide synthesis, *in: Chemistry and Biology of Peptides* (J. Meienhofer, ed.), pp. 221–224, Ann Arbor Science Publishers, Ann Arbor, Michigan.

Suzuki, F., and Koyama, E., 1969, Hydroxylation of proline in collagen model peptides, *Biochim. Biophys. Acta* **177**:154.

Tamburro, A. M., Scatturina, A., and Marchiori, F., 1968, Studi conformazionali su polypeptidi sequenziali-nota III. Sintesi di poli-L-prolil-L-fenilalanil glicina e poli-L-fenilalanil-L-prolilglicina, *Gazz. Chim. Ital.* **96**:638.

Torchia, D. A., 1972, The poly(hydroxy-L-proline) ring conformation determined by proton magnetic resonance, *Macromolecule* **5**:566.

Torchia, D. A., and Lyerla, J. R., 1974, Molecular mobility of polypeptides containing proline as determined by ^{13}C magnetic resonance, *Biopolymers* **13**:97.

Traub, W., 1969, Polymers of tripeptides as collagen models. V. An X-ray study of poly(L-prolyl-glycyl-glycine), *J. Mol. Biol.* **43**:479.

Traub, W., 1974, Some stereochemical implications of the molecular conformation of collagen. *Israel J. Chem.* **12**:435.

Traub, W., and Piez, K., 1971, The chemistry and structure of collagen, *Adv. Protein Chem.* **25**:243.

Traub, W., and Yonath, A., 1966, Polymers of tripeptides as collagen models. I. X-ray studies of poly(L-prolyl-glycyl-L-proline) and related polytripeptides, *J. Mol. Biol.* **16**:404.

Veis, A., and Drake, M. P., 1963, The introduction of intramolecular covalent cross-linkages into ichthyocol tropocollagen with monofunctional aldehydes, *J. Biol. Chem.* **238**:2003.

Verber, D. F., Brady, S. F., and Hirschmann, R., 1972, Some novel amine protecting groups, *in: The Chemistry and Biology of Peptides* (J. Meienhofer, ed.), pp. 315–319, Ann Arbor Science Publishers, Ann Arbor, Michigan.

von der Mark, K., Wendt, P., Rexrodt, F., and Kühn, K., 1970, Direct evidence for a correlation between amino acid sequence and cross striation pattern collagen, *FEBS Lett.* **11:**105.

Von Hippel, P. H., and Harrington, W. F., 1959, Enzymic studies of the gelatin → collagen-fold transition, *Biochim. Biophys. Acta* **36:**427.

Von Hippel, P. H., and Harrington, W. F., 1960, The structure and stabilization of the collagen macromolecule, *Brookhaven Symp. Biol.* **13:**213.

Walton, A. G., 1974, Synthetic polypeptide models of collagen. Structure and function, *in: Structure of Biopolymers, Proceedings of the 26th Symposium of the Colston Research Society,* (E.D.T. Atkins and A. Keller, eds.) pp. 139–148, Butterworth, London.

Wendt, P., von der Mark, K., Rexrodt, F., and Kühn, K., 1972, The covalent structure of collagen. The amino acid sequence of the 112-residues amino-terminal part of peptide α1-CB6 from calf-skin collagen, *Eur. J. Biochem.* **30:**169.

Wieland, T., and Bernhard, H., 1953, Über Peptid-synthesin. 7. Mitt. Di Und Tri-peptid Verbindungen des Triphenols, *Ann.* **582:**218.

Winklmair, D., Engel, J., and Ganser, V., 1971, A cooperative transition between two helical conformations in a linear system: Poly-L-proline I ⇆ II. II. Kinetic studies, *Biopolymers* **10:**721.

Wolman, Y., Gallop, P. M., and Patchornik, A., 1961, Peptide synthesis via oxidation of hydrazides, *J. Am. Chem. Soc.* **83:**1263.

Yamashiro, D., Noble, R. L., and Li, C. H., 1972, New protecting groups for amino acids in solid phase peptide synthesis, *in: Chemistry and Biology of Peptides* (J. Meienhofer, ed.), pp. 197–202, Ann Arbor Science Publishers Inc., Ann Arbor, Michigan.

Index

GPSR Compliance

The European Union's (EU) General Product Safety Regulation (GPSR) is a set of rules that requires consumer products to be safe and our obligations to ensure this.

If you have any concerns about our products, you can contact us on ProductSafety@springernature.com

In case Publisher is established outside the EU, the EU authorized representative is:

Springer Nature Customer Service Center GmbH
Europaplatz 3
69115 Heidelberg, Germany

Batch number: 10761588

Printed by Printforce, the Netherlands